The Produce Contamination Problem

Food Science and Technology International Series

Series Editor

Steve L. Taylor
University of Nebraska – Lincoln, USA

Advisory Board

Ken Buckle
The University of New South Wales, Australia

Mary Ellen Camire
University of Maine, USA

Roger Clemens
University of Southern California, USA

Hildegarde Heymann
University of California – Davis, USA

Robert Hutkins
University of Nebraska – Lincoln, USA

Ron S. Jackson
Quebec, Canada

Huub Lelieveld
Bilthoven, The Netherlands

Daryl B. Lund
University of Wisconsin, USA

Connie Weaver
Purdue University, USA

Ron Wrolstad
Oregon State University, USA

A complete list of books in this series appears at the end of this volume.

The Produce Contamination Problem

Causes and Solutions

Second Edition

Edited by

Karl R. Matthews
*Rutgers, The State University of New Jersey
Department of Food Science, New Brunswick, NJ, USA*

Gerald M. Sapers
*USDA, Agricultural Research Service, (Retired),
Wyndmoor, PA, USA*

Charles P. Gerba
*University of Arizona, Department of Soil,
Water and Environmetal Science Tucson, AZ, USA*

AMSTERDAM • BOSTON • HEIDELBERG • LONDON
NEW YORK • OXFORD • PARIS • SAN DIEGO
SAN FRANCISCO • SINGAPORE • SYDNEY • TOKYO

Academic Press is an imprint of Elsevier

Academic Press is an imprint of Elsevier
525 B Street, Suite 1800, San Diego, CA 92101-4495, USA
32 Jamestown Road, London NW1 7BY, UK
225 Wyman Street, Waltham, MA 02451, USA

Second Edition

Notice
No responsibility is assumed by the publisher for any injury and/or damage to persons or
property as a matter of products liability, negligence or otherwise, or from any use or
operation of any methods, products, instructions or ideas contained in the material herein.
Because of rapid advances in the medical sciences, in particular, independent verification of
diagnoses and drug dosages should be made

British Library Cataloguing in Publication Data
A catalogue record for this book is available from the British Library

Library of Congress Cataloging-in-Publication Data
A catalog record for this book is available from the Library of Congress

For information on all Academic Press publications
visit our web site at store.elsevier.com

ISBN-13: 978-0-12-404611-5

Printed and bound in the United States

14 15 16 17 18   10 9 8 7 6 5 4 3 2 1

Working together
to grow libraries in
developing countries

www.elsevier.com • www.bookaid.org

Contents

v

PART 3   CONTAMINATION AVOIDANCE PRE AND POSTHARVEST

PART 4   TECHNOLOGY FOR REDUCTION OF HUMAN PATHOGENS IN FRESH PRODUCE

CHAPTER 17   Disinfection of Contaminated Produce with Conventional Washing and Sanitizing Technology

CHAPTER 18   Advanced Technologies for Detection and Elimination of Bacterial Pathogens 433

Preface

The premise of the book is that once human pathogen contamination of fresh produce occurs, it is extremely difficult to reduce pathogen levels sufficiently with currently available technologies based on washing with sanitizing agents to assure microbiological safety. Outbreaks since the first edition of this book was published have been attributed to consumption of fresh commodities including but not limited to cantaloupes, mangoes, tomatoes, seed sprouts, and salad greens. Globally extensive research, published in thousands of scientific papers and documents, has focused on the microbiological safety of fresh produce. Disinfection methods whether conventional or based on new innovative technology fail to reduce pathogen loads on produce to levels consistent with product safety.

Produce decontamination methods are surprisingly similar in the level of reductions in pathogens they achieve. Methods that are effective, irradiation and high pressure, are either not accepted by a large segment of consumers or not practical for all types of fresh produce. The intrinsic interaction of microbes with plant tissues, internalization, biofilm formation, and association with inaccessible sites contributes to the insufficient reduction in microbial populations to assure product safety. The key to improving the microbial safety of fresh products is the development of wiser strategies to avoid human pathogen contamination of the products rather than focusing on disinfection technologies with limited efficacy.

Escherichia coli O157:H7 and *Salmonella* continue to be associated with outbreaks of illness linked to the consumption of fresh produce. Recently, a large outbreak in Europe was linked to consumption of fenugreek sprouts contaminated with *Escherichia coli* O104:H4. The serotype was not commonly associated with foodborne illness outbreaks. The pathogen was particularly virulent, causing high morbidity and mortality. In 2013 in the U.S. a large *Cyclospora* outbreak was attributed to an imported salad mix. The outbreak was unusual given the pathogen involved and the size. The specific sources, workers, irrigation water, soil amendments, flooding of fields, of the pathogens remain elusive.

The success of the first edition encouraged us to move forward with a second edition, with contributors who are experts in the area of food safety and produce production, harvesting, packing, and fresh-cut processing, to provide a critical problem-oriented look at produce contamination and its avoidance.

The book is organized into four sections. New chapters have been added and each chapter revised to include the latest information. In the first section, the scope and sources of contamination are covered. Chapters focus on microbial interaction with plant tissue and the limitations of present decontamination methods. Five chapters focus on major sources of contamination – manure, air, water, and wildlife – and examine where and how during crop production, harvesting, packing, or fresh-cut processing these sources might contaminate fresh produce.

In the second section, commodities associated with outbreaks (leafy vegetables, melons, tomatoes, tree fruits and nuts, and berries) are each examined to determine what intrinsic characteristics or production practices make them especially vulnerable to contamination. A chapter on seed sprouts was added to the second edition; the issues surrounding the microbial safety of this commodity group are underscored by the large outbreak in Europe linked to contaminated seed sprouts.

Chapters in the third section provide international perspective on produce contamination issues, focusing on outbreak trends, marketing and distribution practices, produce imports and exports, governmental agencies and regulations concerned with produce safety, and avoidance of contamination through application of Good Agricultural and Manufacturing Practices and guidance documents.

In the fourth section, technology for reduction of human pathogens in fresh produce is examined. Current technology for produce disinfection by washing and application of sanitizing agents is described. The prospects for technological advances in rapid detection and inactivation of microbial contaminants on produce are examined. The book ends with a chapter summarizing conclusions and recommendations for reduction in the risk of human pathogen contamination of fresh produce.

I am grateful to my coeditors, Dr. Gerald M. Sapers and Dr. Charles P. Gerba, for their many contributions and insight, which made this edition distinctive. Although only our names appear on the cover of the book, many people have made important contributions to it. We acknowledge the subject experts whose insights regarding produce contamination contribute to making each chapter of this book a well-written, comprehensive, and up-to-date examination of their respective topics.

We express our gratitude to all of these people as well as Carrie Bolger and Nancy Maragioglio at Elsevier for the enthusiastic support of this project and their great patience in dealing with our difficulties in meeting major deadlines.

Karl R. Matthews, Editor

Contributors

J. Fernando Ayala-Zavala
Centro de Investigación en Alimentación y Desarrollo, A.C., Hermosillo, Sonora, Mexico

Jerry A. Bartz
Department of Plant Pathology, University of Florida, Gainesville, FL, USA

John Brooks
USDA-ARS, Genetics and Precision Agriculture Unit, Mississippi State, MS, USA

Alejandro Castillo
Department of Animal Science, Kleberg Center, College Station, TX, USA

Benjamin J. Chapman
4-H Youth Development and Family & Consumer Sciences, NC Cooperative Extension Service, North Carolina State University, Raleigh, NC, USA

Larry Clark
NWRC Headquarters, Wildlife Services, APHIS-USDA, Fort Collins, CO, USA

Kofitsyo S. Cudjoe
Section for Bacteriology – Food and GMO, National Veterinary Institute, Oslo, Norway

Michael P. Doyle
The University of Georgia, Center for Food Safety, Griffin, GA, USA

Jeffrey A. Farrar
Associate Commissioner for Food Protection, US Food and Drug Administration, Center for Food Safety and Applied Nutrition, USA

Leticia Felix-Valenzuela
Centro de Investigación en Alimentación y Desarrollo, A.C., Hermosillo, Sonora, Mexico

Charles P. Gerba
Department of Soil, Water and Environmental Science, University of Arizona, Tucson, AZ, USA

Jack Guzewich
US Food and Drug Administration, Center for Food Safety and Applied Nutrition

Casey J. Jacob
Diagnostic Medicine/Pathobiology, Kansas State University, Manhattan, KS, USA

Gro S. Johannessen
Section for Bacteriology – Food and GMO National Veterinary Institute, Oslo, Norway

Susanne E. Keller
Institute for Food Safety and Health, Bedford Park, IL, USA

Kalmia E. Kniel
Department of Animal and Food Science, University of Delaware, Newark, DE, USA

Miguel Angel Martinez-Tellez
Centro de Investigación en Alimentación y Desarrollo, A.C., Hermosillo, Sonora, Mexico

Massimiliano Marvasi
Department of Plant Pathology, University of Florida, Gainesville, FL, USA

Veronica Mata-Haro
Centro de Investigacion en Alimentacion y Desarrollo, AC, Hermosillo, Sonora, Mexico

Karl R. Matthews
Department of Food Science, Rutgers-The State University of New Jersey, New Brunswick, NJ, USA

Patricia D. Millner
Environmental Microbial Safety, Beltsville, MD, USA

Brendan A. Niemira
USDA-ARS-ERRC, Wyndmoor, PA, USA

Douglas A. Powell
Vice-President of Communications, IEH Laboratories & Consulting Group, Seattle, WA, USA

Daniel H. Rice
Director, Food Laboratory, NY State Dept. of Agriculture and Markets, Albany, NY, USA

Channah Rock
Department of Soil, Water and Environmental Science, University of Arizona, University of Arizona, Tucson, AZ, USA

M. Ofelia Rodriguez-Garcia
Departamento de Farmacobiología, CUCEI, Universidad de Guadalajara, Guadalajara, Mexico

Gerald M. Sapers
Eastern Regional Research Center, Agricultural Research Service, U.S. Department of Agriculture (Retired), Wyndmoor, PA, USA

Adrienne E.H. Shearer
Department of Animal & Food Sciences, University of Delaware, Newark, DE, USA

Barbara Smal
Department of Food Science, University of Guelph, Guelph, Ontario, Canada

Max Teplitski
Department of Plant Pathology, University of Florida, Gainesville, FL, USA

Keith Warriner
FSQA Program Director, Food Science Bldg 38, University of Guelph, Montreal, Canada

Sima Yaron
Technion – Israel Institute of Technology, Biotechnology and Food Engineering, Hafia, Israel

Howard Q. Zhang
USDA-ARS-ERRC, Wyndmoor, PA, USA

PART

Produce Contamination: Scope and Sources

1

Scope of the Produce Contamination Problem

1

Gerald M. Sapers[1], Michael P. Doyle[2]

[1]*Eastern Regional Research Center, Agricultural Research Service, U.S. Department of Agriculture (Retired), Wyndmoor, PA,* [2]*The University of Georgia, Center for Food Safety, Griffin, GA, USA*

CHAPTER OUTLINE

Introduction

Produce-associated outbreaks—a new problem?

For decades, concerns regarding the microbiological safety of foods have focused largely on the animal products responsible for outbreaks of *E. coli* O157:H7 from ground beef; salmonellosis from poultry, meats, eggs, and dairy products; and listeriosis from soft cheeses and processed meats. Outbreaks of botulism were associated with canned vegetables, but fresh fruits and vegetables generally were considered to be safe, except in countries where the combination of endemic gastrointestinal diseases, unsafe agricultural practices, and poor sanitation resulted in traveler's diarrhea

The Produce Contamination Problem. http://dx.doi.org/10.1016/B978-0-12-404611-5.00001-4

and other illnesses acquired by consumption of locally grown fresh produce. U.S. produce packers and the fresh-cut industry have long believed that their products were made safe by the use of a triple-wash technology using chlorinated water or other approved sanitizing agents.

In recent years, however, this picture has changed dramatically due to an increase in the number of outbreaks of foodborne illnesses associated with fresh and fresh-cut fruits and vegetables. Many large outbreaks, involving widely consumed commodities such as apple cider, cantaloupe melons, raspberries, bagged lettuce and spinach, tomatoes, green onions, and sprouts, have been reported during the past decade (Brackett, 1999; Beuchat, 2002; WHO, 2008). This increase may be due in part to greater consumption of fresh produce in response to the recommendations of health and nutrition professionals. Increased consumption has translated into increased production and distribution of fresh produce, but the growth of produce packing and fresh-cut processing facilities with regional or national distribution capabilities has exposed more consumers to products that may have been contaminated on a single processing line or at a single farm. Additionally, to meet increased demand for out-of-season items, sourcing of fresh produce has become a global endeavor, involving some growing locations where the potential for human pathogen contamination of fruits and vegetables may be high. Furthermore, with better methods for identifying and tracking foodborne outbreaks, the local and state health departments and the Centers for Disease Control and Prevention (CDC) have become better at detecting produce-associated outbreaks, many of which would not have been recognized previously, or the source not identified.

Consequences of produce-associated outbreaks

Pathogen contamination of fresh produce has important public health consequences. Not only are there more cases of illness from produce-associated outbreaks, but highly vulnerable population groups—the very young, the old, and the immunocompromised—are often affected. For these individuals, the severity of foodborne illnesses can be much greater, if not life-threatening, and there may be serious long-term consequences to health. An indirect health-related consequence is the reduced intake of beneficial nutrients from fruits and vegetables by individuals who consume less fresh produce because of concern about acquiring a foodborne illness.

The economic consequences of produce-associated outbreaks are substantial, including the medical costs and lost income of patients, and the costs of damage control (disposal of unmarketable products, product recalls, cleanups, and retrofitting) and lost production time incurred by the affected produce packer/processor. In addition, there are the costs associated with litigation, awards from successful lawsuits, and long-term damage to the company's reputation, reflected by reduced sales of fresh produce. A history of outbreaks can be damaging to an entire segment of the produce industry (e.g., spinach, lettuce, sprouts, green onions, cantaloupes, and tomatoes) or to a production area (e.g., the Salinas Valley of California), resulting in increased costs for compliance with government-mandated changes in production and processing practices and in reduced sales of products nationwide. The estimated cost to tomato growers from the 2008 multistate *Salmonella* Saint-paul outbreak (over 1400 cases reported) was approximately $200 million (Anon., 2008). This outbreak was originally attributed to contaminated tomatoes, but subsequent investigation implicated jalapeño peppers as the major vehicle, with serrano peppers also as a vehicle, and tomatoes as a possible vehicle (CDC, 2008a). The overall economic cost to the produce industry could be a generalized reduction in sales and consumption of fresh fruits and vegetables due to reduced confidence in their safety.

Key aspects of the produce contamination problem
Characteristics of produce-associated outbreaks

Data compiled by the CDC provide insight into trends in the prevalence, size, and causes of produce-related outbreaks (CDC, 2000; CDC, 2006a; CDC, 2008b) and the Outbreak Online Database for 2003 to 2010 (CDC, 2012a). Between 1993 and 1997, the prevalence of outbreaks associated with fresh fruits and vegetables, as reported by the CDC in summary tables for each year, was erratic with no upward trend (Table 1.1). However, there was an abrupt increase in the prevalence of produce-associated outbreaks between 1998 and 2002, perhaps in part because of a change in surveillance and/or reporting methodology (CDC, 2006a). Since then, the number of outbreaks has remained at a high level but with considerable fluctuation from year

Table 1.1 Number of Reported Foodborne-Disease Outbreaks and Cases Associated with Fruits and Vegetables, U.S.[1]

Year[2]	Outbreaks[3]	Cases
1993–1994	29	5,524
1995–1996	22	6,114
1997–1998	59	2,604
1999–2000	124	4,301
2001–2002	111	4,347
2003–2004	65	3,065
2005–2006	91	4,160
2007–2008	92	5,142
2009–2010	59	2,074

[1]Data from summary tables reported by the CDC surveillance reports (2000, 2006) and CDC Foodborne Outbreak Online Database (FOOD) (http://www.cdc.gov/foodborneoutbreaks/).
[2]Data sets for each two-year period are pooled.
[3]Entries represent outbreaks associated with individual produce items or combinations where each component is a specified item of produce. Items designated by the CDC as "salad" or "salad bar," without additional designations (e.g., lettuce-based) to exclude the presence of a major non-produce component such as chicken or pasta, are not included in this tabulation.

to year. Likewise, the number of cases has fluctuated greatly, often because of the occurrence of a small number of very large outbreaks in a single year. It is not clear yet whether the marked decreases in outbreaks and cases in 2009 to 2010 represent the beginning of a downward trend.

Similarly, it is difficult to determine whether the number of outbreaks associated with fresh produce is increasing or decreasing relative to outbreaks associated with non-produce vehicles. CDC data from 2006 to 2008 (CDC, 2009; CDC, 2010; CDC, 2011) indicate that the number of produce-associated outbreaks ranks behind outbreaks linked to fish, poultry, and beef, all of which are in the range of 25 to 47 outbreaks per year. Our compilation of produce related outbreaks, based on CDC data for the same years (CDC, 2012a) falls within a similar range (42–54). However, when the basis of comparison is the number of cases each year, outbreaks in the two produce categories specified by CDC (leafy vegetables and fruits and nuts) sometimes result in more and sometimes fewer cases than are associated with non-produce outbreaks. Any trends in such comparisons are obscured by the occurrence of a small number of very large outbreaks each year.

The number of outbreaks associated with specific human pathogens during 2003 to 2010 is shown in Table 1.2. Norovirus, *Salmonella*, and *E. coli* O157:H7 were responsible for most of the outbreaks; however, the number of outbreaks and cases for each agent varied from year to year, and in each year, single large outbreaks were associated with other pathogens (e.g., hepatitis A in 2003, *Cryptosporidium* in 2004, and *Cyclospora* in 2005). Interestingly, no produce-associated outbreaks were attributed to *Listeria monocytogenes* during 2003 to 2007 (CDC, 2008b). However, two small outbreaks from *L. monocytogenes* on sprouts were reported in 2008 (CDC, 2012a), an outbreak of listeriosis (10 cases) occurred in 2010 from contaminated diced celery (Gaul et al., 2013), and a larger outbreak (146 cases) of listeriosis occurred in 2011 from contamination of whole cantaloupes (CDC, 2011a). In 2011, a large *S.* Agona outbreak (106 cases) was linked to imported papayas (CDC, 2011b), and an outbreak of *E. coli* O157:H7, linked to Romaine lettuce, resulted in 60 cases (CDC, 2011c). In Germany, a devastating outbreak (852 cases) of Shiga toxin-producing *E. coli* O104 (STEC O104:H4) was linked to sprouts, probably due to contaminated fenugreek seeds; some cases in the U.S. were attributed to travel in Germany (CDC, 2011d). In the U.S., an outbreak of *S.* Enteritidis, linked to alfalfa sprouts, resulted in 25 cases (CDC, 2011e). Alfalfa sprouts also were involved in another *Salmonella* outbreak resulting in 140 cases (CDC, 2011f). In 2012, the CDC reported large outbreaks (270 cases) from *S.* Typhimurium and *S.* Newport, linked to cantaloupe (CDC, 2012b). Smaller outbreaks were attributed to Shiga toxin-producing *E. coli* O26 in raw clover sprouts (29 cases) (CDC, 2012c) and *S.* Bredeney in raw peanuts and peanut butter (41 cases) (CDC, 2012d).

CDC data reported for 1998 to 2010 reveal that the prevalence of outbreaks is greater for vegetables than for fruits (CDC, 2000; 2006; 2008b; 2012a). An examination of outbreak data for 2003 to 2010 (Table 1.3) reveals that the principal problem commodities were green salads and lettuce, other leafy vegetables or herbs, sprouts, tomatoes, melons, unpasteurized juice, fruit salad, and nut and nut products.

Table 1.2 Human Pathogens Involved in Reported Outbreaks Associated with Fruits and Vegetables[1]

Year	Pathogen	Outbreaks	Cases
2003–2004	Norovirus	23	1,003
	Salmonella	14	883
	E. coli O157:H7	7	395
	Campylobacter jejuni	2	22
	Shigella sonnei	2	62
	Cryptosporidium parvum	2	356
	Hepatitis A	1	935
2005–2006	Norovirus	31	1,201
	Salmonella	16	607
	E. coli O157:H7	11	450
	Cyclospora cayetanensis	2	606
	Hepatitis A	1	40
	Staphylococcus aureus	1	35
2007–2008	Norovirus	53	1,388
	Salmonella	21	3,312
	E. coli O157:H7	9	172
	Shigella sonnei	2	116
	Cyclospora cayetanensis	2	62
	Listeria monocytogenes	2	40
	Bacillus cereus	1	25
	Hepatitis A	1	22
2009–2010	Salmonella	25	1,183
	Norovirus	24	755
	E. coli O157:H7	5	69
	E. coli O145	1	31
	Clostridium perfringens	1	19
	Cyclospora cayetanensis	1	8
	Hepatitis A	1	5
	Staphylococcus aureus	1	4

[1]See footnotes for Table 1.1.

Many of these commodities are vulnerable to contamination because they grow on or close to soil where exposure to human pathogens may occur. The number of cases and their distribution among commodities varies from year to year. In recent years, major produce-related outbreaks have been caused by *Salmonella* contamination of tomatoes (FDA, 2004, 2006a), cantaloupes (CDC, 2012b), and orange juice (FDA, 2005a), *E. coli* O157:H7 contamination of fresh-cut lettuce (FDA, 2006b, 2007) and bagged spinach (CDC, 2006b; CDC, 2012e), *Cyclospora* contamination of basil (FDA, 2005b), and hepatitis A contamination of green scallions from Mexico

Table 1.3 Items of Fresh Produce Most Frequently Implicated in Outbreaks of Foodborne Disease[1]

Year	Produce Item	Outbreaks	Cases
2003–2004	Green salads, lettuce, other leafy greens	32	653
	Vegetable salads, other vegetables	9	1,349
	Fruit salad, other fruits	7	129
	Melons	5	312
	Sprouts	5	82
	Tomatoes	3	122
	Unpasteurized juice	2	356
	Nuts and nut products	2	62
2005–2006	Green salads, lettuce, other leafy greens	50	1,731
	Fruit salad, other fruits	17	636
	Vegetable salads, other vegetables	12	342
	Tomatoes	11	524
	Melons	4	67
	Unpasteurized juice	2	178
	Nut and nut products	2	815
	Sprouts	1	4
2007–2008	Green salads and lettuce	41	1,038
	Fruit salad, other fruits	15	443
	Vegetable salads, other vegetables	10	280
	Sprouts	9	188
	Tomatoes	6	1,643
	Melons	6	763
	Unpasteurized juice	3	44
	Nut and nut products	1	714
2009–2010	Green salads and lettuce	16	613
	Vegetable salads, other vegetables	11	278
	Fruit salad, other fruits	10	256
	Sprouts	8	493
	Tomatoes	6	107
	Unpasteurized juice	5	249
	Melons	2	70
	Nut and nut products	1	8

[1]See footnotes for Table 1.1.

(CDC, 2003). Several of the outbreaks linked to leafy greens were traced to farms in the Central Valley and Salinas Valley regions of California.

Prevalence of produce contamination with human pathogens

The sporadic nature of produce-related outbreaks is suggestive of localized contamination events, which makes systematic study of contamination sources difficult. One approach to assessing the magnitude of the problem is to obtain data on the prevalence of produce contamination for different commodities and growing locations. Both the FDA and USDA have conducted large-scale studies of selected commodities to determine the prevalence of contamination. The FDA's testing of imported produce (FDA, 2001b) revealed a relatively high prevalence of *Salmonella* and *Shigella* contamination on culantro (50%), cilantro (9%), cantaloupe (7.3%), celery (3.6%), parsley (2.4%), lettuce (1.7%), and scallions (1.7%), all of which are grown in close contact with soil. Testing of domestic produce (FDA, 2003a) revealed a lower prevalence of contamination (total 1.1%) than was found with imported produce (total 4.4%). Domestically grown scallions (3.2%) and cantaloupe (3.1%) had the highest prevalence of contamination, while contamination levels of cilantro, parsley, and lettuce were each about 1%.

A USDA study of selected produce items, sampled at wholesale and distribution centers (USDA, 2004), revealed a much lower prevalence of contamination. *Salmonella* spp. were detected only on lettuce (0.14%), while *E. coli* with a virulence factor was detected on Romaine lettuce (1.34%), leaf lettuce (1.25%), and on cantaloupe, celery, and tomatoes at prevalence levels less than 0.2%.

Other studies of fresh and fresh-cut produce, grown either organically or conventionally, revealed a very low prevalence or absence of human pathogen contamination (Riordan et al., 2001; Sagoo et al., 2001; Anon, 2002; Phillips and Harrison, 2005; Johnston et al., 2006; Mukherjee, 2006; Dallaire et al., 2006; Danyluk et al., 2007; Bobe et al., 2007). However, Heisick et al. (1989) reported a high prevalence of *L. monocytogenes* contamination (26–30%) on potatoes and radishes at retail markets. Castillo et al. (2006) reported high prevalence levels of *Salmonella* (14–20%) and *Shigella* (6–17%) in freshly squeezed orange juice and on fresh oranges collected at public street markets and street booths in Guadalajara, Mexico. Gorski et al. (2011) determined the prevalence of *S. enterica* in the environment in and around Monterey, California. Positive results were obtained for samples of water (7.1%), wildlife (4.2%), and soil (2.6%), but 261 samples of preharvest lettuce and spinach were negative for *Salmonella*. A much greater prevalence of human pathogens in produce was reported in studies performed in India (Ansingkar and Kulkarni, 2010) and Ethiopia (Guchi and Ashenafi, 2010). Mandrell (2009) has reviewed other studies of the prevalence of human pathogens on fresh produce.

These results suggest that contamination levels of most fresh produce by enteric pathogens may be too low for broadly focused prevalence studies to provide helpful guidance in identifying primary sources of contamination. This represents an important gap in our understanding of produce contamination.

Microbial attachment and survival on produce surfaces

When human pathogens come in contact with produce in the crop production environment, they can rapidly attach and strongly adhere to produce surfaces (Sapers et al., 1999; Liao and Sapers, 2000; Ukuku and Fett, 2006). Some pathogens also can form resistant biofilms on plant surfaces (Carmichael et al., 1999; Annous et al., 2005). These topics have been reviewed (Carmichael et al., 1999; Ukuku et al., 2005; Mandrell et al., 2006; Doyle and Erickson, 2008; Ryser et al., 2009) and are discussed in greater detail in Chapter 2.

The extent to which attached human pathogens survive and proliferate on produce surfaces, both prior to and following harvest, is dependent on the type of pathogen, characteristics of the produce, and microbial attachment site (Carlin and Nguyen-The, 1994; Zhuang et al., 1995; Chancellor et al., 2006; Ukuku and Fett, 2006). Important environmental factors include temperature (Zhuang et al., 1995; Duffy, 2005a), humidity (Stine et al., 2005b; Fonseca, 2006; Iturriaga et al., 2007), nutrient availability (Carmichael et al., 1999), and interactions with epiphytic microbes (Francis and O'Beirne, 1998; Aruscavage et al., 2006; Cooley et al., 2006) and plant pathogens (Wells and Butterfield, 1997). Pathogen survival is greater in porous or broken tissue than on smooth tissue (Wei et al., 1995; Janes et al., 2002), and growth can occur in wounds (Wei et al., 1995; Beuchat and Scouten, 2004; Fatemi et al., 2006). Pathogens also can become internalized within plant tissues via attachment and infiltration at pores and cut edges, especially when present in cell populations of greater than a million (Bartz and Showalter, 1981; Bartz, 1982; Seo and Frank, 1999; Solomon et al., 2002a, 2000b).

Studies with tomatoes and cantaloupes, inoculated with human pathogens or surrogates, have revealed that as the time interval between inoculation and washing with sanitizing agents increases from one hour to several days, the efficacy of the sanitizer treatments in reducing pathogen populations decreases significantly (Ukuku and Sapers, 2001; Ukuku et al., 2001; Sapers and Jones, 2006). Microbial internalization, and/or biofilm formation, occurring during this interval between contamination and washing, may be contributing factors. If pathogen contamination of produce occurs preharvest or during harvest, sufficient time may elapse before washing in the packing or processing facility to enable development of these protective factors and thereby reduce the efficacy of sanitizer treatments.

Potential sources of produce contamination
Preharvest sources

Foodborne outbreak investigations have helped in the identification of sources of produce contamination. Such investigations can be characterized into five phases: surveillance/detection, epidemiology, environmental/traceback, regulatory/ enforcement, and prevention/research. In theory, outbreak investigators should be able to genetically match laboratory-confirmed pathogens from ill individuals with laboratory-confirmed pathogens from epidemiologically implicated foods, and thereby identify where and how the contamination occurred. However, this process is often complicated by variability in diagnostic testing procedures and delays in reporting of results and in conducting epidemiologic investigations. Additionally, many perishable foods with a short shelf-life, such as fresh produce, may no longer be available for laboratory testing. Traceback can be complicated by poor record-keeping and commingling of products from different growers throughout the food chain from production to consumption. Hence, identifying the specific source of an outbreak at the farm or field level is often not possible. A more detailed presentation of the difficulties encountered in foodborne outbreak investigations is provided in Chapter 3.

Field studies conducted in crop production locations, packinghouses, and processing facilities, and studies with model systems, have revealed some potential sources of produce contamination (WHO, 2008). These are described in greater detail in Chapters 4 through 8. The ultimate source of enteric pathogens is usually the feces of domesticated animals, wildlife, or humans. Field studies have revealed potential sources of human pathogens in farm environments associated with animal production (Rodriguez et al., 2006; McAllister et al., 2006; Doane et al., 2007), fecal contamination from wildlife (Rice et al., 1995; Wallace et al., 1997; Kullas et al., 2002; Hamilton et al., 2006a; Yan et al., 2007), composted manure (Islam et al., 2004; Ingham et al., 2005), soil (Gagliardi et al., 2003; Johannessen et al., 2005), runoff (Muirhead et al., 2006), irrigation water (Steele et al., 2005; Stine et al., 2005a; Duffy et al., 2005b; Hamilton et al., 2006b; Espinoza-Medina et al., 2006), and the hands of packinghouse workers (Espinoza-Medina et al., 2006). Whether these contamination sources represent an actual food safety hazard will depend on the extent to which human pathogens in the farm or packinghouse environment are able to contact produce surfaces, attach, survive environmental stresses and exposure to sanitizing agents, and then multiply to a population level sufficient to cause illness. These are questions that need to be addressed by researchers to allow identification of the most effective intervention(s), be it at the farm, packinghouse, processing line, or elsewhere in the produce-handling continuum.

Contamination during packing

Studies by Duffy et al. (2005) have revealed that packing equipment may be a source of human pathogen contamination of fresh produce. Gagliardi et al. (2003) implicated process water used for cooling and washing of melons as a source of contamination. Garcia et al. (2006) attributed *E. coli* contamination of apples used for cider production to microbial buildup in dump tanks, when the sanitizer/wash solution was not adequately replenished, and to inadequate cleaning and sanitizing of scrubbers, spray nozzles, and conveyors. Keller et al. (2002) also found that bacteria proliferate in an apple cider mill when equipment is inadequately sanitized, thereby resulting in cider contamination.

Contamination during fresh-cut processing

It is well-established that conventional cleaning and sanitizing treatments applied to fresh produce generally reduce attached pathogen populations by only 90 to 99%

(1- to 2-log reduction), likely due to pathogen survival in protected sites or biofilms on produce surfaces or to neutralization of the sanitizer by the organic load of the process water (WHO, 2008; UGA, 2011; Holvoet et al., 2012; Chapters 2 and 17). This is true both for uncut and fresh-cut commodities. However, exposure of cut produce, especially leafy vegetables, to contaminated wash water increases the risk of bacterial attachment at cut surfaces, subsequent internalization (Seo and Frank, 1999; Solomon et al., 2002b), and proliferation of the human pathogens during product handling, storage, and distribution. Similarly, contamination of fresh-cut cantaloupes with human pathogens, by transfer from the rind surface to the flesh during cutting (Ukuku and Sapers, 2001), could result in extensive growth during storage and distribution of the fresh-cut product under conditions of temperature abuse.

Detection of *L. monocytogenes* in fresh-cut apples, which resulted in a product recall in 2001 (FDA, 2001a), provides evidence of a contamination risk associated with the use of browning inhibitors and other processing aids in fresh-cut processing. If not frequently refreshed, substantial amounts of nutrients, leached from cut produce, can build-up in such solutions, thereby making them suitable media for the proliferation of environmental contaminants such as *L. monocytogenes*. Additionally, this pathogen can grow, albeit slowly, at the low temperatures of fresh-cut processing rooms and under conditions of temperature abuse during product distribution and retailing.

Many studies have revealed that human pathogens can survive and grow on fresh-cut produce (Steinbruegge et al., 1988; Fernandez-Escartin et al., 1989; Carlin et al., 1995). Human pathogen survival and growth on fresh-cut produce is affected by many of the factors discussed earlier, especially temperature (Kallander et al., 1991; Piagentini et al., 1997; Farber et al., 1998), interaction with the indigenous microflora (Carlin et al., 1996; Francis and O'Beirne, 1998), nutrient availability, and use of controlled or modified atmospheres for storage or packaging (Berrang et al., 1989; Abdul-Raouf et al., 1993; Omary et al., 1993; Kakiomenou et al., 1998). Improvements in plant sanitation and maintenance of the cold chain, from the packing or processing plant through distribution and retailing to the consumer, are important prerequisites to reducing pathogen contamination of fresh-cut produce.

Gaps in our understanding of produce contamination
Current state of knowledge

With produce-related outbreaks frequently in the news, and the public health and economic costs so high, why does this problem continue in spite of the large research effort carried out by government, academia, and the private sector to improve food safety? Means of detecting and tracking human pathogens in the food supply continue to improve. Much is known about the foodborne pathogens responsible for produce-related outbreaks, their ability to attach to fresh fruits and vegetables, and the efficacy of various conventional and new disinfection technologies. However, many challenges remain; for example, the low infectious dose of *E. coli* O157:H7, the limited efficacy of many approved sanitizers to kill pathogens on produce surfaces, the resistance of pathogens to cleaning and disinfection when in biofilms, and the limitations that outbreak investigators have in tracking a contamination event to a specific location and source, as discussed in Chapter 3.

What we don't know

We know how to identify and subtype the strain of the causative agent of an outbreak and link human isolates to food isolates, but the challenge is to readily identify the actual source of the pathogen or contamination event. Perhaps this is because the source is a flock of birds or a meandering feral pig, both random and unpredictable events. Perhaps the event is a dust storm conveying desiccated manure from a distant feedlot to a produce farm, again a random occurrence, but this should be more predictable and a risk to avoid. In order to address the problem of produce-associated outbreaks of foodborne illness, we need a better understanding of the contamination process, including transmission of pathogens in aerosols and water, survival of pathogens in manure and soil, mode of contact between human pathogens and produce surfaces, extent of pathogen adhesion and/or entrapment at the attachment site, opportunities for pathogen internalization, opportunities for biofilm formation, and the role of environmental conditions such as temperature, humidity, rainfall, and wind velocity. These factors are discussed in Chapters 2, 5, and 6.

Also to be considered is information regarding agricultural practices, hygienic behavior of farm workers, risks associated with field packing and hydrocooling operations, proximity to potential contamination sources (exposed irrigation canals; nearby pastures, feedlots, or flyways; presence of human pathogens in soil; and scat of local populations of deer, rodents, amphibians, feral pigs, and other wildlife). These contributing factors are discussed in Chapters 4 through 8. Special attention is needed for those commodities most frequently associated with major outbreaks— what makes them more vulnerable to contamination with human pathogens? These problem commodities are addressed in Chapters 9 through 14.

Developing effective interventions

Based on the foundation of improved understanding of the major routes of produce contamination, and of the ability of pathogens to survive and grow on produce, more effective interventions must be developed to reduce the potential for produce contamination. These interventions would be incorporated into guidance documents, HACCP (Hazard Analysis of Critical Control Points) plans, and updated good agricultural and manufacturing practices, making them more effective in preventing produce contamination. Also needed are more effective regulatory actions, applied not only in processing facilities but also at the farm level, to reduce the risk of contamination and exclude contaminated produce from the marketplace. Since many types of fresh produce are sourced internationally, regulation of produce safety should be addressed in global terms. These topics are discussed in Chapters 15 and 16.

Finally, we must consider the promise and limitations of technology in providing means of rapid detection of human pathogens in fresh produce, identification of contamination sources, and disinfection of contaminated produce to reduce the risk of foodborne illness. In recognizing the limitations of conventional technology for produce disinfection, researchers can design more efficacious antimicrobial treatments (see Chapter 17). The prospects for advanced technological solutions are addressed in Chapter 18.

The primary purposes of this book are to address what is known about contamination of fresh produce by human pathogens, and to identify those interventions that may be applied to reduce the risk of contamination. In this second edition of *The Produce Contamination Problem*, we have added separate chapters on airborne contamination, and contamination of seeds and sprouts, to strengthen our coverage of these important topics. Using this information, specific gaps in our understanding of these topics can be identified and used to set an agenda for prioritized research that will enable production of safer produce. Chapter 19 summarizes the state of our knowledge, provides recommendations for development of more effective interventions, and examines policy issues that can influence improvements in the microbiological safety of fresh produce.

In summary, there are deficiencies in the current state of knowledge of human pathogen contamination of fresh produce; the survival and proliferation of microbial contaminants during packing, processing, storage, distribution, and marketing of produce; and the efficacy of conventional interventions; all of which contribute to the problem of produce-associated outbreaks. Improvements in our understanding of sources of produce contamination coupled with implementation of more efficacious food safety interventions are needed to achieve greater success in reducing the occurrence of such outbreaks.

References

Abdul-Raouf, U.M., Beuchat, L.R., Ammar, M.S., 1993. Survival and growth of *Escherichia coli* O157:H7 on salad vegetables. Appl. Environ. Microbiol. 59, 1999–2006.

Annous, B.A., Solomon, E.B., Cooke, P.H., 2005. Biofilm formation by *Salmonella* spp. on cantaloupe melons. J. Food Safety 25, 276–287.

Anon, 2002. Results of 4th Quarter National Survey 2002. European Commission Coordinated Programme for the Official Control of Foodstuffs for 2002. Bacteriological safety of pre-cut fruit & vegetables, sprouted seeds and unpasteurized fruit & vegetable juices from processing and retail premises. www.fsai.ie/surveillance/food_safety/microbiological/4th Quarter2.pdf (accessed 11.06.07.).

Anon., 2008. FDA criticized over its response to Salmonella outbreak. Baltimore Sun: August 1 www.baltimoresun.com/news/health/bal-te.fda01aug01,0,3276708.story (accessed 15.10.08.).

Ansingkar, V., Kulkarni, N., 2010. Incidences of endophytic human pathogens in fresh produce. WebmedCentral MICROBIOLOGY 1 (12):WMC001299.

Aruscavage, D., Lee, K., Miller, S., et al., 2006. Interactions affecting the proliferation and control of human pathogens on edible plants. J. Food Sci. 71, R89–R99.

Bartz, J.A., 1982. Infiltration of tomatoes immersed at different temperatures to different depths in suspensions of *Erwinia carotovora* subsp. *carotovora*. Plant Dis. 66, 302–306.

Bartz, J.A., Showalter, R.K., 1981. Infiltration of tomatoes by aqueous bacterial suspensions. Phytopathol. 71, 515–518.

Beuchat, L.R., 2002. Ecological factors influencing survival and growth of human pathogens on raw fruits and vegetables. Microb. Infect. 4, 413–423.

Beuchat, L.R., Scouten, A.J., 2004. Factors affecting survival, growth, and retrieval of *Salmonella* Poona on intact and wounded cantaloupe rind and stem scar tissue. Food Microbiol. 21, 683–694.

Berrang, M.E., Brackett, R.E., Beuchat, L.R., 1989. Growth of *Listeria monocytogenes* on fresh vegetables stored under controlled atmosphere. J. Food Prot. 52, 702–705.

Bobe, G., Thede, D.J., Ten Eyck, T.A., et al., 2007. Microbial levels in Michigan apple cider and their association with manufacturing practices. J. Food Prot. 70, 1187–1193.

Brackett, R.E., 1999. Incidence, contributing factors, and control of bacterial pathogens in produce. Postharvest. Biol. Technol. 15, 305–311.

Carlin, F., Nguyen-The, C., 1994. Fate of *Listeria monocytogenes* on four types of minimally processed green salads. Lett. Appl. Microbiol. 18, 222–226.

Carlin, F., Nguyen-The, C., da Silva, A.A., 1995. Factors affecting the growth of *Listeria monocytogenes* on minimally processed fresh endive. J. Appl. Bacteriol. 78, 636–646.

Carlin, F., Nguyen-The, C., Morris, C.E., 1996. Influence of background microflora on *Listeria monocytogenes* on minimally processed fresh broad-leaved endive (*Cichorium endivia* var. *latifolia*). J Food Prot. 59, 698–703.

Carmichael, I., Harper, I.S., Coventry, M.J., et al., 1999. Bacterial colonization and biofilm development on minimally processed vegetables. J. Appl. Microbiol. Symp. Suppl. 85, 45S–51S.

Castillo, A., Villarruel-López, A., Navarro-Hidalgo, V., et al., 2006. *Salmonella* and *Shigella* in freshly squeezed orange juice, fresh oranges and wiping cloths collected from public markets and street booths in Guadalajara, Mexico: Incidence and comparison of analytical routes. J. Food Prot. 69, 2595–2599.

CDC, 2003. Hepatitis A outbreak associated with green onions at a restaurant—Monaca, Pennsylvania, 2003. MMWR 52 Dispatch, 1–3.

CDC, 2000. Surveillance for foodborne disease outbreak-United States, 1993–1997. MMWR 49 (SS01), 1–51. Mar, 17.

CDC, 2006a. Surveillance for foodborne disease outbreak-United States, 1998–2002. MMWR 55 (SS10), 1-34.Nov, 10.

CDC, 2006b. Ongoing multistate outbreak of *Escherichia coli* serotype O157:H7 infections associated with consumption of fresh spinach—United States. September 2006. MMWR 55 (Dispatch), 1–2.

CDC, 2008a. Outbreak of *Salmonella* serotype Saintpaul infections associated with multiple raw produce items-United States, 2008. MMWR 57 (34), 929–934.

CDC, 2008b. Outbreak surveillance data. Annual listing of foodborne disease outbreaks, United States, 1990-2006. www.cdc.gov/foodborneoutbreaks/outbreak_data.htm (accessed 15.10.08.).

CDC, 2009. Surveillance for foodborne disease outbreaks-United States, 2006. MMWR 58, 609–615.

CDC, 2010. Surveillance for foodborne disease outbreaks-United States, 2007. MMWR 59, 973–979.

CDC, 2011. Surveillance for foodborne disease outbreaks-United States, 2008. MMWR 60, 1197–1202.

CDC, 2011a. Multistate outbreak of listeriosis linked to whole cantaloupes from Jensen Farms, Colorado. Final update, December 8, 2011. http://www.cdc.gov/listeria/outbreaks/cantalo upes-jensen-farms/index.html (accessed 17.07.12.).

CDC, 2011b. Multistate outbreak of human *Salmonella* Agona infections linked to whole, fresh imported papayas. Final update, August 29, 2011. http://www.cdc.gov/salmonella/a gona-papayas/index.html (accessed 17.07.12.).

CDC, 2011c. Investigation Announcement: multistate outbreak of *E. coli* O157:H7 infections linked to romaine lettuce. http://www.cdc.gov/ecoli/2011/ecoliO157/romainelettuce/1207 11/index.html (accessed 17.07.12.).

CDC, 2011d. Investigation update: outbreak of shiga toxin-producing *E. coli* O104 (STEC O104:H4) infections associated with travel to Germany. Final update, July 8, 2011. http://www.cdc.gov/ecoli/2011/ecoliO104/index.html (accessed 17.07.12.).

CDC, 2011e. Investigation update: multistate outbreak of human *Salmonella* Enteritidis infections linked to alfalfa sprouts and spicy sprouts. Final update, July 6, 2011. http://www.cdc.gov/salmonella/sprouts-enteritidis0611/070611/index.html (accessed 17.07.12.).

CDC, 2011f. Investigation update: multistate outbreak of human *Salmonella* I 4,[5],12:i:- infections linked to alfalfa sprouts. Final update, Feb. 10, 2011. http://www.cdc.gov/salm onella/i4512i-/021011/index.html (accessed 17.07.12.).

CDC, 2012a. Foodborne Outbreak Online Database (FOOD). http://www.cdc.gov/foodborne outbreaks/ (accessed 09.10.12.).

CDC, 2012b. Multistate outbreak of *Salmonella* Typhimurium and *Salmonella* Newport infections linked to cantaloupe. Posted Sept. 13, 2012. http://www.cdc.gov/salmonella/typhimurium-cantaloupe-08-12/index.html (accessed 25.09.12.).

CDC, 2012c. Multistate outbreak of Shiga toxin-producing *Escherichia coli* O26 infections linked to raw clover sprouts at Jimmy John's restaurants. http://www.cdc.gov/ecoli/2012/O26/index.html (accessed 17.07.12.).

CDC, 2012d. Multistate outbreak of *Salmonella* Bredeney infections linked to peanut butter manufactured by Sunland, Inc. http://www.cdc.gov/salmonella/bredeney-09-12/index.html (accessed 15.11.12.).

CDC, 2012e. Multistate outbreak of Shiga toxin-producing *Escherichia coli* O157:H7 infections linked to organic spinach and spring mix blend (final update). http://www.cdc.gov/ecoli/2012/O157H7-11-12/index.html (accessed 13.12.12.).

Chancellor, D.D., Tyagi, S., Bazaco, M.C., et al., 2006. Green onions: Potential mechanism for hepatitis A contamination. J. Food Prot. 69, 1468–1472.

Cooley, M.B., Chao, D., Mandrell, R.E., 2006. *Escherichia coli* O157:H7 survival and growth on lettuce is altered by the presence of epiphytic bacteria. J. Food Prot. 69, 2329–2335.

Dallaire, R., LeBlanc, D.I., Tranchant, C.C., et al., 2006. Monitoring the microbial populations and temperatures of fresh broccoli from harvest to retail display. J. Food Prot. 69, 1118–1125.

Danyluk, M.D., Jones, T.M., Abd, S.J., et al., 2007. Prevalence and amounts of *Salmonella* found on raw California almonds. J. Food Prot. 70, 820–827.

Doane, C.A., Pangloili, P., Richards, H.A., 2007. Occurrence of *Escherichia coli* O157:H7 in diverse farm environments. J. Food Prot. 70, 6–10.

Doyle, M.P., Erickson, M.C., 2008. Summer meeting 2007—the problems with fresh produce: an overview. J. Appl. Microbiol. 105, 317–330.

Duffy, E.A., Cisneros-Zevallos, L., Castillo, A., et al., 2005. Survival of *Salmonella* transformed to express green fluorescent protein on Italian parsley as affected by processing and storage. J. Food Prot. 68, 687–695.

Duffy, E.A., Lucia, L.M., Kells, J.M., et al., 2005. Concentrations of *Escherichia coli* and genetic diversity and antibiotic resistance profiling of *Salmonella* isolated from irrigation water, packing shed equipment, and fresh produce in Texas. J. Food Prot. 68, 70–79.

Espinoza-Medina, I.E., Rodríguez-Leyva, F.J., Vargas-Arispuro, I., et al., 2006. PCR identification of *Salmonella*: Potential contamination sources from production and postharvest handling of cantaloupes. J. Food Prot. 69, 1422–1425.

Farber, J.M., Wang, S.L., Cai, Y., et al., 1998. Changes in populations of *Listeria monocytogenes* inoculated on packaged fresh-cut vegetables. J. Food Prot. 61, 192–195.

Fatemi, P., LaBorde, L.F., Patton, J., et al., 2006. Influence of punctures, cuts, and surface morphologies of Golden Delicious apples on penetration and growth of *Escherichia coli* O157:H7. J. Food Prot. 69, 267–275.

FDA, 2001. Enforcement report. Recalls and field corrections: foods—Class I. Recall number F-535-1. 20 August 2001Sliced apples in poly bags. www.fda.gov/bbs/topics/ENFORCE/2001/ENF00708.html (accessed 01.04.02.).

FDA, 2001. FDA Survey of Imported Fresh Produce. US Food and Drug Administration Center for Food Safety and Applied Nutrition, Office of Plant and Dairy Foods and Beverages January 30, 2001. http://www.fda.gov/Food/FoodSafety/Product-SpecificInformation/FruitsVegetablesJuices/GuidanceComplianceRegulatoryInformation/ucm118891.htm (accessed 21.12.12.).

FDA, 2003, FDA Survey of Domestic Fresh Produce. U.S. Department of Health and Human Services, US Food and Drug Administration Center for Food Safety and Applied Nutrition Office of Plant and Dairy Foods and Beverages January 2003. http://www.fda.gov/Food/FoodSafety/Product-SpecificInformation/FruitsVegetablesJuices/GuidanceComplianceRegulatoryInformation/ucm118306.htm (accessed 21.12.12.).

FDA, 2004. FDA investigates certain Roma tomatoes as source of foodborne illness outbreaks in Pennsylvania, Ohio, and Mid-Atlantic states. FDA Statement Food and Drug Administration, US Department of Health and Human Services: July 23, 2004. http://www.fda.gov/NewsEvents/Newsroom/PressAnnouncements/2004/ucm108332.htm (accessed 21.12.12.).

FDA, 2005a. FDA issues nationwide health alert on Orchid Island unpasteurized orange juice products. FDA Statement Food and Drug Administration, US Department of Health and Human Services: July 8, 2005. http://www.fda.gov/NewsEvents/Newsroom/PressAnnouncements/2005/ucm108457.htm (accessed 21.12.12.).

FDA, 2005b. FDA works to trace source of foodborne illness in Florida. FDA News Food and Drug Administration, US Department of Health and Human Services: June 3, 2005. http://www.fda.gov/NewsEvents/Newsroom/PressAnnouncements/2005/ucm108439.htm (accessed 21.12.12.).

FDA, 2006a. FDA notifies consumers that tomatoes in restaurants linked to *Salmonella* Typhimurium outbreak. FDA News Food and Drug Administration, US Department of Health and Human Services: November 3, 2006. http://www.fda.gov/NewsEvents/Newsroom/PressAnnouncements/2006/ucm108782.htm (accessed 21.12.12.).

FDA, 2006b. Update: FDA narrows investigation of E. coli O157:H7 outbreak at Taco Bell restaurants. FDA News Food and Drug Administration, US Department of Health and Human Services: December 13, 2006. http://www.fda.gov/NewsEvents/Newsroom/PressAnnouncements/2006/ucm108803.htm (accessed 21.12.12.).

FDA, 2007. FDA and States closer to identifying source of *E. coli* contamination associated with illnesses at Taco John's restaurants. FDA News Food and Drug Administration, US Department of Health and Human Services: January 12, 2007. www.fda.gov/bbs/topics/NEWS/2007/NEW01546.html (accessed 12.01.07.).

Fernandez Escartin, E., Castillo Ayala, A., Saldana Lozano, J., 1989. Survival and growth of *Salmonella* and *Shigella* on sliced fresh fruit. J. Food Prot. 52, 471–472.

Fonseca, J.M., 2006. Postharvest quality and microbial population of head lettuce as affected by moisture at harvest. J. Food Sci. 71, M45–M49.

Francis, G.A., O'Beirne, D., 1998. Effects of the indigenous microflora of minimally processed lettuce on the survival and growth of *Listeria innocua*. Int. J. Food Sci. Technol. 33, 477–488.

Gagliardi, J.V., Millner, P.D., Lester, G., et al., 2003. On-farm and postharvest processing sources of bacterial contamination to melon rinds. J. Food Prot. 66, 82–87.

Garcia, L., Henderson, J., Fabri, M., et al., 2006. Potential sources of microbial contamination in unpasteurized apple cider. J. Food Prot. 69, 137–144.

Gaul, L.K., Faraq, N.H., Shim, T., et al., 2013. Hospital-acquired listeriosis outbreak caused by contaminated diced celery. Clin. Infect. Dis. 56, 20–26.

Gorski, L., Parker, C.T., Liang, A., et al., 2011. Prevalence, distribution, and diversity of *Salmonella enterica* in a major produce region of California. Appl. Environ. Microbiol. 77, 2734–2748.

Guchi, B., Ashenafi, M., 2010. Microbial load, prevalence and antibiograms of *Salmonella* and *Shigella* in lettuce and green peppers. Ethiop. J. Health Sci. 20, 41–48.

Hamilton, M.J., Yan, T., Sadowsky, M.J., 2006. Development of goose- and duck-specific DNA markers to determine sources of *Escherichia* coli in waterways. Appl. Environ. Microbiol. 72, 4012–4019.

Hamilton, A.J., Stagnitti, F., Premier, R., et al., 2006. Quantitative microbial risk assessment models for consumption of raw vegetables irrigated with reclaimed water. Appl. Environ. Microbiol. 72, 3284–3290.

Heisick, J.E., Wagner, D.E., Nierman, M.I., et al., 1989. *Listeria* spp. found on fresh market produce. Appl. Environ. Microbiol. 55, 1925–1927.

Holvoet, K., Jacxsens, L., I Sampers, et al., 2012. Insight into the prevalence and distribution of microbial contamination to evaluate water management in the fresh produce processing industry. J. Food Prot. 75, 671–681.

Ingham, S.C., Fanslau, M.A., Engel, R.A., et al., 2005. Evaluation of fertilization-to-planting and fertilization-to-harvest intervals for safe use of noncomposted bovine manure in Wisconsin vegetable production. J. Food Prot. 68, 1134–1142.

Iturriaga, M.H., Tamplin, M.L., Escartín, E.F., 2007. Colonization of tomatoes by *Salmonella* Montevideo is affected by relative humidity and storage temperature. J. Food Prot. 70, 30–34.

Islam, M., Doyle, M.P., Phatak, S.C., et al., 2004. Persistence of enterohemorrhagic *Escherichia coli* O157:H7 in soil and leaf lettuce and parsley grown in fields treated with contaminated manure composts or irrigation water. J. Food Prot. 67, 1365–1370.

Janes, M.E., Cobbs, T., Kooshesh, S., et al., 2002. Survival differences of *Escherichia coli* O157:H7 strains in apples of three varieties stored at various temperatures. J. Food Prot. 65, 1075–1080.

Johannessen, G.S., Bengtsson, G.B., Heier, B.T., et al., 2005. Potential uptake of *Escherichia coli* O157:H7 from organic manure into Crisphead lettuce. Appl. Environ. Microbiol. 71, 2221–2225.

Johnston, L.M., Jaykus, L.-A., Moll, D., et al., 2006. A field study of the microbiological quality of fresh produce of domestic and Mexican origin. Int. J. Food Microbiol. 112, 83–95.

Kallander, K.D., Hitchins, A.D., Lancette, G.A., et al., 1991. Fate of *Listeria monocytogenes* in shredded cabbage stored at 5 and 25 °C under a modified atmosphere. J. Food Prot. 54, 302–304.

Kakiomenou, K., Tassou, C., Nychas, G.-J., 1998. Survival of *Salmonella enteritidis* and *Listeria monocytogenes* on salad vegetables. World J. Microbiol. Biotech. 14, 383–387.

Keller, S.E., Merker, R.I., Taylor, K.T., et al., 2002. Efficacy of sanitation and cleaning methods in a small apple cider mill. J. Food Prot. 65, 911–917.

Kullas, H., Coles, M., Rhyan, J., et al., 2002. Prevalence of *Escherichia coli* serogroups and human virulence factors in feces of urban Canada geese (*Branta canadensis*). Int. J. Environ. Health Res. 12, 153–162.

Liao, C.-H., Sapers, G.M., 2000. Attachment and growth of *Salmonella* Chester on apple fruits and in vivo response of attached bacteria to sanitizer treatments. J. Food Prot. 63, 876–883.

Mandrell, R.E., Gorski, L., Brandl, M.T., 2006. Attachment of microorganisms to fresh produce. In: Sapers, G.M., Gorny, J.R., Yousef, A.E. (Eds.), Microbiology of Fruits and Vegetables. CRC Press, pp. 375–400.

Mandrell, R.E., 2009. Enteric human pathogens associated with fresh produce. In: Fan (Ames, Iowa:, X. (Ed.), Microbial safety of fresh produce. Blackwell Publishing. sources, transport, and ecology.

McAllister, T.A., Bach, S.J., Stanford, K., et al., 2006. Shedding of *Escherichia coli* O157:H7 by cattle fed diets containing monensin or tylosin. J. Food Prot. 69, 2075–2083.

Muirhead, R.W., Collins, R.P., Bremer, P.J., 2006. Interaction of *Escherichia coli* and soil particles in runoff. Appl. Environ. Microbiol. 72, 3406–3411.

Mukherjee, A., Speh, D., Jones, A.T., et al., 2006. Longitudinal microbiological survey of fresh produce grown by farmers in the upper Midwest. J. Food Prot. 69, 1928–1936.

Omary, M.B., Testin, R.F., Barefoot, S.F., et al., 1993. Packaging effects on growth of *Listeria innocua* in shredded cabbage. J. Food Sci. 58, 623–626.

Phillips, C.A., Harrison, M.A., 2005. Comparison of the microflora on organically and conventionally grown spring mix from a California processor. J. Food Prot. 68, 1143–1146.

Piagentini, A.M., Pirovani, M.E., Güemes, D.R., et al., 1997. Survival and growth of *Salmonella* Hadar on minimally processed cabbage as influenced by storage abuse conditions. J. Food Sci. 62, 616–618; 631.

Rice, D.H., Hancock, D.D., Besser, T.E., 1995. Verotoxigenic *E. coli* O157 colonization of wild deer and range cattle. Vet. Rec. 137, 524.

Riordan, D.C., Sapers, G.M., Hankinson, et al., 2001. A study of U.S. orchards to identify potential sources of *Escherichia coli* O157:H7. J. Food Prot. 64, 1320–1327.

Rodriguez, A., Pangloli, P., Richards, H.A., et al., 2006. Prevalence of *Salmonella* in diverse environmental farm samples. J. Food Prot. 69, 2576–2580.

Ryser, E.T., .Hao, J., . Yan, Z., 2009. Internalization of pathogens in produce. Chapter 3. In: Fan, X. (Ed.), Microbial safety of fresh produce, pp. 55–80. IFT Press Wiley-Blackwell.

Sagoo, S.K., Little, C.L., Mitchell, R.T., 2001. The microbiological examination of ready to eat organic vegetables from retail establishments in the United Kingdom. Lett. Appl. Microbiol. 33, 434–439.

Sapers, G.M., Jones, D.M., 2006. Improved sanitizing treatments for fresh tomatoes. J. Food Sci. 71, M252–M256.

Sapers, G.M., Miller, R.L., Mattrazzo, A.M., 1999. Effectiveness of sanitizing agents in inactivating *Escherichia coli* in Golden Delicious apples. J. Food Sci. 64, 734–737.

Seo, K.H., Frank, J.F., 1999. Attachment of *Escherichia coli* O157:H7 to lettuce leaf surface and bacterial viability in response to chlorine treatment as demonstrated by using confocal scanning laser microscopy. J. Food Prot. 62, 3–9.

Solomon, E.B., Potenski, C.J., Matthews, K.R., 2002. Effect of irrigation method on transmission to and persistence of *Escherichia coli* O157:H7 on lettuce. J. Food Prot. 65, 673–676.

Solomon, E.B., Yaron, S., Matthews, K.R., 2002. Transmission of *Escherichia coli* O157:H7 from contaminated manure and irrigation water to lettuce plant tissue and its subsequent internalization. Appl. Environ. Microbiol. 68, 397–400.

Steele, M., Mahdi, A., Odumeru, J., 2005. Microbial assessment of irrigation water used for production of fruit and vegetables in Ontario, Canada. J. Food Prot. 68, 1388–1392.

Steinbruegge, E.G., Maxcy, R.B., Liewen, M.B., 1988. Fate of *Listeria monocytogenes* on ready to serve lettuce. J. Food Prot. 51, 596–599.

Stine, S.W., Song, I., Choi, C.Y., et al., 2005. Application of microbial risk assessment to the development of standards for enteric pathogens in water used to irrigate fresh produce. J. Food Prot. 68, 913–918.

Stine, S.W., Song, I., Choi, C.Y., et al., 2005. Effect of relative humidity on preharvest survival of bacterial and viral pathogens on the surface of cantaloupe, lettuce, and bell peppers. J. Food Prot. 68, 1352–1358.

UGA, 2011. Pathogen contamination during harvesting, packing, or fresh-cut processing of leafy greens, A systems approach for produce safety. University of Georgia Center for Food safety. http://www.ugacfs.org/producesafety/Pages/Basics/FateProcessing.html.

Ukuku, D.O., Fett, W.F., 2006. Effects of cell surface charge and hydrophobicity on attachment of 16 *Salmonella* serovars to cantaloupe rind and decontamination with sanitizers. J. Food Prot. 69, 1835–1843.

Ukuku, D.O., Sapers, G.M., 2001. Effect of sanitizer treatments on *Salmonella* Stanley attached to the surface of cantaloupe and cell transfer to fresh-cut tissues during cutting practices. J. Food Prot. 64, 1286–1291.

Ukuku, D.O., Pilizota, V., Sapers, G.M., 2001. Influence of washing treatment on native microflora and *Escherichia coli* population of inoculated cantaloupes. J. Food Safety 21, 31–47.

Ukuku, D.O., Liao, C.-H., Gembeh, S., 2005. Attachment of bacterial human pathogens on fruit and vegetable surfaces. In: Ukuku, D., Imam, S., Lamikanra, O. (Eds.), Produce degradation, pathways and prevention. CRC Press, pp. 421–440.

USDA, 2004. Microbiological Data Program Progress Update and 2002 Data Summary. http://www.ams.usda.gov/AMSv1.0/getfile?dDocName=MDPSUMM02.

Wallace, J.S., Cheasty, T., Jones, K., 1997. Isolation of Verocytotoxin-producing *Escherichia coli* O157 from wild birds. J. Appl. Microbiol. 82, 399–404.

Wei, C.I., Huang, T.S., Kim, J.M., et al., 1995. Growth and survival of *Salmonella* Montevideo on tomatoes and disinfection with chlorinated water. J. Food Prot. 58, 829–836.

Wells, J.M., Butterfield, J.E., 1997. *Salmonella* contamination associated with bacterial soft rot of fresh fruits and vegetables in the marketplace. Plant Dis. 81, 867–872.

WHO, 2008. Microbial hazards in fresh leafy vegetables and herbs. Meeting Report. Microbial risk assessment series 14, World Health Organization, Food and Agriculture Organization of the United Nations. ftp://ftp.fao.org/docrep/fao/011/i0452e/i0452e00.pdf.

Yan, T., Hamilton, M.J., Sadowsky, M.J., 2007. High-throughput and quantitative procedure for determining sources of *Escherichia coli* in waterways by using host-specific DNA marker genes. Appl. Environ. Microbiol. 73, 890–896.

Zhuang, R.-Y., Beuchat, L.R., Angulo, F.J., 1995. Fate of *Salmonella* Montevideo on and in raw tomatoes as affected by temperature and treatment with chlorine. Appl. Environ. Microbiol. 61, 2127–2131.

Microbial Attachment and Persistence on Plants

Sima Yaron

Faculty of Biotechnology and Food Engineering, Technion – Israel Institute of Technology, Haifa, Israel

CHAPTER OUTLINE HEAD

Introduction

The number of outbreaks of foodborne illness arising from the consumption of fresh and fresh-cut produce has risen dramatically over the last two decades (Sivapalasingam et al., 2004), but this increase became more moderate in the last few years. From 1990 to 2005, fresh produce products were associated with 713 outbreaks in the U.S., resulting in 34,049 cases of illness. In addition, the average number of illnesses per a produce-related outbreak was significantly higher than those from other foods (Anon, 2007). In 2009 to 2010 fresh produce products were associated with 70 outbreaks caused by identified bacteria or viruses, resulting in 2,327 cases of illness, while foods originated from animals like poultry, beef, pork, eggs, dairy products, and seafood caused, together, 149 outbreaks with 4,615 cases. The

The Produce Contamination Problem. http://dx.doi.org/10.1016/B978-0-12-404611-5.00002-6

pathogen-commodity pairs responsible for most illnesses were *Salmonella* in eggs (2,231 illnesses), *Salmonella* in sprouts (493 illnesses), and *Salmonella* in vine-stalk vegetables (422 illnesses), while the pathogen-commodity pairs that caused the most hospitalizations were *Salmonella* in vine-stalk vegetables (88 hospitalizations), *E. coli* O157 in beef (46 hospitalizations), and *Salmonella* in sprouts (41 hospitalizations) (CDC, 2013).

Although the reasons behind the increase in prevalence of outbreaks associated with fresh fruit and vegetables are somewhat unclear, several factors most likely play an important role. First, the increase may, in part, be due to improved surveillance of produce commodities and outbreaks. Second, the consumption of fresh produce has increased significantly. From 1976 to 2009, U.S. consumption of fruits and vegetables increased by 3% and 13%, respectively, and the annual per capita consumption of fruits and vegetables increased by 8.4%, reaching 675 lbs. However, the increased consumption of fresh foods does not fully explain the increased incidence of outbreaks associated with these commodities. The incidence of foodborne outbreaks associated with leafy greens, for example, increased by 39% between 1996 and 2005, but the consumption of leafy greens increased by only 9% (Herman et al., 2008). Third, a significant shift was observed from processed fruit and vegetables to fresh consumption. In 2009, 46% of total fruit and vegetables consumption was in fresh forms (Cook, 2011). The growth in consumption of fresh foods was paralleled with an exponential growth in consumption of convenience foods such as fresh-cut fruits and bagged salads, which are more conducive to microbial growth and spoilage than the whole produce from which they are derived (Brandl, 2008). In the U.S., the weekly sales of prepared fruits and vegetables increased from $1298 and $567 per store in 2005, respectively, to $1587 and $804 in 2010 (Padera, 2010). Fourth, the produce industry has become increasingly global, with large volumes of produce being imported into the U.S. The increase in global trade and simultaneously the increase in the complexity of the supply chain for fresh produce make oversight difficult.

Analysis of identified outbreaks associated with produce from 1990 to 2007 showed that contamination of about a fifth of the products occurred on the farm, while about 80% of these outbreaks were associated with improper handling after leaving the farm (Anon, 2010). Indeed, fresh produce is grown in agricultural settings that are prone to contamination by microbial pathogens. Plants do not normally harbor enteric pathogens, but zoonotic bacterial pathogens are easily transferred from other sources. Preharvest sources of pathogenic microorganisms include soil, manure (or compost), irrigation water, water used for pesticide application, insects, and wild or domestic animals. Postharvest sources include human handling, harvesting and transport equipment, animals, dust, wash water, packing-shed equipment, improper storage, and other potential sources of cross-contamination.

Until recently, it was thought that enteric pathogens such as *Escherichia coli* O157:H7 and non-typhoidal *Salmonella enterica* survive poorly in the harsh environment encountered on plant surfaces, where microorganisms must survive sunlight, desiccation, nutrient limitation, and drastic temperature fluctuations, but recent research has shown this not to be the case. Enteric pathogens have been demonstrated to persist in a variety of agricultural settings including water, soils, manure, the plant rhizosphere, and even on exposed (foliar) plant surfaces (Brandl, 2006; Heaton and Jones, 2007). As a result of outbreaks occurring since the mid-1990s, the survival and dissemination of foodborne pathogens in agricultural environments has been studied in detail. More recently, the intimate interactions between enteric pathogens and plant tissue have begun to be scrutinized.

This chapter will discuss the attachment of foodborne pathogens to plant tissue and the mechanisms the pathogens employ to persist on/in the plants. Although a variety of organisms (bacteria, viruses, parasites, etc.) have been implicated in outbreaks arising from produce, this chapter will focus primarily on *E. coli* O157:H7, *S. enterica,* and *Listeria monocytogenes* because of the frequency of outbreaks associated with these pathogens and the depth to which they have been studied.

Ecological niches and introduction into the plant environment

The identification of routes of plant contamination by enteric pathogens is crucial to the design of intervention strategies to prevent contamination from taking place (Brandl and Sundin, 2013). *S. enterica, E. coli* O157:H7, and *L. monocytogenes* are normally found in the gastrointestinal tracts of warm-blooded farm animals. *E. coli* O157:H7 is traditionally associated with ruminant animals such as cattle, sheep, and goats (Erickson and Doyle, 2007). *S. enterica* is found frequently in poultry, but also in pigs, cattle, goats, waterfowl, rodents, and insects (D'Aoust, 1998). Both of these organisms are introduced into the plant environment by dissemination from their animal hosts. *L. monocytogenes* is found in cattle, birds, and fish, but is most often detected in soil, silage, and various aqueous environments. In a recent survey conducted on five farms in New York state *L. monocytogenes* prevalence was high among water samples, –27.6% (48/174), particularly among samples of surface water (Strawn et al., 2013).

At least 80 million dry tons of solid manure are generated annually by the beef, dairy, swine, and poultry industries in the U.S. (Edwards and Someshwar, 2000). Land application remains the most common and economic method to dispose and recycle this huge quantity of animal feces. Animal manures are used widely as fertilizers. They increase the amounts of inorganic compounds and organic matter in the soil, but also enrich the microbial load and diversity. Although the majority of microorganisms contained in manure are not pathogenic to humans, zoonotic pathogens have the potential to be transported from the manure to water, soil, food, and other areas of the environment. Thus proper composting is essential to ensure that pathogens from manure do not directly interact with growing plants. The carriage rate of pathogens in their animal reservoirs is not quite clear. For example, estimates of the prevalence of *E. coli* O157:H7 in cattle range from less than 1% to upward of 25% (Elder et al., 2000). Carriage rate studies differ in the type of animal surveyed (age), feeding regimens, and type of samples obtained (swabs, grabs, fecal pats, etc.), so conclusions as to the carriage rate are difficult to draw. Regardless of the carriage rate, infected cattle are known to shed anywhere between 10^1 and 10^7 cfu of *E. coli* O157:H7 per gram of feces (Besser et al., 2001). Given that typical cattle excrete 20 to 50 kg of feces per day, this provides a large "inoculum" of *E. coli* O157:H7 for the farm environment. The presence of "super-shedders," a few cattle in a herd that shed greater than 10^4 cfu/g feces, also present a significant source of *E. coli* O157:H7 in the produce growing environment (Matthews et al., 2006).

The fate of pathogens from manure depends on many variables, including the level of pathogen shedding by animals, conditions, and duration of manure storage, extraneous microbial interactions within stored manure, and interactions with water, soil, plants, and insects (Ziemer et al., 2010). A number of researchers have investigated the survival of *E. coli* O157:H7, *S. enterica,* and *L. monocytogenes* in manure from various animals, under different conditions such as temperature or aeration, presence of different manure amendments, and at a range of manure-to-soil ratios (Duffy, 2003). There is contrary information on the survival of human pathogens in manure representing on-farm conditions, as well as the fate of pathogens from manure in soil, water, and plants. While some reports have indicated that pathogens in manure do not survive long after they are applied to the soil, other studies contradict this, indicating longer survival periods in soil and water (Ziemer et al., 2010). Kudva et al. found that *E. coli* O157:H7 survived for more than 21 months in ovine manure at levels ranging up to 10^6 cfu/g manure (Kudva et al., 1998). Aeration of the ovine manure pile greatly reduced the survival time. Experiments with artificially inoculated bovine feces have also confirmed the survival of *E. coli* O157:H7 for greater than 40 days, dependent on initial inoculum and holding temperature (Wang et al., 1996). Additional work states that *S. enterica* may be persistent for longer durations than *E. coli* in bovine manure when kept under constant temperature and moisture conditions (Himathongkham et al., 1999). *L. monocytogenes* survived in the soil following manure spreading to land for even a longer period of time (Nicholson et al., 2005). Collectively, these studies indicate that enteric pathogens can survive for long periods of time in animal manures and composts or in soil fertilized with manure, and therefore remain in close proximity to growing crops.

Pathogens in manure may transfer to water either directly or through runoff. Studies have documented an increase in the levels of bacterial pathogens in water sources immediately after heavy rains (Cooley et al., 2007). Contaminated water has been implicated in several outbreaks arising from produce (such as tomato, 2005–2006 or shredded lettuce, 2006). Extensive laboratory research has demonstrated that enteric pathogens originating in manure can survive for long periods of time in water (Wang and Doyle, 1998). Sterilized clear water has often been used as a model system (Kolling and Matthews, 2001; Wang and Doyle, 1998), but the utility of these studies is questionable, since the introduction of an organic load (such as manure) greatly increases the survival of *E. coli* O157:H7 in water (Hutchison et al., 2005). Limited availability of good-quality water increases the use of low-quality water including raw or partially treated wastewater with high microbial loads (FAO/WHO, 2008; Jacobsen and Bech, 2012; Levantesi et al., 2011; Suslow, 2010; Tyrrel et al., 2006). *Salmonella* occurrence in various water bodies was reported worldwide with frequency of positive samples ranging from 3 to 100%, and its concentration was up to 10^4 viable cfu/ml (Levantesi et al., 2011).

Several field and greenhouse studies investigated the ability of enteric pathogens to transfer from contaminated water/soil/manure to the plants (Islam et al., 2004; Lapidot and Yaron, 2009; Solomon et al., 2002b). In most of these studies researchers used very high concentrations of pathogens (above 10^5 cfu/g) to quantitatively determine if the pathogen transfers to the plants. The high levels of contamination applied in these experiments (most often because of technical limitation in quantification of low levels of contamination) are usually not realistic in terms of contamination levels that possibly occur in the environment or during processing. Other researchers applied an enrichment step or microscopic analysis to determine the transfer and persistence of the pathogens without quantification of the exact numbers and showed that transfer to the plant occurs with low levels of contamination too (Mootian et al., 2009). Recent development of molecular-based methods reduced the detection limit, and indicated that even irrigation with water containing as little as ~300 cfu/ml results in persistence of *S*. Typhimurium on the plants for at least 2 days (Kisluk et al., 2012; Kisluk and Yaron, 2012).

A recent systematic review of risk factors for contamination of fruits and vegetables with enteric pathogens at the preharvest level has indicated several significant factors including an application of contaminated or non-stabilized manure, the use of spray irrigation with contaminated water, and growing produce on clay-type soil (Park et al., 2012). These findings, coupled with extensive field-based research, clearly indicate that pathogens can survive for long periods in contaminated water and that contaminated manure and water pose a serious threat to growing crops.

Outbreak investigations reveal sources and persistence of pathogens

Since the first large outbreaks in the early 1990s, specific pathogens have repeatedly been implicated in outbreaks arising from the same plants. For example, from 1990 to 2005, 24 outbreaks of *Salmonella* have involved seed sprouts, and 16 have involved melons (Anon, 2007). Melons were also involved in sporadic infections and outbreaks of *L. monocytogenes* (Laksanalamai et al., 2012; Varma et al., 2007). Spinach, lettuce, and other leafy vegetables have been involved in 29 outbreaks of *E. coli* O157:H7. In most outbreaks the source of the pathogen in the field was not identified, but traceback investigations into a few of these outbreaks have revealed details of the mechanisms of how these bacteria are introduced and persist on the plant surface. For example, an in-depth investigation into the origin of the *E. coli* O157:H7 outbreak, linked to spinach in 2006 (CFERT, 2007), revealed the presence of the outbreak strain in cattle feces, surface water, and feral pigs present near the fields where the spinach was grown. These investigations underscore the complexity of the preharvest environment and the ease with which plant tissue can become contaminated with foodborne pathogens.

Attachment of pathogens to plant tissue
The plant surface

Pathogens introduced onto the plant via water, manure, improper handling, or any other vector must attach themselves and proliferate or at least survive on plant surfaces. Most aerial plant surfaces are covered in cuticle, a hydrophobic material composed primarily of fatty acids, waxes, and polysaccharides. The cuticle prevents plant dehydration and also protects the plant from infiltration by microorganisms. The cuticle favors attachment of hydrophobic molecules. However, breaks in the cuticle may expose hydrophilic structures from within (Patel and Sharma, 2010).

Leaf topography is also an important factor in microbial adhesion. The surface roughness of the leaves depends on the nature of the plant and on the age of the leaves. The distribution of the pathogen on the leaf surface is highly heterogeneous. Cracks in the cuticle, or other damages that expose the epidermal cells of the plant surface are often sites at which bacteria colonize. The stomata provide protective niches for the bacteria, and also can serve as a source of nutrients.

In the field the leaves' surface is an inhospitable environment for enteric pathogens. It is subject to large swings in temperature and relative humidity, limited water or nutrient availability, and potential exposure to UV from sunlight. It also contains a large population of foliar microorganisms in large aggregates that may compete with foodborne pathogens for nutrients in this environment (Lindow and Brandl, 2003). Postharvest contamination may result in higher levels of contamination because cut leaves serve a better niche for *E. coli*, *Salmonella*, and *Listeria* (Ells and Truelstrup Hansen, 2006; Patel and Sharma, 2010; Takeuchi et al., 2000).

In addition to aerial parts, enteric pathogens are able to attach to the rhizosphere of different plant hosts, through which the pathogens can become internalized and further move to other parts. Microorganisms tend to attach to the root hairs formed by trichomes to increase the root surface and to the epidermis. Bacteria bind particularly well to ends of roots and wound sites and bind poorly to the root tips (Matthysse and Kijne, 1998). In contrast, *E. coli* strains preferred to attach to the root tips of alfalfa sprouts (Jeter and Matthysse, 2005). Rhizodermis cells secrete a wide range of compounds, including organic acid ions, inorganic ions, phytosiderophores, sugars, vitamins, amino acids, purines, and nucleosides, and thus root exudates can be used as a source of nutrients (Darrah, 1991). Under specific conditions such as in a hydrophonic cultivation system the probability of plant contamination was seven times higher from the roots than from the leaves for *E. coli*, *Salmonella*, and *Listeria* (Koseki et al., 2011).

Factors affecting attachment of pathogens to plant tissue

Attachment of foodborne pathogens to plant surface is one of the earliest steps required for survival through consumption. Bacteria need to establish initial contact with the plant surface to stabilize and survive in the plant environment and probably also to interact with the plant cells. Attachment is also the initial step for the formation of biofilm on the plant tissue. Regardless of the environmental source, recent data indicates that enteric bacteria can attach to growing plant tissue in a relatively rapid fashion, colonize specific microenvironments that may be plant-species specific, coexist with epiphytic bacteria to survive and grow, and persist for significant periods of time (see earlier reviews (Brandl, 2006; Solomon et al., 2006) for exhaustive information). Investigation of factors that affect the attachment of plant pathogens or symbiots to root or leaves surfaces has shown that after bacteria get into contact with plant surfaces, two processes may occur. Initial adhesion occurs during the first few seconds. This is a weak, reversible and unspecific binding that usually depends on physical factors such as hydrophobicity and charge. In the second phase of binding, a strong irreversible attachment may occur (Dunne, 2002). This process requires the synthesis of cell factors such as fimbriae, flagella, and polysaccharides. Moreover, attachment is initially enhanced by chemotaxis and motility. Studies of the attachment of foodborne pathogens to plants indicate that these bacteria use a similar scheme of attachment.

Laboratory experiments with excised plant tissue or intact whole produce indicate that human enteric pathogens attach rapidly. Ukuku and Sapers (2001) demonstrated that *Salmonella* deposited onto cantaloupe melons could not be washed off after just four hours of incubation. Experiments with *E. coli* and lettuce demonstrated irreversible attachment after just a few hours (Beuchat, 1999). Similar results were found with tomatoes (Iturriaga et al., 2003) or parsley (Lapidot et al., 2006) inoculated with *Salmonella*. Liao and Cooke (2001) used a laboratory model consisting of green pepper slices to study attachment of *Salmonella* Chester. They concluded that 30% of the bacterial inoculum firmly attached to the pepper's cut surface within 30 seconds. These firmly attached cells could not be removed by washing or agitation. These results were confirmed by Han et al. (2000) who found that *E. coli* irreversibly attached to green pepper after a short time and were not removed by washing.

Both host plant and the bacterial properties influence the efficacy in which bacteria attach to plants. Attachment to whole cantaloupes, for instance, was highest for *E. coli* and lowest for *L. monocytogenes*, but *Salmonella* exhibited the strongest attachment after storage in the refrigerator for up to 7 days. This difference was attributed to a linear correlation between bacterial cell surface hydrophobicity and surface charge and the strength of bacterial attachment to cantaloupe surfaces (Ukuku and Fett, 2002b). Attachment to basil, lettuce, or spinach leaves differed among *S. enterica* serovars, *S.* Typhimurium, and *S.* Senftenberg, for example, showed higher attachment compared with *S.* Agona or *S.* Arizonae (Berger et al., 2009). The attachment strength of *Salmonella* serovars to cabbage was significantly lower than that to lettuce (Patel and Sharma, 2010).

Some areas on the plant surface are preferred for attachment. Phyllobacteria have been shown to colonize at various sites in and on leaf surfaces, including the base of trichomes, at stomata, epidermal cell wall junctions, as well as in grooves along veins, depressions in the cuticle, and beneath the cuticle (Beattie and Lindow, 1999). It is less clear if human enteric pathogens like *E. coli* O157:H7, *S. enterica*, and *L. monocytogenes* demonstrate similar behavior. *S.* Thompson was shown to attach around stomata of spinach leaves and in cell margins, similar to where native bacterial biofilms and microcolonies were detected (Warner et al., 2008). Other *Salmonella* serovars aggregate near and within the stomata. However, the ability of *Salmonella* to colonize the surface around the stomata was observed only with certain serovars on specific plants (Golberg et al., 2011). A recent study has shown that attachment of *Salmonella* to artificially contaminated lettuce leaves differed in older leaf parts and leaf regions near the petiole. Moreover, higher levels of *S.* Typhimurium were localized close to the petiole. The bacteria displayed higher affinity toward the abaxial side compared to the adaxial side of the leaves (Kroupitski et al., 2011).

Damaged plant tissues are far more susceptible to colonization than undamaged tissue of the same type (Seo and Frank, 1999). Enteric pathogens attach preferentially to cut surfaces, where more nutrients may be available for their growth and survival (Boyer et al., 2007; Takeuchi and Frank, 2000). *E. coli* O157:H7 and *L. monocytogenes* attached in greater numbers to cut lettuce leaves compared to whole leaf surfaces, whereas *S.* Typhimurium attached equally well to both intact and cut surfaces (Takeuchi and Frank, 2000). Fresh-cut produce is therefore at even higher risk in terms of bacterial colonization compared to the whole product from which it was derived. Mechanical damage occurring during packing and transport may also result in favorable conditions for bacterial attachment. Damage of stems of lettuce plants resulted in the release of sugar-containing latex, which supported the growth and rapid increase of *E. coli* O157:H7 populations (Brandl and Amundson, 2008). This latex may also prevent gaseous and liquid sanitizers from penetrating tissues and inactivating bacteria.

Studies investigating the role of specific bacterial factors such as fimbria or flagella have shown contradictory results. While similar levels of attachment to lettuce were observed with live *E. coli* O157:H7, glutaraldehyde-killed *E. coli* O157:H7, and non-biological material (Solomon and Matthews, 2006), in other research specific bacterial genes were shown to be required for attachment on plant tissue, and some of them are also defined as attachment or virulence factors in animals. By developing a mutants-library, researchers identified defective *S.* Newport mutants attenuated in the attachment to alfalfa sprouts. Analysis of the mutants indicated the role of the bacterial cellulose, curli, capsule, and the sigma factor RpoS in attachment to alfalfa sprouts (Barak et al., 2005; Barak et al., 2007). In contrast, *Salmonella* mutants that do not form the main components of the biofilm matrix attached to parsley leaves in the same levels as the wild-type strain, but these mutants were more sensitive to disinfection of the leaves after storage (Lapidot et al., 2006). Deletion of *fliC*, encodes for components of the *S.* Senftenberg flagella, resulted in a significant reduction of adhesion (Berger et al., 2009), and deletion of SirA, a regulatory protein involved in *S. enterica* biofilm formation and expression of virulence genes, reduced the bacterial attachment to spinach leaves and tomatoes as well as to glass and polystyrene (Salazar et al., 2013).

Environmental factors affect the attachment too. The adhesion of pathogens in wash water to fresh cucumber surfaces depends on temperature, and was less extensive at lower temperatures. The effect of dewaxing of fruits on adhesion depends on the bacteria. While adhesion of *Listeria* to dewaxed fruits was higher than to waxed fruits, the opposite was reported for *S.* Typhimurium and *Staphylococcus aureus* (Reina et al., 2002).

Collectively, these studies demonstrate that enteric pathogens attach rapidly and irreversibly to produce surfaces. Attachment depends on plant and bacterial factors as well as on environmental conditions. In addition, attachment levels usually differentiate between intact and damaged plant tissues. Attached pathogens are extremely difficult to remove with current washing or agitation regimens.

Biofilm formation on produce surfaces

Following attachment of the pathogens onto produce surfaces, bacterial pathogens may become entrapped in a biofilm (Annous et al., 2005). Biofilms are defined as "an assemblage of microorganisms adherent to each other and/or to a surface and embedded in a matrix of exopolymers" (Costerton et al., 1999). It is estimated that between 30 and 80% of the total bacterial population existing on plant surfaces are embedded in biofilms (Lindow and Brandl, 2003). The formation of biofilms by plant epiphytic or pathogenic bacteria has long been known (Danhorn and Fuqua, 2007); however, the discovery that enteric pathogens could establish biofilms on plant surfaces was surprising. In the last few years the number of studies on the formation of biofilms by foodborne bacteria on produce surfaces has expanded. *Salmonella, E. coli, Listeria, Campylobacter*, and *Shigella* have been found to form distinct biofilms on the surfaces of produce ranging from tomatoes to melons to parsley (Agle, 2003; Annous et al., 2005; Iturriaga et al., 2007). Studies determined a correlation between the ability to form biofilms and the attachment to fresh produce and survival by showing that strains that show the most biofilm formation *in vitro* are also attached to plants at higher populations, and/or better survive after disinfection (Lapidot et al., 2006; Patel and Sharma, 2010).

Bacterial cells embedded in biofilms are significantly different from their planktonic (free-floating) counterparts in terms of physiology. The formation of biofilms by bacterial cells on plant surfaces is likely a survival strategy to withstand the harsh environment of the plant surface. Similar to biofilms on food-processing surfaces, bacteria embedded within biofilms on plant tissue are more difficult to remove and more resistant to inactivation than their planktonic counterparts (Chmielewski and Frank, 2003). On parsley plants, for instance, resistance to disinfection treatments (i.e., chlorination) was improved in biofilm producer *Salmonella* (Lapidot et al., 2006). These differences allow biofilm-associated cells to survive the harsh environment of the plant surface in the field as well as during harvest, transport, in the presence of hypochlorite or other sanitizers, and storage.

Several environmental conditions have an impact on biofilm production. For example, biofilm formation of *Salmonella* has been reported to be maximal under reduced nutrient availability, aerobic conditions, low osmolarity, and a low temperature (28°C) (Gerstel and Romling, 2003), and all these conditions exist on plant surface rather than the gut environment. Biofilm-producing isolates of *Salmonella*, which were isolated during tomato outbreaks, adhered and attached better to tomato leaflets (Cevallos–Cevallos et al., 2012b). Two main extracellular components play an important role in the biofilm matrix of *Salmonella*: the exopolysaccharide cellulose and curli fimbriae. A screen of *S.* Newport mutants with lower attachment ability to alfalfa sprouts identified genes code for the curli fimbria (AgfB) and RpoS that regulate the production of curli, cellulose, and adhesions that are important for biofilm formation (Barak et al., 2005). The *bcsA* gene, coding to proteins involved in synthesis of cellulose, was also found to be important for attachment to alfalfa sprouts (Barak et al., 2007). In a further study two additional genes, essential in biofilm formation and swarming, were also found as important factors for infection of alfalfa sprouts (Barak et al., 2009).

Cellulose and curli were also involved in transmission of *S.* Typhimurium from irrigation water onto parsley leaves grown in a greenhouse (Lapidot and Yaron, 2009). Other researchers have shown that bacterial cells introduced to the leaf surface have a better chance of surviving when they are deposited on or in aggregates of other bacteria (Monier and Lindow, 2005). These aggregates are characterized by an exopolysaccharide matrix that contains a dense population of bacterial cells. If foodborne pathogens are in these aggregates, this may limit the effectiveness of sanitizer treatments on produce. Indeed, *S.* Thompson and *Pantoea agglomerans* were shown to form aggregates on cilantro leaves (Brandl and Mandrell, 2003). Indigenous microorganisms had a positive effect on the initial attachment and survival of foodborne pathogens, as was shown for *E. coli* O157:H7 on spinach (Carter et al., 2012). *Wausteria paucula* supported the survival of *E. coli* O157:H7 on lettuce leaves and in the rhizosphere (Cooley et al., 2006). The fungal phytopathogens *Cladosporium cladosporiodes* and *Penicillium expansum* promoted the colonization and infiltration of *Salmonella* in wounded cantaloupe tissue (Richards and Beuchat, 2005). These studies demonstrate that indigenous microorganisms may aid to attachment and long-term survival of foodborne pathogens on the plant surface and during washing with sanitizers. However, it is not clear if this positive effect results from embedding in the already existed biofilms and aggregates or from other interactions like lesion and release of nutrients.

Internalization and persistence

Bacterial foodborne pathogens can attach and persist on produce commodities, but many laboratory studies have reported the ability of these bacteria to survive in internal tissues of plants and perhaps even in the plant cells. There exists a wide assortment of locations in which internalized human enteric pathogens may reside within fresh produce, including the vasculature, intercellular tissues and stomata or cracks of the cuticle, as well as entrapped within crevices (Erickson, 2012). Internalization can occur through several mechanisms pre- or postharvest: the immersion of fruits and leaves into bacterial suspensions, uptake of pathogen through plant roots, and the infiltration of these bacterial cells through naturally opened, damaged or cut tissues. Three recent comprehensive reviews have summarized this topic (Deering et al., 2012; Erickson,

2012; Hirneisen et al., 2012). Researchers have shown that enteric bacteria, able to colonize the rhizosphere, are more likely to colonize internal tissues of alfalfa seedlings (Dong et al., 2003). Still, it is unclear if the presence of aggregates or biofilms of food-borne pathogens on produce surfaces makes them more or less likely to internalize to plant tissues, thus more work is needed to truly establish this correlation.

Several factors can influence the opportunity and capability of foodborne pathogens to internalize in plant tissues. The route of potential internalization and uptake of bacterial foodborne pathogens by produce differs depending on the pathogen, produce commodity, and stage of development of the plant or fruit. Several studies on internalization into harvested fruits have been conducted, and evidence of pathogen internalization has been observed when there has been a temperature differential between the fruit and the inoculum. A positive temperature differential (Buchanan et al., 1999) occurs when a warm piece of fruit is immersed in a cooler fluid. This causes contraction of gases in internal spaces within the fruit, resulting in a partial vacuum that draws in some of the fluid through pores in the fruit surface. When the fluid is contaminated or intentionally inoculated with pathogenic bacteria, these bacteria infiltrate into the fruit (Penteado et al., 2004). Laboratory experiments that describe internalization through immersion in inoculum were usually conducted with high levels of bacteria (10^6–10^8 cfu/ml). Mangoes, held in 46°C water for 90 minutes and then immersed in an inoculum of *S.* Enteriditis at 22°C, were more likely to have bacteria internalized near the stem end (83% of samples) than the blossom end (8% of samples) of the fruit (Penteado et al., 2004). Internalization of *S.* Thompson was also observed in tomatoes exposed to an inoculum under a temperature differential. The bacteria were detected within the core tissue segments immediately underneath the stem scars, and internalized population was affected by tomato variety and time of removal of the stems (Xia et al., 2012). Only 2.5% and 3.0% of oranges exposed to inocula of *E. coli* O157:H7 and *Salmonella*, respectively, under a positive temperature differential contained the internalized pathogens (Eblen et al., 2004). Infiltration of *E. coli* O157:H7 was observed infrequently in cores of apples exposed to an inoculum under a positive temperature differential, and this low frequency of internalization was further decreased when cold apples were placed in a cold inoculum (Buchanan et al., 1999). Infiltration into fruits exposed to the inoculum was observed in almonds, pecans, and lettuce even without negative temperature differential (Beuchat and Mann, 2010; Danyluk et al., 2008).

Processing operations related to leafy greens may also encourage bacterial infiltration. Lettuce leaves inoculated with *E. coli* O157:H7 and subsequently vacuum cooled had higher populations (ca. 1-log cfu/g) of the pathogen infiltrated into the leaf than those leaves that were not subjected to vacuum cooling (Li et al., 2008). Cutting fruits or leaves during processing is another route of internalization of foodborne pathogens. *E. coli* O157:H7 cells attached preferentially to cut lettuce edges and penetrated into cut surfaces. The penetration was more efficient when the cut leaves were stored under 21% oxygen at 4°C compared to storage at 7, 25, or 37°C. Moreover, penetration at 4°C was greater under 21% oxygen than 2.7% oxygen (Takeuchi and Frank, 2000; Takeuchi et al., 2001).

Preharvest pathogen internalization was described by three major routes: (i) internalization through natural opening in the plant surface (stomata, sites of lateral root emergence, etc.); (ii) internalization through sites of biological or physical damage; (iii) bacteria are pulled into internal tissues along with water (Deering et al., 2012). The uptake of foodborne pathogens through roots, leaves, or fruits of plants is a topic that has received much scrutiny in the last decade. Previous studies have varied in the methods used to grow plants (in soil, in hydroponic medium, and composted manure), the method of introduction of these pathogens to plants (on roots, in soil, in irrigation water), the growth medium, or whether the pathogens were delivered in combination with bacterial phytopathogens and soil parasites. The stage of plant development (seeds, seedlings, young and mature plants) may also affect the degree and extent of internalization of foodborne pathogens. Previous internalization studies also used surface sterilization techniques that were varied in their effectiveness to ensure that bacteria recovered in these studies were from internalized tissue.

Root uptake studies mainly focused on the internalization of *E. coli* O157:H7 and *Salmonella* to different crops, but there are differences in the degree of invasiveness. Two methods of plant growth, soil and hydrophonic growth, were usually used. The populations of *E. coli* O157:H7 were determined 9 and 49 days after exposure to 10^2 cfu *E. coli* O157:H7/ml of hydroponically grown cress, lettuce, radish, and spinach plants. *E. coli* O157:H7 internalized the plant tissues after 9 days of growth in lettuce, radish, and spinach tissues, but were not detected in internal tissues of cress (Jablasone et al., 2005). Populations of *E. coli* O157:H7 in spinach fell from 2.5 log cfu/g after 9 days to undetectable levels (< 1 log cfu/g) when assayed at 49 days. *Salmonella* populations were also introduced to the same seedling varieties on hydroponic media. After 9 days *Salmonella* was detected in internal tissues of lettuce and radishes (1–1.6 log cfu/g), but not spinach and cress plants (Jablasone et al., 2005). No internalized *Salmonella* cells were observed in any plants examined after 49 days. *L. monocytogenes* did not invade in all these commodities under the same experimental conditions. The low persistence of *E. coli* O157:H7 or *Salmonella* cells in the leaf tissues of hydrophonicly grown plants suggests that cells may be under physiological and nutritional stress in the vasculature of plants or that plant defenses are effective in killing enteric human pathogens.

Conflicted results were obtained when researchers applied inoculated soil, in which high, little, or no internalization was observed. Franz et al. grew lettuce in hydroponic media and in soil and examined the presence of *E. coli* O157:H7 on surface-sterilized root and leaf tissues. While soil-grown seedlings were positive (contained up to 4 log cfu/g), no *E. coli* O157:H7 cells were recovered from internal tissues of lettuce leaves or root tissues grown in hydroponic media. It was suggested that growth in soil induces root damage, which would allow greater internalization (Franz et al., 2007). Other experiments usually showed higher levels of penetration in hydroponically grown plants. Internalization of *E. coli* O157:H7 did not occur when mature spinach plants were exposed to 10^3 or 10^6 cfu/g in pasteurized soils, but was observed sporadically when plants were grown in hydroponic medium inoculated with 10^7 cfu/ml (Sharma and Donnenberg, 2008). Internalized populations of 3.7

and 4.3 log cfu *E. coli* O157:H7/shoot were recovered in this study after 14 and 21 days of growth in hydroponic solution, respectively, and were recovered sporadically when replanted into soil at day 28. It was suggested that bacterial motility in hydroponic solution may provide more opportunity for uptake than in soil (Hirneisen et al., 2012; Sharma et al., 2009). Internalization of foodborne pathogens may be more frequent in hydroponic systems also because there is less competition from rhizospheric bacteria that are present in soil and may have physiological fitness advantages when colonizing root tissue, providing a platform for internalization (Klerks et al., 2007a).

Solomon et al. (2002b) examined lettuce seedlings after exposure to high populations of *E. coli* O157:H7 in contaminated irrigation water or contaminated manure slurry in contact with roots. *E. coli* O157:H7 were recovered from internal lettuce tissue 5 days after exposure to 10^7 cfu *E. coli* O157:H7/ml irrigation water, indicating that root uptake of the pathogen did occur. This work also showed that the uptake of the pathogen through the roots of lettuce seedlings was more frequent when populations of *E. coli* O157:H7 were high. Uptake and internalization of *E. coli* O157:H7 into root and leaf stem tissue in 45-day-old romaine lettuce plants after application of 10^9 cfu/ml to the soil resulted in approximately 10^2 cfu of *E. coli* O157:H7/g in lettuce tissues after 5 days (Solomon and Matthews, 2005). Populations did not change significantly over 5 days after application, indicating that cells survived but did not grow within lettuce tissue. In another study, crisphead lettuce (*Lactuca sativa*) seedlings planted in soil and manure contaminated with *E. coli* O157:H7 (10^4 cfu/g) did not result in internalization of the pathogen in either 3-week-old seedlings or 7-week-old lettuce plants and leaves (Johannessen et al., 2005). *Salmonella* penetrated the upper parts of parsley plants drip-irrigated with water containing high concentrations of *S*. Typhimurium. In this experiment the bacteria survived in the plant for at least 3 weeks (Lapidot and Yaron, 2009). When low levels of *E. coli* O157:H7 ($10–10^4$/ml or g) were used to inoculate soil or irrigation water for growth of lettuce plants 20 to −30% positive plants containing *E. coli* O157:H7 in internal locations were observed. The bacterial levels were very low and detected only after enrichment (Mootian et al., 2009). On the other hand, when spinach, lettuce, and parsley plants were drip-irrigated in the field with contaminated water, internalization of *E. coli* O157:H7 via plant roots was rare, and when it did occur, the pathogen did not persist 7 days later (Erickson et al., 2010).

Several studies indicated that under the same growth conditions higher bacterial inocula leads to higher levels of internalized population, and lower inocula results in little or no internalization. However, it is possible that in low inocula the internalized bacteria are below detection level (Kisluk et al., 2012). To date, most studies were conducted with plants exposed to very high concentrations of the pathogen (>6 log cfu/ml), but following improvement in methods used to detect the pathogens recent studies targeted the low levels as well. Pathogen internalization occurred in 25% of spinach leaves when plants were sprayed with irrigation water containing *E. coli* O157:H7 at 6 log cfu/ml, but did not occur with 4 log cfu/ml (Erickson et al., 2010). Kisluk et al. (2013) has shown that irrigation of parsley results in detection of *Salmonella* on leaves even when water contained 300 cfu/ml. The internalized bacteria were approximately 1.5% of total parsley leaves-associated bacteria (Kisluk and Yaron, 2012).

Bacterial strain and serovar as well as plant type have a significant role in internalization of the pathogen and its survival in the plant and may explain part of the diverse results in different studies. Franz et al. (2007) placed 8-day-old lettuce seedlings in soil inoculated with either *E. coli* O157:H7 or *S.* Typhimurium at 10^7 cfu/ml and grew the plants for 18 days. *E. coli* internalized more readily (3.95 log cfu/g) and at significantly higher levels than populations of *S.* Typhimurium (2.37–2.57 log cfu/g) (Franz et al., 2007). It is possible that *E. coli* O157:H7 and *Salmonella* spp. have different capabilities to internalize to the root tissues. Klerks et al. (2007a, 2007b) demonstrated that lettuce seedlings planted in manure-amended soil inoculated with *S.* Typhimurium or *S.* Enteritidis did not show evidence of internalized bacteria when plants were analyzed after six weeks, however, *S.* Dublin did internalize to the plants, indicating that *S.* Dublin may be more adept at surviving in internal tissues of lettuce plants than other *Salmonella* serovars. Similarly, *S.* Senftenberg and *S.* Typhimurium showed a different ability to colonize basil plants, when plants were drip- and spray-irrigated under the same environmental conditions (Kisluk et al., 2013). The same conditions were also applied to irrigate parsley plants. Much higher *S.* Typhimurium population levels persisted in parsley, and the authors suggested that the basil essential oils, volatile molecules with antimicrobial activity, probably reduce *Salmonella* levels in basil (Kisluk et al., 2013; Kisluk and Yaron, 2012).

In addition to plant type, the plant age at the inoculation as well as environmental factors affect the susceptibility of the plant to internalization by enteric pathogens. Stomata, hydathodes, trichomes, and other surface structures are potential sites of bacterial infiltration. Internalization may be a result of a passive uptake through the movement of water, but recent studies have shown that presence of *Salmonella* near the stomata depends on active motility of the bacteria and on illumination of the leaves (Kroupitski et al., 2009). Since the density of these surface structures varies from species to species and ages of the plants, it is expected to observe differences in internalization (Erickson, 2012), as indeed was seen in the case of internalization of *Salmonella* through the stomata in leaves of different plants (Golberg et al., 2011). Younger lettuce leaves and roots were associated with a greater risk of contamination of both, *Salmonella* and *E. coli* O157:H7 (Brandl and Amundson, 2008; Gorbatsevich et al., 2013). Contrary to these studies, other studies showed that older plants are more susceptible to internalization than younger plants. *S.* Newport was recovered from internal tissues of 33-day-old lettuce plants two days after inoculation, but not in the 17-day-old plants (Bernstein et al., 2007a). The reason for these differences may be related to differences in sensitivity of old and young leaves in each plant. Recently it was shown that on young plants of lettuce, the older leaves supported *E. coli* survival better compared with the young and middle-aged leaves, while on nearly mature plants, pathogen population sizes were significantly higher on the old and young leaves compared with middle-aged leaves (Van der Linden et al., 2013). Differences in *Salmonella* colonization and internalization were observed in different seasons (Golberg et al., 2011; Kisluk and Yaron, 2012). Kisluk and Yaron (2012), for example, have reported that higher levels of *S.* Typhymurium were detected in the phyllosphere of parsley when plants were irrigated with contaminated water during the night compared to irrigation during the morning and during winter compared to other seasons (Kisluk and Yaron, 2012).

Presence of endogenous bacteria in soil or on the plant may affect the internalization of enteric pathogens as well. This effect can increase or decrease internalization of enteric pathogens. Negative interactions include competition on nutrients and binding sites. Positive interactions may be when endogenous bacteria degrade cell-wall polymers and thereby increase nutrient sources, when they produce biofilms, when they suppress the plant defense response, or when they create sites for internalization (Deering et al., 2012). Internalization of *S.* Newport and *E. coli* O157:H7 to *Arabidopsis thaliana* was compared in autoclaved and non-autoclaved soil. Internalization was higher in the autoclaved soil probably because of suppression of adhesion and growth of the pathogens by bacteria such as *Enterobacter absuriae* colonizing the root surface (Cooley et al., 2003). The coinoculation of the phytopathogen *Pseudomonas syringae* with *E. coli* O157:H7 in six-week-old spinach plants did not enhance the survival of the human pathogen when internalized through a vacuum-infiltration process applied to roots of plants (Hora et al., 2005), but *Salmonella* populations on tomato plants increased significantly when it inoculated with *P. syringae* (Meng et al., 2013). Mechanical disruption and infection of roots of spinach plants with nematodes, preceding inoculation of soil with *E. coli* O157:H7 (10^7 cfu), resulted in internalization of *E. coli* O157:H7 in root tissues (24/24 plants) but not in leaves of spinach plants (0/24 plants) (Hora et al., 2005).

Like postharvest routes of internalization, damages in the plant tissue or naturally occurring lesions may serve as location of invasion. *Salmonella* was found in higher rates in commercial produce when produce was damaged by soft-rot pathogens than with healthy produce (Brandl, 2008). When roots of maize grown in *E. coli* O157:H7 contaminated hydroponic solution were decapitated, internalization of the pathogen was observed (Bernstein et al., 2007b). A rubbing of contaminated lettuce leaves increased internalization of *E. coli* O157:H7 compared with leaves that were left untouched (Erickson et al., 2010).

A few studies were done recently to examine intracellular location of foodborne pathogens. Schikora et al. (2008) showed that *S.* Typhimurium is present within root hair cells of *Arabidopsis* already 3 h after infection. After longer incubation time the bacteria were visible also in other rhizodermal cells. *S.* Typhimurium was also found in tobacco cells (Shirron and Yaron, 2011). In these cases the frequency of internalization was very low. Recently micro-colonies of *E. coli* O157:H7 were detected inside the cell wall of epidermal and cortical cells of spinach and *Nicotiana benthamiana* roots (Wright et al., 2013). A further study is warranted to confirm these observations, and to determine the mechanism of invasion and its significance.

Collectively, the results from the presented studies demonstrate that it is possible for plants to internalize bacterial foodborne pathogens, but this phenomenon likely requires a constant high population of the pathogen. Internalization of enteric pathogens may involve passive diffusion through open structures, water uptake by the roots, or damages in the plant surface, but in many cases it seems to be an active process. In addition, a set of conditions encourages uptake of pathogens like the cutting of leaves and fruits or the positive temperature differential that promotes the internalization of bacteria to fruits. Moreover, the hydroponic growing environment may be more likely to encourage uptake of bacteria. Despite differences in the experimental results, evidence indicates that under specific conditions foodborne pathogens are able to enter the plant tissues. Internalized pathogens are believed to pose a distinct threat. Because of their location within the tissues of otherwise healthy fruits and vegetables, internalized pathogens can transit through the food chain undetected. In addition, no method of treatment or sanitation that is currently used in the food industry has been proven capable of inactivating these internalized organisms in foods eaten raw.

Specific interactions of the pathogens with commodities

Because of the repeated implication of lettuce, cantaloupe, and tomatoes as sources of outbreaks, the following section will review research on the attachment and survival of specific pathogens to these commodities.

Escherichia coli O157:H7 and lettuce

In the period from 1990 to 2005 leaf lettuce was linked to more outbreaks than any other singular type of produce (Anon., 2007). This distinction has much to do with the popularity of lettuce, its wide distribution, the use of cut (bagged) lettuce as a salad ingredient, and the physical nature of the plant itself. Lettuce is grown in close proximity to the ground, allowing for extensive contact with surrounding soil. Lettuce leaves are easily damaged by mechanical disruption, making them more conducive as an environment for bacterial attachment and growth. In addition, lettuce is most often consumed raw as opposed to other vegetables that may be cooked. Lettuce appears to be particularly susceptible to *E. coli* O157:H7 contamination. It was reported that 29% of all lettuce-related outbreaks were caused by *E. coli* O157:H7 (Franz and van Bruggen, 2008). However, other highly virulent non-O157 species associated with lettuce outbreaks such as *E. coli* O145 were also reported. Surveys of farm-level and store-purchased lettuce report extremely low levels of the pathogen, yet outbreaks continue to occur (Delaquis et al., 2007).

Beuchat (1999) demonstrated that *E. coli* could attach to cut lettuce tissue when suspended in either bovine manure or 0.1% peptone. Low levels of bacteria (ca. 10^2 cfu/g) introduced onto lettuce leaves survived 15 days of storage at 4°C. Treatment of contaminated leaves with 200 ppm of chlorine was no more effective at inactivating the bacteria than treatment with water alone, and both had a minor effect. Similar results were reported by Delaquis et al. (2002) for lettuce pieces inoculated with *E. coli* O157:H7 and then immediately washed in both cold and warm water containing 100 ppm chlorine. Other studies indicate that *E. coli* O157:H7 (as well as other enteric pathogens) attached preferentially to cut edges of lettuce pieces compared to the intact surface (Takeuchi and Frank, 2000). Seo and Frank (1999) utilized confocal microscopy to visualize cells of *E. coli* attached at stomata and trichomes of cut lettuce plants. These observations led the authors to conclude that attachment sites for *E. coli* were similar to those reported for phytopathogens.

The attachment of *E. coli* to growing lettuce plants under greenhouse or field conditions also has been determined. Collectively, these experiments demonstrate that plants contaminated by direct contact with the organism, whether carried in manure or water, result in persistent contamination (Cooley et al., 2006; Franz et al., 2007; Solomon et al., 2002a; Solomon et al., 2002b). After irrigation with contaminated water the bacteria are located on the surface or in root and leaf tissues. A recent study has shown that *E. coli* O157:H7 was recovered from 100% of irrigated lettuce plants, but after gentamicin treatment aimed to kill unprotected bacteria on the surface, *E. coli* O157:H7 was recovered from 23% of the plants for the leaves and 31% of the plant for the roots. The internalized bacteria were approximately 0.5% of total lettuce-associated bacteria (Wright et al., 2013).

Under field conditions the *E. coli* O157:H7 population in leaves persisted for several months, but decreased over time (Oliveira et al., 2012). Furthermore, environmental *E. coli* isolates such as avian pathogenic *E. coli* (APEC) survived at higher populations and for longer durations compared to *E. coli* O157:H7 (Markland et al., 2012). Solomon et al. (2002a) found that plants spray-irrigated once with water containing 10^7 cfu/ml of *E. coli* O157:H7 remained culture-positive for 20 days. Treatment of spray-irrigated plants with 200 ppm chlorine failed to completely inactivate the organism. Islam et al. (2004) planted lettuce seedlings into manure-fertilized soil that had been inoculated with a nonpathogenic strain of *E. coli* O157:H7. Lettuce plants harvested 77 days after planting were positive for the organism. Similar levels of persistence of *E. coli* in the lettuce rhizosphere and phyllosphere have been reported by other authors using real-time qPCR to quantify these levels (Ibekwe et al., 2004). A recent study indicates that the pathogen survives on lettuce significantly better in fall than in spring (Oliveira et al., 2012).

A number of recent authors have investigated the role of cell-surface charge, presence of divalent cations, hydrophobicity, capsule production, curli production, and other specific bacterial adherence mechanisms and their role in attachment of *E. coli* to lettuce tissue (Boyer et al., 2007; Hassan and Frank, 2003, 2004). Collectively, these studies have shown very little correlation between the presence of cell-surface appendages, charge, or hydrophobicity on the ability of the bacteria to attach to lettuce tissue. The possibility that the interaction between lettuce and surface-associated *E. coli* is based on simple entrapment is supported by studies using live and glutaraldehyde-killed cells of *E. coli,* which demonstrate little difference in the numbers of live and dead bacteria retained on the lettuce surface. Abiotic fluorescent microspheres adhered to lettuce pieces as well as *E. coli* O157:H7 (Solomon et al., 2006). On the other hand, studies have found significant molecular interactions between *E. coli* O157:H7 and plant tissue (Matthysse et al., 2008; Shaw et al., 2008; Torres et al., 2005). A microarray study aimed to examine transcriptional changes in *E. coli* attached to intact lettuce leaves showed that up to 10% of the genes were affected. Genes involved in biofilm modulation and curli production were up-regulated. The largest group of down-regulated genes consisted of those involved in energy metabolism (Fink et al., 2012). Another microarray study showed that after being exposed to lettuce lysate *E. coli* up-regulates multiple virulence and motility genes and shifts a great part of its metabolism to enable it to utilize substrates known to be prevalent in lettuce such as maltose and sucrose, and to survive oxidative stress, osmotic stress, and activity of toxic plant compounds (Kyle et al., 2010). A similar study of *E. coli* attached to the lettuce rhizosphere has shown that genes involved in attachment and biofilm formation, protein synthesis, and stress responses were up-regulated (Hou et al., 2012).

Although no genetic elements have been definitively identified as essential for attachment, genes involved in biofilm or capsule formation such as the curli fibers have a role in attachment of *E. coli* to lettuce leaves (Fink et al., 2012). Mutants deficient in cellulose production, colanic acid, and poly n-acetylglucosamine reduced the ability to bind to alfalfa sprouts (Matthysse et al., 2008), and increased capsule production led to greater attachment of *E. coli* O157:H7 to lettuce (Hassan and Frank, 2004). Mutants deficient in OmpA (adhesion) were also unable to bind to alfalfa as strongly as wild-type cells (Torres et al., 2005). Shaw et al. (2008) demonstrated that *E. coli* O157:H7 cells used the type III secretion system (T3SS) to attach to leaves of arugula, spinach, and lettuce when leaves were immersed in high populations (10^6 cfu/ml) at 37°C.

Altogether, this irreversible attachment and adaptation to the plant environment triggers the expression of virulence factors and enhances the resistance of the pathogen to oxidative compounds and sanitizers, and may reveal the association of fresh cut leafy greens such as lettuce with *E. coli* O157:H7 outbreaks.

Salmonella and tomatoes

Examination of domestic nontyphoid salmonellosis outbreaks associated with agricultural row crops revealed that the majority of these outbreaks are associated with either tomatoes or cantaloupes consumption (Barak et al., 2011). Between 1998 and 2008 at least 12 outbreaks of salmonellosis were attributed to the consumption of tomatoes (Barak et al., 2008). These 12 outbreaks resulted in 1990 culture-confirmed infections; however, since approximately 97.5% of all *Salmonella* infections are not confirmed, these outbreaks may have resulted in almost 79,600 illnesses (CDC, 2007). Outbreaks arising from tomatoes tend to be quite large and widely dispersed, indicating a point source of contamination early in distribution (Greene et al., 2008). Unlike lettuce, tomatoes are not grown in close proximity to the ground and as such, should be less likely to be contaminated with manure or compost. In addition, the tomato is covered with a cuticle, a heterogeneous layer composed mainly of cutin, long-chain fatty-acid derivatives (wax lipids), and secondary metabolites such as flavonoids. The cuticle plays a role in the fruit development and protection against biotic and abiotic stress conditions (Adato et al., 2009). The external surface of a tomato should make it more difficult for bacteria to attach and grow, compared to the high surface area of leafy vegetables. Despite these differences, tomatoes continue to be identified as vehicles for outbreaks. On the other side, the mature tomato fruit contains simple sugars, sugar alcohols, fatty acids, organic acids, and amino acids, which can be utilized by the external bacteria. The amount of each compound depends on the maturity stage of the fruit and differs among cultivars (Carrari et al., 2006).

Much like early work with lettuce, the attachment, survival, and proliferation of *Salmonella* on tomatoes was investigated in laboratory-scale experiments in response to outbreaks linked to tomatoes (Iturriaga et al., 2003; Zhuang et al., 1995). *Salmonella* survived well on tomato surfaces following immersion into contaminated water (Zhuang et al., 1995). Storage at temperatures above 10°C resulted in rapid bacterial growth over a short period of time (Zhuang et al., 1995). The temperature differential between the tomato and the wash water was found to be extremely important since immersion of warmer tomatoes into colder water resulted in infiltration of *Salmonella* into the internal portion of the tomato core. Lastly, chopped ripe tomatoes supported significant growth of *Salmonella* at 20 and 30°C. This raises the possibility that contamination of tomatoes with low levels of *Salmonella* would result in high levels of growth after chopping and holding at higher temperatures. Iturriaga et al. (2003) used scanning electron microscopy (SEM) to visualize the interactions between *S.* Montevideo and tomato surfaces. They found that attachment occurred rapidly and irreversibly and attributed this attachment to bacterial cells associating with the cuticle. These results were confirmed when Iturriaga et al. (2007) found relative humidity as well as temperature to be important to the persistence of *Salmonella* on tomatoes. A study by Shi et al. (2007) raises the possibility that outbreaks of *Salmonella* arising from tomatoes may be serovar-specific since some strains grew better than others on ripe fruits.

Greenhouse-level studies also have been conducted to investigate interactions between *Salmonella* and tomato plants. Studies of tomato plants grown in *Salmonella*-inoculated hydroponic solutions showed that the pathogen was associated with the stems, leaves, and the hypocotyl regions of plants (Guo et al., 2002a). As expected, more consistent *Salmonella* populations were recovered from the hypocotyl region than the stem of leaf tissues, regardless of whether or not roots were cut before placing in hydroponic solution (Guo et al., 2002b). Barak et al. demonstrated contamination incidences ranging from 6.3 to 61.1% of *S. enterica* in the phyllosphere of various tomato cultivars when seeds were planted in contaminated soil (Barak et al., 2008). The authors further hypothesized that since soil-borne *Salmonella* demonstrated poor attachment to tomato seedlings, soil may not be a common route of preharvest contamination. However, experiments conducted by Guo et al. (2002a) demonstrated that *Salmonella* present in soil in contact with the stem scars of mature green tomatoes could survive for at least 14 days. An increase in population by 2.5 log cfu per tomato was observed (dependent on storage conditions), as well as a potential contamination of the internal portions of the fruit. Field-level studies conducted by the same authors indicated that *Salmonella* introduced onto flowers of tomato plants, both prior to or after fruit set or injected into the stem adjacent to the flowers (to mimic physical damage to the plant), could result in contaminated tomatoes at harvest (Guo et al., 2001). These results indicate that *Salmonella* attach well to flowers and can survive through fruit set and be present at harvest (up to 49 days). Furthermore, the harvest of contaminated tomatoes from plants that had received stem inoculation demonstrates systemic movement of the bacteria from the stem into the tissue of the fruit (Guo et al., 2001).

Hintz et al. (2010) indicated that irrigation with contaminated water is a potential source of fruit contamination, however, evidence that *S. enterica* is able to enter tomato plants through contaminated irrigation water remains inconsistent. Results of this study showed that application of *S.* Newport to the root zone via irrigation water can result in the attachment of *Salmonella* to the roots followed by contamination of different tomato plant tissues. Cevallos–Cevallos et al. (2012a) showed that *Salmonella* may even transfer from aerosols produced by rain to tomato fruits. A recent study has confirmed that *Salmonella* is capable of internalizing tomato plants through the roots and blossoms (Zheng et al., 2013), but other researchers were not able to recover *Salmonella* from the plant tissue (Jablasone et al., 2004; Miles et al., 2009).

Attachment and survival of *Salmonella* on tomatoes depend on both, plants and microbial factors. Barak et al. (2011) showed that the level of *Salmonella* population is cultivar-dependent. This difference in *Salmonella* population is not attributed to the initial attachment, but on genetic aspects of the plant that affect the persistence of the pathogen (Barak et al., 2011). In this study, type 1 trichomes were identified as the preferred colonization site on tomato leaves. The phyllosphere contamination led to fruit contamination. A recent study has shown that aside from cultivar type, serovar type and the plant growth stage are also important factors (Zheng et al., 2013).

Like with *E. coli* and lettuce, no genetic elements have been definitively identified as essential for attachment or survival of *Salmonella* on tomatoes. Molecular studies indicate that *Salmonella* genes required for virulence in animals as well as those required for synthesis of the biofilm extracellular matrix also have a role in the colonization of plant tissue (Barak et al., 2005; Barak et al., 2007). Although these studies were conducted using sprouts as opposed to tomatoes, the results warrant discussion. Using an alfalfa sprout colonization assay, 6000 mutants created through transposon mutagenesis were screened for their ability to bind to plant tissue (Barak et al., 2005). Mutants deficient in *rpoS*, the stationary phase sigma factor, as well as *agfD*, a transcriptional regulator, were significantly reduced in their binding ability (approximately sevenfold fewer cells per sprout). Further experiments demonstrated that cellulose, thin aggregative fimbriae, and the O-antigen capsule are also important determinants in the attachment of *Salmonella* to plant tissue (Barak et al., 2007). Other work examining the colonization of *Salmonella* on alfalfa sprouts indicated that strains lacking elements of a TTSS were able to "hypercolonize" alfalfa seedlings, indicating that the TTSS elements may be recognized by plant defenses to limit *Salmonella* colonization (Iniguez et al., 2005).

A recent study also showed that mutants deficient in *ycfR*, *sirA,* and *yigG* reduced the attachment of *Salmonella* serovars to grape tomatoes. The *ycfR* gene encodes a membrane protein that changes the surface properties of the bacterial cell and regulates biofilm formation. This gene was also involved in resistance to chlorination of produce. SirA is a regulatory protein affecting the expression of different virulence factors and YigG is a putative membrane protein with unknown function (Salazar et al., 2013). Noel et al. (2010) screened for *Salmonella* genes that are differentially regulated in tomatoes relative to *Salmonella* grown in a soft agar LB. They identified several genes, including genes associated with protein synthesis and degradation, virulence, transport, attachment, stress response biofilm, and capsule formation, and genes with unknown functions. The FabH gene was upregulated most strongly in immature tomatoes, probably in response to changes in the concentrations of linoleic acid. Deletion of SirA and MotA (involved in motility) modestly increased fitness, and deletion of YihT (involved in synthesis of the capsule) contributed to fitness in green but not ripe tomatoes, but interestingly, known *Salmonella* genes associated with motility and virulence in animals such as *hila, flhDC,* and *fliF* did not contribute to the fitness of the bacteria in the fruit (Marvasi et al., 2013; Noel et al., 2010).

Salmonella and melons

Over the past two decades, melons, mostly cantaloupes, have been implicated in outbreaks of foodborne illness as well as recalls due to positive pathogen detection. Cantaloupes were implicated in at least 11 outbreaks of salmonellosis in the United States and Canada between 1973 and 2003 (Bowen et al., 2006). Most of the outbreaks linked to cantaloupe resulted in over 20 illnesses and covered a wide geographic area, indicating that contamination likely occurred at a single source. From this information, it can be inferred that *Salmonella* attaches to the surface of the cantaloupe, and survives well during transport and storage through sale, preparation, and consumption. The physical characteristics of cantaloupe melons and the manner in which they are grown make it exceedingly difficult to prevent contamination and also to remove attached bacteria once in place. Melons are grown directly on the ground in constant contact with the soil surface. In addition, melon-plant leaves provide a shady environment for growing melons, reducing the likelihood of large swings in temperature or direct sunlight. The melon fruit itself is covered in netting made up of large cracked pieces of cuticle. Materon et al. (2007) investigated the sources of microbial contamination from cantaloupes grown at ten different farms in Texas. Their results supplied evidence that quality of irrigation water is a significant risk of preharvest contamination. On the other hand, results of a field study indicated that root uptake and systemic transport of *Salmonella* from soil, as a consequence of contaminated irrigation water, is highly unlikely to occur. Contamination of the applied *Salmonella* was detected on the rind surface of the melons when the melons were developed in contact with the contaminated soil (Lopez–Velasco et al., 2012).

The behavior of *Salmonella* on melons has been examined in detail by numerous investigators. Laboratory experiments with whole melons indicate that once introduced onto the surface of the cantaloupe, cells of *Salmonella* are impossible to remove completely, regardless of the sanitizer used or the exposure time. Additional experiments demonstrate that the efficacy of sanitizer treatment is reduced when the organism is allowed to reside on the melon surface for an extended period of time (Ukuku, 2004; Ukuku and Sapers, 2001). The authors postulate that attachment of cells to sites inaccessible to sanitizer action within the cantaloupe netting, and the initiation of bacterial biofilm formation, contribute to the minimal efficacy of aqueous sanitizers. Fissures in the cantaloupe netting have been demonstrated to be infiltrated by cells of *Salmonella* (Annous et al., 2004; Annous et al., 2005) and likely provide attachment sites that aid bacterial survival when in contact with aqueous sanitizers. These results contrast with those found for honeydew melons, which are closely related to cantaloupe, yet lack the characteristic surface netting. Five-minute washing treatments in 2.5 or 5.0% hydrogen peroxide were effective at completely eliminating *Salmonella* from the surface of honeydew melons (>3 log reduction), yet were unable to produce the same results on cantaloupe (Ukuku, 2004). Alvarado–Casillas et al. (2007) reported similar results when comparing the efficacy of sanitizers on cantaloupe with the efficacy on bell peppers.

Melons are rich in nutrients, low in acidity (pH 5.2–6.7), and high in water activity (0.97–0.99), thus are capable of supporting the growth of pathogens (Harris et al., 2003). For that reason one concern with the survival of *Salmonella* attached to cantaloupe surfaces is the ability of fresh-cut melon to support the growth of *Salmonella*. Cutting through the rind of a melon harboring *Salmonella* may result in contamination of fresh-cut pieces with the pathogen (Ukuku and Sapers, 2001). Fresh-cut melon on display at grocery stores generally is stored at temperatures above 4°C and *Salmonella* grows extremely well in melon cubes stored at abusive temperatures (Ukuku, 2004).

Greenhouse-level studies with cantaloupe are far less prevalent than for lettuce or tomato. Stine et al. used a growth chamber to investigate the effects of humidity on the survival of *Salmonella* on cantaloupe (Stine et al., 2005). Inactivation rates of *Salmonella* under dry conditions were significantly greater than the inactivation rate under humid conditions. These results suggest that humid conditions may allow *Salmonella* to persist on a melon surface through harvest, transport, storage, and consumption. In terms of the intimate interactions between cells of *Salmonella* and the melon surface, no molecular mechanisms have been identified and investigated in-depth; however, the cell-surface charge and hydrophobicity have been found to play an important role (Ukuku and Fett, 2002b, 2006).

Listeria and melons

Although the majority of the large detected listeriosis outbreaks have been associated with the consumption of deli meats as well as milk and dairy products, a few outbreaks due to consumption of produce such as cabbage, sprouts, celery, and cantaloupe have been reported. *Listeria* has rarely been implicated in whole or fresh-cut produce, probably because most *Listeria* infections occur sporadically, making the identification of the bacterial source difficult (Varma et al., 2007). Yet, the pathogen has been isolated not only from soil, water, and vegetation, but also from raw vegetables and fruits. Data published from different countries show that *Listeria* species can be detected in 4 to 24% of samples of fresh, minimally processed fruit, vegetables, sprouts, and fresh salads. The pathogenic species *L. monocytogenes* was identified in up to 4.8% of the samples. In some commodities it exceeded 100 cfu/gr

(Jadhav et al., 2012; Kovacevic et al., 2013; Little et al., 2007). A case-control study identified "eating melons at a commercial establishment" as a risk factor for sporadic listeriosis (Varma et al., 2007). Indeed, recently, cantaloupe was implicated in a major multistate outbreak of foodborne listeriosis (www.cdc.gov/listeria/outbreaks/cantaloupes-jensen-farms/index.html). This outbreak caused 147 illnesses (99% were hospitalized) in 28 states, including at least 33 deaths and 1 miscarriage, the largest number of fatalities due to an outbreak of foodborne listeriosis in the U.S. Two serotypes, 1/2a and 1/2b, were associated with this outbreak, which were categorized into five pulse-field gel electrophoresis (PFGE) profiles (Laksanalamai et al., 2012).

Since this outbreak the behavior of *Listeria* on melons and particularly on cantaloupe has been examined in detail. *Listeria* is a psychotrophic microorganism, therefore refrigeration slows, but does not inhibit the growth of *Listeria* (Fang et al., 2013). Melons eaten at commercial formation (e.g., restaurants) are usually sliced and probably stored at refrigerated temperatures, but may also be exposed to temperature abuse. *L. monocytogenes* may be present on the exterior of melons and pre-slicing could allow *L. monocytogenes* to multiply. Indeed, fresh-cut pieces prepared from inoculated whole cantaloupes stored at 4°C for 24 h after inoculation were positive for *L. monocytogenes*. After direct inoculation onto fresh-cut pieces, *L. monocytogenes* survived, but did not grow during 15 days' storage at 4°C, but grew fast at 8 and 20°C (Ukuku and Fett, 2002a). Fang et al. (2013) also showed that *L. monocytogenes* can begin exponential growth on cut melons immediately without going through an adjustment period or lag phase, making this pathogen particularly dangerous to consumers. On the other hand, Ukuku et al. (2012) showed that *L. monocytogenes* survived on cantaloupe rinds for up to 15 days' during storage at 4 and 20°C, but population slightly declined.

The low number cases of *Listeria* associated with melons compared with *Salmonella* can be explained by several properties of *Listeria*. *Listeria* is capable of attaching to whole cantaloupes, but comparing to *E. coli* and *Salmonella* its attachment was the lowest (Ukuku and Fett, 2002a), and the population of *Listeria* transferred from melon rinds to fresh cut pieces were below the transferred population described for *Salmonella* (Ukuku et al., 2012). Moreover, native microflora of whole cantaloupes such as yeast and mold inhibited attachment to rind surface as well as survival and growth of *L. monocytogenes* on cantaloupe surfaces and fresh-cut pieces (Ukuku et al., 2004).

Plant defense response to human enteric pathogens

As discussed above, *S. enterica, Listeria,* and *E. coli* attach to the plant surface. *S. enterica* and *E. coli* also show endophytic characteristics. The ability to adapt to the plant environment probably depends on the plant type and the bacteria serovar. Some studies reported that the endophytic bacteria failed to grow. In these studies a decline was observed in the population over time, but still a portion of the population was able to persist on/in the phyllosphere and the rhizosphere of different plants for extended periods (Dreux et al., 2007; Erickson et al., 2010; Kisluk et al., 2013; Kisluk and Yaron, 2012; Moyne et al., 2011). Other studies have indicated that the bacteria are capable of replication on or in the plants (Deering et al., 2011; Gandhi et al., 2001; Schikora et al., 2008). In addition most observations indicated that enteric pathogens colonize different plants without causing disease symptoms (Gandhi et al., 2001; Klerks et al., 2007a; Lapidot and Yaron, 2009). In the few cases in which disease-like symptoms such as wilting and chlorosis appeared in the leaves in response to the attachment of *Salmonella* serovars, it seems that the recognition and the plant response depend on the structure of the bacterial capsules (Berger et al., 2011; Hernandez-Reyes and Schikora, 2013).

Plants apply a range of mechanisms for protection against microorganisms. The plant protection includes local or systemic production of defense enzymes and antimicrobial molecules. Some mechanisms are expressed constitutively (like the production of essential oils with a broad-range of antimicrobial activity) and the others (like secretion of reactive oxygen species) are induced after exposure to the pathogen. To survive in the plant, endophytic bacteria have to escape the diverse plant defense systems, which function at different levels through the plant-bacteria interactions. Plants and mammals have fundamental biological differences that affect their capacity to defend themselves against microorganisms. While mammals use both the adaptive and innate immunity, plants rely mostly on innate immunity. On the other hand, plant cells produce a cell wall that provides an effective barrier to microorganisms (Zhang and Zhou, 2010).

Following penetration of the cell walls either by degradation of the polymers or through wounds, the microorganisms encounter host cells' extracellular surface receptors, the pattern recognition receptors (PRRs), which recognize conserved pathogen associated molecular patterns (PAMPs) and trigger downstream defense signaling pathways. PAMPs include a growing list of conserved molecules in bacteria or fungi such as flagellin, lipopolysaccharides, glycoproteins, cold shock protein (CSP), elongation factor Tu, and chitin (Nurnberger et al., 2004). Recognition of PAMPs by PRRs initiates PAMP-triggered immunity (PTI), which usually prevents microbial growth and halts infection before the microorganism gains a hold in the plant (Chisholm et al., 2006). The triggered pathways include production and secretion of nitric oxide and reactive oxygen species (ROS) such as superoxide and hydrogen peroxide, ionic fluxes (primary Ca^{2+}, K^+ and H^+), activation of mitogen-activated protein kinases (MAPKs), and reinforce the cell wall at sites of infection (Gomez-Gomez et al., 1999; Nurnberger et al., 2004). These compounds inhibit bacteria, but together with different additional signaling molecules may also locally activate programmed cell death (PCD) that generates a physical barrier restraining nutrient availability, due to the rapid dehydration caused by tissue death (de Pinto et al., 2002).

Successful endophytes are able to escape the plant defense by an active suppression of the PTI response. During infection they are able to multiply by secretion of effector proteins that interfere directly with PTI response and alter plant physiology (Zhou and Chai, 2008). In many Gram-negative bacteria the effector proteins are translocated directly to the plant cell through the bacterial T3SSs. In response to the microbial suppression of the primary defenses, plants developed more specialized mechanisms to detect and inhibit the evaded bacteria. These include the effector triggered immunity (ETI), in which plants activate specific plant resistance (R) proteins to inhibit the effector proteins and to suppress microbial growth (Chisholm et al., 2006). This response often results in hypersensitive response (HR), a localized response that results in a rapid death of the bacteria and the infected plant cells with no spread of bacteria to the surrounding tissues. During the HR, plant cells produce and secrete ROS, nitric oxide, different ions as well as molecular signals like jasmonate, salicylic acid, and ethylene (Jones and Dangl, 2006).

Evidence indicates that the similar mechanisms of plant defense against phytopathogens are applied by plants in response to internalization of human enteric pathogens. Many surface components of *E. coli* and *Salmonella* such as lipopolysaccharides, flagella, peptidoglycans, curli, and pili are homologous to PAMPs of phytopathogens. Deletion of these PAMPs from *E. coli* or *Salmonella* usually resulted in better colonization of the interior of plants than the wild-type strains (Iniguez et al., 2005; Seo and Matthews, 2012), suggesting that the plant invasion and colonization process of human enteric pathogens may be similar to some phytopathogens.

S. enterica has two types of flagellin subunits, FliB and FliC. It has been shown that FliC but not FliB or LPS is recognized by *Nicotiana benthamiana* and tomato as PAMP and activates its PTI (Meng et al., 2013). Moreover, an *E. coli* O157:H7 mutant that produces a great amount of exopolysaccharides (EPS) and a thick capsule exhibits a different survival pattern on *Arabidopsis* compared with the wild-type strain. It was suggested that the EPS may mask underlying bacterial surface components from recognition by the plant host, evading the plant defense response. On the other hand, curli, also a major component of the biofilm matrix, can serve as PAMP (Seo and Matthews, 2012). Thus the role of biofilm formation in triggering or evading the plant response has not been elucidated. Many *Salmonella* serovars also produce cellulases (Yoo et al., 2004), but their activity in the plant environment against the cell wall has not been investigated.

Plant hormones, secreted in the course of the plant response, were found to have a role in persistence of enteric pathogens in plants. Ethylene and salicylic acid decreased endophytic colonization of *Salmonella* in alfalfa (Iniguez et al., 2005). Increased *Salmonella* colonization was observed in an *Arabidopsis* mutant with defective production of jasmonic acid compared with plants with active defenses, indicating that the jasmonic acid signaling pathway is of major importance for induction of defense response against *Salmonella* (Schikora et al., 2008). Furthermore, the mitogen-activated protein kinase (MAPK) cascades, commonly associated with the plant immune response, were activated minutes after exposure to *Salmonella* (Schikora et al., 2008).

T3SSs that secrete virulence proteins (effectors) into the host cells play a major role in pathogenicity of many Gram-negative plant and animal pathogens (reviewed in Cornelis and Van Gijsegem, 2000). Since some Gram-negative foodborne pathogens produce T3SSs, several studies pointed to the role of the T3SS in persistence of the human enteric pathogens in plants. The systems are similar except to the differences seen between the short "needles" of typical animal pathogens and the long "pili" of most plant pathogens. The "needles" of the animal pathogens T3SS are considered too short to allow translocation of effectors through the cell wall of plant cells (He et al., 2004). *S. enterica*, for example, produces two distinct T3SSs that are essential for different stages during pathogenesis in mammals. Removal of *Salmonella* TTSS-1 and some effector proteins affected the bacterial colonization on *Medicago truncatula* (Iniguez et al., 2005). While *S.* Typhimurium actively inhibited the defense response of *N. tabacum* plants and cell suspension, a T3SS-1 mutant was not able to suppress this response (Shirron and Yaron, 2011). In another study, mutants of T3SS-1 or T3SS-2 of *S.* Typhimurium had lower proliferation rates and enhanced hypersensitive response-related symptoms in *Arabidopsis* plants (Schikora et al., 2012). The *Salmonella* effector protein SseF triggered hypersensitive response (HR)-like symptoms in *N. benthamiana* leaves when it was expressed and translocated by the plant pathogens. It was suggested that SseF is recognized by an R protein in *N. benthamiana* (Ustun et al., 2012). Currently there is no evidence that *E. coli* and *Salmonella* use their T3SSs to secrete effectors into the plant cells, but there are also no reports that confirm that they lack the ability to do so (Deering et al., 2012). All these studies indicate that functional T3SSs are required for suppression of the plant immune response and survival in the plants, but the mechanisms have not been explained and effectors capable of such suppression are yet to be determined.

Conclusion

Plants can serve as good vehicles for transfer of the pathogen from the environment into the intestinal tracts of a new host. Thus it is reasonable to propose intimate interactions between the bacteria and the plant to ensure that the enteric pathogens fit the plant environment very quickly and survive in plants long enough and in sufficient numbers for infection of a new host. The studies detailed in this chapter indicate a complex and active interaction between enteric pathogens such as *E. coli*, *Salmonella* and *Listeria,* and plants. Pathogens originate from animals, survive well in water, soil, and farm wastes, attach strongly to plant tissue, and survive the harsh environment through consumption. Some bacterial systems, evolved to attach and invade the host or to suppress the immune response in the host, also help the pathogen to survive in the plant, although these pathogens are not completely fit to the plant environment.

The results from the research presented in this chapter provide some answers as to why the public has witnessed an increase in outbreaks arising from produce. The close proximity between animal and vegetable growing operations will continue to present opportunities for outbreaks arising from fresh produce. Although research is ongoing, our current understanding of the ecology, genetics, and physiology of pathogens attached to plant tissue is rudimentary at best. The use of different strains of pathogens, a wide variety of plants, and diverse methods of inoculation and growth limits the ability to compare the results of different studies and to conduct in-depth research of the plant-pathogens interactions. Researchers will continue to probe the interactions between pathogens and plant surfaces with the goal of finding methods to reduce attachment and promote inactivation. However, the ultimate goal of a pathogen-free produce supply will depend more on the vigilance of individual growers, cooperation from industry groups, and a focused effort from government agencies.

References

Adato, A., Mandel, T., Mintz-Oron, S., Venger, I., et al., 2009. Fruit-surface flavonoid accumulation in tomato is controlled by a SlMYB12-regulated transcriptional network. PLoS Genet. 5, e1000777.

Agle, M.E., 2003. *Shigella boydii* 18. Characterization and biofilm formation University of Illinois, Champaign, IL Ph.D. thesis.

Alvarado-Casillas, S., Ibarra-Sanchez, S., Rodriguez-Garcia, O., Martinez-Gonzales, N., et al., 2007. Comparison of rinsing and sanitizing procedures for reducing bacterial pathogens on fresh cantaloupes and bell peppers. J. Food Prot. 70, 655–660.

Annous, B.A., Burke, A., Sites, J.E., 2004. Surface pasteurization of whole fresh cantaloupes inoculated with *Salmonella* Poona or *Escherichia coli*. J. Food Prot. 67, 1876–1885.

Annous, E.B., Solomon, P., Cooke, H., Burke, A., 2005. Biofilm formation by *Salmonella* spp. on cantaloupe melons. J. Food Saf. 25, 276–287.

Anon., 2007. Outbreak alert! Closing the gaps in our federal food-safety net Center for Science in the Public Interest. Washington, DC. Available at www.cspinet.org.

Anon., 2010. Alliance for Food and Farming. Anal. Produce Relat. Foodborne Illn. Outbreaks. http://www.businesswire.com/portal/site/home/permalink/?ndmViewId=news_view&newsId=20100329006735&newsLang=en.

Barak, J.D., Gorski, L., Liang, A.S., Narm, K.E., 2009. Previously uncharacterized *Salmonella enterica* genes required for swarming play a role in seedling colonization. Microbiology 155, 3701–3709.

Barak, J.D., Gorski, L., Naraghi-Arani, P., Charkowski, A.O., 2005. *Salmonella enterica* virulence genes are required for bacterial attachment to plant tissue. Appl. Environ. Microbiol. 71, 5685–5691.

Barak, J.D., Jahn, C.E., Gibson, D.L., Charkowski, A.O., 2007. The role of cellulose and O-antigen capsule in the colonization of plants by *Salmonella enterica*. Mol. Plant Microbe Interact. 20, 1083–1091.

Barak, J.D., Kramer, L.C., Hao, L.Y., 2011. Colonization of tomato plants by *Salmonella enterica* is cultivar dependent, and type 1 trichomes are preferred colonization sites. Appl. Environ. Microbiol. 77, 498–504.

Barak, J.D., Liang, A., Narm, K.E., 2008. Differential attachment to and subsequent contamination of agricultural crops by *Salmonella enterica*. Appl. Environ. Microbiol. 74, 5568–5570.

Beattie, G.A., Lindow, S.E., 1999. Bacterial colonization of leaves: A spectrum of strategies. Phytopathology 89, 353–359.

Berger, C.N., Brown, D.J., Shaw, R.K., Minuzzi, F., et al., 2011. *Salmonella enterica* strains belonging to O serogroup 1,3,19 induce chlorosis and wilting of *Arabidopsis thaliana* leaves. Environ. Microbiol. 13, 1299–1308.

Berger, C.N., Shaw, R.K., Brown, D.J., Mather, H., et al., 2009. Interaction of *Salmonella enterica* with basil and other salad leaves. The ISME J. 3, 261–265.

Bernstein, N., Sela, S., Neder-Lavon, S., 2007a. Assessment of contamination potential of lettuce by *Salmonella enterica* serovar Newport added to the plant growing medium. J. Food Prot. 70, 1717–1722.

Bernstein, N., Sela, S., Pinto, R., Ioffe, M., 2007b. Evidence for internalization of *Escherichia coli* into the aerial parts of maize via the root system. J. Food Prot. 70, 471–475.

Besser, T.E., Richards, D.L., Rice, D.H., Hancock, D.D., 2001. *Escherichia coli* O157:H7 infection of calves: Infectious dose and direct contact transmission. Epidemiol. Infect. 127, 555–560.

Beuchat, L.R., 1999. Survival of enterohemorrhagic *Escherichia coli* O157:H7 in bovine feces applied to lettuce and the effectiveness of chlorinated water as a disinfectant. J. Food Prot. 62, 845–849.

Beuchat, L.R., Mann, D.A., 2010. Factors affecting infiltration and survival of *Salmonella* on in-shell pecans and pecan nutmeats. J. Food Prot. 73, 1257–1268.

Bowen, A., Fry, A., Richards, G., Beuchat, L., 2006. Infections associated with cantaloupe: A public health concern. Epidemiol. Infect. 134, 675–685.

Boyer, R.R., Sumner, S.S., Williams, R.C., Pierson, M.D., et al., 2007. Influence of curli expression by *Escherichia coli* O157:H7 on the cell's overall hydrophobicity, charge, and ability to attach to lettuce. J. Food Prot. 70, 1339–1345.

Brandl, M.T., 2006. Fitness of human enteric pathogens on plants and implications for food safety. Annu. Rev. Phytopathol. 44, 367–392.

Brandl, M.T., 2008. Plant lesions promote the rapid multiplication of *Escherichia coli* O157:H7 on postharvest lettuce. Appl. Environ. Microbiol. 74, 5285–5289.

Brandl, M.T., Amundson, R., 2008. Leaf age as a risk factor in contamination of lettuce with *Escherichia coli* O157:H7 and *Salmonella enterica*. Appl. Environ. Microbiol. 74, 2298–2306.

Brandl, M.T., Mandrell, R.E., 2003. Fitness of *Salmonella enterica* serovar Thompson in the cilantro phyllosphere. Appl. Environ. Microbiol. 68, 3614–3621.

Brandl, M.T., Sundin, G.W., 2013. Focus on food safety: human pathogens on plants. Phytopathology 103, 304–305.

Buchanan, R.L., Edelson, S.G., Miller, R.L., Sapers, G.M., 1999. Contamination of intact apples after immersion in an aqueous environment containing *Escherichia coli* O157:H7. J. Food Prot. 62, 444–450.

Carrari, F., Baxter, C., Usadel, B., Urbanczyk-Wochniak, E., et al., 2006. Integrated analysis of metabolite and transcript levels reveals the metabolic shifts that underlie tomato fruit development and highlight regulatory aspects of metabolic network behavior. Plant Physiol. 142, 1380–1396.

Carter, M.Q., Xue, K., Brandl, M.T., Liu, F., et al., 2012. Functional metagenomics of *Escherichia coli* O157:H7 interactions with spinach indigenous microorganisms during biofilm formation. PloS one 7, e44186.

CDC. Centers for Disease Control, 2007. Multistate outbreaks of *Salmonella* infections associated with raw tomatoes eaten in restaurants-United States, 2005–2006. Morb. Mortal. Wkly. Rep. 56, 909–911.

CDC. Centers for Disease Control, 2013. Foodborne disease outbreak surveillance. http://www.cdc.gov/outbreaknet/surveillance_data.html#reports.

Cevallos-Cevallos, J.M., Gu, G., Danyluk, M.D., Dufault, N.S., et al., 2012a. *Salmonella* can reach tomato fruits on plants exposed to aerosols formed by rain. Int. J. Food Microbiol. 158, 140–146.

Cevallos-Cevallos, J.M., Gu, G., Danyluk, M.D., van Bruggen, A.H., 2012b. Adhesion and splash dispersal of *Salmonella enterica* Typhimurium on tomato leaflets: effects of rdar morphotype and trichome density. Int. J. Food Microbiol. 160, 58–64.

CFERT. California Food Emergency Response Team, 2007. Investigation of an *Escherichia coli* O157:H7 outbreak associated with Dole pre-packaged spinach. California Department of Health Services, Sacramento, CA.

Chisholm, S.T., Coaker, G., Day, B., Staskawicz, B.J., 2006. Host-microbe interactions: shaping the evolution of the plant immune response. Cellule 124, 803–814.

Chmielewski, R.A.N., Frank, J.F., 2003. Biofilm formation and control in food processing facilities. Compr. Rev. Food Sci. Food Saf. 2, 22–32.

Cook, R., 2011. Tracking Demographics and U.S. Fruit and Vegetable Consumption Patterns. http://agecon.ucdavis.edu/people/faculty/roberta-cook/docs/Articles/BlueprintsEoECons umptionCookFinalJan2012Figures.pdf.

Cooley, M., Carychao, D., Crawford-Miksza, L., Jay, M.T., et al., 2007. Incidence and tracking of *Escherichia coli* O157:H7 in a major produce production region in California. PloS One 2, e1159.

Cooley, M.B., Chao, D., Mandrell, R.E., 2006. *Escherichia coli* O157:H7 survival and growth on lettuce is altered by the presence of epiphytic bacteria. J. Food Prot. 69, 2329–2335.

Cooley, M.B., Miller, W.G., Mandrell, R.E., 2003. Colonization of *Arabidopsis thaliana* with *Salmonella enterica* and enterohemorrhagic *Escherichia coli* O157:H7 and competition by *Enterobacter asburiae*. Appl. Environ. Microbiol. 69, 4915–4926.

Cornelis, G.R., Van Gijsegem, F., 2000. Assembly and function of type III secretory systems. Annu. Rev. Microbiol. 54, 735–774.

Costerton, J.W., Stewart, P.S., Greenberg, E.P., 1999. Bacterial biofilms: A common cause of persistent infections. Science 284, 1318–1322.

D'Aoust, J.Y., 1998. *Salmonella* Species. In: Doyle, M.P., Beuchat, L.R., Montville, T.J. (Eds.), Food Microbiology and Frontiers. ASM Press, Washington, DC.

Danhorn, T., Fuqua, C., 2007. Biofilm formation by plant associated bacteria. Annu. Rev. Microbiol. 61, 401–422.

Danyluk, M.D., Brandl, M.T., Harris, L.J., 2008. Migration of *Salmonella* Enteritidis phage type 30 through almond hulls and shells. J. Food prot. 71, 397–401.

Darrah, P.R., 1991. Models of the rhizosphere. I. Microbial population dynamics around a root releasing soluble and insoluble carbon. Plant Soil 133, 187–199.

de Pinto, M.C., Tommasi, F., de Gara, L., 2002. Changes in the antioxidant systems as part of the signaling pathway responsible for the programmed cell death activated by nitric oxide and reactive oxygen species in tobacco Bright-Yellow 2 cells. Plant Physiol. 130, 698–708.

Deering, A.J., Mauer, L.J., Pruitt, R.E., 2012. Internalization of *E. coli* O157:H7 and *Salmonella* spp. in plants: a review. Food Res. Int. 45, 567–575.

Deering, A.J., Pruitt, R.E., Mauer, L.J., Reuhs, B.L., 2011. Identification of the cellular location of internalized *Escherichia coli* O157:H7 in mung bean, *Vigna radiata*, by immunocytochemical techniques. J. Food Prot. 74, 1224–1230.

Delaquis, P., Bach, S., Dinu, L.D., 2007. Behavior of *Escherichia coli* O157:H7 in leafy vegetables. J. Food Prot. 70, 1966–1974.

Delaquis, P., Stewart, S., Cazaux, S., Toivonen, P., 2002. Survival and growth of *Listeria monocytogenes* and *Escherichia coli* O157:H7 in ready-to-eat iceberg lettuce washed in warm chlorinated water. J. Food Prot. 65, 459–464.

Dong, Y., Iniguez, A.L., Ahmer, B.M., Triplett, E.W., 2003. Kinetics and strain specificity of rhizosphere and endophytic colonization by enteric bacteria on seedlings of *Medicago sativa* and *Medicago Truncatula*. Appl. Environ. Microbiol. 69, 1783–1790.

Dreux, N., Albagnac, C., Carlin, F., Morris, C.E., et al., 2007. Fate of *Listeria* spp. on parsley leaves grown in laboratory and field cultures. J. Appl. Microbiol. 103, 1821–1827.

Duffy, G., 2003. Verocytoxigenic *Escherichia coli* in animal faeces, manures and slurries. J. Appl. Microbiol (Suppl.), 94–103.

Dunne Jr., W.M., 2002. Bacterial adhesion: seen any good biofilms lately? Clin. Microbiol. Rev. 15, 155–166.

Eblen, S.B., Walderhaug, M.O., Edelson-Mammel, S., Chirtel, S.J., et al., 2004. Potential for internalization, growth, and survival of *Salmonella* and *Escherichia coli* O157:H7 in oranges. J. Food Prot. 1578–1584.

Edwards, J.H., Someshwar, A.V., 2000. Chemical, physical, and biological characterization of agricultural and forest by-products for land application. In: Power, J.F., Dick, W.A. (Eds.), Land Application of Agricultural, Industrial, and Municipal By-Products. ASA-CSSA-SSSA, Madison, WI, pp. 1–62.

Elder, R.O., Keen, J.E., Siragusa, G.R., Barkocy-Gallagher, G.A., et al., 2000. Correlation of enterohemorrhagic *Escherichia coli* O157:H7 prevalence in feces, hides, and carcasses of beef cattle during processing. Proc. Nation. Acad. Sci. USA 97, 2999–3003.

Ells, T.C., Truelstrup Hansen, L., 2006. Strain and growth temperature influence *Listeria* spp. attachment to intact and cut cabbage. Int. J. Food Microbiol. 111, 34–42.

Erickson, M.C., 2012. Internalization of fresh produce by foodborne pathogens. Annu. Rev. Food Sci. Technol. 3, 283–310.

Erickson, M.C., Doyle, M.P., 2007. Food as a vehicle for transmission of shiga toxin-producing *Escherichia coli*. 70 J. Food Prot. 70, 2426–2449.

Erickson, M.C., Webb, C.C., Diaz-Perez, J.C., Phatak, S.C., et al., 2010. Surface and internalized *Escherichia coli* O157:H7 on field-grown spinach and lettuce treated with spray-contaminated irrigation water. J. Food Prot. 73, 1023–1029.

Fang, T., Liu, Y., Huang, L., 2013. Growth kinetics of *Listeria monocytogenes* and spoilage microorganisms in fresh-cut cantaloupe. Food Microbiol. 34, 174–181.

FAO/WHO, 2008. Food and Agriculture Organization of the United Nations/World Health Organization. Microbiological hazards in fresh fruits and vegetables. Meeting Rep. Microbiol Risk Assess. Series.

Fink, R.C., Black, E.P., Hou, Z., Sugawara, M., et al., 2012. Transcriptional responses of *Escherichia coli* K-12 and O157:H7 associated with lettuce leaves. Appl. Environ. Microbiol. 78, 1752–1764.

Franz, E., van Bruggen, A.H., 2008. Ecology of *E. coli* O157:H7 and *Salmonella enterica* in the primary vegetable production chain. Crit. Rev. Microbiol. 34, 143–161.

Franz, E., Visser, A.A., Van Diepeningen, A.D., Klerks, M.M., et al., 2007. Quantification of contamination of lettuce by GFP-expressing *Escherichia coli* O157:H7 and *Salmonella enterica* serovar Typhimurium. Food Microbiol. 24, 106–112.

Gandhi, M., Golding, S., Yaron, S., Matthews, K.R., 2001. Use of green fluorescent protein expressing *Salmonella* Stanley to investigate survival, spatial location, and control on alfalfa sprouts. J. Food Prot. 64, 1891–1898.

Gerstel, U., Romling, U., 2003. The *csgD* promoter, a control unit for biofilm formation in *Salmonella* Typhimurium. Res. Microbiol. 154, 659–667.

Golberg, D., Kroupitski, Y., Belausov, E., Pinto, R., et al., 2011. *Salmonella* Typhimurium internalization is variable in leafy vegetables and fresh herbs. Int. J. Food Microbiol. 145, 250–257.

Gomez-Gomez, L., Felix, G., Boller, T., 1999. A single locus determines sensitivity to bacterial flagellin in *Arabidopsis thaliana*. Plant J. Cell Mol. Biol. 18, 277–284.

Gorbatsevich, E., Sela, S., Pinto, S.R., Bernstein, N., 2013. Root internalization, transport and in-planta survival of *Salmonella enterica* serovar Newport in sweet basil. Environ. Microbiol. Rep. 5, 151–159.

Greene, S.K., Daly, E.R., Talbot, E.A., 2008. Recurrent multistate outbreak of *Salmonella* Newport associated with tomatoes from contaminated fields, 2005. Epidemiol. Infect. 136, 157–165.

Guo, X., Chen, J., Brackett, R.E., Beuchat, L.R., 2001. Survival of salmonellae on and in tomato plants from the time of inoculation at flowering and early stages of fruit development through fruit ripening. Appl. Environ. Microbiol. 67, 4760–4764.

Guo, X., Chen, J., Brackett, R.E., Beuchat, L.R., 2002a. Survival of *Salmonella* on tomatoes stored at high relative humidity, in soil, and on tomatoes in contact with soil. J. Food Pro. 65, 274–279.

Guo, X., van Iersel, M.W., Chen, J., Brackett, R.E., et al., 2002b. Evidence of association of salmonellae with tomato plants grown hydroponically in inoculated nutrient solution. Appl. Environ. Microbiol. 68, 3639–3643.

Han, Y., Sherman, D.M., Linton, R.H., Nielsen, S.S., et al., 2000. The effects of washing and chlorine dioxide gas on survival and attachment of *Escherichia coli* O157:H7 to green pepper surfaces. Food Microbiol. 17, 521–533.

Harris, L.J., Farber, J.N., Beuchat, L.R., Parish, M.E., et al., 2003. Outbreaks associated with fresh produce: incidence, growth, and survival of pathogens in fresh and fresh-cut produce. Compr. Rev. Food. Sci. Food Saf. 2, 78–141.

Hassan, A.N., Frank, J.F., 2003. Influence of surfactant hydrophobicity on the detachment of *Escherichia coli* O157:H7 from lettuce. Int. J. Food Microbiol. 87, 145–152.

Hassan, A.N., Frank, J.F., 2004. Attachment of *Escherichia coli* O157:H7 grown in tryptic soy broth and nutrient broth to apple and lettuce surfaces as related to cell hydrophobicity, surface charge, and capsule production. Int. J. Food Microbiol. 96, 103–109.

He, S.Y., Nomura, K., Whittam, T.S., 2004. Type III protein secretion mechanism in mammalian and plant pathogens. Biochim et Biophys Acta. 1694, 181–206.

Heaton, J.C., Jones, K., 2007. Microbial contamination of fruit and vegetables and the behaviour of enteropathogens in the phyllosphere: A review. J. Appl. Microbiol. 104, 613–626.

Herman, K., Ayers, T.L., Lynch, M., 2008. Foodborne disease outbreaks associated with leafy greens, 1973–2006. International Conference on Emerging Infectious Diseases, March 16–19, 2008, Atlanta, GA.

Hernandez-Reyes, C., Schikora, A., 2013. *Salmonella*, a cross-kingdom pathogen infecting humans and plants. FEMS Microbiol. Lett.

Himathongkham, S., Bahari, S., Riemann, H., Cliver, D., 1999. Survival of *Escherichia coli* O157:H7 and *Salmonella* Typhimurium in cow manure and cow manure slurry. FEMS Microbiol. Lett. 178, 251–257.

Hintz, L.D., Boyer, R.R., Ponder, M.A., Williams, R.C., et al., 2010. Recovery of *Salmonella enterica* Newport introduced through irrigation water from tomato (*Lycopersicum esculentum*) fruit, roots, stems and leaves. Hort Sci. 45, 675–678.

Hirneisen, K.A., Sharma, M., Kniel, K.E., 2012. Human enteric pathogen internalization by root uptake into food crops. Foodborne Pathog. Dis. 9, 396–405.

Hora, R., Warriner, K., Shelp, B.J., Griffiths, M.W., 2005. Internalization of *Escherichia coli* O157:H7 following biological and mechanical disruption of growing spinach plants. J. Food Prot. 68, 2506–2509.

Hou, Z., Fink, R.C., Black, E.P., Sugawara, M., et al., 2012. Gene expression profiling of *Escherichia coli* in response to interactions with the lettuce rhizosphere. J. Appl. Microbiol. 113, 1076–1086.

Hutchison, M.L., Walters, L.D., Moore, A., Avery, S.M., 2005. Declines of zoonotic agents in liquid livestock wastes stored in batches on-farm. J. Appl. Microbiol. 99, 58–65.

Ibekwe, M.A., Watt, P.M., Shouse, P.J., Grieve, C.M., 2004. Fate of *Escherichia coli* O157:H7 in irrigation water on soils and plants as validated by culture method and real-time PCR. Can. J. Microbiol. 50, 1007–1014.

Iniguez, A.L., Dong, Y., Carter, H.D., Ahmer, B.M., et al., 2005. Regulation of enteric endophytic bacterial colonization by plant defenses. Mol. Plant Microbe. Interact. 18, 169–178.

Islam, M., Doyle, M.P., Phatak, S.C., Millner, P., et al., 2004. Persistence of enterohemorrhagic *Escherichia coli* O157:H7 in soil and on leaf lettuce and parsley grown in fields treated with contaminated manure composts or irrigation water. J. Food Prot. 67, 1365–1370.

Iturriaga, M.H., Escartin, E.F., Beuchat, L.R., Martínez-Peniche, R., 2003. Effect of inoculum size, relative humidity, storage temperature, and ripening stage on the attachment of *Salmonella* Montevideo to tomatoes and tomatillos. J. Food Prot. 66, 1756–1761.

Iturriaga, M.H., Tamplin, M.L., Escartin, E.F., 2007. Colonization of tomatoes by *Salmonella* Montevideo is affected by relative humidity and storage temperature. J. Food Prot. 70, 30–34.

Jablasone, J., Brovko, L.Y., Griffiths, M.W., 2004. A research note: the potential for transfer of *Salmonella* from irrigation water to tomatoes. J. Sci. Food Agric. 84, 287–289.

Jablasone, J., Warriner, K., Griffiths, M., 2005. Interactions of *Escherichia coli* O157:H7, *Salmonella* Typhimurium, and *Listeria monocytogenes* plants cultivated in a gnotobiotic system. Int. J. Food Microbiol. 99, 7–18.

Jacobsen, C.S., Bech, T.B., 2012. Soil survival of *Salmonella* and transfer to freshwater and fresh produce. Food Res. Int. 45, 557–566.

Jadhav, S., Bhave, M., Palombo, E.A., 2012. Methods used for the detection and subtyping of *Listeria monocytogenes*. J. Microbiol. Methods 88, 327–341.

Jeter, C., Matthysse, A.G., 2005. Characterization of the binding of diarrheagenic strains of *E. coli* to plant surfaces and the role of curli in the interaction of the bacteria with alfalfa sprouts. MPMI 18, 1235–1242.

Johannessen, G.S., Bengtsson, G.B., Heier, B.T., Bredholt, S., et al., 2005. Potential uptake of *Escherichia coli* O157:H7 from organic manure intro crisphead lettuce. Appl. Environ. Microbiol. 71, 2221–2225.

Jones, J.D., Dangl, J.L., 2006. The plant immune system. Nature 444, 323–329.

Kisluk, G., Hoover, D., Kniel, K., Yaron, S., 2012. Quantification of low and high levels of *Salmonella enterica* serovar Typhimurium on leaves. LWT - Food Sci. Technol. 45, 36–42.

Kisluk, G., Kalily, E., Yaron, S., 2013. Resistance to essential oils and survival of *Salmonella enterica* serovars in growing and harvested basil. Environ. Microbiol. In press.

Kisluk, G., Yaron, S., 2012. Presence and persistence of *Salmonella enterica* serotype Typhimurium in the phyllosphere and rhizosphere of spray-irrigated parsley. Appl. Environ. Microboil. 78, 4030–4036.

Klerks, M.M., Franz, E., van Gent-Pelzer, M., Zijlstra, C., et al., 2007a. Differential interaction of *Salmonella enterica* serovars with lettuce cultivars and plant-microbe factors influencing the colonization efficiency. ISME J. 1, 620–631.

Klerks, M.M., van Gent-Pelzer, M., Franz, E., Zijlstra, C., et al., 2007b. Physiological and molecular responses of *Lactuca sativa* to colonization by *Salmonella enterica* serovar Dublin. Appl. Environ. Microbiol. 73, 4905–4914.

Kolling, G.L., Matthews, K.R., 2001. Examination of recovery in vitro and in vivo of nonculturable *Escherichia coli* O157:H7. Appl. Environ. Microbiol. 67, 3928–3933.

Koseki, S., Mizuno, Y., Yamamoto, K., 2011. Comparison of two possible routes of pathogen contamination of spinach leaves in a hydroponic cultivation system. J. Food Prot. 74, 1536–1542.

Kovacevic, M., Burazin, J., Pavlovic, H., Kopjar, M., et al., 2013. Prevalence and level of *Listeria monocytogenes* and other *Listeria* sp. in ready-to-eat minimally processed and refrigerated vegetables. World J. Microbiol. Biotechnol. 29, 707–712.

Kroupitski, Y., Golberg, D., Belausov, E., Pinto, R., et al., 2009. Internalization of *Salmonella enterica* in leaves is induced by light and involves chemotaxis and penetration through open stomata. Appl. Environ. Microbiol. 75, 6076–6086.

Kroupitski, Y., Pinto, R., Belausov, E., Sela, S., 2011. Distribution of *Salmonella* Typhimurium in romaine lettuce leaves. Food Microbiol. 28, 990–997.

Kudva, I.T., Blanch, K., Hovde, C.J., 1998. Analysis of *Escherichia coli* O157:H7 survival in ovine or bovine manure and manure slurry. Appl. Environ. Microbiol. 64, 3166–3174.

Kyle, J.L., Parker, C.T., Goudeau, D., Brandl, M.T., 2010. Transcriptome analysis of *Escherichia coli* O157:H7 exposed to lysates of lettuce leaves. Appl. Environ. Microbiol. 76, 1375–1387.

Laksanalamai, P., Joseph, L.A., Silk, B.J., Burall, L.S., et al., 2012. Genomic characterization of *Listeria monocytogenes* strains involved in a multistate listeriosis outbreak associated with cantaloupe in US. PloS One 7, e42448.

Lapidot, A., Romling, U., Yaron, S., 2006. Biofilm formation and the survival of *Salmonella* Typhimurium on parsley. Int. J. Food Microbiol. 109, 229–233.

Lapidot, A., Yaron, S., 2009. Transfer of *Salmonella enterica* serovar Typhimurium from contaminated irrigation water to parsley is dependent on curli and cellulose, the biofilm matrix components. J. Food Prot. 72, 618–623.

Levantesi, C., Bonadonna, L., Briancesco, R., Grohmann, E., et al., 2011. *Salmonella* in surface and drinking water: Occurrence and water-mediated transmission. Food Res. Int. doi:10.1016/j.foodres.2011.06.037.

Li, H., Tajkarimi, M., Osburn, B.I., 2008. Impact of vacuum cooling on *Escherichia coli* O157:H7 infiltration into lettuce tissue. Appl. Environ. Microbiol. 74, 3138–3142.

Liao, C.H., Cooke, P.H., 2001. Response to trisodium phosphate treatment of *Salmonella* Chester attached to fresh-cut green pepper slices. Can. J. Microbiol. 47, 25–32.

Lindow, S.E., Brandl, M.T., 2003. Microbiology of the phyllosphere. Appl. Environ. Microbiol. 69, 1875–1883.

Little, C.L., Taylor, F.C., Sagoo, S.K., Gillespie, I.A., et al., 2007. Prevalence and level of *Listeria monocytogenes* and other *Listeria* species in retail pre-packaged mixed vegetable salads in the UK. Food Microbiol. 24, 711–717.

Lopez-Velasco, G., Sbodio, A., Tomas-Callejas, A., Wei, P., et al., 2012. Assessment of root uptake and systemic vine-transport of *Salmonella enterica* sv. Typhimurium by melon (*Cucumis melo*) during field production. Int. J. Food Microbiol. 158, 65–72.

Markland, S.M., Shortlidge, K.L., Hoover, D.G., Yaron, S., et al., 2012. Survival of Pathogenic *Escherichia coli* on Basil, Lettuce, and Spinach. Zoonoses Public Health.

Marvasi, M., Cox, C., Xu, Y., Noel, J., et al., 2013. Differential regulation of *Salmonella* Typhimurium genes involved in O-antigen capsule production and their role in persistence within tomatoes. Mol plant-microbe interactions: MPMI.

Materon, L., Martinez-Garcia, M., McDonald, V., 2007. Identification of sources of microbial pathogens on cantaloupe rinds from pre-harvest to post-harvest operations. World J. Microbiol. Biotechnol. 23, 1281.

Matthews, L., Low, J.C., Gally, D.L., Pearce, M.C., et al., 2006. Heterogeneous shedding of *Escherichia coli* O157:H7 in cattle and its implications for control. Proc. Natl. Acad. Sci. USA 103, 547–552.

Matthysse, A.G., Deora, R., Mishra, M., Torres, A.G., 2008. Polysaccharides cellulose, poly-B-1,6-N-acetyl-D-glucosamine, and colanic acid are required for optimal binding of *Escherichia coli* O157:H7 strains to alfalfa sprouts and K-12 strains to plastic but not for binding to epithelial cells. Appl. Environ. Microbiol. 74, 2384–2390.

Matthysse, A.G., Kijne, J.W., 1998. Attachment of *Rhizobiaceae* to plant cells. In: Spaink, H.P., Kondorosi, A., Hooykaas, P.J.J. (Eds.), The *Rhizobiaceae*: molecular biology of model plant associated bacteria Kluwer. Dordrecht, pp. 235–249.

Meng, F., Altier, C., Martin, G.B., 2013. *Salmonella* colonization activates the plant immune system and benefits from association with plant pathogenic bacteria. Environ. Microbiol.

Miles, J.M., Sumner, S.S., Boyer, R.R., Williams, R.C., et al., 2009. Internalization of *Salmonella enterica* serovar Montevideo into greenhouse tomato plants through contaminated irrigation water or seed stock. J. Food Prot. 72, 849–852.

Monier, J.M., Lindow, S.E., 2005. Aggregates of resident bacteria facilitate survival of immigrant bacteria on leaf surfaces. Microbial. Ecol. 49, 343–352.

Mootian, G., Wu, W.H., Matthews, K.R., 2009. Transfer of *Escherichia coli* O157:H7 from soil, water, and manure contaminated with low numbers of the pathogen to lettuce plants. J. Food Prot. 72, 2308–2312.

Moyne, A.L., Sudarshana, M.R., Blessington, T., Koike, S.T., et al., 2011. Fate of *Escherichia coli* O157:H7 in field-inoculated lettuce. Food Microbiol. 28, 1417–1425.

Nicholson, F.A., Groves, S.J., Chambers, B.J., 2005. Pathogen survival during livestock manure storage and following land application. Bioresour. Technol. 96, 135–143.

Noel, J.T., Arrach, N., Alagely, A., McClelland, M., et al., 2010. Specific responses of *Salmonella enterica* to tomato varieties and fruit ripeness identified by in vivo expression technology. PloS One 5, e12406.

Nurnberger, T., Brunner, F., Kemmerling, B., Piater, L.L., 2004. Innate immunity in plants and animals: striking similarities and obvious differences. Immunol. Rev. 198, 249–266.

Oliveira, M., Vinas, I., Usall, J., Anguera, M., et al., 2012. Presence and survival of *Escherichia coli* O157:H7 on lettuce leaves and in soil treated with contaminated compost and irrigation water. Int. J. Food Microbiol. 156, 133–140.

Padera, B., 2010. Fresh cut produce in the USA. Perishables group. http://www.perishablesgroup.com/dnn/LinkClick.aspx?fileticket=mFzB9BJhR0w=.

Park, S., Szonyi, B., Gautam, R., Nightingale, K., et al., 2012. Risk factors for microbial contamination in fruits and vegetables at the preharvest level: a systematic review. J. Food Prot. 75, 2055–2081.

Patel, J., Sharma, M., 2010. Differences in attachment of *Salmonella enterica* serovars to cabbage and lettuce leaves. Int. J. Food Microbiol. 139, 41–47.

Penteado, A.L., Eblen, B.S., Miller, A.J., 2004. Evidence of *Salmonella* internalization into fresh mangos during simulated postharvest insect disinfestation procedures. J. Food Prot. 67, 181–184.

Reina, L.D., Fleming, H.P., Breidt Jr, F., 2002. Bacterial contamination of cucumber fruit through adhesion. J. Food Prot. 65, 1881–1887.

Richards, G.R., Beuchat, L.R., 2005. Metabiotic associations of molds and *Salmonella* Poona on intact and wounded cantaloupe rind. Int. J. Food Microbiol. 97, 329–339.

Salazar, J.K., Deng, K., Tortorello, M.L., Brandl, M.T., et al., 2013. Genes *ycfR, sirA* and *yigG* contribute to the surface attachment of *Salmonella enterica* Typhimurium and Saintpaul to fresh produce. PloS One 8, e57272.

Schikora, A., Carreri, A., Charpentier, E., Hirt, H., 2008. The dark side of the salad: *Salmonella* Typhimurium overcomes the innate immune response of *Arabidopsis thaliana* and shows an endopathogenic lifestyle. PloS One 3, e2279.

Schikora, M., Neupane, B., Madhogaria, S., Koch, W., et al., 2012. An image classification approach to analyze the suppression of plant immunity by the human pathogen *Salmonella* Typhimurium. BMC Bioinform. 13, 171.

Seo, K.H., Frank, J.F., 1999. Attachment of *Escherichia coli* O157:H7 to lettuce leaf surface and bacterial viability in response to chlorine treatment as demonstrated using confocal scanning laser microscopy. J. Food Prot. 62, 3–9.

Seo, S., Matthews, K.R., 2012. Influence of the plant defense response to *Escherichia coli* O157:H7 cell surface structures on survival of that enteric pathogen on plant surfaces. Appl. Environ. Microbiol. 78, 5882–5889.

Sharma, M., Donnenberg, M., 2008. A novel approach to investigate internalization of *Escherichia coli* O157:H7 in lettuce and spinach. Fresh Express Produce Safety Research Conference, Monterey, CA.

Sharma, M., Ingram, D.T., Patel, J.R., Millner, P.D., et al., 2009. A novel approach to investigate the uptake and internalization of *Escherichia coli* O157:H7 in spinach cultivated in soil and hydroponic medium. J. Food Prot. 72, 1513–1520.

Shaw, R.K., Berger, C.N., Feys, B., Knutton, S., et al., 2008. Enterohemorrhagic *Escherichia coli* exploits EspA filaments for attachment to salad leaves. Appl. Environ. Microbiol. 74 2980–2914.

Shi, X., Namvar, A., Kostrzynska, M., Hora, R., et al., 2007. Persistence and growth of different *Salmonella* serovars on pre- and postharvest tomatoes. J. Food Prot. 70, 2725–2731.

Shirron, N., Yaron, S., 2011. Active suppression of early immune response in tobacco by the human pathogen *Salmonella* Typhimurium. PloS One 6, e18855.

Sivapalasingam, S., Friedman, C.R., Cohen, L., Tauxe, R.V., 2004. Fresh produce: A growing cause of outbreaks of foodborne illness in the United States, 1973 through 1997. J. Food Prot. 67, 2342–2353.

Solomon, E.B., Brandl, M.T., Mandrell, R.E., 2006. Biology of foodborne pathogens on produce. In: Matthews, K.R. (Ed.), Microbiology of Fresh Produce. ASM Press, Washington, DC.

Solomon, E.B., Matthews, K.R., 2005. Use of fluorescent microspheres as a tool to investigate bacterial interactions with growing plants. J. Food Prot. 68, 870–873.

Solomon, E.B., Matthews, K.R., 2006. Interaction of live and dead *Escherichia coli* O157:H7 and fluorescent microspheres with lettuce tissue suggests bacterial processes do not mediate adherence. Lett. Appl. Microbiol. 42, 88–93.

Solomon, E.B., Potenski, C.J., Matthews, K.R., 2002a. Effect of irrigation method on transmission to and persistence of *Escherichia coli* O157:H7 on lettuce. J. Food Prot. 65, 673–676.

Solomon, E.B., Yaron, S., Matthews, K.R., 2002b. Transmission of *Escherichia coli* O157:H7 from contaminated manure and irrigation water to lettuce plant tissue and its subsequent internalization. Appl. Environ. Microbiol. 68, 397–400.

Stine, S.W., Song, I., Choi, C.Y., Gerba, C.P., 2005. Effect of relative humidity on preharvest survival of bacterial and viral pathogens on the surface of cantaloupe, lettuce, and bell peppers. J. Food Prot. 68, 1352–1358.

Strawn, L.K., Fortes, E.D., Bihn, E.A., Nightingale, K.K., et al., 2013. Landscape and meteorological factors affecting prevalence of three food-borne pathogens in fruit and vegetable farms. Appl. Environ. Microbiol. 79, 588–600.

Suslow, T.V., 2010. Produce Safety Project Issue Brief: Standards for Irrigation and Foliar Contact Water. http://www.producesafetyproject.org/reports?id=0007.

Takeuchi, K., Frank, J.F., 2000. Penetration of *Escherichia coli* O157:H7 into lettuce tissues as affected by inoculum size and temperature and the effect of chlorine treatment on cell viability. J. Food Prot. 63, 434–440.

Takeuchi, K., Hassan, A.N., Frank, J.F., 2001. Penetration of *Escherichia coli* O157:H7 into lettuce as influenced by modified atmosphere and temperature. J. Food Prot. 64, 1820–1823.

Takeuchi, K., Matute, C.M., Hassan, A.N., Frank, J.F., 2000. Comparison of the attachment of *Escherichia coli* O157:H7, *Listeria monocytogenes, Salmonella* Typhimurium, and *Pseudomonas fluorescens* to lettuce leaves. J. Food Prot. 63, 1433–1437.

Torres, A.G., Jeter, C., Langley, W., Matthysse, A.G., 2005. Differential binding of *Escherichia coli* O157:H7 to alfalfa, human epithelial cells, and plastic is mediated by a variety of surface structures. Appl. Environ. Microbiol. 71, 8008–8015.

Tyrrel, S.F., Knox, J.W., Weatherhead, E.K., 2006. Microbiological water quality requirements for salad irrigation in the United Kingdom. J. Food Prot. 69, 2029–2035.

Ukuku, D.O., 2004. Effect of hydrogen peroxide treatment on microbial quality and appearance of whole and fresh-cut melons contaminated with *Salmonella*. Int. J. Food Microbiol. 95, 137–146.

Ukuku, D.O., Fett, W., 2002a. Behavior of *Listeria monocytogenes* inoculated on cantaloupe surfaces and efficacy of washing treatments to reduce transfer from rind to fresh-cut pieces. J. Food Prot. 65, 924–930.

Ukuku, D.O., Fett, W., Sapers, G.M., 2004. Inhibition of *Listeria monocytogenes* by native microflora of whole cantaloupe. J. Food Saf. 24, 129–146.

Ukuku, D.O., Fett, W.F., 2002b. Relationship of cell surface charge and hydrophobicity to strength of attachment of bacteria to cantaloupe rind. J. Food Prot 65, 1093–1099.

Ukuku, D.O., Fett, W.F., 2006. Effects of cell surface charge and hydrophobicity on attachment of 16 *Salmonella* serovars to cantaloupe rind and decontamination with sanitizers. J. Food Prot. 69, 1835–1843.

Ukuku, D.O., Olanya, M., Geveke, D.J., Sommers, C.H., 2012. Effect of native microflora, waiting period, and storage temperature on *Listeria monocytogenes* serovars transferred from cantaloupe rind to fresh-cut pieces during preparation. J. Food Prot. 75, 1912–1919.

Ukuku, D.O., Sapers, G.M., 2001. Effect of sanitizer treatments on *Salmonella* Stanley attached to the surface of cantaloupes and cell transfer to fresh-cut tissues during cutting practices. J. Food Prot. 64, 1286–1291.

Ustun, S., Muller, P., Palmisano, R., Hensel, M., et al., 2012. SseF, a type III effector protein from the mammalian pathogen *Salmonella enterica*, requires resistance-gene-mediated signalling to activate cell death in the model plant *Nicotiana benthamiana*. The New Phytol. 194, 1046–1060.

Van der Linden, I., Cottyn, B., Uyttendaele, M., Vlaemynck, G., et al., 2013. Survival of Enteric Pathogens During Butterhead Lettuce Growth: Crop Stage, Leaf Age, and Irrigation. Foodborne Pathog. Dis.

Varma, J.K., Samuel, M.C., Marcus, R., Hoekstra, R.M., et al., 2007. *Listeria monocytogenes* infection from foods prepared in a commercial establishment: a case-control study of potential sources of sporadic illness in the United States. Clin. Infect. Dis. 44, 521–528.

Wang, G., Doyle, M.P., 1998. Survival of enterohemhorrhagic *Escherichia coli* O157:H7 in water. J. Food Prot. 61, 662–667.

Wang, G., Zhao, T., Doyle, M.P., 1996. Fate of enterohemorrhagic *Escherichia coli* O157:H7 in bovine feces. Appl. Environ. Microbiol. 62, 2567–2570.

Warner, J.C., Rothwell, S.D., Keevil, C.W., 2008. Use of episcopic differential interference contrast microscopy to identify bacterial biofilms on salad leaves and track colonization by *Salmonella* Thompson. Environ. Microbiol. 10, 918–925.

Wright, K.M., Chapman, S., McGeachy, K., Humphris, S., et al., 2013. The Endophytic Lifestyle of *Escherichia coli* O157:H7: Quantification and Internal Localization in Roots. Phytopathology 103, 333–340.

Xia, X., Luo, Y., Yang, Y., Vinyard, B., et al., 2012. Effects of tomato variety, temperature differential, and post-stem removal time on internalization of *Salmonella enterica* serovar Thompson in tomatoes. J. Food Prot. 75, 297–303.

Yoo, J.S., Jung, Y.J., Chung, S.Y., Lee, Y.C., et al., 2004. Molecular cloning and characterization of CMCase gene (*celC*) from *Salmonella* Typhimurium UR. J. Microbiol. 42, 205–210.

Zhang, J., Zhou, J.M., 2010. Plant immunity triggered by microbial molecular signatures. Molecular Plant 3, 783–793.

Zheng, J., Allard, S., Reynolds, S., Millner, P., et al., 2013. Colonization and Internalization of *Salmonella enterica* in Tomato Plants. Appl. Environ. Microbiol. 79, 2494–2502.

Zhou, J.M., Chai, J., 2008. Plant pathogenic bacterial type III effectors subdue host responses. Curr. Opin. Microbiol. 11, 179–185.

Zhuang, R.Y., Beuchat, L.R., Angulo, F.J., 1995. Fate of *Salmonella* Montevideo on and in raw tomatoes as affected by temperature and treatment with chlorine. Appl. Environ. Microbiol. 61, 2127–2131.

Ziemer, C.J., Bonner, J.M., Cole, D., Vinje, J., et al., 2010. Fate and transport of zoonotic, bacterial, viral, and parasitic pathogens during swine manure treatment, storage, and land application. J. Anim. Sci. 88, E84–94.

Identification of the Source of Contamination

3

Jeffrey A. Farrar[1], Jack Guzewich[2]

[1]*Associate Commissioner for Food Protection, U.S. Food and Drug Administration, Center for Food Safety and Applied Nutrition,* [2]*U.S. Food and Drug Administration, Center for Food Safety and Applied Nutrition (Retired)*

CHAPTER OUTLINE

Introduction

State and local health departments make up the backbone of the system to detect and investigate foodborne outbreaks and illnesses in the United States. Numerous types of expertise are required in these investigations including knowledge and experience in surveillance systems, epidemiological investigations, laboratory

The Produce Contamination Problem. http://dx.doi.org/10.1016/B978-0-12-404611-5.00003-8
2009 Published by Elsevier Inc.

methods, bacterial and viral ecology, water engineering, and environmental investigations. Some health departments have staffing, experience, communication capabilities, training, and interest in some of these areas. However, few agencies have standardized, documented, and practiced procedures and performance standards for the entire foodborne outbreak investigation process. Even fewer have dedicated, highly trained, multidisciplinary teams that work together throughout the entire investigative process. Without a multidisciplinary, multiagency approach, using trained and experienced experts throughout, foodborne outbreak investigations will remain unable to consistently and accurately identify the causes of the outbreaks and develop effective prevention measures to reduce the likelihood of recurrence (FDA, 2009).

This chapter will provide an overview of the entire foodborne outbreak investigative process as a means of identifying the source of contamination but will focus primarily upon the environmental phase of the investigation.

Overview: phases of a foodborne outbreak investigation

Foodborne outbreak investigations can be roughly characterized into five phases: surveillance/detection, epidemiologic, environmental/traceback, regulatory/enforcement, and prevention/research. This characterization assumes that laboratory diagnostics are part of each of these five phases. These separate, yet overlapping and related phases require very different knowledge, methods, expertise, and legal authority. However, investigators involved in all phases must interact and work together throughout the entire process if the complete story of an outbreak is to be revealed and the risk of additional outbreaks is to be minimized.

Surveillance and detection

The first phase, surveillance and detection, provides a signal to public health investigators that an unexpected number of similar illnesses occurred during a certain time-period. These signals can be from a variety of active and passive surveillance efforts. Consumers may report illnesses to local health jurisdictions after eating at a restaurant or attending an event. Medical care providers may report diagnoses of foodborne pathogens to local or state public health agencies, local or state laboratories may notice a higher than expected number of similar pathogens during a specific time period, and emergency room care providers may notify local public health agencies of an unusual number of individuals with similar symptoms (e.g., bloody diarrhea). National surveillance systems, such as PulseNet, have proven very effective in identifying small outbreaks composed of a genotypically similar group of pathogens from a number of geographically widespread, sporadic cases that previously were not connected. However, there remains significant variability among states and local jurisdictions regarding the promptness and thoroughness of surveillance and reporting systems for foodborne illnesses and outbreaks.

Epidemiologic

Once the surveillance system has detected an unusual occurrence of disease, an epidemiologic investigation may be initiated. This stage attempts to identify the specific vehicle responsible for the illnesses, and to characterize the ill individuals in terms of age, gender, race, and location. Laboratories attempt to determine the agent in patient specimens concurrent with the epidemiologic investigation. Ill individuals are interviewed, and food histories are collected along with travel history and visits or exposures to known vectors such as animal petting zoos. Epidemiologists use well-described methods such as case control or cohort studies to statistically compare foods eaten by ill individuals to foods consumed by well individuals in order to identify a specific food vehicle. Even without laboratory confirmed vehicles, statistical associations derived from these investigations can provide compelling evidence of what food or foods caused the illnesses. However, epidemiologic investigations may not always be successful in identifying a specific food. Recall of specific food items consumed may dim after a couple of weeks, or multiple, unrelated food items may be identified.

The time required to complete an epidemiologic investigation varies but frequently requires days or weeks of repeated telephone or in-person interviews followed by descriptive and detailed data analysis. However, as in all other phases of a foodborne outbreak investigation, rapid completion is critical, especially when there is a reasonable likelihood of ongoing exposures to contaminated foods. Similar to the demand by public health agencies for increased industry documentation of best growing, processing, and shipping practices, public health agencies are beginning to address expectations of written performance standards and standard operating procedures for foodborne outbreak investigations (Anon., 2002). Epidemiological investigations of foodborne outbreaks can result in findings suggestive of two possible broad sources of exposure: point source and multiple or continuing sources (e.g., tomatoes from the eastern shore of Virginia). In point source exposures, ill individuals are found to have a common event or place of exposure, for example eating at a single restaurant or attending the same catered event prior to onset of symptoms. Some point source exposure events may manifest as very low numbers of ill individuals that occur over extended periods of time or that reappear from time to time. These types of foodborne outbreaks warrant special attention to environmental or food worker sources of contamination within the food facility. These point source exposures must be carefully assessed during the environmental investigation to determine the probability of food workers as the most likely source of contamination. Even with cultures, serology, detailed reviews of work logs and daily job responsibilities, preparation of product work flow charts, and one-on-one interviews of food workers without managers present, there may not be sufficient information to definitively rule out food workers. However, without such comprehensive, standardized efforts, public health agencies will continue to be criticized for the lack of effort, and further investigation efforts at manufacturers or farms may not be considered practical.

Multiple source exposures, such as illnesses associated with multiple, unrelated restaurants in multiple states, strongly suggest that the source of contamination is a widely distributed product and that contamination likely occurred prior to the retail facility. In these situations, measuring and recording routine practices and procedures at retail facilities may still be useful to identify risk factors that may contribute to or detract from the survival or growth of pathogens (e.g., refrigeration temperatures, pH, water activity, times, and temperatures at each stage, and whether the product was washed and how it was washed).

When specific foods are implicated in epidemiological investigations, ill individuals are further interviewed to determine precisely when and where they purchased or consumed the implicated food, to provide as much specific information about the food as possible including brand, type of product, size, grade, color, and whether they have any remaining opened or unopened product. This specific information is critical for tracebacks and environmental investigations. Without this specific information, further investigation may not be possible. Therefore, the highest priority should be assigned to dispatching staff to households of ill individuals to collect any remaining opened or unopened product implicated in the investigation. Asking family members to deliver remaining products only invites delays in initiating valuable laboratory testing efforts. If a specific retail food facility is determined to be associated with illnesses and the epidemiological investigation has not yet identified or is unable to identify a specific food, investigators should collect samples of all foods deemed to be a possible cause of the outbreak. Samples can be refrigerated or frozen until decisions are made on the epidemiological study.

Knowing that lettuce or tomatoes purchased from store A are associated with illness is important information but is of marginal utility in determining the ultimate source and cause of contamination and in preventing additional exposures to contaminated foods. Many stores carry numerous types and sizes of lettuce and tomatoes from multiple suppliers and sources. Standardized commodity-specific questionnaires should be developed in advance for foods implicated in previous outbreaks, and these questionnaires could be posted on websites by federal agencies for use by state and local agencies in subsequent multistate/multijurisdiction foodborne outbreak investigations.

Increasingly, epidemiologists attempt to confirm whether consumers may have utilized "club cards" in the purchase of contaminated foods. This objective data can be used to verify specific product information and purchase dates and thus can provide valuable information during investigations. Retailers generally provide club-card data to public health investigators once permission to share the information is given by the consumer. These contacts, mechanisms, and standardized questionnaires for obtaining permission to use club-card information should be developed in advance by state agencies and shared with local agencies. Having multiple counties or states contacting a retail food chain asking for the same or similar information for different consumers only creates confusion, delays, and duplication of effort, and creates doubt regarding the ability of agencies to coordinate their efforts. Communication should include clarification of roles and responsibilities of participating agencies.

Providing frequent, current, detailed updates to those involved in the environmental phase of the investigation, especially during outbreaks with a reasonable probability of ongoing exposures, will allow regulators to better focus limited resources.

Environmental

When a specific food item is identified in the epidemiologic investigation, the third phase, the environmental investigation, attempts to determine how, where, when, and why the contamination occurred so that further exposures can be reduced or eliminated, and effective prevention measures can be implemented. Before initiating environmental investigations, it is critical that the results of the epidemiologic investigation be carefully reviewed and discussed to understand the inherent strengths and weaknesses. Ideally, outbreak investigators would be able to genetically match laboratory confirmed pathogens from ill individuals with laboratory confirmed pathogens from epidemiologically implicated foods and identify how and where the contamination occurred. However, this ideal investigation is more often the exception rather than the rule. Because of the inherent delays and variability in incubation periods, diagnostic testing, reporting of results, and in completing epidemiologic investigations, many perishable foods with a relatively short shelf-life may no longer be available for laboratory testing. Many fields growing these crops will have been replanted with other crops. Additionally, environmental investigations may take months of work, consuming huge amounts of resources and resulting in tremendous economic damage to firms and industries. Therefore, it is critical that regulatory agencies have trained, experienced epidemiologists and statisticians available to review and discuss epidemiological findings with communicable disease agencies before beginning the environmental investigation.

Although epidemiological studies are extremely powerful tools; analytical mistakes, incomplete investigations, biases in questionnaires/interviews, or purely chance associations with incorrect foods can result in erroneous conclusions with enormous consequences. A previous epidemiological investigation of illnesses in multiple states incorrectly linked illnesses to consumption of domestic strawberries, although imported raspberries were eventually identified as the vehicle responsible for the illnesses. This erroneous conclusion, based on an incomplete epidemiological investigation, resulted in incorrect preventive guidance to consumers, possibly resulting in additional exposures, extraordinary financial losses to the domestic strawberry industry, massive costs to regulators in the investigation of the wrong commodity, significant delays in implementing correct control measures, and a loss of credibility for the public health community. Although it is critical that foodborne outbreak investigations be quickly completed, it is more important that the findings are correct.

Occasionally, decisions to initiate tracebacks and environmental investigations must be made with only preliminary or limited epidemiological findings. For example, outbreaks with a strong possibility of ongoing exposures to pathogens such as *E. coli* O157:H7 may require immediate actions with less than complete or definitive epidemiological findings.

Once a decision is made to initiate an environmental investigation, detailed written practices and procedures should be implemented. Receiving, storage, and food preparation practices should be carefully measured and documented. Factors such as time, temperature, water activity, and pH within food manufacturing and retail food facilities may contribute to the growth or survival of pathogens. Previous investigations of green onion-associated hepatitis A outbreaks revealed that some retail facilities stored green onions in containers with water without thoroughly cleaning and sanitizing the container between lots of green onions. This practice may have contributed to the spread of contamination to multiple batches. Accurate, thorough descriptions along with objective measurements using recently calibrated equipment, where appropriate, are essential.

Environmental investigations have been erroneously referred to as tracebacks. Tracebacks of contaminated foods attempt to document each stop or location in the farm-to-fork continuum. Tracebacks are one important part of environmental investigations; however, this terminology does not accurately convey the scope and complexity of this phase of the outbreak investigation.

Previously, evaluations of food facilities implicated in foodborne outbreaks were composed of routine inspections of food facilities, using established, entrenched regulatory inspection forms, checklists, and mindsets. Not surprisingly, this approach often yielded little in the way of useful information as to the cause of the outbreak and even less useful information regarding effective prevention measures. Findings of the lack of a suitable shield on fluorescent lights may have been documented as violations of existing law or regulations, but this finding had little to do with how *Salmonella* or *Shigella* was introduced, survived, or grew in or on a specific food, or which prevention measures could be implemented to reduce the risk of recurrence. The term "environmental investigation" includes tracebacks but also includes methodical, scientific reviews of each point in the farm-to-table continuum to assess opportunities for introduction, survival, and growth of pathogens.

Many regulatory agencies have recently adopted a more scientific, methodical, investigative approach to environmental and traceback investigations. Without this approach, foodborne outbreak investigations will continue to yield few clues as to the causes of outbreaks and prevention measures for reducing the probability of recurrence.

In order to better understand where and how contamination occurred and whether pathogens could have been introduced, survived, or could have grown during food preparation or food manufacturing conditions, investigators must have extensive knowledge of specific pathogens, their ecology, and the effect of pH, time, and temperature and other factors on the organism. When the source of the contamination is determined to be prior to the retail food facility, investigators must use the same concepts as in the retail food facility to understand and document growing, harvesting, processing, and shipping practices associated with contaminated foods and food ingredients and must complete a scientific assessment of the probability of introduction, survival, and growth of the organism at each point.

Specifics of environmental investigations

Environmental investigations must include evaluations of each step in the farm-to-fork continuum to look for opportunities for introduction, survival, or growth of the pathogen. In essence, the investigators attempt to rule out contamination at each

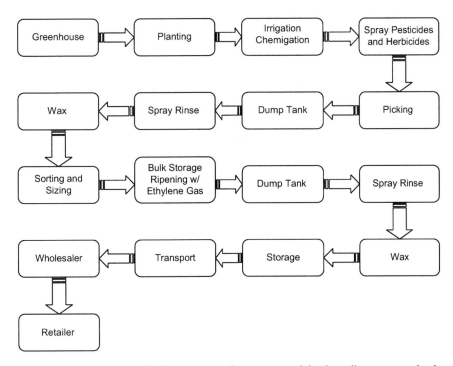

Critical information such as pH, times, temps, volume, water activity, ingredients may need to be included to ensure a thorough understanding of the entire food preparation/food processing event.

FIGURE 3.1

Generic example of a food process flow chart.

step in the process. Flow charts (Figure 3.1), capturing very specific measurements such as time, temperature, pH, amounts of ingredients, and size of containers are a highly recommended tool in environmental investigations. These charts can provide food scientists with objective data to better determine whether food preparation or processing procedures could have contributed to the outbreak, and more importantly, whether additional preventive steps or barriers should be implemented. Additionally, descriptions and documentation of cleaning and sanitation procedures and chemicals are important.

Environmental investigations of retail, wholesale, and food manufacturing facilities are identical in concept. Thorough assessment and documentation of the entire process at each step is critical. Each type of facility could be segmented into three broad areas: incoming ingredients, food processing/food preparation, and outgoing product. Previous environmental investigations of retail food and food processing facilities were often conducted by a single individual. However, due to the scope, complexity, and specialized needs of these investigations, a single individual is likely insufficient to complete an environmental investigation promptly and thoroughly, even with single retail food facilities. In large processing facilities, several individuals with targeted training and experience are often necessary, each assigned to lead one of the three segments identified earlier, in order to complete the investigation in a thorough, timely manner.

These investigations cannot be completed thoroughly and accurately without having appropriate equipment. Most environmental investigators have access to obvious equipment such as digital thermometers, stop watches, and digital cameras. However, many have not traditionally utilized specialized equipment such as needle probes for thermocouple thermometers to measure the temperature of hamburgers while cooking, oven probes to monitor the temperature of products throughout the cooking cycle, oxidation-reduction potential meters, water activity meters, and waterproof data loggers that measure the temperature of dishwashers and flume water. Additional areas of expertise and equipment needed may include tracer dyes or compounds to assess cross-connections and flow of water or wastewater, air-stream monitoring and air sampling devices, and water sampling and concentration devices. Digital photographs and videos can often convey in a few seconds conditions or practices that may take pages to explain in writing. However, investigators should ensure that they have legal authority to take photographs prior to initiating the investigation.

Where possible, investigators are strongly encouraged to visually observe "routine" food preparation/food processing procedures and record objective measurements of preparation and processing practices, as well as routine cleaning and sanitation procedures. Often, what is written in procedures manuals is not what actually occurs in the kitchen or in the food processing facility. Even with the inevitable bias introduced during physical observation by investigators, valuable clues often are obtained by simply observing and documenting a process from start to finish.

Food contamination may have been confirmed by an epidemiologic association between consumption of a food and illness or by laboratory recovery of an agent in a food or by observations of conditions during manufacturing, preparation, or serving that likely lead to contamination. In any of these cases, there may still be additional contaminated food available to consume. Investigators must work quickly with the firm responsible for processing/serving/selling the food to have any remaining food removed from distribution, collect samples where appropriate, determine distribution, and alert the public in all areas where the food may still be available. Many of the produce outbreaks in recent years have taught us that consumers will keep fresh-produce items in their homes much longer than the shelf-life assumed by growers, packers, and retailers. Therefore, the public can still be at risk for exposure long after the shelf-life on the bag has passed. However, in some fresh-produce outbreaks, by the time the epidemiology and food testing steps have been completed and a specific food has been implicated, there truly is no possibility of the food item remaining in distribution, and consequently, no risk of further exposure for the public. In this case, regulatory and public health agencies must evaluate the benefits of issuing consumer-level advisories. In some situations, previously unknown cases may be identified by public notifications.

Frequently, environmental sampling of retail and food processing facilities is given only a minimal amount of effort, not surprisingly resulting in negative results. For example, many environmental investigations report collecting 5 to 10 environmental samples in a restaurant or 15 to 20 samples in a large food processing facility. The number of samples collected in any single facility should be maximized to provide the best opportunity for recovering pathogens, if they are there. Frequently, sampling efforts are determined by figures that laboratories provide regarding the number of samples they can process. Investigators are often instructed not to collect more samples than their laboratory can handle. However, this limitation should not be allowed to drive the environmental investigation. Managers should identify and have in place options to process additional environmental samples once the local or state capacity is exceeded. In some situations, samples from defined areas within a facility can be composited to lessen the total number of samples yet still allow for a realistic opportunity to detect pathogens that may be present in defined areas.

Although there is no precise numerical guideline for the number of samples to be collected in these investigations, 10 to 12 samples from a retail or food processing facility is likely to be an insufficient number of samples to have a reasonable probability of detecting pathogens if they are present. Investigators should think in terms of a minimum of 50 to 100 samples per retail facility and even larger numbers from food processing facilities. Portable coolers or freezers, large enough in size to hold the perishable samples, should be provided. Written procedures for adequately packaging samples should include necessary steps to prevent spillage or leaking of samples during transit to laboratories.

Even though many firms may have completed one or more cleaning and sanitation processes prior to environmental sampling, this should not deter investigators from collecting environmental samples. Media exists to help neutralize the effects of cleaning and sanitizing chemicals. Additionally, pathogens can survive and grow in difficult-to-clean places in food facilities. However, where possible, samples should be collected prior to cleaning and sanitation steps.

Within the food facility, investigators should be strongly encouraged to consider the need for disassembling, or have food facility staff disassemble equipment used to prepare, slice, or process implicated foods and then to take sufficient numbers of samples from the equipment. Do not accept arguments from firms or investigative staff that "taking apart that piece of equipment would take too long." Screwdrivers and wrenches to dissemble equipment should be part of a standard environmental investigation "go kit," and investigators should be encouraged to collect samples from both easy access locations (e.g., floor drains, cleaning rags, mops, brooms, vacuum cleaners, wheels of forklifts/carts) and from hard-to-reach places (e.g., undersides of cutting boards bolted to tables, gears of conveyor belts) as cleaning and sanitation processes may not reach pathogens in these locations. Regulatory actions and possible legal challenges to findings from environmental investigations are more effectively addressed when proper, documented chain-of-custody procedures are followed for all samples (see Figure 3.2). Where appropriate, control samples such

Department of Health Services
Food and Drug Branch

EVIDENCE/SAMPLE TAG

IS# 138071007 A

Received From	Delivered To	Date	Time
A+A Fish Co.	P. Smith	7/10/07	1000 Hrs
P. Smith	J. Jones - Lab	7/10/07	1430 Hrs
J. Jones	Sample Locker	7/10/07	1445 Hrs
J. Jones/Sample Locker	B. Thomas	7/11/07	0800 Hrs
B. Thomas	Sample Locker	7/11/07	1300 Hrs

Comments:

DHS 8067 (7/05)

Department of Health Services
Food and Drug Branch

EVIDENCE/SAMPLE TAG

IS# 138071007A

Received From	Delivered To	Date	Time
Sample Locker/B. Thomas	M. Johnson	7/12/07	1000 Hrs
M. Johnson	Disposal	7/12/07	1300 Hrs

Comments:

DHS 8067 (7/05)

FIGURE 3.2 Example of a chain of custody form/tag that can be physically attached to samples.

as unused whirl pack bags and sterile gloves should also be collected. Supervisors should emphasize to investigators that early and frequent communication with the laboratory must occur during the investigation.

Some public health agencies routinely avoid interviewing food workers simply because "we know what they will say" or because "we won't be able to tell if the food worker got it from the food or introduced it into the facility." Obviously, without effort in this area, potentially useful information might be lost to the investigation. Other agencies are more aggressive with one-on-one interviews, and in collecting stool cultures or serum samples from food workers in implicated facilities. Public health agencies are encouraged to make a concentrated effort to try to rule out food worker contamination at the point of service. Without these efforts to assess the probability of food worker contamination in outbreaks with a single point of service, further traceback efforts and environmental investigations may not be initiated. Additionally, certain patterns of illnesses may signal a strong need to test food workers and/or conduct more intensive environmental sampling efforts. For example, one or two cases of a specific PFGE matching pathogen may be loosely associated with a specific food facility during a two- to three-month period. Then, three to four months later, one or two additional matching cases may appear, again with some association

Table 3.1 Hypothetical Traceback from Joe's Steakhouse to World Distribution, Providing Dates of Receipt at Each Node and Volume of Product Received

Example of a Completed Time Line

	Receipt Date									Event	
DATE	5/06	5/07	5/08	5/09	5/10	5/11	5/12	5/13	5/14	5/15	5/16
At Joe's Steak House (POS) Daily inventory	3	0	4	5	1	0	0	0	3	2	0
From XYZ Produce	0	5	0	4	0	4	2	5	0	4	0
From Nations Foods	2	0	2	0	2	0	0	0	0	0	2
At XYZ Produce no inventory											
From Zenith Fresh	0	40	40	30	25	35	50	0	45	35	35
From Best Produce	0	0	10	5	5	10	5	55	10	10	5
From Superior Vegetables	50	10	0	0	0	0	0	0	0	0	0
At Zenith Fresh no inventory											
From New Products	450	500	0	0	0	0	250	300	200	300	0
From Fresh-N-Fast Grower- State F	0	0	550	450	400	500	200	300	300	200	450
At Best Produce no inventory											
From World Dist. Country X	0	0	950	0	0	0	900	0	0	0	950

with the same food facility. This pattern may suggest a need for intensive employee assessment efforts along with intensive environmental sampling.

Traceback investigations

Traceback investigations are frequently the first step in an environmental investigation (FDA, 2001). Distribution patterns of foods vary significantly by commodity, season of the year, and type of food. For example, tomatoes destined for dicing at a single processing facility may include product from multiple growers and multiple distributors in multiple states. Following processing, this commingled product may be shipped to dozens of customers who may resell to one or more middlemen until the product reaches the end user (see Table 3.1).

To ensure accuracy in traceback investigations and prevent wasted resources, investigators should carefully select a subsample of those with confirmed illnesses for a traceback investigation. It is not necessary, nor is it logistically feasible, to trace product from every ill case. The subsampling of confirmed ill individuals should give preference to those with the most definitive recall of when and where they purchased or consumed the contaminated product and to those with singular points of purchase or exposure. Objective data such as club-card records, credit-card bills and receipts, and personal calendars provide additional support to an individual's recall of events that occurred weeks before. This subsample, where appropriate, should also include individuals who purchased or consumed product from different states, retail chains, restaurants, or different brands of product. This provides regulators with a higher degree of confidence in the ability of the traceback to produce the correct answer. This process is referred to as **triangulation**.

Determining the time period of interest is one of the first, and frequently one of the most difficult challenges in traceback investigations. Most frequently, this time period is derived from estimates based upon a sample of the confirmed cases with the most definitive and, ideally, singular recall of exposure to the implicated foods. Unless there are laboratory confirmed samples from implicated products with definitive lot codes or use-by dates, this analysis generally begins with the earliest and latest known exposure/consumption dates from confirmed cases. Then, minimum and maximum incubation periods are included along with the maximum consumable shelf-life under proper refrigeration of the product and minimum and maximum shipment times.

A narrowly defined time period of interest for tracebacks is preferred and allows investigators to more efficiently utilize limited resources. Excessively broad time periods may involve dozens of suppliers, wholesalers, distributors, brokers, and even more farms of origin, making these efforts less likely to be successful or even to be attempted. However, past investigations have demonstrated the need to err on the side of a slightly broader time range to avoid repeated visits to the same facilities as new cases are discovered.

Conducting tracebacks by telephone is tempting because they are quick and are not as resource intensive as in-person tracebacks. However, telephone tracebacks will frequently give you the wrong results very quickly, and thus, are strongly discouraged except in extraordinary circumstances. In-person, on-site tracebacks have repeatedly revealed critical information that would likely not have been obtained by a telephone call. Additionally, these records are necessary for regulators to take enforcement actions when appropriate. Where possible, investigators should personally visit each point in the food-to-table continuum, carefully review and obtain legible copies of all records of incoming and outgoing products during the time period of interest, and review processing practices. Portable, high-speed copiers or scanners can be invaluable when large volumes of records are involved or when business owners do not want to release original copies for photocopying by investigators.

Occasionally, regulators will receive calls from communicable disease staff asking for assistance in conducting an epidemiological traceback. In some epidemiological investigations, a specific ingredient may not be discernable (e.g., tacos). Tracebacks of multiple ingredients in these implicated foods may provide additional insight into which ingredient distribution pattern may best match the geographic distribution of ill individuals. However, all efforts to conduct a thorough and prompt epidemiological investigation should be exhausted first before these requests are made.

Two types of traceback diagrams are currently used to document traceback investigations: a reference traceback flow diagram (Figure 3.3) and a traceback spreadsheet (Table 3.2). These diagrams not only help document findings but also provide investigators with visual depictions of common growers, shippers, processors, or retailers to better discern relationships and commonalities that may exist among the subsample of ill individuals. These traceback diagrams also allow regulators to narrow the focus of the investigation to certain shipment or production dates using purchase dates, exposure dates, incubation periods, daily inventories, and shipping times described earlier. These diagrams should include basic information such as amounts of implicated product shipped/received, lot codes, and dates shipped/received. This information is frequently obtained from invoices, shipping manifest records, and daily inventory and ordering records that are provided when ordering, picking up, and delivering food products.

In addition to collecting and reviewing records, investigators should interview knowledgeable individuals at each food facility to confirm information provided in records. Not infrequently, paper records may not list deviations from usual procedures. For example, retail food facilities may tell suppliers that they only want a specific type, brand, color, or size of food in each shipment (e.g., 3 lb bags of iceberg and romaine blend with carrots from manufacturer X). However, product

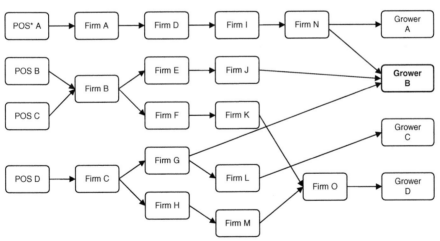

*POS = Point of Service

The boxes and lines that are in bold lettering indicate the implicated product's distribution pathway that links all of the points-of-service (POS) outbreaks to the implicated grower. Tracebacks were initiated at four points of service in response to a multistate outbreak of illnesses that were epidemiologically associated with a specific food product. In the traceback related to POS A, Grower A and B were implicated. In the tracebacks initiated at POS B and POS C, Growers B and D were both implicated. None of these distributors or subdistributors were identified as a common supplier to all four points of service. Therefore, none of these were considered as the most likely source of the contamination. Although a total of four growers were implicated, the only one in common to all of the four tracebacks is Grower B. The only grower assigned for an environmental/farm investigation was Grower B. Distributors/subdistributors G and E were also assigned as a lower priority investigation to document the opportunities for survival and growth during these interim steps. Dates of shipment and receipt, along with volume of shipments can be included in each box to better identify the most likely source.

FIGURE 3.3 Example of multistate traceback flow diagram.

Table 3.2 Example of a Traceback Spreadsheet

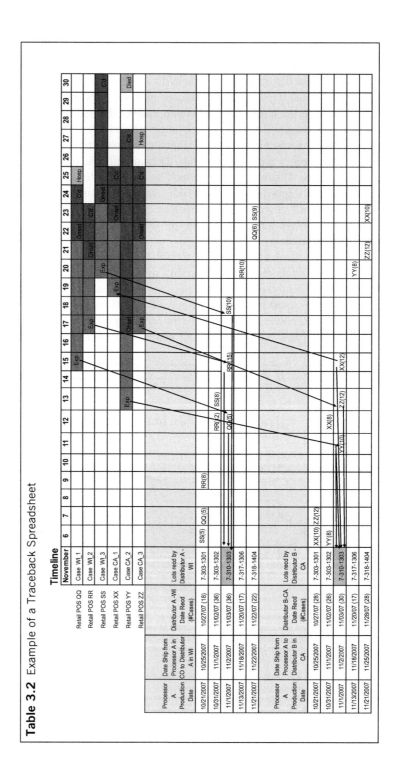

shortages occur on given days, and suppliers may substitute products from the same or different manufacturers when the originally requested product is not available (e.g., 3 lb bags of iceberg with red leaf lettuce, spring mix, and radicchio from manufacturer Y). Failure to take into account this relatively simple, frequently occurring transaction could result in thousands of hours of investigative time spent pursuing the incorrect commodity or ingredient. Food manufacturers, distributors, and retailers should require suppliers to note any substitutions clearly on shipping manifests. Investigators should collect accurate information on the precise delivery times and preparation times for implicated meals at retail facilities. These times may also assist investigators in narrowing the scope of the traceback and environmental investigation.

Tracebacks of foodborne outbreaks within a single local jurisdiction may be done by local officials. However, these investigations often result in findings of suppliers located in other local or state jurisdictions or even in other countries that are outside the jurisdictional authority of the original local agency. Additionally, outbreaks originating in one local jurisdiction often quickly expand to other counties or states. Close coordination of all tracebacks with appropriate state and federal regulators is critical to avoid duplication of effort, to ensure consistency, to ensure that adequate authority exists for collection of records, and to ensure that the highest priorities are addressed first. Multijurisdictional outbreaks are best coordinated by state or federal agencies. State and federal agencies should provide clear direction and protocols, along with frequent feedback to local agencies who wish to assist in the traceback and environmental investigations (FDA, 2001).

Regulatory/Enforcement

The regulatory/enforcement phase of the investigation may be done concurrently with, or more preferably after, the initial environmental investigation phase. However, it is critical that the two phases be viewed and implemented separately. The knowledge, skills, and abilities for those involved in the environmental investigation are very different from those in the regulatory/enforcement phase. Attempting to use the same investigator for both purposes at the same time will likely result in an incomplete job in both areas.

Most retail and food manufacturing facilities have limited resources to respond to multiple requests from investigators and can become overwhelmed quickly by the demands associated with outbreak investigations. The highest priority must always be implementation of measures necessary to identify the scope of the distribution of the implicated product and prevent further exposures. Routine inspections of retail and wholesale food facilities should not be placed before information gathering and implementation of measures needed to control the outbreak.

The regulatory/enforcement phase collects evidence of conditions that may have been in violation of statutes, regulations, or guidance documents. Previous inspection findings should also be reviewed. These inspectional findings then need to be discussed to determine if regulatory or enforcement action is warranted.

Prevention/Research

Although investigators involved in the environmental phase of foodborne outbreak investigations have made significant advances in developing standardized methods and in utilizing highly skilled and experienced investigators, the precise mechanism of how pathogens come into contact with fresh produce remains unclear. This lack of understanding has pointed out the need for quick, prioritized research projects so that effective prevention steps can be identified and implemented. Instead of taking three or four years from the request for proposal from a granting agency to a completed scientific paper, federal agencies could consider new approaches to this research including teams of dedicated "high priority/quick turnaround" researchers to address urgent issues identified during the investigation.

Many believe that the contamination first occurs in the field through contaminated water, air, or animal or human feces coming in direct contact with the plants. This contamination, once it is on or within the plant material, is not likely to be eliminated by subsequent washing/sanitizing of the produce and may be further distributed through mixing of small contaminated batches with larger lots. Applied research to confirm or disprove these assumptions and to identify interventions and priorities for their implementation is critically needed. Federal agencies should publish yearly lists of research priorities for commodities involved in recurring outbreaks. These lists can be used by granting agencies and by industry to evaluate and fund produce-related research efforts.

Training needs for environmental investigators of retail, food processing facilities, packing sheds, and farms

Training to emphasize an integrated, multidisciplinary team of laboratory, communicable disease, and environmental health staff is important. Training courses such as the NEHA/CDC Epi Ready course emphasize this team approach, building upon the skills, expertise, and authorities each team member brings to the investigation. Core competencies in environmental investigations of retail and food processing facilities may be established and would include areas such as:

- Basic epidemiology of foodborne outbreak investigations
- Basic microbiology/ecology of common foodborne pathogens
- Aseptic sampling and chain of custody procedures
- Basic traceback and environmental investigation procedures
- Documenting the flow of food and identifying opportunities for contamination, growth, survival, and destruction
- Basic interviewing techniques
- Legal aspects of investigations

Standardized food worker questionnaires, in appropriate languages, can be developed for interviews with food workers and should include questions about specific responsibilities on the dates in question, deviations from these responsibilities, and symptoms consistent with the illness under investigation during the time period of interest. All food workers involved in preparing implicated foods or working with implicated ingredients should be interviewed individually where possible. These interviews should be conducted without the manager or supervisor present. If necessary, translators should be provided during these interviews.

Resuming operations

At some point in time following a foodborne outbreak at a retail or food processing facility, questions will begin to arise regarding what the facility must do or should do to resume operations. Although these requirements may vary by pathogen and by local, state, or federal requirements, the following guidelines may be useful:

- Assess the need to dispose of all foods and ingredients that may have been cross-contaminated.
- Consider the need for one or more "exceptional" cleaning and sanitation cycles using steam and approved cleaning and sanitizer compounds and processes followed by procedures to verify the effectiveness of the cleaning and sanitation.
- Determine whether the firm should be required to implement an ongoing cleaning and sanitation effectiveness monitoring program. Monitoring programs may incorporate ATP devices to give instant feedback on levels of organic matter present after cleaning and sanitizing. Written procedures should also include parameters for frequency of sampling, number of samples, location of sampling, and steps to take if baseline levels are exceeded.
- Discuss the need for requiring the food facility to retrain and provide regular ongoing training to staff, in their native language if necessary, by an approved trainer in basic food safety practices and in specific food preparation and processing procedures.
- Consider whether additional preventive measures such as implementing, modifying, or revalidating HACCP plans for a processing facility; establishing new time or temperature requirements; establishing new guidance for minimizing bare-hand contact and excluding ill food workers; providing improved access to handwashing stations; color coding of items such as cutting boards for meat and poultry, equipment, and hardhats for personnel with access to one area of a food processing facility; and reconfiguring food preparation or processing areas, which may reduce the risk of recurrence.
- Determine whether increased inspections of the food facility for some time period are appropriate, and identify specific areas for more intensive review.
- Provide clear expectations of food facilities along with relevant timelines and expected outcomes (preferably in writing) to the responsible individual to avoid confusion.

Farm investigations

When environmental investigation findings suggest that the point of contamination may have occurred prior to the retail or food processing facility, farm investigations may be initiated. Farm investigations require specialized expertise and training in such areas as sanitary surveys of agricultural wells, hydrogeological connections between surface water and underground aquifers, legal requirements and jurisdictions for irrigation water quality, recycled water treatment processes, manure composting processes and requirements, wildlife identification from feces or tracks, wildlife habitat and ranges, wildlife trapping and sampling procedures, environmental (water, air, soil, sediment) sampling procedures, along with crop production, harvesting, and cooling practices for different commodities (FDA, 2005).

Evidence to date suggests that the point of contamination for most produce-associated outbreaks originates on the farm during growing and harvesting. The agents that have been involved are in two categories: those most commonly associated with animals (e.g., zoonotic including *E. coli* O157:H7 and *Salmonella*) and those associated with humans (e.g., *Cyclospora, Shigella,* and hepatitis A).

Epidemiologists and microbiologists refer to the place where pathogens are normally found in nature as the reservoir. If an investigator is visiting a farm implicated in an outbreak with an agent that has a human reservoir, the emphasis has to be on identifying how the pathogen moved from humans to the food such as through human feces, or worker hands or water contaminated with human feces. If the investigator is visiting a farm implicated in an outbreak with an animal reservoir, the investigation needs to focus on how the pathogen moved from domestic animals or wildlife to the produce. This would most likely occur through indirect contact with animal feces through water spread or dust spread or direct contact with the animal or its feces.

Farm environmental investigations begin with a thorough inventory of the farm operation including a map of the farm layout; the source and species of seed/seedlings and when they are planted; fertilizers used and how they are applied; pesticides/herbicides used and the water used to dilute them and how they are applied; how and when crops are irrigated and the source of water used; how and when the crop is harvested, and how the crop is transported. Since a farm investigation happens after the crop has been harvested, this information will have to be obtained through an interview with the grower. The field portion of the investigation should include a walk around the field looking for animal tracks and feces and evidence of human feces. The source(s) of water should be evaluated. If it is a surface source, like a stream, pond, or reservoir, conduct a sanitary survey of the watershed looking for opportunities for fecal contamination by humans or animals, depending on the agent of concern. For example, could there have been a discharge of raw sewage into the stream upstream of an irrigation intake due to overload of a municipal system during heavy rain? If the water source is a well, determine if it has been constructed to reduce opportunities for contamination. Is the well properly sealed? Is it down hill from potential sources of contamination? Is it likely to be under the influence of nearby surface water?

Identify nearby domestic animals including cattle. Contamination can spread from the animals to the crops through runoff from fields or feedlots, from waste lagoons and possibly from blowing dust. Document items such as distance of field from possible contamination sources such as feedlots and dairies and prevailing wind direction. Identify wildlife species near or in the fields. This can include birds (and the proximity of fields to flyways or frequent foraging locations), mammals, reptiles, and amphibians. These animals can be attracted to the crops or to water in the fields or near the fields. Determine if manure was used for soil augmentation. When was the manure applied? From what type of animals was the manure? If manure was used, had it been composted? Obtain records that document the composting process.

If the agent is one that may have a human reservoir, the health of the farm workers should be evaluated where possible. Determine if farm workers have been ill with gastroenteritis by speaking with the workers or their supervisors and with local public health officials or clinics that the workers may have visited. How was worker health monitored by supervisors? Were workers ever excluded due to illness? Determine worker practices for harvesting the crop. Did they wash their hands after breaks and after using the toilet? What is the availability of toilet and handwashing facilities in the field? Were gloves worn in the field by workers who would have handled the crop? Since the implicated field may have been harvested weeks before, observations of the same harvest crew in a different location may be helpful where possible.

Examine equipment that could have come in contact with the crop such as bins, ladders, knives, gloves, and boots. Is the equipment clean and cleanable? How was cleaning/sanitizing verified when it was being used? Was water applied to the commodity during or immediately after harvest? What was the source of the water? Document the training fieldworkers have received for sanitation and hygiene, and the language of instruction. Are practices such as harvesting products from the ground (apples dropped from the tree or from the packing shed conveyor belt) or harvesting slightly decayed products in use? Are chemicals such as browning inhibitor solutions recycled to save money?

Collect appropriate environmental samples such as soil, water, crop, and feces in any location that may account for contamination of the crop. It is appropriate to collect a large number of samples in a highly suspect or implicated field. Contamination is likely to be low level and sporadic, so a large number of samples is needed to increase the likelihood of finding the agent. Samples should be tested for the agent of concern and for generic *E. coli* as an indicator of possible fecal contamination.

Packinghouse investigations

Past investigations of packinghouses, which are often unenclosed facilities, have documented that the water supply or wildlife could be likely sources of contamination or could result in spreading contamination (Figures 3.4 and 3.5). Water in packinghouses may be used for washing, fluming, or cooling produce. Water sources used in packinghouses should meet standards for potable water or be from a public

FIGURE 3.4 This photo shows an apple bin drenching apparatus in a packing shed with a pigeon roosting over the equipment and pigeon feces beside the equipment.

FIGURE 3.5 The drenching machine contains an open reservoir of recirculating water in the bottom of the machine used to introduce preservatives/chemicals onto the fruit.

water supply. This water should be disinfected by a method that maintains a residual of disinfectant throughout the processing period and should include frequent, if not real-time, monitoring of disinfection levels. Examine records for this disinfection, and have operators demonstrate their manual testing while you observe.

Many packing sheds have only a roof and concrete floor, making them attractive locations for birds and other wildlife. Investigators should carefully examine the packinghouse for evidence of animal feces, nests, and tracks. The use of appropriate equipment, such as ultraviolet lights of specific wavelengths and observations of the facility at night, may be helpful in this effort.

Evaluate worker health and hygiene similar to efforts made for field workers. A review of time cards and one-on-one interviews with relevant workers to identify responsibilities during the time period of interest are critical. Evaluate the cleanliness of equipment along with actual cleaning and sanitation procedures, and measures to determine the efficacy of cleaning. Determine the training field workers have received for sanitation and hygiene.

Vacuum cooler/hydrocooler investigation

Various postharvest processes exist to quickly decrease the temperature of produce harvested from the field and thus prolong shelf-life. These include forced chilled air, hydrocoolers, and vacuum coolers. Hydrocoolers flush large volumes of recirculating, chilled water over the produce in wax-coated shipping boxes. This process could serve as the source of contamination or could spread contamination; therefore, adequate water quality, along with monitoring of the disinfection levels of the water and cleaning and sanitation processes, must be maintained at all times to prevent cross-contamination. Similarly, some produce is cooled using vacuum pressurized equipment with or without overhead sprays to partially replenish water removed during the vacuum cooling process. During the investigation, consideration should be given to collecting environmental samples from equipment and thoroughly reviewing and documenting cleaning, sanitation, and disinfection procedures.

Fresh-cut produce processor investigations

A significant percentage of produce outbreaks have been linked to fresh-cut produce. Investigations of implicated fresh-cut processors have not identified obvious sources of contamination at these processors, but investigators have noted that the washing and fluming steps could result in spreading contamination from one or a few contaminated items to a much larger volume or number of items. Similarly, the commingling of cut produce could spread contaminants from one or a few contaminated items to a large number of final packages of produce. Water used in fresh-cut operations should meet potable water standards. Infrequent monitoring of disinfection levels of water may allow a "slug" of contamination to pass through the system. A deliberate, careful assessment of the water quality, disinfection procedures, temperature, and disinfection monitoring system is required. Worker health and hygiene also need to be evaluated.

Environmental investigations at packinghouses, hydrocoolers and fresh-cut processors should include collecting "library" samples from implicated lot codes if available, collecting water samples, documenting the type and frequency of water clarity/disinfection monitoring by processors, and sampling processing equipment surfaces and other locations that could harbor microorganisms.

Intentional contamination

Investigators of foodborne disease outbreaks always need to keep in mind the possibility that the contamination event was intentional. Communication with appropriate local or state law enforcement agencies should occur immediately if intentional contamination is suspected. Some of the features that might tip off investigators to intentional contamination include an unusual agent not previously associated with the vehicle involved, an unusually high attack rate, or an unusually severe disease when compared to past experience. Other information that might indicate intentional contamination include knowledge of existing threats to the food supply; knowledge of previous threats by or issues with employees at the facility or farm; evidence of unauthorized personnel in fields, packing plants, or fresh-cut processors; evidence of unauthorized supplies or equipment in the area where the food was produced or processed; and illness in food or processing plant workers similar to that in outbreak cases.

Lessons learned

Lessons learned from past produce outbreak investigations include the following:

- Using rotating, inexperienced, untrained investigators on environmental investigations of retail food facilities, farms, packing sheds, and food processors has significantly compromised the effectiveness of past investigations. Investigations need to use a team approach, incorporating persons with expertise and training in epidemiology, microbiology, water quality, animal/wildlife science and regulatory investigations that are used to working with each other. Sending out different investigators on follow-up visits wastes time as these new investigators are not familiar with the setting involved or the past findings, thus requiring them to take additional time to come up to speed before they can provide useful input.
- Produce outbreak investigations usually involve multiple government agencies at the local, state, and federal levels. Coordination of these entities is always a challenge and can result in duplication and gaps in efforts as well as communication problems. One example of the kinds of confusion that can occur is when multiple agencies independently visit the grocery store or restaurant where implicated produce was purchased or served and ask for the same information multiple times. Another example is when regulators have to return to a facility numerous times to collect additional information. Even though it may not be possible to collect all information at the initial visit, careful and methodical planning and daily debriefings can help reduce the number of return visits required. Communication and coordination among investigating agencies throughout the process is critical.
- If the produce implicated in an outbreak was served at a single restaurant, it is necessary to try to determine if the produce became contaminated there. Many times, local health departments conduct routine regulatory inspections

of the establishment rather than an epidemiologically based, detailed food preparation review to assess the likelihood that contamination did not occur at the restaurant. These traditional inspection approaches to illness or outbreak investigations are entirely inadequate and should no longer be used as a primary investigative approach. Detailed, well-documented, science-based investigations, as described previously, are imperative. Additionally, making no effort to assess food worker contamination at the facility may result in erroneous conclusions or an inability to pursue the investigation further.

- Dispatching staff to collect remaining implicated product from ill-consumer households is critical. Advising consumers to "drop off" remaining product is not acceptable. Freezing remaining positive product after testing at state or local labs may prove helpful for further analysis.
- Assessing whether incoming ingredient volume is approximately equal to outgoing finished product volume may be useful in determining if there are gaps in paperwork or processing.

Recommendations

The following guidelines are recommended:

- Local, state, and federal agencies must look for opportunities to improve speed and quality of all phases of the investigative process. For example, population-based comparisons of the number of annual reported foodborne outbreaks may help identify geographic areas (states, counties) that need additional training or infrastructure support to ensure rapid and complete reporting. Additionally, a state-based food complaint system that receives, compiles, and analyzes information from local agencies and then forwards these data to a national database may help identify outbreaks earlier.
- Public health agencies should develop written performance standards and standard operating procedures and encourage monitoring of these standards during all phases of foodborne outbreak investigations including reporting and surveillance systems.
- Standardized commodity-specific questionnaires should be developed for those commodities previously implicated in more than one outbreak. These questionnaires should include very detailed information needed for tracebacks that may be specific to that commodity. These should be posted on websites by federal agencies for use by state and local agencies in foodborne outbreak investigations.
- Contacts, standardized questionnaires, and procedures for obtaining permission from ill individuals to use club-card information should be developed in advance by state agencies.
- Understanding the strengths and weaknesses of epidemiological studies is critical for regulatory agencies, yet many regulatory agencies do not have these resources. Regulatory agencies should consider hiring experienced epidemiologists to quickly review and discuss epidemiological findings before beginning

the environmental investigation. These trained, experienced epidemiologists can also help regulatory staff move away from the traditional inspection approach to a more scientific approach.

- Agencies with statutory authority for the specific types of facilities should develop specially trained, multidisciplinary teams to investigate foodborne outbreaks at retail food facilities, food processing facilities, distributors, brokers, and farms using epidemiological concepts to assess opportunities for introduction, survival, and growth of pathogens. These teams should not be simultaneously tasked with regulatory inspections.
- Federal agencies, in consultation with state and local agencies and industry, should publish yearly lists of research priorities for commodities involved in recurring foodborne outbreaks.
- Food manufacturers, distributors, and retailers should move quickly to improve the traceability of all food products. Although some efforts appear to be underway in the produce area, these efforts may take a decade or longer to result in significant improvements in traceability. Although longer term, industry-wide efforts are critical, incremental gains can be made in the short term. For example, buyers can require suppliers to note any substitutions clearly on shipping manifests.
- State and federal agencies should provide clear direction and protocols to local agencies who wish to assist in the traceback and environmental investigations.
- Environmental sampling at retail food facilities, food processors, and farms should be reexamined to require much larger numbers of samples. New methods for on-farm sampling (concentrating large volumes of water, sediment sampling, soil sampling, etc.) are needed.
- Time is critical during fresh produce-related investigations. Outbreaks involving illnesses or exposures in multiple counties in a single state should be aggressively coordinated by appropriate state agencies. Similarly, outbreaks with illnesses or exposures in multiple states should be aggressively coordinated by appropriate federal agencies.

Editor's Note

The following section contains excerpts from the FDA website concerning the FDA Food Safety Modernization Act (FSMA) which may impact investigative responsibilities and procedures.

The FDA Food Safety Modernization Act (FSMA), the most sweeping reform of our food safety laws in more than 70 years, was signed into law by President Obama on January 4, 2011. It aims to ensure the U.S. food supply is safe by shifting the focus from responding to contamination to preventing it (www.fda.gov/Food/G uidanceRegulation/FSMA/default.htm). Section 105 of the Food Safety Modernization Act directs the FDA to set science-based standards for the safe production and harvesting of fruits and vegetables that the Agency determines minimize the risk of serious adverse health consequences or death. The FDA proposes to set standards associated with identified routes of microbial contamination of produce, including (1) agricultural water; (2) biological soil amendments of animal origin; (3) health and hygiene; (4) animals in the growing area; and (5) equipment, tools, and buildings. The proposed rule includes additional provisions related to sprouts. Details of the FSMA Proposed Rule for Produce are summarized in *Fact Sheets on the Subparts of the FSNA proposed Rule for Produce: Standards for the Growing, Harvesting, Packing, and Holding of produce for Human Consumption* (www.fda.gov/Food/FoodSafety/FSMA/ucm334552.htm). It will be necessary for the FDA to provide oversight, ensure compliance with requirements, and respond effectively when problems emerge. For the first time, FDA has been given an inspection mandate. Building a new food safety system based on prevention will take time, and FDA is creating a process for getting this work done (www.fda.gov/Food/GuidanceRegulation/FSMA/ucm257978.htm).

References

Anonymous, 2002. National assessment of epidemiological capacity in food safety. Council of State and Territorial Epidemiologists. www.cste.org/pdffiles/fsreportfinal.pdf.

FDA, 2001. Guide to traceback of fresh fruits and vegetables implicated in epidemiological investigations. www.fda.gov/ora/Inspect_ref/igs/epigde/epigde.html.

FDA, 2005. Guide to produce farm investigations. www.fda.gov/ora/inspect_ref/igs/farminvestigation.html.

FDA, 2009. CIFOR outbreak guidelines document. www.fda.gov/ora/fed_state/NFSS/Default.htm.

Manure Management

4

Patricia D. Millner

U.S. Department of Agriculture, Beltsville Agricultural Research Center, Beltsville, MD

Introduction

Animal manure is a well-recognized potential source of a wide variety of infectious agents (Table 4.1) that can cause disease in humans, directly or indirectly, particularly through consumption of contaminated water or food (Burger, 1982; Cole et al., 1999; Feachem et al., 1981; Guan and Holley, 2003; Spencer and Guan, 2004; Strauch, 1991; Strauch and Ballarini, 1994). Foodborne illness outbreaks involving fresh fruits and vegetables over the past decade have heightened concerns about contamination of produce from fugitive enteric pathogens at the primary field production level. Possible contamination sources of concern include wildlife and domestic farm animals, insect vectors, runoff from pasture and rangeland grazing or feedlots, contaminated surface water, and manure-based soil amendments. The illness outbreaks and the ensuing industry, consumer, and government responses have increased overall awareness of the potential for foodborne pathogen contamination at the field level and the possible linkage to manure pathogens introduced into a complex landscape environment. Characteristics of the different types, virulence, fate, and transport responses of manure pathogens are essential inputs for ultimate use in quantitative microbial risk assessments within a livestock and fresh produce agroenvironment.

The Produce Contamination Problem. http://dx.doi.org/10.1016/B978-0-12-404611-5.00004-X
2009 Published by Elsevier Inc.

Table 4.1 Some Pathogenic Microorganisms of Public Heath Concern in Manure and Their Animal Sources

Bacteria	Potential Animal Source(s)
Campylobacter coli and *C. jejuni*	Cattle, sheep, swine, poultry, goats, wildlife
Bacillus anthracis	Cattle, sheep, swine, wildlife
Brucella abortus	Cattle, sheep, goats, wildlife
Escherichia coli patho- & toxigenic strains	Cattle, sheep, swine, wildlife, birds, water fowl
Leptospira spp.	Cattle, swine, horses, dogs, rodents, wildlife
Listeria monocytogenes	Cattle
Mycobacterium bovis	Cattle
Mycobacterium avium paratuberculosis	Cattle
Salmonella spp.	Cattle, sheep, swine, poultry, goats
Yersinia enterocolitica	Swine
Viruses	
Avian – Swine influenza (USDA, 2008)	Poultry, swine, wild birds, water fowl
Hepatitis E	Swine[a]
Parasites	**Disease**
Protozoa	
Balatidium coli	**Pigs, swine, guinea pigs, other mammals**
Cryptosporidium parvum	**Cattle, sheep, swine, amphibians, reptiles, birds, water fowl**
Giardia spp.	**Cattle, sheep, swine, amphibians, reptiles, birds, water fowl**
Toxoplasma spp.	**Felines, warm-blooded animals**
Helminths	
Ascaris suum	**Swine**
Taenia spp.	**Cattle, swine**
Trichuris trichiura	**Swine**

[a]*Herremans, M., Vennema, H., Bakker, J., van der Veer, B., Duizer, E., Benne, C. A. et al. (2007). Swine-like hepatitis E viruses are a cause of unexplained hepatitis in the Netherlands. J. Viral Hepat. 14,140–146.*

While concern was increasing about manure pathogens inadvertently coming into contact with fresh produce in the field in some production regions, other regions were concerned about the potential for environmental pollution resulting from inadequate handling, storage, stabilization, and land use of animal manure. In response to the last concern, a variety of manure treatment technologies were being developed and evaluated to reduce the potential for environmental overloads of nutrients, pathogens, and air emissions from concentrated animal production facilities and land application of animal manure. Unlike many traditional means of manure management in which pathogen reduction occurs mainly by default rather than intentionally, the new manure management technologies (Burton and Turner, 2003; Williams, 2003) included pathogen destruction as an integral and critical element of the process.

In the United States and many other countries, domestic and municipal sewage sludge is subject to regulated use practices, multicriteria treatment processes designed for pathogen destruction (USEPA, 1994, 2000b), pathogen testing (USEPA, 2003), and storage guidelines (USEPA, 2000a). In the United States, federal regulations establish a standard set of treatment and pathogen limits that correspond to a range of land application circumstances and subsequent landscape, public access, and private agricultural crop and livestock contact situations for biosolids (treated sewage sludge). Individual states may impose additional requirements beyond those established by the U.S. Environmental Protection Agency. In contrast, no federal or state regulations specify pathogen reduction or testing for animal manure prior to land application. However, revised U.S. Clean Water Act regulations emphasize managing land application of manure to reduce input of nutrient, microbial, and other pollutants to surface waters. In such cases, microbial pollution is measured by the traditional indicator bacterial group, fecal coliforms. This is accomplished by limiting application rates or hauling manure from one region to another where soil nutrients are not already overloaded and waterways are not impaired, and therefore, manure application is not severely limited by nutrient management requirements and Total Maximum Daily Load limits. Manure handling and application strategies have been developed to mitigate hauling costs; however, these can lead to ammonia volatilization, which can impair air quality and lead to state regulations on emissions, as has occurred in some jurisdictions.

Although state regulations and guidance, based on Natural Resource Conservation Service (NRCS) guidance (USDA, 1999) are available to assist producers in handling and managing animal manure stockpiles and storage, they emphasize engineering (USDA, 1999) and nutrient management, with some recent attention to air-quality impacts (USDA, 2008) rather than pathogen aspects. Some states also have fact sheets on application-season timing; however, these focus on logistics relative to manure accumulation in confinement facilities, soil conditions, and crop needs, not on pathogen survival and persistence. This focus primarily results from the fact that major use of animal manure is on major grain, forage, and fiber crops, rather than fresh produce.

For confined, rather than grazed or pastured animal production, treatment generally involves the initial collection and removal of manure-urine (slurry) from the animal housing units, with subsequent storage in lagoons followed by spraying onto fields. Alternatively, for solid manure, it may be stacked in high piles, sometimes under roofs and on stabilized surfaces to prevent uptake of precipitation and leaching of nutrients and runoff. It may be further treated by composting or several other means if equipment is available. As newer technologies are implemented, nutrient stabilization, pathogen reduction, and volatile organic emissions reductions will be realized during treatment and with land application.

Manure use on crops

Animal manure use as a fertilizer on crop production land has been practiced for millennia worldwide. Although this practice has generally served the farming community's needs and represents an acceptable form of resource conservation on various grain, bean, and cotton crops, it distributes surviving pathogens across large areas (Bicudo and Goyal, 2003). Use of animal manure in primary production of fruit and vegetable crops is a clear hazard and increases the likelihood of contamination by enteric pathogens that survive in the manure-based inputs. Foodborne illness outbreaks with contaminated produce (24%) were calculated to nearly equal those associated with meats (29%) in the United States between 1990 and 1998.

During this period several outbreaks involving produce were reported from small, organic gardens in which raw manure had recently been applied (Cieslak et al., 1993; Guan and Holley, 2003; Nelson, 1997). Organic production relies on animal manures, crop rotation and residues, nitrogen-fixing legumes, composts, and mineral rock powders to maintain soil quality and provide plant nutrients; cultivation, cultural controls, and biocontrols are used to manage insects, weeds, and other pests. Current USDA organic certification regulations require producers to use thermophilic conditions to compost manure, or if raw manure is used, then harvest cannot occur before 90 to 120 days postapplication (USDA, 2000). Neither the USDA National Organics Program (NOP) nor the National Agricultural Statistics Service maintain specific records on the number of certified organic or conventional farms using manure or manure-based products in production of fresh produce. The NOP requires organic farmers and food handlers to meet a uniform organic standard and makes certification mandatory for operations with organic sales exceeding $5000. The NOP implements the regulations through third-party certifiers that it audits. Approximately 50 state and private certification programs in the United States and over 40 foreign programs have been accredited by the NOP.

Many organic growers use manure-based inputs, and many conventional growers also use such inputs, not just for their fertilizer value, but also for their benefit in building and maintaining soil quality. The steady increase in organic food production and distribution worldwide involves adherence to a variety of safety standards (Cooper et al., 2007). Records show that outbreaks with fresh produce have occurred with organically as well as conventionally raised products. Clearly, both methods of production involve similar types of inputs (i.e., seeds, transplants, water, fertilizer, cover crops or previous green crop residue incorporation into soil, employees, etc.). Consequently, the nonpreferential contamination of fresh produce outbreaks across organic and conventional sources suggests that actual on-site conditions and practices, rather than marketing-based labels (like "organic"), are the critical determinants of the sanitary condition of fresh produce. However, small growers and backyard garden enthusiasts may require continued information regarding good agricultural practices for use of self-prepared composts.

In addition to direct manure application to land, runoff from animal grazing areas or fenced lots to primary fresh market crop production can increase the risk of pathogen contamination to fresh produce crops. Over the past several decades, the economics and efficiencies of animal husbandry have led to an increase in the animal density per unit area within livestock and poultry production facilities in the United States and abroad. Such intensive production conditions now used for broilers, layers, turkeys, swine, beef, and dairy animals generate major quantities of manure within relatively limited landscape areas. Appropriate use of these manures as fertilizers requires calculation of the nutrient content, particularly nitrogen and phosphorus, relative to crop needs and existing soil test values to avoid use of excessive amounts that may lead to pollution of surface and groundwater by nutrients (nitrate, phosphorous), organic matter, sediments, pathogens, and other materials (Al-Kaisi et al., 1998a, 1998b; Davis et al., 1997). Consequently, several advanced manure management systems have been developed to handle very large volumes of manure from intensive livestock and poultry operations (Vanotti et al., 2003, 2005b; Williams, 2003). Development has focused not only on nutrients, but also on disinfection of pathogens in liquid and solid-phase materials (Vanotti et al., 2005a). Although these developments targeted swine manure, the technologies are applicable to dairy systems using liquid collection schemes.

Survival of pathogens in manure

Stressors, such as fluctuations and extremes in temperature, moisture, pH, UV irradiation, and nutrient availability, along with biological pressures from competition, predation, parasitism, toxins, and inhibitory substances, contribute to the natural attenuation of microbial populations in the environment and in manured soil. Microorganisms exhibit a considerable degree of variability relative to their tolerance to these stressors. In addition, viruses and parasites (protozoans and helminths) are dependent on host organisms, sometimes very specific ones, for reproduction, and therefore once shed in manure or other excretions, will not reproduce in the open environment. In contrast, bacteria, such as *E. coli* O157:H7, *Salmonella,* and *Listeria monocytogenes,* are not host dependent, and thus their populations respond dynamically up and down over time to changes in environmental conditions.

Bacteria

The complexity and uncertainties involved in predicting the fate (partitioning, growth, die-off rates) and transport of manure pathogens in the innumerable variety of agricultural situations limit extensive direct application of traditional decay and transfer functions, and likely require alternative and new approaches to modeling such as those suggested in a review of this subject (Pachepsky et al., 2006). Some of the reported variations in fate of manure pathogens relative to environmental and situational conditions are reviewed in this section.

Survival of *E. coli* O157:H7 in stockpiled, raw sheep manure at 21 months contrasts with its elimination after four months in parallel stockpiles that were aerated (oxygen content not reported) by periodic mixing, and survival for only 47 days in bovine manure stockpiles that were periodically mixed (Kudva et al., 1998). However, freezing (–20°C) and cold storage of bovine manure (4 and 10°C) prolonged survival up to 100 days, whereas increasing temperatures shorten survival: 70, 56, and 59 days at 5, 22, and 37°C (Wang et al., 1996). E. coli O157:H7 in laboratory-incubated, manured field soil showed a steady population decline to undetectable levels within 165 days at 15°C and within 231 days at 21°C, whereas E. coli O157:H7 persisted in corresponding samples of this manured soil in which competing microbial factors were absent because the amended soils were autoclaved prior to inoculation (Jiang et al., 2002).

In field studies during late autumn and winter in the United Kingdom, generic E. coli in cattle, sheep, and swine manure survived on grassed areas for very long periods (up to six months), and in at least one case, up to 162 days, when initial populations ranged from 4.31 to 5.34 $\log_{10}$ cfu/g respectively (Avery et al., 2004). Average D-values for the cattle, sheep, and swine manure for E. coli were calculated as 38, 36, and 26 days for the test conditions (Avery et al., 2004). In addition, E. coli O157:H7 has been shown to survive up to 28 days in significant numbers on farm structural surfaces that have contacted manure (Williams et al., 2005).

However, plowing and harrowing of soil amended with naturally contaminated pig slurry effectively and rapidly (i.e., immediately) reduced populations of E. coli and Salmonella DT104 on a clay soil (Boes et al., 2005). In contrast, harrowing only, or surface application to winter wheat stands only, or injection in winter wheat stands only, prolonged survival of E. coli to 21 days, and Salmonella to 7 days.

Interest in the potential contamination of leafy greens and herbs by E. coli O157:H7 led to a study of organic iceberg lettuce production using composted bovine manure, solid manure, as well as manure slurry as the fertilizer. Results showed that no bacterial pathogens, not E. coli O157:H7, Salmonella spp., or L. monocytogenes were recovered from the lettuce, even through E. coli O157:H7 was present in all the manure amendments applied to the soils (Johannessen et al., 2004). The authors concluded that further research is needed to resolve how contamination of the lettuce was avoided. The absence of E. coli and Enterococcus spp. from interior or exterior portions of potato skins on tubers harvested 214 days after soil was amended with raw or composted manure containing both of these bacteria, has also been reported (Entry et al., 2005).

Recent reviews of bacterial pathogen survival in animal manures (Bicudo and Goyal, 2003; Guan and Holley, 2003) clearly show that Salmonella spp. survive in some situations for up to 60 days in a variety of nonthermophilic manure systems. In contrast, Salmonella spp. and Ascaris suum ova were destroyed completely after exposure for 24 hours in swine manure biogas digesters operated at 55°C (Plym-Forshell, 1995). However, salmonellae survived 35 days and 60% of Ascaris ova survived up to 56 days in the mesophilic manure pit where the digested manure from the biogas unit was stored. These results show the effectiveness of manure disinfection processes that involve thermophilic temperatures, and the extended survival periods of bacterial and parasitic manure pathogens in mesophilic, facultative storage units even when substantial nutrients have been depleted by the prior digestion.

Ammonia is typically generated by most stored manures; however, its effect on pathogens in the manure had not been specifically evaluated until recently. In disinfection tests with bovine manure, results show that *Salmonella* was destroyed more rapidly and cost-effectively by gaseous ammonia generated from addition of 0.5% aqueous ammonia than from addition of 2% w/w urea to 12% total solids manure slurry (Ottoson et al., 2008). *C. jejuni* was the bacterium most resistant to the anaerobic digestion of cattle slurry (supplemented with pig, hen, and potato waste) at 28°C, followed by *S. enterica* Typhimurium and *Y. enterocolitica* (Kearney et al., 1993).

Simulated seasonal temperature sequence effects on die-off of *Salmonella* serovars (Agona, Hadar, Heidelberg, Montevideo, Oranienburg, and Typhimurium) were greatest in the first week of the winter-summer sequence (–18, 4, 10, 25°C) as compared with the spring-summer sequence (4, 10, 25, 30°C) in a 180-day study with 5 $\log_{10}$cfu/g inoculated directly to moist clay and loamy sand soils, with and without fresh swine manure slurry (Holley et al., 2006). Total die-off (no detection in enrichments) of inocula was more rapid (160 days) in spring-summer temperature sequences regardless of manure, incorporation, or soil moisture content (60–80%) of winter-summer sequence treatments (160 days). By considering the calculated decimal reduction times of 30 days or more for 90% reduction of salmonellae in the application treatments, and common estimated slurry concentrations of 3.0 to 600.0 salmonellae/ml, and application rates of slurry to land, 25 g/kg, the authors concluded that a 30-day delay between field application of manure in spring or fall and use of treated land would minimize risk of environmental contamination and uptake by animals of *Salmonella* (Holley et al., 2006). However, they do not specify the different types of crops for which this 30-day delay would be appropriate, whether the 30-day delay refers to application and planting date or to harvest date, nor do they address other pathogens and their survival within this 30-day period. Such a recommendation also is not consistent with recent leafy green marketing agreement requirements that prohibit use of raw, untreated manures, and the 90- to 120-day required delay between application and harvest for U.S. certified organic producers.

A four-state study of environmental and herd-level risk factors associated with *Salmonella* prevalence in dairy cows, including conventional and organic farms, identified major contributing risk factors as access to surface water, *Salmonella*-positive manure storage, land application of manure slurry or spray irrigation, and cows eating or grazing in fields where manure was surface-applied rather than soil-incorporated within the same growing season (Fossler et al., 2005a, 2005b). In contrast, free-range rearing conditions, sometimes used on organic farms, were found to be slightly beneficial in reducing *Salmonella* spp. but not *Campylobacter* spp. or *L. monocytogenes* contamination in chicken flocks (Esteban et al., 2008).

Animal management practices at organic and conventional farms that focus not only on manure management, but also on measures to control access of wildlife to the housing units and water troughs, has been repeatedly identified as a critical point in maintaining farm hygiene relative to several major zoonotic pathogens, including *E. coli* O157:H7 and other shiga-toxing positive serotypes, *Salmonella* spp., *L. monocytogenes, Campylobacter* spp., *Cryptosporidium,* and *Giardia* (Castellan et al., 2004; Meerburg et al., 2006; Murinda et al., 2004). Identification of on-farm pathogen reservoirs and vectors can aid development and use of farm-specific pathogen reduction programs.

Protozoan and helminthic parasites

Several eukaryotic parasites in manure are characteristically more resistant to the range of environmental stressors encountered in various agricultural situations and treatment technologies than are viruses or most nonspore-forming bacteria (Bowman, 2009; Fayer and Ungar, 1986; Robertson et al., 1992; Tzipori and Widmer, 2008). Ova of the parasitic helminths, *Ascaris lumbricoides,* and *A. suum* are particularly persistent because the outer shell is resistant to most environmental stressors that adversely affect other groups of microorganisms. Thus, in evaluation of efficacy of treatment technologies, *Ascaris* has been used as a conservative benchmark for microbial destruction.

Environmental stressor effects on the infectivity of *Cryptosporidium* oocysts showed that in water at 4°C, infectivity is maintained for two to six months (Tzipori, 1983). Oocysts also tolerate a wide variety of common disinfectants such as sodium hydroxide, sodium hypochlorite, and benzylkonium chloride without significant loss in infectivity (Campbell et al., 1982). However, ammonia but not pH (Fayer et al., 1996; Jenkins et al., 1998), desiccation, and very extreme temperatures (e.g., freeze-drying, freezing, or 30 min at 65°C) completely eliminated viability and infectivity of the oocysts (Tzipori, 1983). Moist heat at 55°C for 15 to 20 minutes, such as present in the thermophilic phase and core of a composting mass or in a thermophilic digestor, destroyed oocyst infectivity in calf feces and intestinal contents (Anderson, 1985). These temperatures are easily achievable in thermophilic manure treatment technologies that are operated properly. Process management is key to ensuring that pathogens are destroyed, even when a treatment technology with the capacity to meet the pathogen destruction criteria is used.

With controlled-environment laboratory studies, *Cryptosporidium parvum* oocysts were reported to be more resistant to degradation than *Giardia muris* cysts (Olson et al., 1999) in soil, water, and cattle manure. *Giardia* cysts were infective for only one week at 4 and 25°C, whereas *Cryptosporidium* remained infective for eight weeks at 4°C, and four weeks at 25°C. At −4°C, *Giardia* was noninfective within one week, but *Cryptosporidium* remained infective for more than 12 weeks. Field studies in the summer in the United Kingdom have shown that *Cryptosporidium* oocysts in swine manure on grassy fields were reportedly reduced by 1-log during both eight- and 31-day periods, whereas similar 1-log reductions for *Salmonella, E. coli* O157, *L. monocytogenes,* and *C. jejuni* averaged only 1.86, 1.70, 2.80, and 1.86 days, respectively (Hutchison et al., 2005).

In addition, recent advances in molecular characterization of species and genotypes of *Cryptosporidium* and *Giardia* show that such approaches are essential to ensure accurate identifications of organisms in environmental transmission studies (Fayer et al., 2008; Santin et al., 2008). Molecular epidemiological studies strongly indicate that *C. parvum* is the major species pathogenic to both cattle and humans, and that certain genotypes and subtypes predominate in calves worldwide (Xiao et al., 2007). In contrast, molecular data suggest that zoonotic transmission is not as prevalent in the epidemiology of giardiasis (Xiao and Fayer, 2008). In a study of 14 farms from seven eastern U.S. states, *C. parvum, C. bovis,* and *C. andersoni* were found on two, six, and eight farms, and infected 0.4, 1.7, and 3.7% of the 541 cows, respectively (Fayer et al., 2007). Low prevalence of *Cryptosporidium* overall, and for each of the previous species individually, in mature cows in this study was very highly significant ($p \leq 0.0001$), compared with young cattle including those previously examined on most of the same farms. The very low level of infection of mature cows with *C. parvum* suggests that field practices be developed and used to manage manure and potential runoff for young cattle likely to shed this pathogen. In the western United States, a study involving more than 5200 fecal samples, from 22 sites in seven states, showed that fresh fecal material from feedlot systems contained about 1.3 to 3.6 *C. parvum* oocysts/g feces, or about 2.8×10^4 to 1.4×10^5 oocysts/animal-day (Atwill et al., 2006).

Clearly, data are needed to evaluate the effectiveness of various field management practices designed to reduce pathogen loading and transport off-site within large-scale animal production areas. Runoff from cattle grazing areas as well as feeding and holding/resting lots may have off-site impacts if appropriate catchment and diversion measures are not available or are incorrectly implemented. Data on field management strategies to reduce prevalence and amounts of pathogens off-site is accumulating. For example, in Canada, a field study with and without vegetative filter strips (VFS), with three slope conditions (1.5, 3.0, and 4.5%) and two 44-minute rainfall intensities (25.4 and 63.5 mm/h) showed that VFS were very effective regardless of slope in reducing *C. parvum* oocysts in surface runoff (Trask et al., 2004). Total recovery of oocysts in runoff from the VFS ranged from 0.6 to 1.7% and 0.8 to 27.2% with low and high rate rainfall, respectively, whereas oocyst recovery from non-VFS sites ranged from 4.4 to 14.5% and 5.3 to 59%, from low- and high-rate rainfall, respectively (Trask et al., 2004).

Viruses

Limited data are available in publicly accessible databases documenting viral bioburdens in animal manures in the United States. Hepatitis E virus (HEV), a nonenveloped virus that is relatively more environmentally stable than enveloped viruses, was recovered from 15 of 22 swine manure pit and 3 of 8 manure lagoon samples in a multifarm survey, although no HEV was recovered from drinking- or surface-water samples on the 28 farm sites studied (Kasorndorkbua et al., 2005). The presence of exotic Newcastle disease (END) after depopulation and decontamination (D & D) of the infected birds in a California outbreak was assessed with emphasis on manure, compost, and manure conveyors (Kinde et al., 2004). At one ranch, END was recovered up to 16 days postdepopulation, but not thereafter; no END was recovered from a second ranch. Further research on avian influenza and END, both enveloped viruses, resulted in development of a real-time reverse transcriptase-PCR method for rapid quality control checks on composts and D & D in the event of exotic disease outbreaks (Guan et al., 2008).

Although bovine enterovirus (BEV) is not a zoonotic pathogen, it has been suggested as a potential indicator of fecal contamination from animals (cattle/deer) and a molecular epidemiological tool; it is a relatively stable, nonenveloped virus (Ley et al., 2002). It is rapidly inactivated by thermophilic (55°C) anaerobic digestion of manure as is bovine parvovirus, (BPV; enveloped) (Monteith et al., 1986). Both viruses are also inactivated by composting for 28 days. However, mesophilic (35°C) anaerobic digestion prolonged BEV and BPV survival to 13 and eight days, respectively, whereas 30 minutes at 70°C only inactivated BEV. Additional research is needed to determine if either of these bovine viruses would be suitable indicators of manure treatment efficacy. African swine fever virus (ASFV; enveloped) and swine vesicular disease virus (SVDV; extraordinarily resistant to desiccation, freezing, and the fermentation and smoking processes used to preserve food) are rapidly inactivated by thermophilic temperatures (≥ 50°C) in any of several treatment technologies (Turner et al., 1999). With other exotic viruses of swine, including foot-and-mouth disease virus, Aujeszky's disease virus, and classical swine fever virus (both enveloped), thermophilic temperatures (≥ 60°C) were found to be essential for rapid viral inactivation (Turner et al., 2000).

A North Carolina state-sponsored study with fresh, swine manures compared virus survival in conventional lagoons to that in five environmentally superior and technologically advanced candidate manure treatments (Costantini et al., 2007; Williams, 2003). Pretreatment manure from all farms had detectable porcine sapoviruses [PoSaVs] and rotavirus A [RV-A], whereas porcine enteric viruses (porcine noroviruses [PoNoVs]) and rotavirus C [RV-C] were present only in some of the farms using the candidate technologies (Costantini et al., 2007). After treatment, only the conventional technology samples contained detectable PoSaV RNA. Candidate farm posttreatment samples with detectable RV-A and RV-C were not infectious by cell culture immunofluorescence assay, nor did they result in clinical signs or seroconversion in inoculated gnotobiotic pigs. Results indicate that the specific environmentally superior manure treatment technologies evaluated would reduce the viral bioburden in treated liquids and solids.

Pastures, lots, and runoff

Runoff can move significant amounts of pathogens from the original site of manure deposition, be it a pasture, pen, or lot (Thurston–Enriquez et al., 2005). Understanding the transport and survival of zoonotic pathogens potentially present in livestock manure and runoff is critical for development of appropriate and effective practical measures to reduce adverse environmental, food safety, and public health impacts that nonpoint source releases can potentially have. Results from several complex studies are beginning to inform this issue.

Release of *E. coli* from fecal cowpats during rainfall was reported to occur primarily as individual bacterial cells (Muirhead et al., 2005); bacterial cells preferentially attach to manure colloids and organic matter and small-size silt and clay particles (Guber et al., 2007). However, variation among strains of *E. coli* in dairy manure were shown to have significantly different attachment affinities for various soil textural fractions (Pachepsky et al., 2008).

An agricultural management area scale study with 10 dairies and ranches showed that fecal coliform concentrations were highly variable both within and between animal loading units (Lewis et al., 2005). Fecal coliform concentrations for pastures ranged from 2.3 to 6.36 $\log_{10}$cfu/100 ml and for dairy lots ranged from 3.29 to 8.22 $\log_{10}$cfu/100 ml, with mean concentrations of 5.08 and 6.5 $\log_{10}$cfu/100 ml for pastures and lots, respectively. The investigators (Lewis et al., 2005) noted that the previously cited results are being used by dairy managers to change on-farm practices, by regulatory agency staff, and by sources of technical and financial assistance. Results from a different set of field study tests over 26 months with a cattle feedlot runoff control-vegetative treatment (VT) system showed that the system effectively reduced environmental risk by containing and removing *E. coli* O157:H7 and *Campylobacter* spp. from feedlot runoff (Berry et al., 2007). In this study, *Cryptosporidium* oocysts and *Giardia* cysts were infrequently isolated, and generic *E. coli* populations in the vegetative treatment soil declined with time, although their presence (12 of 30 samples in VT areas vs. 1 of 30 samples in nonrunoff impact areas) in freshly cut hay from the VT areas indicated the risk of contamination in that region. Both *E. coli* O157:H7 and *Campylobacter* spp. were absent from the baled hay.

In another agricultural management scale study that included 350 storm runoff samples from dairy lots and other high-cattle-use landscapes (Miller et al., 2007, 2008), a California team found 59 and 41% prevalence of *Cryptosporidium* oocysts and *Giardia duodenalis* cysts, respectively, in runoff associated with areas containing calves less than two months old. In contrast, only 10% of runoff samples associated with cattle older than six months were positive for these protozoans. Percent landscape slope, animal stocking number, and density were not as important factors as animal age, intensity, and cumulative amount of precipitation for *Cryptosporidium* oocyst concentrations. Animal age, stocking number, and instantaneous precipitation were major factors in concentrations of *G. duodenalis* cysts in runoff samples. Specific beneficial management practices, notably vegetated buffer strips especially located near calf areas, were associated with reduced runoff loads of these protozoans, whereas cattle exclusion and removal of manure was not. These results support those from other studies cited here that indicate targeted strategies for field management of stock and manure have some potential for reducing manure risk impacts off-site. Additional research is needed to determine if similar field management strategies in land areas adjacent to and surrounding primary fresh produce croplands will reduce fugitive enteric pathogen contamination with the sensitive crop fields.

Another route of off-site transport being examined involves movement of protozoan oocysts through shallow soils via macropores (Harter et al., 2008). A modeling study using packed soil boxes with and without macropores enabled collection of macropore flow data as for that in shallow soils. Macropore flow was shown to be responsible for *C. parvum* transport through the shallow soil to underlying pore spaces when soil bulk density, precipitation, and total shallow subsurface flow rate were taken into account. A risk assessment of oocyst transport was conducted and was determined to be consistent with the reported occurrence of oocysts in springs or groundwater from fractured or karstic rocks protected only by shallow overlying soils (Harter et al., 2008).

Manure treatment technologies

Manure is considered a potential source of pathogens, but this does not mean that every sample of manure will contain all the various types of pathogens that have been reported or that they will be present at maximally reported concentrations. However, manure treatment technologies, just like food processing technologies, are designed to destroy the most resistant types of pathogens likely to be encountered.

Treatment systems for manure from animal housing units have evolved beyond traditional collect, store, land apply approaches and now include processes that aid in protecting soil, water, and air quality (Humenik, 2001; Vanotti et al., 2007, 2008; Westerman and Bicudo, 2005). When manure storage is coupled with managed treatment processes, the result on pathogens essentially acts as a multibarrier system. Some treatment systems address several of these requirements, whereas some are specialized and address only individual factors (USDA, 1999, 2007).

Manure treatment technologies are grouped into two major categories that reflect the primary mechanisms involved in the processes: (1) physico-chemical, or (2) biological. Some systems can be integrated in sequence to meet several treatment criteria in different phases of a system. Some of the physico-chemical approaches to treatment include thermal conversion (combustion, gasification and pyrolysis), solid-liquid separation and filtration, advanced alkaline treatment, and aeration/mixing. Some of the biological approaches to treatment include thermophilic composting (Rynk, 1992), vermicomposting, anaerobic digestion (Bicudo and Goyal, 2003), thermophilic digestion (Ahring et al., 2002; Aitken et al., 2007), autothermal thermophilic aerobic digestion (Layden et al., 2007), sequencing batch reactors (Juteau et al., 2004), and constructed wetlands (Cirelli et al., 2007; Humenik, 2001; Karim et al., 2008).

In general, thermophilic processes, particularly those operated in-vessel, are designed to expose all treated material to extreme lethal temperatures (60–65°C), while still maintaining sufficient metabolic activity by the nonpathogenic bacteria to sustain the process heat (Juteau et al., 2004). Thermophilic composting remains one of the most cost-effective treatment technologies for manure solids; it functions well in a variety of environments. Initial capital as well as operations and maintenance costs are minimal compared with other treatment technologies.

Composting

Use of composting as a cost-effective treatment for manure and the widespread use of compost in large and small primary production systems, even home gardens, warrants special mention of practices that ensure the product will be significantly sanitized of original pathogens. Manure composting as used here refers to the controlled aerobic, thermophilic (self-heating) decomposition of organic matter by microorganisms such that three major objectives are met: (1) nutrient stabilization, (2) pathogen reduction, and (3) odor and vector-attraction reduction. Self-heating of stacked manure without attention to the time and temperature needed for all parts of the stockpile to meet pathogen reduction criteria, nutrient stabilization, and vector-attraction reduction does not adequately meet the requirement for a managed process. Such practice is simply stockpiling, with default self-heating in some parts. If the final compost product were intended for corn, wheat, cotton, or other field crops, this type of stockpiling might be adequate relative to pathogen content and the other two major treatment objectives. However, without adherence to a managed process and all that this involves, including regular temperature monitoring-recording at selected places within the pile, along with pathogen testing, the producer would have no basis for asserting that all parts of the pile were subjected to temperatures that would substantially reduce pathogens. Such unmanaged composting is common among backyard garden composts, and hence the strong caution to avoid inclusion of animal feces and diseased plant material in such piles.

To produce compost that is disinfected for use on crops (such as leafy greens, herbs, carrots, radishes, green onions, strawberries, etc.), in which the harvestable portions will directly contact soil, requires quality control and standards at each step in the composting process. A critical control point approach to the composting process could aid compost producers in meeting stringent requirements for use of manure-based products in primary fresh produce cropping systems. Growers, processors, distributors, and buyers need a science-based quality assurance system for compost inputs supported by validated quality assurance test standards for pathogens (such as *E. coli* O157:H7, *Salmonella* spp., *L. monocytogenes, Campylobacter* spp., and parasites). Compost test standards currently used by the U.S. Composting Council Standards of Testing Assurance program for members reflect USEPA biosolids test methods for fecal coliforms, *Salmonella*, and *Ascaris* ova (Thompson et al., 2002; USEPA, 2003). Test methods for other pathogens need to be validated.

During thermophilic composting, the mass of the pile insulates the core and metabolic heat generated by rapid microbial decomposition of the organic matter cannot escape quickly enough to equalize the temperature to ambient. Thus, the core pile temperatures increase, while surface temperatures remain insufficient to destroy pathogens, unless piles are turned so that the outer mass is exposed to thermophilic core temperatures (Shepherd et al., 2007). Turning is best accomplished early in the process, that is, during the first two to three weeks while maximal thermophilic temperatures are generated. Pile temperature usually declines immediately after turning, but rebounds to 55°C or greater as long as readily decomposable organic matter remains. Microbial respiration rate declines as readily available nutrients are depleted, and maximal temperatures achieved after each turning coincide with the decline in readily available nutrients. With time (e.g., 10–12 weeks or more), the humification begins, and nutrients are further immobilized within the biomass of living and dead microbial cells. Lignin and cellulosic compounds are the slowest to decompose and do not provide readily available carbohydrate for any of the bacteria of concern as foodborne illness or public health pathogens.

Guidance on composting basics, feedstock mix ratios, technology requirements, and conditions for large and small on-farm composting of manure and other organic feedstocks are available from several sources (Christian et al., 1997; Misra et al., 2003) and from many state agricultural extension programs. Critical factors in composting include aeration, nutrients, C:N ratio, moisture, pile structure, pH, temperature, and time (De Bertoldi et al., 1986; Haug, 1993). Aeration, either mechanically with blowers, by turning, or with passive means, is essential to meet the microbial requirement for oxygen needed for aerobic decomposition. The porosity, air flow characteristics, structure, and physical texture of the biomass mixture also impacts pile aeration. Management of the composting mixtures is needed to ensure that the process achieves the target time-temperature criteria for pathogens, and that turning and other operations are conducted according to good manufacturing practices.

Composting formats comprise a range of process and facility types. Some very mechanized and highly managed composting systems can involve frequent turning of windrows or mechanical aeration and biofiltration of static (stationary) piles that are either free-standing or enclosed within a vessel or containment system. The latter may include plastic polymer silage-type tubing, various types of synthetic material covers, and fully enclosed metal containers equipped with air-flow control devices, temperature sensors, feedback controls, and leachate collection systems. These mechanized approaches are useful for large-scale operations that maximize materials throughput and minimize process times. Other less mechanized compost systems, typical of some on-farm approaches (Christian et al., 1997), utilize passively ventilated static piles. Oxygen transfer rates in these passive formats are less efficient than mechanically aerated formats; hence decomposition proceeds somewhat slowly and temperature maxima are not as great or as sustained as for forced aeration systems. Such formats require a greater footprint than mechanized approaches, but they also have lower capital and operating costs. Because static piles are not turned until after the thermophilic phase ends, constructing the pile on a bed of at least 30 cm of woodchips, old hay, or straw and covering the outer surface of the pile with a layer of similar materials, or unscreened coarse textured compost to a depth of 20 to 30 cm (as is done with the aerated static format; Rynk, 1992) provides an insulated zone sufficient to ensure that all the "new" compostable mixture is within a thermophilic zone. This can also provide an absorptive layer for leachate and moderate creation of anaerobic zones from liquid accumulation pockets. Excessively tall piles run the risk of compaction at the base and diminished heating.

Good manufacturing practices for compost that is intended for use in primary fresh produce cropping systems, such as leafy greens and herbs, can best be met when compost producers conduct their operations within a Hazard Analysis Critical Control Point (HACCP) framework. Though not typical of current composting industry practice in the United States, the main elements for critical control point composting are outlined as follows. In the United Kingdom, national regulations stipulate very high thermophilic temperatures (> 60°C) for 3 days in static aerated pile or 14 days in turned windrow composting that includes any catering (i.e., food) wastes (DEFRA, 2004). In the United States, several states and localities use the time-temperatures and operational requirements that apply to biosolids compost (USEPA, 2003). The USEPA requirements of 55°C for three consecutive days in a static aerated pile format (this includes a base and blanket layer 15 to 30 cm thick as insulation to ensure that the new material is well within the thermophilic zone) have been applied by states for all other types of composting, including landscape trimmings, wood, food, papermill sludge, and dissolved air flotation waste to mention a few. Carcass composting has special requirements, and the product is not appropriate for use on primary fresh produce crops. For windrow composting, USEPA requirements stipulate 55°C with five turnings during the two-week period when the thermophilic temperatures are generated. Both formats require testing of final product according to USEPA standards for *Salmonella* and fecal coliforms (USEPA, 2003), plus other microbes as state or industry requirements specify. Material from piles that do not achieve time-temperature or test standards must be recomposted until they meet requirements.

Compost tea is a compost-derived product used by some growers as a foliar spray or soil drench to promote plant growth and protection against phytopathogens (Scheuerell and Mahafee, 2002). Compost tea (CT) production processes that use supplemental nutrients can support regrowth of even a few surviving cells of *E. coli* O157:H7 and *Salmonella* (Ingram and Millner, 2007). Production of CT without supplements avoids growth of these pathogens from initial trace concentrations. Sanitization of CT equipment is essential between batches.

Overall, critical control points suggested for composting operations are associated with each of the major steps in the process as follows.

1 Delivery, Inspection, and Input Preparation. Avoid spread of microorganisms from reception areas and materials to subsequent and especially final product handling areas, notably by delivery and operation vehicles; avoid processing delays that allow microbes to multiply; establish dirty-to-clean areas to avoid cross-contamination; attend to spills in a timely manner; control access of birds and vermin; avoid use of manure delivery trucks for hauling of final product unless equipment is cleaned and sanitized; train employees; monitor adherence to plan.

2 Thermophilic, Decomposition, and Disinfection. Avoid cross-contamination from vehicles, equipment, and containers; maintain the dirty-to-clean sequencing; control access of birds and vermin; periodically calibrate temperature measurement and recording devices; measure and record temperatures according to protocols; troubleshoot temperature failures and segregate material not meeting

time-temperature exposure for retreatment; clean and sanitize equipment; train employees; monitor adherence to plan.

3 Curing, Maturation, and Stabilization. Avoid cross-contamination from vehicles and equipment; turn piles as necessary; maintain the dirty-to-clean sequencing; control access of birds and vermin; clean and sanitize equipment; train employees; monitor adherence to plan.

4 Sampling and Analysis. Avoid cross-contamination during sampling via equipment and personnel; submit samples according to protocols for testing at a certified compost testing laboratory; train employees; monitor adherence to plan.

5 End-Product Preparation. Ensure product is stored properly to avoid recontamination by equipment, personnel or vermin; avoid blending with untreated, unstabilized products; transport in clean vehicles that have been sanitized if previously used for hauling manure or potentially contaminated wastes; maintain analyses on batch-lot production and product deliveries for traceback when needed; advise clientele on appropriate storage and use of product; provide copies of temperature and test records to clientele; caution clientele regarding inappropriate usages.

In contrast to thermophilic composting, vermicomposting (worm composting) cannot be conducted at temperatures sufficient to kill pathogens because the epigeic worms used (*Eisenia foetida* and *Lumbricus rubellus*) do not tolerate high temperatures or excessive amounts of ammonia. Thus, manure solids and urine require pretreatment or significant dilution (50%) before being introduced into the vermicompost bins or containers. A two-step process that involves precomposting by a thermophilic method and subsequent vermicomposting or vice versa, was reported to meet pathogen reduction requirements while yielding a stable, consistent, and nutrient-rich product (Ndegwa and Thompson, 2001).

SUMMARY

Advanced treatment technologies that achieve high standards of nutrient stabilization, pathogen reduction, and reduction of odor and vector attraction are available for animal operations that generate large quantities of liquids or solids. In an era when agriculture is increasingly challenged to protect soil, water, and air resources, while maintaining crop and animal production, there is an urgent need to develop, adapt, and use innovative manure management technologies and practices to reduce pathogen loads in manure prior to land application. Storage systems such as traditional lagoons and high stacks cannot provide predictable pathogen control because the environmental factors that impact microbial survival vary widely in such uncontrolled management schemes. Grazing or free-ranging animal systems present different challenges because manure deposition on rangeland can potentially be dispersed across large drainage areas that will impact distant off-site locations.

Vegetated filter strips and treatment areas that intercept run-off from agricultural drainage channels show promise when coupled with grazing and feedlot livestock management practices in reducing off-site pathogen loads. Information about effectiveness of field measures and practices at the field- and watershed-levels is urgently needed. A HACCP framework for composting operations needs to be developed, used, and evaluated for its performance characteristics relative to reducing the number of incompletely disinfected composts and its cost-effectiveness.

References

Ahring, B.K., Mladenovska, Z., Iranpour, R., Westermann, P., 2002. State of the art and future perspectives of thermophilic anaerobic digestion. Water Sci. Technol. 45, 293–298.

Aitken, M.D., Sobsey, M.D., Van Abel, N.A., Blauth, K.E., Singleton, D.R., Crunk, P.L., et al., 2007. Inactivation of *Escherichia coli* O157:H7 during thermophilic anaerobic digestion of manure from dairy cattle. Water Res. 41, 1659–1666.

Al-Kaisi, M.M., Davis, J.G., Waskom, R.M., 1998a. Fact sheet 1.222. Liquid manure application methods. Colorado State University Extension, Fort Collins, CO.

Al-Kaisi, M.M., Waskom, R.M., Davis, J.G., 1998b. Fact sheet No. 1.223. Liquid manure application to cropland. Colorado State University Agricultural Extension, Fort Collins, CO.

Anderson, B.C., 1985. Moist heat inactivation of *Cryptosporidium* sp. Am. J. Public Health 75, 1433–1444.

Atwill, E.R., Pereira, M.D., Alonso, L.H., Elmi, C., Epperson, W.B., Smith, R., 2006. Environmental load of *Cryptosporidium parvum* oocysts from cattle manure in feedlots from the central and western United States. J. Environ. Qual. 35, 200–206.

Avery, S.M., Moore, A., Hutchison, M.L., 2004. Fate of *Escherichia coli* originating from livestock faeces deposited directly onto pasture. Lett. Appl. Microbiol. 38, 355–359.

Berry, E.D., Woodbury, B.L., Nienaber, J.A., Eigenberg, R.A., Thurston, J.A., Wells, J.E., 2007. Incidence and persistence of zoonotic bacterial and protozoan pathogens in a beef cattle feedlot runoff control vegetative treatment system. J. Environ. Qual. 36, 1873–1882.

Bicudo, J.R., Goyal, S.M., 2003. Pathogens and manure management systems: A review. Environ. Technol. 24, 115–130.

Boes, J., Alban, L., Bagger, J., Mogelmose, V., Baggesen, D.L., Olsen, J.E., 2005. Survival of *Escherichia coli* and *Salmonella* Typhimurium in slurry applied to clay soil on a Danish swine farm. Prev. Vet. Med. 69, 213–228.

Bowman, D.D., 2009. Manure pathogens. McGraw-Hill, New York.

Burger, H.J., 1982. Large-scale management systems and parasite populations prevalence and resistance of parasitic agents in animal effluents and their potential hygienic hazard. Vet. Parasitol. 11, 49–60.

Burton, C.H., Turner, C., 2003. Manure management—Treatment strategies for sustainable agriculture Silsoe Research Institute. Wrest Park: Silsoe, Bedford, UK.

Campbell, I., Tzipori, A.S., Hutchison, G., Angus, K.W., 1982. Effect of disinfectants on survival of *Cryptosporidium* oocysts. Vet. Rec. 111, 414–415.

Castellan, D.M., Kinde, H., Kass, P.H., Cutler, G., Breitmeyer, R.E., Bell, D.D., 2004. Descriptive study of California egg layer premises and analysis of risk factors for *Salmonella enterica* serotype *enteritidis* as characterized by manure drag swabs. Avian. Dis. 48, 550–561.

Christian, A.H., Evanylo, G.K., Pease, J.W., 1997. On-farm composting—A guide to principles, planning, and operations. Virginia Cooperative Extension, Blacksburg, VA.

Cieslak, P.R., Barrett, T.J., Griffin, P.M., Gensheimer, K.F., Beckett, G., Buffington, J., et al., 1993. *Escherichia coli* O157:H7 infection from a manured garden. Lancet 342, 367.

Cirelli, G.L., Consoli, S., Di Grande, V., Milani, M., Toscano, A., 2007. Subsurface constructed wetlands for wastewater treatment and reuse in agriculture: Five years of experiences in Sicily, Italy. Water Sci. Technol. 56, 183–191.

Cole, D.J., Hill, V.R., Humenik, F.J., Sobsey, M.D., 1999. Health, safety, and environmental concerns of farm animal waste. Occup. Med. 14, 423–448.

Cooper, J., Niggli, U., Leifert, C., 2007. Handbook of organic food safety and quality. Woodhead Publishing Ltd, Cambridge, UK.

Costantini, V.P., Azevedo, A.C., Li, X., Williams, M.C., Michel Jr., F.C., Saif, L.J., 2007. Effects of different animal waste treatment technologies on detection and viability of porcine enteric viruses. Appl. Environ. Microbiol. 73, 5284–5291.

Davis, J.G., Andrews, J.E., Al-Kaisi, M.M., 1997. Fact Sheet No. 1.221. Liquid Manure Management. Colorado State University Agricultural Extension, Fort Collins, CO.

De Bertoldi, M., Ferranti, M.P., L'Hermite, P., Zucconi, F., 1986. Compost: Production, Quality and Use. Elsevier Applied Science, London.

DEFRA, 2004. Guidance on the treatment in approved composting or biogas plants of animal by-products and catering waste—Animal By-Products Regulation 2003. www.defra.gov.uk/animalh/by-prods/pdf/compost_guidance.pdf (accessed 17.08.08.).

Entry, J.A., Leytem, A.B., Verwey, S., 2005. Influence of solid dairy manure and compost with and without alum on survival of indicator bacteria in soil and on potato. Environ. Pollut. 138, 212–218.

Esteban, J.I., Oporto, B., Aduriz, G., Juste, R.A., Hurtado, A., 2008. A survey of food-borne pathogens in free-range poultry farms. Int. J. Food Microbiol. 123, 177–182.

Fayer, R., Graczyk, T.K., Cranfield, M.R., Trout, J.M., 1996. Gaseous disinfection of *Cryptosporidium parvum* oocysts. Appl. Environ. Microbiol. 62, 3908–3909.

Fayer, R., Santin, M., Trout, J.M., 2007. Prevalence of *Cryptosporidium* species and genotypes in mature dairy cattle on farms in eastern United States compared with younger cattle from the same locations. Vet. Parasitol. 145, 260–266.

Fayer, R., Santin, M., Trout, J.M., 2008. *Cryptosporidium ryanae n. sp* Cryptosporidiidae: Apicomplexa in cattle (*Bos taurus*). Vet. Parasitol. 156, 191–198.

Fayer, R., Ungar, B.L., 1986. *Cryptosporidium* spp. and cryptosporidiosis. Microbiol. Rev. 50, 458–483.

Feachem, R., Bradley, D.J., Garelick, H., Mara, D.D., 1981. Appropriate technology for water supply and sanitation: Health aspects of excreta and sewage management-a state-of-the art review. The World Bank, Washington, DC.

Fossler, C.P., Wells, S.J., Kaneene, J.B., Ruegg, P.L., Warnick, L.D., Bender, J.B., 2005. Herd-level factors associated with isolation of *Salmonella* in a multi-state study of conventional and organic dairy farms I. *Salmonella* shedding in cows. Prev. Vet. Med. 70, 257–277.

Fossler, C.P., Wells, S.J., Kaneene, J.B., Ruegg, P.L., Warnick, L.D., Eberly, L.E., 2005. Cattle and environmental sample-level factors associated with the presence of *Salmonella* in a multi-state study of conventional and organic dairy farms. Prev. Vet. Med. 67, 39–53.

Guan, J., Chan, M., Ma, B., Grenier, C., Wilkie, D.C., Pasick, J., 2008. Development of methods for detection and quantification of avian influenza and Newcastle disease viruses in compost by real-time reverse transcription polymerase chain reaction and virus isolation. Poult. Sci. 87, 838–843.

Guan, T.Y., Holley, R.A., 2003. Pathogen survival in swine manure environments and transmission of human enteric illness—A review. J. Environ. Qual. 32, 383–392.

Guber, A.K., Pachepsky, Y.A., Shelton, D.R., Yu, O., 2007. Effect of bovine manure on fecal coliform attachment to soil and soil particles of different sizes. Appl. Environ. Microbiol. 73, 3363–3370.

Harter, T., Atwill, E.R., Hou, L., Karle, B.M., Tate, K.W., 2008. Developing risk models of *Cryptosporidium* transport in soils from vegetated, tilted soilbox experiments. J. Environ. Qual. 37, 245–258.

Haug, R.T., 1993. The Practical Handbook of Compost Engineering. Lewis Publishers, Boca Raton, FL.

Holley, R.A., Arrus, K.M., Ominski, K.H., Tenuta, M., Blank, G., 2006. *Salmonella* survival in manure-treated soils during simulated seasonal temperature exposure. J. Environ. Qual. 35, 1170–1180.

Humenik, F.J., 2001. Lesson 25—Manure treatment options. MidWest Plan Service, Ames, IA.

Hutchison, M.L., Walters, L.D., Avery, S.M., Moore, A., 2005. Decline of zoonotic agents in livestock waste and bedding heaps. J. Appl. Microbiol. 99, 354–362.

Ingram, D.T., Millner, P.D., 2007. Factors affecting compost tea as a potential source of *Escherichia coli* and *Salmonella* on fresh produce. J. Food Prot. 70, 828–834.

Jenkins, M.B., Bowman, D.D., Ghiorse, W.C., 1998. Inactivation of *Cryptosporidium parvum* oocysts by Ammonia. Appl. Environ. Microbiol. 64, 784–788.

Jiang, X., Morgan, J., Doyle, M.P., 2002. Fate of *Escherichia coli* O157:H7 in manure-amended soil. Appl. Environ. Microbiol. 68, 2605–2609.

Johannessen, G.S., Frøseth, R.B., Solemdal, L., Jarp, J., Wasteson, Y., Rørvik, L.M., 2004. Influence of bovine manure as fertilizer on the bacteriological quality of organic Iceberg lettuce. J. Appl. Microbiol. 96, 787–794.

Juteau, P., Tremblay, D., Ould-Moulaye, C.B., Bisaillon, J.G., Beaudet, R., 2004. Swine waste treatment by self-heating aerobic thermophilic bioreactors. Water Res. 38, 539–546.

Karim, M.R., Glenn, E.P., Gerba, C.P., 2008. The effect of wetland vegetation on the survival of *Escherichia coli, Salmonella typhimurium*, bacteriophage MS-2 and polio virus. J. Water Health 6, 167–175.

Kasorndorkbua, C., Opriessnig, T., Huang, F.F., Guenette, D.K., Thomas, P.J., Meng, X.J., et al., 2005. Infectious swine hepatitis E virus is present in pig manure storage facilities on United States farms, but evidence of water contamination is lacking. Appl. Environ. Microbiol. 71 7831–7387.

Kearney, T.E., Larkin, M.J., Levett, P.N., 1993. The effect of slurry storage and anaerobic digestion on survival of pathogenic bacteria. J. Appl. Bacteriol. 74, 86–93.

Kinde, H., Utterback, W., Takeshita, K., McFarland, M., 2004. Survival of exotic Newcastle disease virus in commercial poultry environment following removal of infected chickens. Avian Dis. 48, 669–674.

Kudva, I.T., Blanch, K., Hovde, C.J., 1998. Analysis of *Escherichia coli* O157:H7 survival in ovine or bovine manure and manure slurry. Appl. Environ. Microbiol. 64, 3166–3174.

Layden, N.M., Mavinic, D.S., Kelly, H.G., Moles, R., Bartlett, J., 2007. Autothermal thermophilic aerobic digestion (ATAD)—Part I: Review of origins, design, and process operation. J. Environ. Eng. Sci. 6, 665–678.

Lewis, D.J., Atwill, E.R., Lennox, M.S., Hou, L., Karle, B., Tate, K.W., 2005. Linking on-farm dairy management practices to storm-flow fecal coliform loading for California coastal watersheds. Environ. Monit. Assess. 107, 407–425.

Ley, V., Higgins, J., Fayer, R., 2002. Bovine enteroviruses as indicators of fecal contamination. Appl. Environ. Microbiol. 68, 3455–3461.

Meerburg, B.G., Jacobs-Reitsma, W.F., Wagenaar, J.A., Kijlstra, A., 2006. Presence of *Salmonella* and *Campylobacter* spp. in wild small mammals on organic farms. Appl. Environ. Microbiol. 72, 960–962.

Miller, W.A., Lewis, D.J., Lennox, M., Pereira, M.G.C., Tate, K.W., Conrad, P.A., Atwill, E.R., 2007. Climate and on-farm risk factors associated with *Giardia duodenalis* cysts in storm runoff from California coastal dairies. Appl. Environ. Microbiol. 73, 6972–6979.

Miller, W.A., Lewis, D.J., Pereira, M.D.G., Lennox, M., Conrad, P.A., Tate, K.W., Atwill, E.R., 2008. Farm factors associated with reducing *Cryptosporidium* loading in storm runoff from dairies. J. Environ. Qual. 37, 1875–1882.

Misra, R.V., Roy, R.N., Hiraoka, H., 2003. On-farm composting. Food and Agriculture Organization of the United Nations, Rome.

Monteith, H.D., Shannon, E.E., Derbyshire, J.B., 1986. The inactivation of a bovine enterovirus and a bovine parvovirus in cattle manure by anaerobic digestion, heat treatment, gamma irradiation, ensilage and composting. J. Hyg. (Lond.) 97, 175–184.

Muirhead, R.W., Collins, R.P., Bremer, P.J., 2005. Erosion and subsequent transport state of *Escherichia coli* from cowpats. Appl. Environ. Microbiol. 71, 2875–2879.

Murinda, S.E., Nguyen, L.T., Nam, H.M., Almeida, R.A., Headrick, S.J., Oliver, S.P., 2004. Detection of sorbitol-negative and sorbitol-positive Shiga toxin-producing *Escherichia coli, Listeria monocytogenes, Campylobacter jejuni*, and *Salmonella* spp. in dairy farm environmental samples. Foodborne Pathog. Dis. 1, 97–104.

Ndegwa, P.M., Thompson, S.A., 2001. Integrating composting and vermicomposting in the treatment and bioconversion of biosolids. Bioresour. Technol. 76, 107–112.

Nelson, H., 1997. The contamination of organic produce by human pathogens in animal manures. Available on-line at http://eap.mcgill.ca//SFMC_1.htm. Ecological Agriculture Projects, Faculty of Agricultural and Environmental Sci, McGill Univ. (Macdonald Campus), Ste-Anne-de-Bellevue: Canada (accessed 17.08.08.).

Olson, M.E., Goh, J., Phillips, M., Guselle, N., McAllister, T.A., 1999. *Giardia* cyst and *Cryptosporidium* oocyst survival in water, soil, and cattle feces. J. Environ. Qual. 28, 1991–1996.

Ottoson, J., Nordin, A., von Rosen, D., Vinneras, B., 2008. *Salmonella* reduction in manure by the addition of urea and ammonia. Bioresour. Technol. 99, 1610–1615.

Pachepsky, Y.A., Sadeghi, A.M., Bradford, S.A., Shelton, D.R., Guber, A.K., Dao, T., 2006. Transport and fate of manure-borne pathogens: Modeling perspective. Agric. Water Manag. 86, 81–92.

Pachepsky, Y.A., Yu, O., Karns, J.S., Shelton, D.R., Guber, A.K., van Kessel, J.S., 2008. Strain-dependent variations in attachment of *E. coli* to soil particles of different sizes. Intl. Agrophys. 22, 61–66.

Plym-Forshell, L., 1995. Survival of salmonellas and *Ascaris suum* eggs in a thermophilic biogas plant. Acta. Vet. Scand. 36, 79–85.

Robertson, L.J., Campbell, A.T., Smith, H.V., 1992. Survival of *Cryptosporidium parvum* oocysts under various environmental pressures. Appl. Environ. Microbiol. 58, 3494–3500.

Rynk, R., 1992. On-Farm Composting Handbook, NRAES No. 54. NRAES (Natural Resource, Agricultural, and Engineering Service) Cooperative Extension, Ithaca, NY.

Santin, M., Trout, J.M., Fayer, R., 2008. A longitudinal study of cryptosporidiosis in dairy cattle from birth to 2 years of age. Vet. Parasitol. 155, 15–23.

Scheuerell, S., Mahafee, W., 2002. Compost tea: Principles and prospects for plant disease control. Compost. Sci. Util. 10, 313–338.

Shepherd Jr., M.W., Liang, P., Jiang, X., Doyle, M.P., Erickson, M.C., 2007. Fate of *Escherichia coli* O157:H7 during on-farm dairy manure-based composting. J. Food Prot. 70, 2708–2716.

Spencer, J.L., Guan, J., 2004. Public health implications related to spread of pathogens in manure from livestock and poultry operations. Methods Mol. Biol. 268, 503–515.

Strauch, D., 1991. Survival of pathogenic micro-organisms and parasites in excreta, manure and sewage sludge. Rev. Sci. Tech. 10, 813–846.

Strauch, D., Ballarini, G., 1994. Hygienic aspects of the production and agricultural use of animal wastes. J. Vet. Med. Ser. B. 41, 176–228.

Thompson, W., Legge, P., Millner, P., Watson, M., 2002. Test methods for the examination of composts and composting, CD-ROM. U.S. Composting Council. http://tmecc.org/tmecc/ (accessed 17.08.08.).

Thurston-Enriquez, J.A., Gilley, J.E., Eghball, B., 2005. Microbial quality of runoff following land application of cattle manure and swine slurry. J. Water Health 3, 157–171.

Trask, J.R., Kalita, P.K., Kuhlenschmidt, M.S., Smith, R.D., Funk, T.L., 2004. Overland and near-surface transport of *Cryptosporidium parvum* from vegetated and nonvegetated surfaces. J. Environ. Qual. 33, 984–993.

Turner, C., Williams, S.M., Burton, C.H., Cumby, T.R., Wilkinson, P.J., Farrent, J.W., 1999. Pilot scale thermal treatment of pig slurry for the inactivation of animal virus pathogens. J. Environ. Sci. Health B. 34, 989–1007.

Turner, C., Williams, S.M., Cumby, T.R., 2000. The inactivation of foot and mouth disease, Aujeszky's disease and classical swine fever viruses in pig slurry. J. Appl. Microbiol. 89, 760–767.

Tzipori, S., 1983. Cryptosporidiosis in animals and humans. Microbiol. Rev. 47, 84–96.

Tzipori, S., Widmer, G., 2008. A hundred-year retrospective on cryptosporidiosis. Trends. Parasitol. 24, 184–189.

USDA, 1999. Agricultural waste management field handbook, National Engineering Handbook Part 651 US Department of Agriculture. Natural Resource Conservation Service, Washington, DC www.wsi.nrcs.usda.gov/products/W2Q/AWM/handbk.html.

USDA, 2000. National organic program, 7 CFR Part 205. The Organic Foods Production Act of 1990, as amended (7 U.S.C. 6501–6522) 65 FR 80637, Dec. 21, 2000 U.S. Department of Agriculture, Agricultural Marketing Service, Washington, DC.

USDA, 2007. Animal waste management software and user's guide U.S. Department of Agriculture, Natural Resource Conservation Service, Washington, DC http://www.wsi.nrcs.usda.gov/products/W2Q/AWM/pgrm.html#DOCUMENTS.

USDA, 2008. NRCS air quality and atmospheric change fact sheets—Air quality and animal operations U.S. Department of Agriculture, Natural Resource Conservation Service, Washington, DC www.airquality.nrcs.usda.gov/Documents/.

USEPA, 1994. A plain English guide to the EPA Part 503 biosolids rule, EPA/832/R-93/003. Office of Wastewater Management U.S. Environmental Protection Agency, Washington, DC.

USEPA, 2000. Guide to field storage of biosolids and other organic by-products used in agriculture and for soil resource management, EPA/832/B-00–007. Office of Wastewater Management U.S. Environmental Protection Agency, Washington, DC.

USEPA, 2000. Standards for the use or disposal of sewage sludge. Code of Federal Regulations (USEPA CFR), 40CFR 503 Title 40, Parts 425 to 699 [Revised as of July 1, 2000]. Office of Wastewater Management U.S. Environmental Protection Agency, Washington, DC 754–786.

USEPA, 2003. Environmental regulations and technology control of pathogens and vector attraction in sewage sludge, EPA/625/R-92–013. Office of Research and Development, National Risk Management Research Laboratory U.S. Environmental Protection Agency, Cincinnati, OH.

Vanotti, M.B., Hunt, P.G., Szogi, A.A., Humenik, F.B., Millner, P.D., Ellison, A.Q., 2003. Solids separation, nitrification-denitrification, soluble phosphorus removal, solids processing system. N. C. Anim. Waste Manag. Workshop 30–35.

Vanotti, M.B., Millner, P.D., Hunt, P.G., Ellison, A.Q., 2005. Removal of pathogen and indicator microorganisms from liquid swine manure in multi-step biological and chemical treatment. Bioresour. Technol. 96, 209–214.

Vanotti, M.B., Rice, J.M., Ellison, A.Q., Hunt, P.G., Humenik, F.J., Baird, C.L., 2005. Solid-liquid separation of swine manure with polymer treatment and sand filtration. Trans. Amer. Soc. Agric. Engin. 48, 1567–1574.

Vanotti, M.B., Szogi, A.A., Hunt, P.G., Millner, P.D., Humenik, F.J., 2007. Development of environmentally superior treatment system to replace anaerobic swine lagoons in the USA. Bioresour. Technol. 98, 3184–3194.

Vanotti, M.B., Szogi, A.A., Vives, C.A., 2008. Greenhouse gas emission reduction and environmental quality improvement from implementation of aerobic waste treatment systems in swine farms. Waste Manag. 28, 759–766.

Wang, G., Zhao, T., Doyle, M.P., 1996. Fate of enterohemorrhagic *Escherichia coli* O157:H7 in bovine feces. Appl. Environ. Microbiol. 62, 2567–2570.

Westerman, P.W., Bicudo, J.R., 2005. Management considerations for organic waste use in agriculture. Bioresour. Technol. 96, 215–221.

Williams, A.P., Avery, L.M., Killham, K., Jones, D.L., 2005. Persistence of *Escherichia coli* O157 on farm surfaces under different environmental conditions. J. Appl. Microbiol. 98, 1075–1083.

Williams, C.M., 2003. Development of environmentally superior technologies, year 3, progress report for technology determinations per agreements between the Attorney General of North Carolina and Smithfield Foods and Premium Standard Farms, and Frontline Farmers, reporting period July 26, 2002–July 25, 2003. North Carolina State University Animal and Poultry Waste Management Center North Carolina State University, Raleigh, NC.

Xiao, L., Fayer, R., 2008. Molecular characterization of species and genotypes of *Cryptosporidium* and *Giardia* and assessment of zoonotic transmission. Int. J. Parasitol. 38, 1239–1255.

Xiao, L., Zhou, L., Santin, M., Yang, W., Fayer, R., 2007. Distribution of *Cryptosporidium parvum* subtypes in calves in eastern United States. Parasitol. Res. 100, 701–706.

Bioaerosol Contamination of Produce: Potential Issues from an Unexplored Contaminant Route

5

John P. Brooks[1], Charles Gerba[2]

[1]*Genetics and Precision Agriculture Unit, USDA-ARS, Mississippi State, MS,* [2]*Department of Soil, Water and Environmental Science, University of Arizona, Tucson, AZ*

CHAPTER OUTLINE HEAD

Introduction

The presence of foodborne pathogens on fresh fruits and vegetables has and always will be a cause for concern (Doyle and Erickson, 2008; Hamilton et al., 2006; Blumenthal et al., 2000; Berger et al., 2010). Given the recent trend towards consumption of fresh fruits and vegetables, the need for understanding and addressing all potential routes of contamination is more important than ever (Figure 5.1). Scallan et al. (2011) estimates approximately 37.2 million infections occur nationwide, with approximately 9.4 million attributed to foodborne infections. The majority of foodborne infections are attributed to viruses, specifically norovirus, with *Salmonella* spp. and *Campylobacter* spp. a distant second (Scallan et al., 2011). The fresh leaf lettuce outbreak of 1995 ushered in an era of fresh produce foodborne contamination (Ackers et al., 1998) and possibly, equally as important, the interest of the news media and the public. The public chastising of fresh crop producing companies, whether deserved or not, has led the industry to increase all efforts to protect their product from all potential contaminant sources. Despite these efforts, since the turn of the century, a seemingly large number of outbreaks and/or food recalls have occurred for both fresh food crops and other food products, leading to increased research efforts and monitoring. Oftentimes, recalls are due to vigilant

The Produce Contamination Problem. http://dx.doi.org/10.1016/B978-0-12-404611-5.00005-1
2014 Published by Elsevier Inc.

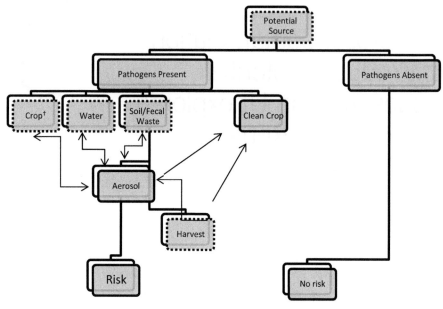

FIGURE 5.1

Potential for aerosol generation from multiple sources during the crop growth and harvest. Each dotted box contributes to aerosols, with aerosols contributing back to the contamination of each respective contaminant route/source.

monitoring efforts and more effective, sensitive techniques capable of detecting low pathogen numbers on or in various food matrices. When an outbreak occurs, the news media and hence the public begin the search for answers and seek to assess blame and responsibility for the contamination. Oftentimes, despite the efforts of industry, government, and academic agencies, the answer is multi-faceted with many potential pathogen contamination routes acting either singularly as the source or interconnected in an indecipherable matrix of contamination. Obvious routes of infection involve food handlers or the use of land-applied fecal wastes, though as has been seen with recent outbreaks, the sources are rarely obvious.

In the case of fresh food crops, it is thought that the majority of foodborne contamination occurs at the field level possibly through: 1) wildlife contamination (CDHS, 2007); 2) human error (USFDA, 1998); or 3) contaminated irrigation water (Keraita et al., 2007). One possibility, not previously investigated, is the aerosol transport and subsequent contamination of fresh vegetables and fruits with foodborne pathogens. It is well known, among plant pathologists, that bacterial and fungal plant pathogens utilize air pathways to disseminate and infect plants (Harrison, 1980; Brown and Hovmoller, 2002). The question is, do these same pathways exist for foodborne pathogens? To date, no known human pathogens have been traced back to crops contaminated via aerosol drift; this shouldn't be taken to mean that this process doesn't exist, but rather the sensitivity of current field and epidemiological monitoring tools aren't up to the task of accurately assessing the impact of aerosols and food crop contamination. We can, however, predict the potential for aerosol contamination of fresh food crops, using the information we currently have regarding the dynamics of aerosol transport, and use this as an informative discussion towards food protection in a potentially new arena. To understand why aerosols, specifically bioaerosols, can be difficult to predict with regard to food crops, this chapter will first seek to define bioaerosols and our understanding regarding aerosol transport of foodborne bacteria and viruses. A review of the available literature will then be conducted to assess the potential for crop contamination.

As discussed in other chapters, enteric pathogens may contaminate the surfaces and internal tissues of crops such as spinach, lettuce, tomatoes, and peppers (Brandl, 2008; Berger, et al., 2010). Consumption of fresh produce has been linked to nation-wide outbreaks of enteric illness (USFDA, 2006; Hanning et al., 2009; Warriner et al., 2009; Wendel et al., 2009). Scallan et al. (2011) estimated that most bacterial food-borne outbreaks have been associated with either *Salmonella* spp., *Campylobacter jejuni*, *Escherichia coli* O157:H7, or *Listeria monocytogenes*. Recent data have suggested *Salmonella* spp. and *E. coli* O157:H7 as the primary etiological agents most often associated with fresh food crop outbreaks. There are many potential reservoirs for each of these pathogens including: nearby manure or biosolids land-application operations, concentrated animal feeding operations (CAFO), feral animals, and con-taminated irrigation water. These reservoirs can all serve as aerosol sources, under varying circumstances, which will be discussed below.

Aerosols

Aerosols can be defined as aerosolized particles; more specifically bioaerosols refer to biological particles that have been aerosolized. In this case, bioaerosol can refer to aerosolized cells (e.g., *Legionella pneumophila* in an air vent), viruses (e.g., a cough containing influenza), spores (e.g., *Aspergillus* fungal spores following a rain event), or biological cell remnants (e.g., endotoxin or peptidoglycan). Typical sizes range from 0.5 to 5 μm for smaller-sized particles (often referred to as inhalable particles); however, larger-sized particles can be associated with sizes up to 30 μm and typically settle out of the air stream or are ingested when exposed to humans. Environmental bioaerosols rarely exist as single cells or viruses, most are associated with inorganic particulates, such as soil, dust, or water droplets. As would be expected, the type of bioaerosol, and hence its composition, is entirely dependent on the source (Brooks et al., 2004). Bioaerosols generated by land application of manure or biosolids can be composed of a mixture of soil, dry plant material, and the land-applied waste residual (Brooks et al., 2005a). Bioaerosols attached to other particles, also known as biologi-cal rafts, often enhance survival over singular cells or particles (Brooks et al., 2004). Dust carriage is largely responsible for carriage of bioaerosols across oceans (Jones and Harrison, 2004), and for the carriage of fungal plant diseases across continents (Brown and Hovmoller, 2002). That being said, the drier the soil, dust, or vegetation particle, the further it will travel; moist particles, which are dense, tend to settle out of the atmosphere more quickly. It is also well known that dry conditions tend to promote bacterial and viral inactivation, therefore only the most recalcitrant of bioaerosols are capable of traveling long distances, hence why fungal and bacterial spores are often cited as traveling across continents and oceans.

Several factors affect the fate and transport of pathogens, including relative humidity, temperature, wind speed, ultraviolet radiation (UV), oxygen radicals, and the aerosolization process. Ideally, a mid-high humidity, low temperature, and minimal UV environment would be necessary to facilitate the contamination of fresh food crops. Consider that the fresh food crops in question are grown in the western U.S., hence drier conditions with higher temperatures; the chance for bioaerosol survival may be low (discussed below). Relative humidity is known to drastically affect the survival of aerosolized foodborne pathogens and indicators (Brooks et al., 2004). Aerosolized *E. coli* has been shown to survive under a low- to mid-range relative humidity level under controlled laboratory conditions (Cox, 1966), while some Gram-positive bacteria demonstrated the opposite reaction (Theunissen et al., 1993). Some naked-capsid viruses (e.g., enteric viruses) can be inactivated at low relative humidity levels (Mohr, 2001). Bacterial inactivation through dehydration and desiccation result from conformational changes to the cell wall (e.g., peptidoglycan layer and phospholipid bilayer). Generally speaking, thicker peptidoglycan layers lead to longer persistence under conditions of desiccation. Viruses, due to their relative simplicity, are not as affected by changes to humidity and temperature (Brooks et al., 2004). It is well known that UV damage affects DNA replication through production of thymine dimers; generally speaking, both viruses and bacteria are negatively affected by UV and increased solar radiation (Brooks et al., 2004). This is important to note, as previously stated, the majority of fresh food crops tend to be grown in environments where solar radiation is high and sunny days are typical for more than two-thirds of the year.

Three factors dominate aerosol transport, including: 1) aerosol release or emission; 2) dispersion in three dimensions; and 3) deposition rate (Brooks et al., 2004). Typically, the release of a bioaerosol involves a "violent," at least at a cellular or molecular level, dispersal event. Wind speed can provide the energy necessary to break bonds between the bioaerosol and surface, initiating an emission event (Jones and Harrison, 2004). Jones and Harrison (2004) reviewed the literature and determined that wind speed may be more effective in generating aerosols from plant tissue surfaces than from soil. Mechanical forces (e.g., farm equipment) are equally as important to initiate the aerosol emission release (Brooks et al., 2005a, b). The agitation forces involved in, for instance, a front-end loader dropping solid manure into a hopper, may generate more aerosols than the actual land application event (Brooks et al., 2005a). Likewise, the mechanical agitation involved in spray irrigation of effluent (e.g., reel guns or central-pivot irrigation) provides the energy necessary to generate aerosols, though much of this is a function of the liquid effluent, which is more readily aerosolized than solid material (Brooks et al., 2004). These mechanical forces are important, particularly when referring to the aerosol dispersal of enteric pathogens, which are often tightly associated with host cellular or fecal debris

(Chetochine et al., 2006). Rain forces may be a significant disruptive force, capable of aerosolizing fecal droppings and other pathogen sources located near crops (Butterworth and McCartney, 1991; Cevallos-Cevallos, et al., 2012a, b). Once bioaerosols are aerosolized, they are subject to transport processes consisting of prevailing air currents, convection, diffusion, and eventual gravitational settling (Brooks et al., 2004; Jones and Harrison, 2004). Smaller particles (<5 μm) are highly subjected to these processes, while large particles tend to settle out of the air cycle and are deposited more quickly (Brooks et al., 2004). Once an aerosolized particle is near a surface, in this case the surface of a leaf or fruiting body, attractive forces aid in surface attachment, and these forces are dominated by low energy bonds such as Van der waals and electrostatic forces (Brooks et al., 2004).

Potential for crop contamination – sources of bioaerosols
Spray-irrigation and rain-induced aerosols

Fecal droppings, as a result of wild hogs, is one of the presumed causes of the 2006 fresh spinach outbreak (CDHS, 2007). Fischer et al. (2001) demonstrated that wildlife, such as deer, are capable of harboring *E. coli* O157:H7 in both field and experimental trials. Though natural carriage only demonstrated low pathogen levels, the study demonstrated that infected deer were capable of horizontally transferring the pathogen from infected to non-infected deer. Likewise, the presence of fecal materials (e.g., manure) from infected domesticated livestock could provide the vehicle for pathogen delivery (Hutchison et al., 2005). That being said, the presence of the infected cattle or deer may not be enough to yield a contamination event and subsequent outbreak; a mechanical agitation event is necessary to transfer the pathogen load from the soil surface to the crop leaf and/or fruiting body (Cevallos-Cevallos, 2012b).

Rain has long been associated with pathogen contamination events (Curriero, et al., 2001). Rain-induced runoff and subsequent contamination of irrigation water or groundwater is a fairly well-known mechanism for horizontal and vertical bacterial and viral transport (Curriero et al., 2001; Brooks et al., 2012). However, aerosolization of foodborne pathogens via this route has been largely ignored. There is precedent for it though as plant-specific pathogens such as tobamovirus utilize a rain-induced aerosolization route to infect new hosts (Fillhart et al., 1997). Likewise, Harrison (1980) discussed the aerosolization of *Erwinia carotovora* in a field from naturally infected potato crops following a rain event, and were typically more often detected when conditions were more favorable for detection, such as high humidity, low temperature, or nightfall. Graham and Harrison (1975) showed that larger rain droplets were more effective than smaller ones at generating aerosols from potato stems infected with *E. carotovora*. One potential difference between these studies and foodborne pathogens is the biological amplification of *Erwinia* spp. (a natural plant pathogen), as opposed to a foodborne pathogen, most likely not propagating itself to high levels within the infected plant or fruiting body tissue. Even under these ideal conditions (e.g., rain, propagation, etc.), Graham et al. (1977) calculated that only 0.003% of the available infecting bacteria ($\sim 10^7$ cfu/g^{-1} of infected tissue) were capable of aerosolization, which assuming 2% of leaves were infected only yields approximately 10^8 aerosolized *Erwinia* cells ha^{-1} (or 10,000 cfu/m^{-3} air going approximately 10 m high). In these cases, low aerosolization rates pose no problem to the pathogen since it is still near its natural host, thus enabling it to propagate; this would not be the case for a foodborne pathogen. In the case of a foodborne pathogen, at most only a small portion of that 1 ha field may be contaminated and at levels orders of magnitude below 10^7 per gram of plant tissue. The work of Butterworth and McCartney (1991) corroborates this by demonstrating that rain-induced aerosolization of leaf surface-inoculated *Pseudomonas syringae*, *Klebsiella planticola*, and *Bacillus subtilis* produced detectable levels only up to 20 cm from the target. The study also demonstrated the inefficiency associated with aerosol dispersion, which was less than 5% of the total inoculated bacteria (Butterworth and McCartney, 1991).

Aerosolization appears to be an inefficient phenomenon even when considering deliberate aerosolization from point sources such as spray-irrigation of inoculated water (Brooks et al., 2005b). This may be due to the low survival associated with aerosolized Enterobactericae, as shown by Teltsch et al. (1980), or due to immediate dilution effects once in the air environment. Considering that many of the most well-known foodborne pathogens are Gram-negative bacteria, it appears that the aerosolization process is not well suited for these pathogens. That being said, rain can theoretically aerosolize fecal pathogens located in soil or fecal material atop of soil (Cevallos-Cevallos et al., 2012b). Take, for instance, the *E. coli* O157:H7 contamination on fresh spinach. It has been hypothesized that feral swine contributed to the contamination by tracking and depositing contaminated feces, either their own or cattle feces on their hooves, onto soil near the implicated spinach fields (Jay et al., 2007; CDPH, 2007). In theory, a rain event could aerosolize contaminated feces and any offending pathogen, a theory that has been tested (Cevallos–Cevallos et al., 2012b). Though the study was artificial (e.g., petri dish with a liquid inoculum as the source), Cevallos–Cevallos et al. (2012b) demonstrated that *Salmonella* could aerosolize when impacted with a 60 mm h^{-1} rain intensity, reaching heights of 85 cm, which was adequate to attach to tomato skin. Unfortunately, the effect of UV and high-temperature variations were not simulated, and the use of a liquid *Salmonella* aliquot created an ideal environment for aerosolization. The presence of high organic matter fecal wastes would most likely decrease aerosolization, given the binding properties of some fecal materials (Chetochine et al., 2006). The amount of agitation needed to aerosolize deposited fecal material would most likely exceed that provided by rain; additionally, the heavier the rainfall (with higher energy), may serve to scrub the air, actually suppressing aerosolization (Cevallos–Cevallos et al., 2012b) and enhancing horizontal runoff (Brooks et al., 2012). To date no other studies have been conducted regarding this method of aerosolization of foodborne pathogens, particularly using fecal material in the field or controlled greenhouse experiments. It goes without stating that this is a field that requires more investigation, particularly under natural field conditions. Natural, non-inoculated field situations will be difficult to measure, given the random combination of these systems (e.g., field, contamination, rain event, etc.).

Assuming that a 24-hour aerosol surveillance program was put into place throughout a field, the bioaerosol levels in the field would most likely be minimum and therefore, difficult to detect. As has been discussed, even under induced aerosolization, low aerosol levels are produced when compared to source pathogen levels (Brooks et al., 2005a). In all likelihood, a fecal dropping in the field subjected to rain-induced aerosolization will only have an immediate impact on the surrounding vegetation, which could prove difficult for monitoring efforts, but nonetheless could prove problematic.

Irrigation of contaminated water has been hypothesized as a potential crop contamination route, either through spray, furrow, or surface-irrigation. Many studies have been conducted regarding irrigation as a form of crop foodborne pathogen delivery (Stine et al., 2005; Soderstrom et al., 2008; Greene et al., 2008; Pachepsky et al., 2011). As with rain aerosolization, no documented evidence suggests this as a means for delivery of foodborne pathogens to food crops, only that it was presumed to have occurred, but to date no study has measured the aerosolized pathogen *in situ*. Stine et al. (2005) demonstrated that contaminated furrow-irrigation water was unable to contaminate bell pepper (~30 cm from soil surface), indicating that minimal aerosolization occurred. In fact, the majority of contamination through surface-irrigation will most likely arise from direct contact, as shown with lettuce and cantaloupe (e.g., crops that touch the soil) (Song et al., 2006; Stine et al., 2005). Doyle and Erickson (2008) reviewed the literature and considered crop proximity to the soil (e.g., potato, lettuce, etc.) to be one of the top reasons for field contamination, in part due to direct contact with irrigation water. Likewise, Doyle and Erickson (2008) considered the possibility of aerosol dispersion of pathogens as a means for contaminating aerial plant tissue, though aside from spray-irrigation of contaminated water, the authors do not cite evidence of aerosolized furrow surface-irrigation. It is, however, possible that furrow- or flood-irrigation can generate significant aerosols as demonstrated by a recent study conducted by Paez–Rubio et al. (2005). Though, in that study, the researchers investigated a flood-irrigated (domestic wastewater) field under various field conditions, particularly high and low wind speeds. High wind speeds produced a 6-order magnitude difference in aerosolized enteric bacteria when compared to the low wind-speed conditions (Paez–Rubio et al., 2005). Though this study and others like it most likely represent a "worst case scenario" and likely do not represent actual fresh food crop-irrigation scenarios, they do demonstrate precedent for this phenomenon. As with rain-induced aerosolization of fecal droppings, this is an area of research where many inroads have yet to be made.

On the other hand, overhead spray-irrigation is consistently considered a top source of potential contamination (Berger et al., 2010). Studies conducted by Islam et al. (2004), Solomon et al. (2003), and Hamilton et al. (2006) all demonstrated the microbiological risks associated with spray-irrigation of contaminated water. In all studies, the use of either furrow or subsurface drip-irrigation systems reduced the risks substantially. Overhead spray-irrigation can be considered a form of aerosolization, but more realistically, in the field, it is more akin to direct contact with the contaminating source, as the droplets generated by the sprayer directly fall to the plant tissue. Consider the use of overhead spray irrigators, applying water directly to the edible leaf tissue of a lettuce or spinach crop; in this case, it's less about the aerosolization of the pathogens and more about direct application of the pathogen to the tissue surface. There is evidence to suggest injured crop tissue is more susceptible to this means of contamination (Barker–Reid et al., 2009), most likely a result of the direct inoculation from overhead.

That being said, spray systems, such as these, are known to generate a fine mist, in addition to large droplets (Dungan, 2011); thus, irrigation drift should be considered a possibility when large-scale center-pivot irrigation systems are used. The drift could be capable of long downwind distance travel (Brooks et al., 2004), and given that damaged crops may be more susceptible to this form of contamination, aerosol drift arising from spray-irrigation should not be ignored. This may be more of a concern when adjoining farms have varying levels of food safety concerns (e.g., a non-food crop farm spray irrigating contaminated water adjacent to a baby spinach farm). Though this form of pathogen delivery has yet to be studied, with regard to fresh food crops, it goes without stating that aerosols generated by contaminated water, delivered by center pivot-irrigation is a possible source of contamination, at least within 500 m of the source (Bausum et al., 1982). Of course, this all depends on the microbial quality of the water in question; the data regarding center-pivot irrigation systems and aerosols were all collected using minimally treated manure or municipal effluent with high levels of indicators and pathogens (Brooks et al., 2004).

Land application of manure/biosolids and CAFO aerosols

Perhaps one of the more researched areas regarding downwind transport of pathogen-laden aerosols is land application of waste residuals (e.g., manure, municipal biosolids, etc.). Specifically, the land application of municipal biosolids and wastewater has been more thoroughly researched in recent years (Brooks et al., 2004, 2005a, b; Paez–Rubio et al., 2007). Additionally, the generation of aerosols from downwind of concentrated animal feeding operations (CAFOs) has received more attention as well (Hutchison et al., 2008; Chinivasagam et al., 2009; Brooks et al., 2010; Dungan et al., 2010; Dungan, 2011). Though, admittedly, these operations and studies have little to do with fresh food crop contamination, their implications should be taken into account as another potential source for aerosol contamination. To consider these as sources, one has to assume that these operations are near fresh food crop operations; the 2006 spinach *E. coli* O157:H7 outbreak was in part due to the proximity of an animal-feeding operation to the spinach operation (Jay et al., 2007). For the most part, fresh or minimally treated manure/municipal biosolids are not used near or on fresh food crop lands (Brooks et al., 2012). This is in part due to design and regulations, but with more urban sprawl and less land available to dispose the millions of tons of manure/biosolids, the eventuality suggests that land application of waste residuals will take place near fresh food crop operations.

As with some of the previous sections, there is no evidence to suggest land application of waste residuals contributes to aerosolized pathogens that contaminate food crops; likewise, no studies have been conducted to test this hypothesis. As presented with the previous sections, we will use the available literature to suggest the possibility for these aerosols reaching fresh food crops. There are a handful of reviews which present current information regarding bioaerosols generated during land application of biosolids and manure (Brooks et al., 2004; Pillai 2007; Dungan, 2010). Similarly, aerosols generated by CAFOs have been studied and reviewed (Dungan, 2011). The majority of reviews suggest that while production of bioaerosols is common place during these operations, the bioaerosols produced are largely kept to the application site. Transport distances have been reported between 200 and 400 m downwind of the source (Bausum et al., 1982; Camann et al., 1988) for wastewater spray application. Dungan (2011) reviewed spreading of animal wastes and reported that liquid spray resulted in culturable bacteria above background levels up to 200 m downwind of the application site (Boutin et al., 1988). On the other hand, biosolids have demonstrated travel distances up to 165 m during disk incorporation (i.e., soil dust generation) (Low et al., 2008). The small-sized particles generated during liquid spray account for the long travel distances; therefore, these operations would tend to increase risks to adjacent fields. Solid residuals, such as biosolids, will travel longer distances when combined with soil or other dry particles, which can act as biological rafts, protecting and carrying the bioaerosol further downwind. Regardless of residual, once aerosolized, bioaerosol concentrations appear to quickly dilute at distances greater than 200 m downwind of the application site. Any food crop operation sharing a fence line with an application site needs to be aware of the potential for aerosol transport offsite, and would be prudent to avoid harvesting during the land-application operation.

Stationary locations, constantly generating bioaerosols, such as indoor and open-air CAFOs, may pose a significant risk to nearby cropping operations. On the other hand, land-application scenarios occur on a field, at most, twice per year (Brooks et al., 2005b); needless to say, this significantly reduces exposures and contamination risks. Dungan (2011) reviewed the literature pertaining to CAFO-generated bioaerosols, and demonstrated that as with other scenarios, 200 m appears to be the furthest distance at which bioaerosols were above background. However, in these scenarios, the bioaerosol plume is constantly generated with ebbs and flows occurring throughout the day and night depending on animal activity and growth stage (Dungan, 2011). These locations pose a more significant risk to fields located nearby, as in all likelihood, a bioaerosol plume may be in the field, depending on wind direction. As mentioned in other sections, this hypothesis has not been tested or verified. Likewise, many fresh food crop land owners are actively reducing these types of risk by ensuring fields are not located near CAFOs. It is important to state, that for the most part, in studies of bioaerosols originating from land application of residuals or CAFOs, rarely are true pathogens (e.g., *Salmonella* spp., *Campylobacter* spp.) detected; in fact, most studies detect pathogen indicators, which are typically present in the residual at levels at least 1000 times greater than human enteric pathogens. Thus, pathogens are rarely detected at these 200 m downwind distances and therefore risks are even lower when considering pathogens (Brooks et al., 2012). Take, for instance, a series of risk assessments conducted by Brooks et al. (2012), in which separate exposures to aerosols and fresh food crops combined with land application of waste residuals yielded significant risks during immediate, short-term exposures. Particularly, for food crops, only exposures that were considered an immediate safety bypass (e.g., circumvention of land-application safety measures by using high pathogen level wastes) generated significant risks. Risks produced, as a result of merely following stipulated regulations and recommendations for crop harvesting and use of manure/biosolids, would be low. In the case of a bioaerosol contaminating a food crop, one could assume that the contaminant source level would be far below that of an immediate safety bypass, as described above. If one were to assume a pathogen aerosolization event at 200 m from the crop site, coupled with dilution, inactivation, and settling on plant tissue, the risks would be low, as the crop would then be harvested and washed prior to public exposure. However, unforeseen circumstances can always occur, and in actuality, most outbreaks occur as a result of unforeseen biological circumstances. For instance, the pathogen may land on damaged tissue, which may promote internalization (Barker–Reid et al., 2009). Likewise, the pathogen may land on soil, with high organic matter and moisture, thus promoting active growth (Zaleski et al., 2005), which can in-turn lead to a high-risk scenario.

Harvesting and mechanical aerosolization

Harvest and postharvest contamination is the final opportunity for a pathogen to contaminate plant tissue. Postharvest contamination in the food-meat industry has been well studied (Pearce et al., 2006; Posch et al., 2007). Mechanical agitation during the slaughter process is known to generate aerosols at significant levels, which in turn provides a means for horizontal transmission. Pearce et al. (2006) and Posch et al. (2007) determined that aerosol levels could reach as high as 10^4 *Campylobacter* spp. during poultry slaughtering and 10^3 cfu/m^{-3} air (aerobic bacteria) during swine processing, respectively. The commercial preparation of fresh food crops is not as intensive as with meat products, nor is the product as contaminated. Potentially, though, a pathogen on the surface of plant tissue can be aerosolized during harvest. Perombelon and Lowe (1973) described the aerosolization of *Erwinia carotovora* during the mechanical beating of potato stems during harvest. As was the case with raindrop-associated aerosols, the propagation of the pathogen within the plant tissue accounted for high numbers present in the tissue, which would then lead to high rates of aerosolization. As discussed above, this most likely wouldn't occur with foodborne pathogens; though some propagation has been suggested to occur, most likely the levels would not reach that of a true plant pathogen and its host.

Conclusions

The increase in the reported number of pathogen outbreaks and recalled fresh food crops has led the media, public, and researchers to attempt to discover all possible crop-contamination routes. Despite an increase in reports, perhaps due to more active vigilance, our nation's fresh food supply is relatively safe, which consequentially helps to negatively highlight a recall or outbreak. Because these outbreaks occur so infrequently (relative to the number of products consumed on an annual basis) and investigations are conducted from a retrospective view-point, pinpointing a pathogen source is difficult. Oftentimes, only speculative "educated hypotheses" are the best we can muster. The aerosolization of pathogens, whether through rain-induced aerosolization, irrigation, land application of residual wastes, or harvesting processes is likely the most difficult to pinpoint pathogen source for fresh food crops. Aerosol detection techniques and the ability to separate an aerosol route from other more obvious contaminant routes make the identification task a more nuanced scientific investigation. This is coupled with the fact that in nearly all the presented examples, the aerosol route of contamination needs to occur in conjunction with a pathogen source, which itself occurred as a result of a contamination route (e.g., feral animal depositing fecal matter or contaminated irrigation water). Separating the two contamination routes and demonstrating which one is the primary, may be difficult. As discussed above, the contamination of fresh food crops with aerosolized pathogens is an area of research that has been discussed, but very little empirical data exists to suggest this as a potential issue. However, there is precedent for this transport route, particularly, when one examines the transfer of plant pathogens in the field. It can be assumed that foodborne pathogens could exploit these opportunities as well. Particularly, the use of surface or overhead irrigation, with contaminated water, appears to be the most prominent source for these aerosols. Maintaining a high microbiological standard for the water, which is used for these practices, is the most likely primary critical control point. Removing or reducing the presence of feral animals from entering crop lands will reduce other potential routes. Perhaps the most controllable potential sources is the presence of nearby CAFOs or residual waste land application. Proper knowledge and planning for these operations will avoid these potential aerosol routes. That being said, this area is ripe for research opportunities and should be investigated with the vigor that other areas of foodborne pathogens on fresh food crops have been investigated.

References

Ackers, M.L., Mahon, B.E., Leahy, E., Goode, B., Damrow, T., Hayes, P.S., et al., 1998. An outbreak of *Escherichia coli* O157:H7 infections associated with leaf lettuce consumption. J. Infect. Dis. 177, 1588–1593.

Barker-Reid, F., Harapas, D., Engleitner, S., Kreidl, S., Holmes, R., Faggian, R., 2009. Persistence of *Escherichia coli* on injured iceberg lettuce in the field, overhead irrigated with contaminated water. J. Food Prot. 72, 458–464.

Bausum, H.T., Schaub, S.A., Kenyon, K.F., Mitchell, J., 1982. Comparison of coliphage and bacterial aerosols at a wastewater spray irrigation site. Appl. Environ. Microbiol. 43, 28–38.

Berger, C.N., Sodha, S.V., Shaw, R.K., Griffin, P.M., Pink, D., Hand, P., et al., 2010. Fresh fruit and vegetables as vehicles for the transmission of human pathogens. Environ. Microbiol. 12, 2385–2397.

Blumenthal, U.J., Duncan Mara, D., Peasey, A., Ruiz-Palacios, G., Stott, R., 2000. Guidelines for the microbiological quality of treated wastewater used in agriculture: recommendations for revising WHO guidelines. Bull. WHO 78, 1104–1116.

Boutin, P., Torre, M., Serceau, R., Rideau, P.J., 1988. Atmospheric bacterial contamination from land spreading of animal wastes: evaluation of the respiratory risk for people nearby. Agric. Engng. Res. 39, 149–160.

Brandl, M.T., 2008. Plant lesions promote the rapid multiplication of *Escherichia coli* O157:H7 on postharvest lettuce. Appl. Environ. Microbiol. 74, 5285–5289.

Brooks, J.P., Gerba, C.P., Pepper, I.L., 2004. Aerosol emission, fate, and transport from municipal and animal wastes. J. Residuals Sci. Technol. 1, 13–25.

Brooks, J.P., McLaughlin, M.R., Gerba, C.P., Pepper, I.L., 2012. Land application of manure and class b biosolids: An occupational and public quantitative microbial risk assessment. J. Environ. Qual. 41, 2009–2023.

Brooks, J.P., Tanner, B.D., Josephson, K.L., Gerba, C.P., Haas, C.N., Pepper, I.L., 2005a. A national study on the residential impact of biological aerosols from the land application of biosolids. J. Appl. Microbiol. 99, 310–322.

Brooks, J.P., Tanner, B.D., Gerba, C.P., Haas, C.N., Pepper, I.L., 2005b. Estimation of bioaerosol risk of infection to residents adjacent to a land applied biosolids site using an empirically derived transport model. J. Appl. Microbiol. 98, 397–405.

Brooks, J.P., McLaughlin, M.R., Scheffler, B., Miles, D.M., 2010. Microbial and antibiotic resistant constituents associated with biological aerosols and poultry litter within a commercial poultry house. Sci. Total Environ. 408, 4770–4777.

Brown, J.K.M., Hovmoller, M.S., 2002. Aerial dispersal of the global and continental scales and its impact on plant disease. Science 297, 537–541.

Butterworth, J., McCartney, H.A., 1991. The dispersal of bacteria from leaf surfaces by water splash. J. Appl. Bacteriol. 71, 484–496.

Camann, D.E., Moore, B.E., Harding, H.J., Sorber, C.A., 1988. Microorganism levels in air near spray irrigation of municipal wastewater: the Lubbock infection surveillance study. J. Water pol. Control Fed. 60, 1960–1969.

CDPH, 2007. Investigation of an *Escherichia coli* O157:H7 outbreak associated with Dole pre-packaged spinach. http://www.cdph.ca.gov/pubsforms/Documents/fdb%20eru%20Spnch%20EC%20Dole032007wph.pdf (accessed 19.05.13.).

Cevallos-Cevallos, J.M., Gu, G., Danyluk, M.D., van Bruggen, A.H.C., 2012a. Adhesion and splash dispersal of *Salmonella enterica* Typhimurium on tomato leaflets: effects of rdar morphotype and trichome density. Int. J. Food Microbiol. 160, 58–64.

Cevallos-Cevallos, J.M., Gu, G., Danyluk, M.D., Dufault, N.S., van Bruggen, A.H.C., 2012b. Salmonella can reach tomato fruits on plants exposed to aerosols formed by rain. Int. J. Food Microbiol. 158, 140–146.

Chetochine, A.S., Brusseau, M.L., Gerba, C.P., Pepper, I.L., 2006. Leaching of phage from class B biosolids and potential transport through soil. Appl. Environ. Microbiol. 72, 665–671.

Chinivasagam, H.N., Tran, T., Maddock, L., Gale, A., Blackall, P.J., 2009. The aerobiology of the environment around mechanically ventilated broiler sheds. J. Appl. Microbiol. 108, 1657–1667.

Cox, C.S., 1966. The survival of *Escherichia coli* atomized into air and into nitrogen from distilled water and from solution of protecting agents as a function of relative humidity. J. Gen. Microbiol. 43, 383–399.

Curriero, F.C., Patz, J.A., Rose, J.B., Lele, S., 2001. The association between extreme precipitation and waterborne disease outbreaks, in the United States, 1948–1994. Am. J. Public Health 91, 1194–1199.

Doyle, M.P., Erickson, M.C., 2008. Summer meeting 2007—the problems with fresh produce: an overview. J. Appl. Microbiol. 105, 317–330.

Dungan, R.S., Leytem, A.B., Verwey, S.A., Bjoneberg, D.L., 2010. Assessment of bioaerosols at a concentrated dairy operation. Aerobiologia 26 171–184.

Dungan, R.S., 2011. Fate and transport of bioaerosols associated with livestock operations and manures. J. Anim. Sci. 88, 2693–3706.

Fillhart, R.C., Bachand, G.D., Castello, J.D., 1997. Airborne transmission of tomato mosaic Tobamovirus and its occurrence in red spruce in the northeastern United States. Can. J. For. Res. 27, 1176–1181.

Fischer, J.R., Zhao, T., Doyle, M.P., Goldberg, M.R., Brown, C.A., Sewell, C.T., et al., 2001. Experimental and field studies of *Escherichia coli* O157:H7 in white-tailed deer. Appl. Environ. Microbiol. 67, 1218–1224.

Graham, D.C., Quinn, C.E., Bradley, L.F., 1977. Quantitative studies on the generation of aerosols of *Erwinia carotovora* var. *atroseptica* by simulated raindrop impaction on blackleg-infected potato stems. J. Appl. Bacteriol. 43, 413–424.

Graham, D.C., Harrison, M.D., 1975. Potential spread of *Erwinia* spp. in aerosols. Phytopathology 65, 739–741.

Greene, S.K., Daly, E.R., Talbot, E.A., Demma, L.J., Holzbauer, N., Patel, N.J., et al., 2008. Recurrent multistate outbreak of *Salmonella* Newport associated with tomatoes from contaminated fields, 2005. Epidemiol. Infect. 136, 157–165.

Hamilton, A.J., Stagnitti, F., Premier, R., Boland, A.M., Hale, G., 2006. Quantitative microbial risk assessment models for consumption of raw vegetables irrigated with reclaimed water. Appl. Environ. Microbiol. 72, 3284–3290.

Hanning, I.B., Nutt, J.D., Ricke, S.C., 2009. Salmonellosis outbreaks in the United States due to fresh produce: source and potential intervention measures. Foodborne Pathog. Dis. 6, 635–648.

Harrison, M.D., 1980. Aerosol dissemination of bacterial plant pathogens. Ann. N. Y. Acad. Sci. 353, 94–104.

Hutchison, M.L., Walters, L.D., Moore, T., Thomas, D.J., Avery, S.M., 2005. Fate of pathogens present in livestock wastes spread onto fescue plots. Appl. Environ. Microbiol. 71, 691–696.

Hutchison, M.L., Avery, S.M., Monaghan, J.M., 2008. The air-borne distribution of zoonotic agents from livestock waste spreading and microbiological risk to fresh produce from contaminated irrigation sources. J. Appl. Microbiol. 105, 848–857.

Islam, M., Morgan, J., Doyle, M.P., Phatak, S.C., Millner, P., Jian, X., 2004. Fate of *Salmonella enterica* serovar Typhimurium on carrots and radishes grown in fields treated with contaminated manure composts or irrigation water. Appl. Environ. Microbiol. 70, 2497–2502.

Jay, M.T., Cooley, D., Carychao, D., Wiscomb, G.W., Sweitzer, R.A., Crawford-Miksza, L., et al., 2007. *Escherichia coli* O157:H7 in feral swine near spinach fields and cattle, central California coast. Emerg. Infect. Dis. 13, 1908–1911.

Jones, A.M., Harrison, R.M., 2004. The effects of meteorological factors on atmospheric bioaerosol concentrations – a review. Sci. Total Environ. 326, 151–180.

Keraita, B., Konradsen, F., Drechsel, P., Abaidoo, R.C., 2007. Effect of low-cost irrigation methods on microbial contamination of lettuce irrigated with untreated wastewater. Trop. Med. Int. Health 12, 15–22.

Low, S.Y., Paez-Rubio, T., Baertsch, C., Kucharski, M., Peccia, J., 2008. Off-site exposure to respirable aerosols produced during the disk-incorporation of class B biosolids. J. Environ. Engng. 133, 987–994.

Mohr, A.J., 2001. Fate and transport of microorganisms in air. In: Hurst, C.J., Crawford, R.L., Knudsen, G.R., McInerney, M.J., Stetzenbach, L.D. (Eds.), Manual of Environmental Microbiology, second ed. ASM Press, Washington, DC.

Pachepsky, Y., Shelton, D.R., McLain, J.E.T., Patel, J., Mandrell, R.E., 2011. Irrigation waters as a source of pathogenic microorganisms in produce: a review. Adv. Agron. 113, 73–138.

Paez-Rubio, T., Viau, E., Romero-Hernandez, S., Peccia, J., 2005. Source bioaerosol concentration and rRNA gene-based identification of microorganisms aerosolized at a flood irrigation wastewater reuse site. Appl. Environ. Microbiol. 71, 804–810.

Paez-Rubio, T., Ramrui, A., Sommer, J., Xin, H., Anderson, J., Peccia, J., 2007. Emission rates and characterization of aerosols produced during the spreading of dewatered class B biosolids. Environ. Sci. Technol. 41, 3537–3544.

Pearce, R.A., Sheridan, J.J., Bolton, D.J., 2006. Distribution of airborne microorganisms in commercial pork slaughter processes. Int. J. Food Microbiol. 107, 186–191.

Perombelon, M.C.M., Lowe, R., 1973. Bacterial soft rot and blackleg potato. Rep. Scott. Hort. Res. Inst. 1972, 52–53.

Pillai, S.D., 2007. Bioaerosols from land-applied biosolids: issues and needs. Water Environ. Res. 79, 270–278.

Posch, J., Feirei, G., Wuest, G., Sixi, W., Schmidt, S., Haas, D., et al., 2007. Transmission of *Campylobacter* spp. in a poultry slaughterhouse and genetic characterization of the isolates by pulsed-field gel electrophoresis. Br. Poult. Sci. 47, 286–293.

Scallan, E., Hoekstra, R.M., Angulo, F.J., Tauxe, R.V., Widdowson, M.A., Roy, S.L., et al., 2011. Foodborne illness acquired in the United Sates-Major pathogens. Emerg. Infect. Dis. 17, 7–15.

Soderstrom, A., Osterberg, P., Lindqvist, A., Jonsson, B., Lindberg, A., Blide-Ulander, S., et al., 2008. A large *Escherichia coli* O157 outbreak in Sweden associated with locally produced lettuce. Foodborne Pathog. Dis. 5, 339–349.

Solomon, E.B., Pang, H.J., Matthews, K.R., 2003. Persistence of *Escherichia coli* O157:H7 on lettuce plants following spray irrigation with contaminated water. J. Food Prot. 66, 2198–2202.

Song, I., Stine, S.W., Choi, C.Y., Gerba, C.P., 2006. Comparison of crop contamination by microorganisms during subsurface drip and furrow irrigation. J. Environ. Engr. 132, 1243–1248.

Stine, S.W., Song, I., Choi, C.Y., Gerba, C.P., 2005. Application of microbial risk assessment to the development of standards for enteric pathogens in water used to irrigate fresh produce. J. Food Prot. 68, 913–918.

Teltsch, B., Kedmi, S., Bonnet, L., Borenzstajn-Rotem, Y., Katzenelson, E., 1980. Isolation and identification of pathogenic microorganisms at wastewater-irrigated fields: ratios in air and wastewater. Appl. Environ. Microbiol. 39, 1183–1190.

Theunissen, H.J.J., Lemmens-Den Toom, N.A., Burggraaf, A., Stolz, E., Michel, M.F., 1993. Influence of temperature and relative humidity on the survival of *Chlamydia pneumoniae* in aerosols. Appl. Environ. Microbiol. 59, 2589–2593.

USFDA, 1998. Guide to minimize microbial food safety hazards for fresh fruits and vegetables. U.S. Department of Health and Human Services, Food and Drug Administration.

USFDA, 2006. FDA statement on foodborne *E. coli* O157:H7 outbreak in spinach. USFDA, Washington, D.C. Available at: http://www.fda.gov/bbs/topics/NEWS/2006/NEW01489.html (accessed 19.05.12.).

Warriner, K., Huber, A., Namvar, A., Fan, W., Dunfield, K., 2009. Recent advances in the microbial safety of fresh fruits and vegetables. Adv. Food Nutr. Res. 57, 155–208.

Wendel, A.M., Johnson, D.H., Sharapov, U., Archer, J.R., Monson, T., Koschmann, C., et al., 2009. Multistate outbreak of *Escherichia coli* O157:H7 infection associated with consumption of packaged spinach, August-September 2006: the Wisconsin investigation. Clin. Infect. Dis. 48, 1079–1086.

Zaleski, K.J., Josephson, K.L., Gerba, C.P., Pepper, I.L., 2005. Potential regrowth and recolonization of Salmonellae and indicators in biosolids and biosolids-amended soil. Appl. Environ. Microbiol. 71, 3701–3708.

Water Quality

Charles P. Gerba, Channah Rock

Department of Soil, Water and Environmental Science, University of Arizona, Tucson, AZ

CHAPTER OUTLINE

Introduction

Water not only plays an essential role in the growth of produce, but also in cleaning and hygienic uses pre- and postharvest. Ensuring the microbial quality of this water is critical to prevent the contamination of the produce by waterborne enteric pathogens. Water has always played a key role as a vehicle for the transmission of pathogens transmitted by the fecal-oral route. For more than a century it has been recognized that fecally contaminated water, either used for drinking or irrigation of food crops, can result in the transmission of enteric pathogens. Conventional treatment of drinking water (filtration and disinfection) has eliminated diseases such as cholera and typhoid in the developed world; however, waterborne outbreaks still occur because of treatment or distribution system failures. This is because raw water sources are always subject to contamination by animals and sewage discharge. At the same time that modern drinking water treatment began, the use of raw sewage for food crop irrigation was banned or severely restricted. However, in arid regions with limited water resources, guidelines for the use of treated waste water for food crops traditionally eaten raw have been developed, although in the developing world they are seldom enforced.

Several produce outbreaks have been known or suspected to have arisen from contamination in the field, suggesting contamination by irrigation or during handling (Dentinger et al., 2001; CDC, 2003; Nyard et al., 2008; Soderstrom et al., 2008; Crawford et al., 2011; Gelting et al., 2011). Perhaps more significant may be the low-level transmission of viruses and protozoan parasites by food from contaminated irrigation water. Quantitative microbial risk assessment has suggested that low levels of virus in irrigation water can result in a significant level of risk of infection to consumers (Petterson et al., 2001). Stine et al. (2005) estimated that less than one hepatitis A virus per 10 liters in irrigation water could result in a risk exceeding 1:10,000 per year, considering the efficiency of transfer of the virus to crop and survival until harvest time. The 1:10,000 risk of infection per year is currently the acceptable level used by the U.S. Environmental Protection Agency (EPA) for drinking water (Regli et al., 1991).

Contamination of produce may also occur through the use of contaminated water to apply pesticides, fertilizers, washing, hydrocooling, handwashing, and icing.

Irrigation water

The largest user of fresh water in the world is agriculture, with more than 70% being used for irrigation. About 240 million hectares, 17% of the world's cropland, are irrigated, producing one-third of the world's food supply (Shanan, 1998). Nearly 70% of this area is in developing countries. Irrigation with sewage or sewage-contaminated waters in developing countries is fairly common and usually not regulated. Although guidelines for waste water reuse have been developed by the World Health Organization (WHO, 2006), their application in developing countries will remain difficult, due to inadequate institutional capability and general lack of financial resources.

The World Health Organization estimates that 10% of the world's population consumes food that is irrigated with untreated waste water (WHO, 2006), amounting to about 20 million hectares of crop land (Scott et al., 2004). Significant irrigation with waste water of food crops occurs near the major cities in Peru and Bolivia in Latin America (Scott et al., 2004). In Pakistan sewage is used directly for irrigation of vegetables commonly eaten raw (Ensink, 2004). In many other parts of the world river water contaminated with untreated or disinfected sewage is used for food crop irrigation (Knox et al., 2010; Drechsel et al., 2011; Kittigul et al., 2012; Castro–Rosas et al., 2012; Park et al., 2012).

Nationally, Arizona ranked third in 2011 in the production of fresh market vegetables and second to only California in the production of head lettuce, leaf and romaine lettuce, broccoli, cauliflower, spinach, cantaloupes, and honeydews. Leafy greens alone represent approximately $652 million in farm income and are the highest value crop in the state of Arizona (Arizona Farm Bureau, 2011). All these crops are grown almost entirely by irrigated agriculture. It is thus surprising that we know little about the microbial quality of this water. Most studies have dealt with the occurrence and fate of enteric pathogens in reclaimed water used for irrigation and not the quality of surface waters currently in use. Little data exists on the occurrence of pathogens in irrigation waters, which do not intentionally receive sewage discharges.

There are few published studies on the quality of non-reclaimed waste water used as an irrigation source (Steele and Odumeru, 2004). Irrigation agriculture requires approximately two acre-feet of water per acre of growing crops. The frequency and volume of application must be carefully programmed to compensate for deficiencies in rainfall distribution and soil moisture content during the growing season. Rivers and streams are tapped by large dams and then diverted into extensive canal systems. Ground water is pumped from wells into canals (which places the water at risk from surface contamination), and catchments are constructed to trap storm-water runoff. Because water availability is often critical, little attention is given to the microbial quality of the irrigation runoff. In water-short areas, available sources are subject to contamination by sewage discharge from small communities (unplumbed housing along canals in developing countries is common), cattle feedlot drainage, animals grazing along canal embankments, storm-water events, and return irrigation water (non-infiltrated water from the field being irrigated is returned to the irrigation channels) (Table 6.1). Since irrigation channels are frequently small, these occurrences of pollution discharge can result in rapid deterioration of water quality.

Currently, no microbial indicator standards exist for irrigation waters used for produce production in the U.S. It has been suggested by the produce industry that the bathing water standard guideline (126 *E. coli*/100 ml) established by the EPA be used (EPS, 1973). This guideline was developed from epidemiological studies of bather exposure in recreational waters and has no direct relationship to risk associated with infection or illness rates that might result from produce irrigation waters. Thus, there is currently no scientific basis for its use in irrigated agriculture.

Table 6.1 Factors Influencing Water Quality in Man-Made Irrigation Channels

Factor	Remarks
Rainfall	Rain can act to resuspended sediments containing microbes; wash in fecal matter from bank sides and enhance drainage.
Storm-water drainage	In some areas canals may receive storm-water flows from streets and grazing land.
Return flows	Water remaining from flood-field irrigation may be returned to canals and used to irrigate other fields.
Water fowl	Large numbers of water fowl may contribute bacterial pathogens such as *Campylobacter*.
Animal occurrence bank sides	Animal grazing may occur on bank sides of irrigation channels. In some areas the bank sides are used as urban parks and pet feces may be washed into canals during rainfall events.
Urban areas	Runoff during storms may be diverted to irrigation canals.
Channel size and depth	The impact of storm events or other contamination events are greater in smaller channels because of less water for dilution.
Distance of canal from water source	The greater the distance from the water source, the greater the chance for contamination.
Recreational use of source water	Bathers can contribute significant amounts of pathogens if the lakes or other bodies of water are used for recreational purposes.
Season	Temperature controls survival of pathogens. Greater UV light intensity may adversely affect survival.

Water quality standards for irrigation water

Traditionally bacterial indicators of fecal contamination have been used to assess fecal contamination of water. This is because of the difficulty, cost, and time needed to detect waterborne pathogens. Coliform, fecal coliform (thermotolerant coliform), and *Escherichia coli* have been used for this purpose. Since coliform and fecal coliform bacteria can originate from non-fecal sources *E. coli* is a more specific indictor of fecal contamination. Most of the research on enteric pathogen contamination of vegetables and fruits during production has been done to evaluate the safety of reclaimed waste water irrigation. Many states in the United States have standards for the treatment of reclaimed water to be used for food crop irrigation (Asano, 1998), and the World Health Organization has also made recommendations (WHO, 2006). The state of California requires advanced physical-chemical treatment and extended disinfection to produce virus-free effluent. A total coliform standard less than 2/100 mL must also be met (Asano, 1998). The state of Arizona had a virus standard of one plaque-forming unit per 40 L and *Giardia* cysts of one per 40 L in addition to a fecal coliform standard of 25 per 100 mL (Rose and Gerba, 1991). Currently, in the state of Arizona, class A+ water (treated waste water that has less than 10 mg/L of total nitrogen and no detectable fecal coliform bacteria) can be used for irrigation of food crops by drip- or furrow-irrigation. Spray-irrigation is not allowed, and the food crop must not be allowed to come in contact with the water product (ADEQ, 2013). Although these standards for the use of reclaimed waste water exist for food crops eaten raw in the United States, irrigation using reclaimed water for crop-irrigation is seldom practiced. In developing countries, raw or partially treated waste water is often used to irrigate crops, especially in arid regions.

One of the few early studies conducted on irrigation waters documented the wide range in microbial quality of this water (Geldreich and Bordner, 1971). The wide variation was attributed to the discharge of domestic sewage into streams from which the irrigation water was obtained. This study was conducted in the western United States (Wyoming, Utah, and Colorado). Median values of fecal coliform bacteria ranged from 70 to 450,000/100 ml. Based on results obtained for *Salmonella* occurrence in the same waters, they recommend a fecal coliform standard for irrigation waters of 1000/100 mL.

Guidelines for the microbial water quality of surface water tend to be more lenient than those for waste water because of the belief that enteric viruses and other human pathogens are less likely to be present or less numerous (Steele and Odumeru, 2004). The criteria range from less than 100 to less than 1000 fecal coliforms per 100 mL. Guidelines also include criteria for *Escherichia coli* and fecal streptococcus (Steele and Odumerus, 2004).

Recently the Arizona and California Leafy Greens Marketing Agreements (LGMA) were adapted in an effort to minimize the risk of contamination of leafy greens similar to those established for recreational waters, which are based on *E. coli* concentrations. These standards were based on epidemiological studies, which indicated that the probability of gastrointestinal infection from recreational bathers in freshwater was 1.9% when *E. coli* concentrations averaged 126/100 ml (Kay and Wyuer, 1992). The primary agent causing the illness was unknown, but believed to be viral-based symptomology. How appropriate such a standard is to produce crops remains to be determined. At least one study has suggested that this standard has no correlation between *E. coli* carrying toxin-producing genes in irrigation source water (Shelton et al., 2011).

Without comprehensive data on the relative occurrence of pathogens and indicator bacteria in irrigation waters, it is difficult to say if such standards are truly reflective of risk to produce crops. Given the greater survival of enteric viruses and protozoan parasites in water, current suggested *E. coli* standards may not be reflective of these waterborne pathogens. Also, a scientific basis for sampling location and frequency is not yet available. Won et al. (2013) in their studies of the spatial-temporal variations of *E. coli* in reservoirs and irrigation canals found that a single water sample was not reflective of the quality of the water over time, and that rainfall events had a major impact on the water quality.

Occurrence of pathogens in irrigation water

The microbial quality of irrigation water depends on the source of the water and contamination as it is transmitted through the delivery system. Sources of human enteric pathogens may involve sewage discharges into source water, septic tanks, and recreational bathers, for example (Table 6.1). In the United States, disinfection of waste water effluents is required before discharge into surface waters, greatly reducing the risks from enteric bacterial and viral pathogens. However, this is not a common practice in much of the world, including Europe. Humans are believed to be the only significant source of enteric viruses in water, although hepatitis E virus may be the exception. Zoonotic pathogens may not only originate from domestic animals, but also from wild animals such as migrating water fowl.

Although ground water is often considered a microbially safe source for irrigation water, it can also be a source of waterborne pathogens. In a study of ground water used for irrigation of vegetables in Thailand, 27% of the samples yielded enteric viruses (Cheong et al., 2009). Studies in the United States have indicated that 8 to 31% of the ground waters may contain viruses (Abbaszadegan et al., 2003; Borchardt et al., 2003). Because of their small size and long-term survival, enteric viruses are more likely to contaminate ground water. However, at least one produce outbreak involving *E. coli* O157:H7 has been attributed to the use of contaminated ground water used for irrigation (Gelting et al., 2011). Pathogens in ground water may originate from septic discharges; leaking sewer lines; or infiltration from lakes, rivers, or oxidation ponds.

Several outbreaks have been linked to the use of contaminated irrigation water. In a multistate outbreak of *Salmonella* Newport involving tomatoes, the strain involved in the outbreak was detected in the water used to irrigate the fields in which the tomatoes were grown (Greene et al., 2008). Irrigation water was also believed to be involved in another *Salmonella* outbreak involving lettuce grown in Italy (Nygard et al., 2008). In Sweden, an outbreak of *E. coli* 0157:H7 associated with locally grown lettuce was traced to the use of river water contaminated by local farm animals (Soderstrom et al., 2008). It was suggested in an outbreak of *E. coli* O145 that contamination of romaine lettuce was associated with contaminated irrigation water (Taylor, 2011). A multi-state outbreak of *E. coli* O157:H7 from spinach was believed due to the use of contaminated ground water (Gelting et al., 2011).

Several studies have reported the occurrence of enteric pathogens in irrigation water (Table 6.2). *Salmonella* and the enteric protozoan parasites *Giardia, Cryptosporidium,* and microsporidia have been reported in irrigation waters used for produce. Recently an outbreak of gastroenteritis associated with microsporidium contaminated cucumbers occurred in Sweden, where preharvest contamination was suggested as the likely source of this pathogen (Decrene et al., 2012). Human adenoviruses, enteroviruses, rotaviruses, and noroviruses have also been detected in both surface and ground waters used for irrigation of produce (Table 6.2). *Salmonella* has many potential animal sources including both warm- and cold-blooded animals (Garcia-Villanova et al., 1987). A study of irrigation waters used for tomato farms in the mid-Atlantic region of the United States found *Salmonella* in 2/13 samples or ~15% (Micallef et al., 2012). In a more extensive study in Arizona *Salmonella* was detected in 30/112 (26.7%) (Kayed, 2004). *Campylobacter* and toxigenic *E. coli* have also been reported in irrigation waters. In a large irrigation system in Yaqui Mexico not receiving sewage discharges, *E. coli* was detected in only 4% of the samples with a range of 1 to 22,325/100 ml; coliphage were detected in 30% of the samples, *Giardia, Cryptosporidium*, and enteroviruses were also detected at low levels (Gortares–Moroyoqui et al., 2011).

It is likely that any irrigation water from surface sources or that passes through open channels can be expected to contain waterborne pathogens at one time or another, given the potential large number of animal sources. In developed countries, where sewage treatment is required, risks are most likely related to environmental events, such as rainfall, where runoff into water sources can dramatically increase the level of pathogens into the water. While numerous studies have been conducted on the fate and transport of pathogens in watersheds, similar studies have yet to be done on man-made irrigation systems. Predictive models for microbial water quality would be useful, but accurate predictions may be difficult given the heterogeneity of inputs and factors that influence fate and transport (Pachepsky et al., 2011).

Contamination of produce during irrigation

The likelihood of the edible parts of a crop becoming contaminated by contaminated irrigation water depends upon a number of factors including growing location, type of irrigation application, and nature of the produce surface (Table 6.3). If the edible part of the crop grows on or near the soil surface, it is more likely to become contaminated than a fruit growing in the aerial parts of a plant (Manshadi et al., 2013). Some produce surfaces are furrowed or have other sources that may

Table 6.2 Occurrence of Pathogens in Irrigation Water

Study	Location	Results	Remarks
Robertson and Gjerde, 2001	Norway	*Salmonella* detected in irrigation water used for bean spouts	
Thurston–Enriquez et al., 2002	Arizona and Central America	*Giardia, Cryptosporidium* and microsporidia detected	
Okafo et al., 2003	Nigeria	*Salmonella* detected in 2 to 14% of irrigation water used for produce production	
Gannon et al., 2004a	Western Canada	*E. coli* 0157:H7 and *Salmonella* detected	
Chaidez et al., 2005	Western Mexico	*Giardia* detected in 48% of samples and *Cryptosporidium* in 50%	
Duffy et al., 2005	Texas	*Salmonella* detected in irrigation water	
Esponoza–Medina et al., 2006	Brazil	*Salmonella* in 23.5% of irrigation water used to irrigate cantaloupe	*Salmonella* also detected in 9.1% of ground water samples and 4.8% of chlorinated water samples
Van Zyl et al., 2006	South Africa	Human rotaviruses detected in 14% of irrigation water samples used on raw vegetables	Rotavirus also detected on 1.4% of vegetable samples Materon et al., 2007
Materon et al., 2007	Southern Texas	*Salmonella* detected in river water used to irrigate cantaloupes	Washing melons with chlorine containing water did not eliminate the *Salmonella*
Izumi et al., 2008	Japan	*E. coli* 0157:H7 in irrigation water used to irrigate persimmons	*Salmonella* also isolated in pesticide spray water made from irrigation water
Mota et al., 2009	Sinaloa, Mexico	*Giardia* geometric mean 32.94/100 liters; *Cryptosporidium* 82.34/100 liters	Used for produce irrigation
Gortares et al., 2011	Sonora, Mexico	Enteroviruses average 0.05 MPN/100 liters; *Giardia* 3.5/100 liters; *Cryptosporidium* 1.78/100 liters	Used for produce irrigation
Kokkinos et al., 2012	Europe	Adenoviruses detected in 28% of samples; norovirus 4%	Used for production of leafy greens

Table 6.3 Factors Affecting the Contamination of the Edible Parts of Plants during Irrigation

Growing location of the edible portion of the plant
- distance from the soil or water surface

Frequency of irrigation
- number of days last irrigated before harvest

Surface of the edible portion
- smooth
- webbed
- rough

Type of irrigation method
- furrow or flood
- sprinkler
- drip
  - surface
  - subsurface

retain water (e.g., tomato vs. cantaloupe). There are four distinct methods of irrigation: sprinkler systems, gravity-flow systems (flood-irrigation), drip or trickle methods, and subsurface-irrigation. The number of farms in the United States in 2011 is estimated at 2.2 million, with the land in farms totaling 917 million acres (Arizona Department of Agriculture, 2011). According to the 2008 Farm and Ranch Irrigation Survey, the vast majority of water applied to horticultural crops in 2008 was applied by sprinkler-irrigation. The second most common method was drip-, trickle-, or low-flow-irrigation (USDA, 2008). In a recent survey in the United Kingdom, it was found that overhead irrigation was the predominant method for fruits and vegetables (Tyrrel et al., 2006). Worldwide, less than 1% of the total irrigated area is believed to be irrigated by drip-irrigation (Postel et al., 2001). However, drip-irrigation is growing rapidly around the world, covering almost 3,000,000 hectares (Postel et al., 2001).

The type of irrigation method can greatly influence the degree of crop contamination. For example, the degree of PRD-1 virus transference to lettuce was found to be 4.4%, 0.02%, and 0.00039% for spray-, furrow-, and drip-irrigation (Stine et al., 2005a, 2005b; Choi et al., 2004). The reduction in contamination by bacteria (*Escherichia coli*) on lettuce by use of subsurface drip- versus flood-irrigation was 99.9%; however, it was only 99% for virus (PRD-1) (Song et al., 2006). Stine et al. (2005a) compared surface and subsurface irrigation as sources of contamination of cantaloupe, iceberg lettuce, and bell peppers when the water was seeded with coliphage PRD-1 under field-growing conditions in Arizona. Coliphage was detected on both the lettuce and cantaloupe, but not on the bell peppers.

Oron et al. (1995) applied irrigation water containing up to 1000 plaque-forming units/mL of a vaccine strain of poliovirus to tomato plants by subsurface drip irrigation in an outdoor setting in Israel. Some virus was detected in the leaves of the plants, but not in the fruits. The authors stated that the high content of the virus in the water might explain the occurrence of the virus in the leaves. No virus was detected in plants irrigated with waste water containing the same level of virus. The authors suggested that this might be due to the interaction of the virus with particulate or soluble matter present in the waste water, but absent in the irrigation water, preventing their entrance into the roots.

Alum et al. (2011) studied the effectiveness of drip irrigation in the control of viral contamination of salad crops (tomato and cucumber) in a greenhouse in potted plants. The plants were irrigated with secondary effluent using surface drip and sub-surface irrigation. Irrigation water was periodically seeded with coliphages MS-2, PRD-1, poliovirus type 1, adenovirus 40, and hepatitis A virus. Surface-irrigation always resulted in the surface contamination of both the aboveground and the underground parts of the plants. In lettuce it was observed that only the outer leaves of the plant became contaminated. No contamination of the plants occurred when subsurface drip-irrigation was used. No systemic uptake of viruses was observed in any of the crops.

Choi et al. (2004) assessed viral contamination of lettuce by surface and sub-surface drip irrigation using coliphage MS-2 and PRD-1. A greater number of coliphages was recovered from the lettuce in the subsurface plots as compared to those in the furrow-irrigated plots. Shallow drip-tape installation and preferential water paths through cracks on the soil surface appeared to be the main causes of high viral contamination. In subsurface drip-irrigation, penetration of the irrigation water to the soil surface led to direct contact with the lettuce stems. Thus, drip-tape depth can influence the probability of produce contamination. Greater contamination by PRD-1 was observed, which might be due to its longer survival time.

A number of recent studies have also documented the contamination of produce by bacterial pathogens from spray-irrigation. Kim et al. (2012) found that contamination of the edible parts of lettuce was greater from spray-irrigation than flood-irrigation. Oliveria et al. (2012) found similar results evaluating contamination of lettuce using *E. coli* 0157:H7. However, most studies seem to indicate that the enteric bacteria die off at a fairly rapid rate on the surface of the produce after spray-irrigation due to desiccation, sunlight, and other environmental factors (Oliveria et al., 2012). For example, Ingram et al. (2011) found that after spray-irrigation the *E. coli* 0157:H7 decreased in titer by 99.9% within a 24-hour period. Survival is greater on roots and other more protected areas of the crops, thus persistence is likely crop-specific. It has been shown that soil splash is another way that produce can become contaminated either by spray-irrigation or rainfall events (Jacobsen et al., 2012; Monaghan and Hutchison, 2012; Cevallos–Cevallos et al., 2012). Pathogens usually survive longer in soil than on plants (Wood et al., 2010), and so contamination of edible parts of the plants can occur sometime after irrigation with contaminated water.

Root uptake of enteric pathogens and internalization has been an area of a great deal of recent research with varying results due to differences in experimental design, systems tested, pathogens, and the types of crops studied (Hirneisen et al., 2012). This is most easily demonstrated when plants are grown hydroponically (Wei et al., 2011). However, the internalization and contamination is usually only demonstrated when very high numbers of organisms are present in the irrigation water (~>1,000,000/ml).

Such high levels of bacterial or viral pathogens are unlikely to occur in even sewage-contaminated irrigation waters. Risk assessments are needed to determine if any significant risks exist from internalization of pathogens from irrigation waters.

Survival of pathogens on produce in the field

Studies on the survival of viruses on produce postharvest indicate that little inactivation occurs because of the low temperatures of storage (Seymour and Appleton, 2001). Tierney et al. (1977) found that poliovirus survived on lettuce for 23 days after flooding of outdoor plots with waste water. The virus persisted in the soil for two months during the winter and two to three days during the summer months. Sadovski et al. (1978) spiked waste water and tap water used to irrigate cucumbers with a high titer of poliovirus. They were able to detect the virus on the cucumbers grown with either (1) surface drip-irrigation or (2) the soil and drip lines covered with polyethylene sheets, although the virus was detected only occasionally on the cucumbers irrigated with plastic covering the soil, right after irrigation. Viruses on the soil-irrigated cucumbers survived for at least 8 days after irrigation.

Hepatitis A virus and coliphage PRD-1 survival on growing produce was found to be similar under high and low humidity conditions (Stine et al., 2005). In general, the inactivation rates of these viruses were lower than those of *E. coli* 0157:H7, *Shigella sonnei*, and *Salmonella enterica* on cantaloupe, lettuce, and bell peppers. The hepatitis A virus was reduced about 90% after 14 days, indicating that they could survive from an irrigation event to harvest time.

Survival of bacterial pathogens on the crop appears to be less than the soil, although this may be crop-specific (Wood et al., 2010). *E. coli* 0157:H7 appears to be limited to a few days or weeks on plant surfaces, but persist much longer in the soil (Wood et al., 2012; Ingram et al., 2011). However, this is related to the degree of contamination. Solomon et al. (2003) found that if lettuce was spray-irrigated with 10,000 *E. coli* 0157:H7/ml the bacterium could be detected after 30 days. *Salmonella* and *E. coli* are also capable of growth on produce surfaces preharvest under certain conditions (Stine et al., 2005a). Islam et al. (2004) observed survival of *Salmonella* for weeks on carrots and radishes after irrigation with artificially contaminated irrigation water. Also using artificially contaminated irrigation water containing *Giardia* cysts and *Cryptosporidium* oocysts, Armon et al. (2002) readily contaminated vegetables during flood-irrigation. Of all the vegetables studied, the highest prevalence of oocysts occurred on zucchini.

Other sources

Use of fecally contaminated water for application of pesticides or in wash water may lead to produce contamination (Stine et al., 2011). The use of untreated water to make up pesticide spray has been suspected as a source of enteric pathogens, and Izumi et al. (2008) reported the isolation of *Salmonella* in pesticide spray used on persimmons, in which irrigation water was used to dilute the pesticide before application. The quality of process water impacts the effectiveness of washing (Allende et al., 2008). Keraita and Drechsel (2004) reported that wash water used for produce in Ghana was an important source of contamination with enteric bacteria. Even low amounts of contaminated product in a batch impacted the safety of the entire lot that passed through a washing tank. The risk of cross-contamination is not necessarily eliminated by using large quantities of water (López–Gálvez et al., 2009). That is why it is important to use a sanitizing agent in the process water to kill microbes before they attach or become internalized in the produce, avoiding cross-contamination. However, this is only effective with proper amounts of sanitizer and maintenance of process equipment (e.g., washing baths) (Gil et al., 2009; Holvoet et al., 2012).

A hydrocooler was found to be a source of melon rinds with fecal coliforms and fecal enterococci (Gagliardi et al., 2003). The use of ice for hydrocooling after harvesting may be another source of contamination (Cannon et al., 1991).

Finally, it has been observed that use of contaminated hand wash water, resulting from multiple users by different individuals, can result in hand contamination by enteric bacteria (Ogunsola and Adesiji, 2008).

SUMMARY AND CONCLUSIONS

The role water plays in the contamination of produce has been little studied, despite its potential significance. A recent review of preharvest risk factors of microbial contamination of fruits and vegetables concluded that the existing literature suggests that reducing microbial contamination of irrigation water and soil are the most promising targets for prevention and control of produce contamination (Oark et al., 2012). Information on the occurrence of bacterial indicators of fecal contamination and pathogens in irrigation water and potential sources of irrigation water contamination is needed so the risks can be better defined. We need to understand the ecology of enteric pathogens and indicator bacteria in terms of transport and survival of enteric pathogens in complex irrigation delivery systems to better define the risks to produce. Meaningful standards for indicator bacteria that better assess the risk of produce contamination and risk of infection to the consumer need to be developed. Irrigation methods and the type of produce affect the degree of contamination. Spray-irrigation of produce is common, and this offers the greatest potential for contamination of the edible parts of the produce. Although the percentage of pathogen transfer from contaminated water to produce by some types of irrigation methods (e.g., drip-irrigation) may be low, risks can still be considered significant because of the low numbers of some enteric pathogens, such as viruses, necessary to cause infection (Peterson and Ashbolt, 2001). Environmental conditions, such as temperature and humidity, may determine the survival of viruses on produce surfaces. Limited studies suggest that enteric viruses can survive on produce longer than enteric bacteria, and that such viruses introduced on produce surfaces at the time of irrigation can survive through harvesting. Similar data needs to be developed for foodborne protozoan parasites.

References

Abbaszadegan, M., LeChevallier, M., Gerba, C.P., 2003. Occurrence of viruses in US ground water. J. Am. Water Works Assoc. 95, 107–120.

Allende, A., Selma, M.V., López-Gálvez, F., Villaescusa, R., Gil, M.I., 2008. Impact of wash water quality on sensory and microbial quality, including Escherichia coli cross contamination, of fresh-cut escarole. J. Food Prot. 71, 2514–2518.

Alum, A., Enriquez, C., Gerba, C.P., 2011. Impact of drip irrigation method, soil, and virus type on tomato and cucumber contamination. Food Environ. Virol. 3, 78–85.

Arizona Farm Bureau, 2011. Arizona Agricultural Statistics Bulletin. Phoenix, AZ http://www.nass.usda.gov/Statistics_by_State/Arizona/Publications/Bulletin/index.asp.

ADEQ. Arizona Department of Environmental Quality, Reclaimed Water Classifications and Permits. Accessed June 13, 2013: http://www.azdeq.gov/environ/water/permits/reclaimed.html#class.

Armon, R., Gold, D., Brodsky, M., Oron, G., 2002. Surface and subsurface irrigation with effluents of different qualities and presence of Cryptosporidium oocysts in soil and on crops. Water Sci. Technol. 49, 115–122.

Asano, T., 1998. Waste water and reclamation reuse. Technomic Publishing, Lancaster, P.A..

Borchardt, M.A., Bertz, P.D., Spencer, S.K., Battigelli, D.A., 2003. Incidence of enteric viruses in ground water from household wells in Wisconsin. Appl. Environ. Microbiol. 69, 1172–1180.

Cannon, R.O., Hirschhorn, J.R.B., Rodeheaver, D.C., et al., 1991. A multistate outbreak of Norwalk virus gastroenteritis associated with consumption of commercial ice. J. Infect. Dis. 164, 860–863.

Castro-Rosas, J., Cerna-Cortes, J.F., Mendez-Reyes, E., et al., 2012. Presence of faecal coliforms, Escherichia coli and diarrheagenic E. coli pathotypes in ready-to-eat salads, from an area where crops are irrigated with sewage water. Int. J. Food Microbiol. 156, 176–180.

CDC. Centers for Disease Control and Prevention, 2003. Hepatitis A outbreak associated with green onions at a restaurant—Monaca, Pennsylvania, (2003). MMWR 52, 1155–1157.

Cevallos-Cevallos, J.M., Danyluk, D., Gy, G., Vallad, G.E., van Bruggen, A.H.C., 2012. Dispersal of Salmonella typhimurium by rain splash onto tomato plants. J. Food Prot. 75, 472–479.

Chaidez, C., Soto, M., Gortares, P., Mena, K., 2005. Occurrence of Cryptosporidium and Giardia in irrigation water and its impact on fresh produce industry. Int. J. Environ. Health Res. 15, 339–345.

Cheong, S., Lee, C., Song, S.W., et al., 2009. Enteric viruses in raw vegetables and groundwater used for irrigation in South Korea. Appl. Environ. Microbiol. 75, 7745–7751.

Choi, C., Song, I., Stine, S., et al., 2004. Role of irrigation and wastewater: Comparison of subsurface irrigation and furrow irrigation. Water Sci. Technol. 50, 61–68.

Crawford, W.M., 2011. Escherichia coli O145 outbreak associated with romaine lettuce: environmental assessment. Presented at International Association for Food Protection Annual Meeting, August 2, 2011. Milwaukee, WI.

Decrene, V., Lebband, M., Botero-Kleiven, S., Gustavsson, A.M., Lofdahl, M., 2012. First reported foodborne outbreak associated with microsporidia, Sweden, October 2009. Epidemiol. Infect. 140, 519–527.

Dentinger, C., Bower, W.A., Nainan, O.V., et al., 2001. An outbreak of hepatitis A associated with green onions. J. Infect. Dis. 183, 1273–1276.

Drechsel, P., Scott, C.A., Raschid-Sally, L., Redwood, M., Bahri, A., 2011. Wastewater irrigation and health. Earthscan, London.

Duffy, E.A., Lucia, L.M., Kells, J.M., et al., 2005. Concentrations of *Escherichia coli* and genetic diversity and antibiotic resistance profiling *Salmonella* isolated from irrigation water, packing shed equipment, and fresh produce in Texas. J. Food Prot. 68, 70–79.

EPA, 1973. *Water quality criteria.* Environmental Protection Agency. DC Ecological Research Series, EPA R3-73-033, Washington.

Ensink, J., Mahmood, T., van der Hoek, W., et al., 2004. A nation-wide assessment of wastewater use in Pakistan: An obscure activity or a vitally important one? Water Policy. 6, 1–10.

Esponoza-Medina, I.E., Rodriguez, F.J., Vargas-Arispuro, I., Islas-Osuna, M.A., 2006. PCR identification of *Salmonella*: Potential contamination sources from production and postharvest handling of cantaloupes. J. Food Prot. 69, 1422–1425.

Gagliardi, J.V., Millner, P.D., Lester, G., Ingram, D., 2003. On-farm and postharvest processing sources of bacterial contamination to melon rinds. J. Food Prot. 66, 82–87.

Gannon, V.P.J., Graham, T.A., Read, S., et al., 2004. Bacterial pathogens in rural water supplies in southern Alberta, Canada. J. Toxicol. Environ. Health A. 67, 1643–1653.

Garcia-Villanova, B., Ruiz, C., Espinar, A., Bolanos, S., Carmona, M.J., 1987. A comparative study of strains of *Salmonella* isolated from irrigation waters, vegetables and human infections. Epidemiol. Infect. 98, 271–276.

Geldreich, E.E., Bordner, R.H., 1971. Fecal contamination of fruits and vegetables during cultivation and processing for market. J. Milk. Food Technol. 34, 184–198.

Gelting, R.J., Baloch, M.A., Zarate-Bermudez, M.A., Selman, C., 2011. Irrigation water issues potentially related to the 2006 multistate *E. coli* O157:H7 outbreak associated with spinach. Ag. Water Manag. 98, 1395–1402.

Gil, M.I., Selma, M., Lopez-Galvez, Allende, A., 2009. Fresh-cut sanitation and wash water disinfection: problems and solutions. Int. J. Food. Microbiol. 134, 37–45.

Greene, S.K., Daly, E.R., Talbot, E.A., et al., 2008. Recurrent multistate outbreak of *Salmonella* Newport associated with tomatoes from contaminated fields, 2005. Epidemiol. Infect. 136, 157–165.

Gortares-Moroyoqui, P., Castro-Espinoza, L., Naranjo, J.S., Karpiscak, M.M., Freitas, R.J., Gerba, C.P., 2011. Microbiological water quality in a large irrigation system: El Valle de Yaqui, Sonora, Mexico. J. Environ. Sci. Health Part. A. 46, 1708–1712.

Hirneisen, K.A., Sharma, M., Kniel, K.E., 2012. Human enteric pathogen internalization by root uptake into food crops. Food Borne. Path. Dis. 9, 306–405.

Holvoet, K., Jacxsens, L., Sampers, I., Uytteddaele, M., 2012. Insight into the prevalence and distribution of microbial contamination to evaluate water management in the fresh produce processing industry. J. Food Prot. 75, 671–681.

Ingram, D.T., Patel, J., Sharma, M., 2011. Effect of repeated irrigation with water containing varying levels of total organic carbon on the persistence of *Escherichia coli* O157:H7 on baby spinach. J. Food Prot. 74, 709–717.

Islam, M., Morgan, J., Doyle, M.P., et al., 2004. Fate of *Salmonella enterica* serovar Typhimurium on carrots and radishes grown in fields treated with contaminated manure composts or irrigation water. Appl. Environ. Microbiol. 70, 2497–2502.

Izumi, H., Tsukada, Y., Poubol, J., Hisa, K., 2008. On-farm sources of microbial contamination of persimmon fruit in Japan. J. Food Prot. 71, 52–59.

Jacobsen, C.S., Bech, T.B., 2012. Soil survival and transfer to freshwater and fresh produce. Food Res. Int. 45, 557–566.

Kay, D., Wyuer, M., 1992. Recent epidemiological research leading to standards. In: Kay, D. (Ed.), Recreational water quality management, volume 1 coastal waters. Ellis Harwood, Chichester, UK, pp. 129–156.

Kayed, D., 2004. Microbial quality of irrigation water used in the production of fresh produce in Arizona University of Arizona. Tucson, AZ Ph.D. Dissertation.

Keraita, B.N., Drechsel, P., 2004. Agricultural use of untreated urban wastewater in Ghana. In: Scott, C.A., Faruqui, N.I., Raschid-Sally, L. (Eds.), Wastewater use in irrigated agriculture. Commonwealth Agricultural Bureau International Publishing, Wallingford, UK, pp. 101–112.

Kim, S.R., Lee, S.H., Kim, W.I., et al., 2012. Effects of medium, soil, and irrigation contaminated with *Escherichia coli* and *Bacillus cereus* on the microbiological safety of lettuce. Korean. J. Hort. Sci. Technol. 30, 442–448.

Kittigul, L., Panjangampathana, A., Pombubpa, K., et al., 2012. Detection and genetic characterization of norovirus in environmental water samples in Thailand. Southeast Asian J. Trop. Med. Public Health 43, 323–332.

Knox, J.W., Tyrrel, S.F., Dacache, A., Weatherhead, E.K., 2010. A geospatial approach to assessing microbiological water quality risks associated with irrigation abstraction. Water Environ. J. 25, 282–289.

Kokkinos, P., Kozyra, I., Lazie, S., et al., 2012. Harmonized investigation of the occurrence of human enteric viruses in the leafy green vegetable supply chain in three European countries. Food Environ. Virol. 4, 179–191.

López-Gálvez, F., Allende, A., Selma, M.V., Gil, M.I., 2009. Prevention of Escherichia coli cross-contamination by different commercial sanitizers during washing of fresh-cut lettuce. Int. J. Food Microbiol. 133, 167–171.

Manshadi, F.D., Karpiscak, M., Gerba, C.P., 2013. Enteric bacterial contamination and survival on produce during irrigation with dairy wastewater. J. Water Reuse Desalination 3, 102–110.

Materon, L.A., Martinez-Garcia, M., McDonald, V., 2007. Identification of sources of microbial pathogens on cantaloupe rinds from pre-harvest to post-harvest operations. Wld. J. Microbiol. Biotechnol. 23, 1281–1287.

Micallef, S.A., Rosenberg-Goldstein, R.E., Gerge, A., et al., 2012. Occurrence and antibiotic resistance of multiple *Salmonella* serotypes recovered from water, sediment and soil on mid-Atlantic tomato farms. Environ. Res. 114, 31–39.

Monaghan, J.M., Hutchison, M.L., 2012. Distribution and decline of human pathogenic bacteria in soil after application in irrigation water and the potential for soil-splash-mediated dispersal onto fresh produce. J. Appl. Microbiol. 112, 1007–1019.

Mota, A., Mena, K.D., Soto-Beltran, M., et al., 2009. Risk assessment of *Cryptosporidium* and *Giardia* in water irrigating fresh produce in Mexico. J. Food Prot. 72, 2184–2188.

Nyard, K., Lassen, J., Vold, L., et al., 2008. Outbreak of *Salmonella* Thompson infections linked to imported rucola lettuce. Foodborne Pathog. Dis. 5, 165–173.

Ogunsola, F.T., Adesiji, Y.O., 2008. Comparison of four methods of hand washing in situations of inadequate water supply. West. Afr. J. Med. 27, 24–28.

Okafo, C.N., Umoh, V.J., Galadima, M., 2003. Occurrence of pathogens on vegetables harvested from soils irrigated with contaminated streams. Sci. Total Environ. 311, 49–56.

Oliveria, M., Vinas, I., Usall, J., Anguera, M., Abadias, M., 2012. Presence and survival of *Escherichia coli* O157:H7 on lettuce leaves and in soil treated with contaminated compost and irrigation water. Internl. J. Food Microbial. 156, 133–140.

Oron, G., Goemans, M., Manor, Y., Feyen, J., 1995. Poliovirus distribution in the soil-plant system under reuse of secondary wastewater. Water Res. 29, 1069–1078.

Pachepsky, Y., Shelton, D.R., McLain, J.E.T., Oatel, J., Mandrell, A., 2011. Irrigation waters as a source of pathogenic microorganism in produce: a review. Adv. Agronomy 113, 75–141.

Park, S., Szonyi, S., Raju, G., Nightingale, K., et al., 2012. Risk factors for microbial contamination in fruits and vegetables at the preharvest level: a systematic review. J. Food Prot. 75, 2055–2081.

Petterson, S.R., Teunis, P.F., Ashbolt, N.J., 2001. Modeling virus inactivation on salad crops using microbial count data. Risk Anal. 21, 1097–1108.

Postel, S., Polak, P., Gonzales, F., Keller, J., 2001. Drip irrigation for small farmers: a new initiative to alleviate hunger and poverty. Water Int. 26, 3–13.

Regli, S., Rose, J.B., Haas, C.N., Gerba, C.P., 1991. Modeling the risk from *Giardia* and viruses in drinking water. J. Am. Water Works. Assoc. 88, 76–84.

Robertson, L.J., Gjerde, B., 2001. Occurrence of parasites on fruits and vegetables in Norway. J. Food Prot. 64, 1793–1798.

Rose, J.B., Gerba, C.P., 1991. Assessing potential health risks from viruses and parasites in reclaimed water in Arizona and Florida, USA. Water Sci. Technol. 23, 2091–2098.

Sadovski, A.Y., Fattal, B., Goldberg, D., Katzenelson, E., Shuval, H.I., 1978. High levels of microbial contamination of vegetables irrigated with wastewater by the drip method. Appl. Environ. Microbiol. 36, 824–830.

Seymour, I.J., Appleton, H., 2001. Foodborne viruses and fresh produce. J. Appl. Meas. 91, 759–773.

Scott, C.A., Faruqui, L., Raschid-Sally, L., 2004. Wastewater use in irrigated agriculture: management challenges in developing countries. In: Scott, C.S., Faruqui, N.L., Raschid-Sally (Eds.), Wastewater use in irrigated agriculture: Confronting the livelihood and environmental realities. Commonwealth Agricultural Bureau International Publishing, Wallingford, UK, pp. 1–10.

Shanan, L., 1998. Irrigation development: Proactive planning and interactive management. In: Bruins, H., Harvey, R, L. (Eds.), The arid frontier. Kluwer Academic Press, London.

Shelton, D.R., Karns, J.S., Coppock, C., et al., 2011. Relationship between eae and six virulence genes in an agricultural watershed: implications for irrigation water standards and leafy green commodities. J. Food Prot. 74, 8–23.

Soderatorm, A., Osterberg, P., Lindqvist, A., et al., 2008. A large *Escherichia coli* 0157 outbreak in Sweden associated with locally produced lettuce. Foodborne. Pathog. Dis. 5, 339–349.

Solomon, E.B., Pang, H., Matthews, K.R., 2003. Persistence of *Escherichia coli* O157:H7 on lettuce plants following spray irrigation with contaminated water. J. Food Prot. 66, 2198–2202.

Song, I., Stine, S.W., Choi, C.Y., Gerba, C.P., 2006. Comparison of crop contamination by microorganisms during subsurface drip and furrow irrigation. J. Environ. Eng. 132, 1243–1248.

Steele, M., Odumeru, J., 2004. Irrigation water as source of foodborne pathogens on fruits and vegetables. J. Food Prot. 67, 2839–2849.

Stine, S.W., Song, I., Pimentel, J., et al., 2005. The effect of relative humidity on pre-harvest survival of bacterial and viral pathogens on the surface of cantaloupe, lettuce, and bell pepper were studied. J. Food Prot. 68, 1352–1358.

Stine, S.W., Song, I., Choi, C.Y., Gerba, C.P., 2005. Application of microbial risk assessment to the development of standards for enteric pathogens in water used to irrigate fresh produce. J. Food Prot. 68, 1352–1358.

Stine, S.C., Song, I., Choi, C.Y., Gerba, C.P., 2011. Application of pesticide sprays to fresh produce: a risk assessment for hepatitis A and *Salmonella*. Food Environ. Virol. 3, 86–91.

Tierney, J.T., Sullivan, R., Larkin, E.P., 1977. Persistence of poliovirus 1 in soil and on vegetables grown in soil previously flooded with inoculated sewage sludge or effluent. Appl. Environ. Microbiol. 33, 109–113.

Thurston-Enriquez, J.A., Watt, P., Dowd, S.C., et al., 2002. Detection of protozoan parasites and microsporidia in irrigation waters used for crop production. J. Food Prot. 65, 378–382.

Tyrrel, S.F., Knox, J.W., Weatherhead, E.K., 2006. Microbiological water quality requirements for salad irrigation in the United Kingdom. J. Food Prot. 69, 2029–2035.

USDA, 2008. Farm and Ranch Irrigation Survey, Census of Agriculture. United State Department of Agriculture http://www.agcensus.usda.gov/Publications/Irrigation_Survey/.

Van Zyl, W.B., Page, N.A., Grabow, W.O., et al., 2006. Molecular epidemiology of group a rotaviruses in water sources and selected raw vegetables in southern Africa. Appl. Environ. Microbiol. 72, 4554–4560.

Wei, J., Yan, J., Sims, T., Kniel, K.E., 2011. Internalization f murine norovirus 1 by *Lactuca sativa* during irrigation. Appl. Environ. Microbiol. 7, 2508–2512.

Won, G., Kline, T.R., Lejeune, J.T., 2013. Spatial-temporal variations of microbial water quality in surface reservoirs and canals used for irrigation. Ag. Water Manag. 116, 73–78.

Wood, J.D., Bezanson, G.S., Gordon, R.J., Jamieson, R., 2010. Population dynamics of *Escherichia coli* inoculated by irrigation into the phyllosphere of spinach grown under commercial conditions. Int. J. Food Microbiol. 143, 198–204.

World Health Organization (WHO), 2006. Guidelines for the safe use of wastewater, excreta and greywater, Vol. 2. Wastewater use in agriculture. WHO, Geneva, Switzerland.

Disease Risks Posed by Wild Birds Associated with Agricultural Landscapes

Larry Clark

U.S. Department of Agriculture, Animal and Plant Health Inspection Service, Wildlife Services,
National Wildlife Research Center, Fort Collins, CO

CHAPTER OUTLINE

The Produce Contamination Problem. http://dx.doi.org/10.1016/B978-0-12-404611-5.00007-5
2014 Published by Elsevier Inc.

Introduction

There are over 1,400 cataloged human pathogens, with approximately 62% classified as zoonotic (Taylor et al., 2001). While most evidence of direct transmission of pathogens to humans involves domestic and companion animals, the reservoir for most zoonoses is wildlife, yet there are relatively few well-documented cases for the direct involvement of transmission from wildlife to humans (Kruse et al., 2004). In part, this absence of evidence reflects the mobility of wildlife, the difficulty in accessing relevant samples, and the smaller number of studies focused on characterizing wildlife pathogens relative to the human and veterinary literature (McDiarmid, 1969; Davies et al., 1971; Hubalek, 2004). Because humans generally do not have direct contact with wild birds, exposure to pathogens is via indirect routes, i.e., environmental. This indirect exposure route (sapro-zoonotic) makes identifying the wild-bird source of the pathogen all the more difficult. Thus, most assessments implicating birds in carriage or transmission of pathogens of zoonotic importance are based on reasonable inference. The first step in this process is documentation that birds are hosts or can carry or transmit the pathogen. The second step involves a demonstration that the bird species involved is associated with agricultural food production. Usually the evidence provided is actual census information, behavioral observations, or evidence of the presence of feces. The third step involves demonstration that a pathogen or parasite that could originate from a bird associated with agricultural production or processing is the causative agent of human disease. The best evidence would be genetic, but even other diagnostic methods would suffice. Unfortunately this last step is rarely documented in the literature. However, with better diagnostic technologies and better understanding of the disease ecology, it is feasible that documenting the actual risks posed by wildlife to human health will become easier, and we will be better able to identify control points for pathogen management originating from wildlife.

Most of the evidence in the literature focuses on relatively few commensal wildlife species in urbanized environments, or at best, general wildlife surveillance and monitoring efforts (Tsiodras et al., 2008). In the absence of any compelling direct evidence, this review summarizes the circumstantial evidence, relying mostly on the characterization of host range of pathogens, similarities of virulence traits of animal and human pathogens, and habitat-use patterns of wild birds in agricultural and urban landscapes. Nonetheless, the material presented here does represent a solid circumstantial case for the potential of wild birds to contaminate field crops and act as agents for the transmission of pathogens to humans. More directed studies will be needed to form a more informed assessment as to what actual risks wild birds pose to field crop contamination with human pathogens, and by implication, to human health. Finally, this review also briefly covers mitigation efforts that might be undertaken to reduce risks of pathogen transmission by wild birds.

Bird species commonly associated with agriculture

There are two competing views regarding the value of birds to agriculture. The first view values the ecological services that birds may provide as consumers of pest insects and rodents (Mouysset et al., 2011; Power, 2011; Wenny et al., 2011; Benayas and Bullock, 2012). To capitalize on this service, a great deal of effort has been placed on research and management programs such as conservation reserves and other local habitat management to encourage beneficial species to associate themselves with agricultural production (Berges et al., 2010; Vickery et al., 2009). A full discussion of this topic is beyond the scope of this review. A competing view is the potential damage birds might cause to agriculture either from crop damage and depredation (Gebhardt et al., 2011; Kale et al., 2012; Merkens et al., 2012), or as an agent facilitating the contamination of production or processing of crops with pathogens (Bach and Delaquis, 2009). At the present time no bioeconomic or risk assessment models have comprehensively integrated these benefits and costs.

Wild birds, and especially migratory species, can become long-distance vectors for a wide range of microorganisms (Benskin et al., 2009; Hughes et al., 2009; Francesca et al., 2012; Crowder et al., 2013). Moreover, many bird species incorporate agricultural fields into their habitat use patterns. However, for the purpose of this review, focus is limited to only a few groups of birds: gulls (Charadriformes), waterfowl (Anseriformes), pigeons and doves (Columbiformes), and selected passerine birds (Passeriformes) such as blackbirds, crows, starlings, and sparrows. These groups of birds tend to have high use patterns of agricultural habitats, they are abundant, and they have closer commensal relationships with human activity. These species also tend to be abundant and gregarious; hence they provide greater opportunity and capacity to contribute larger fecal loads to the environments they use.

Contamination of produce can occur via many routes, e.g., at the field level during the growing season, during harvesting, postharvest handling, processing, shipping, marketing, or in the home (Beuchat and Ryu, 1997). Wild birds are most likely to be involved in contamination while the crops are in the field, and perhaps at field-side processing and storage facilities. Moreover, the likelihood that birds are responsible for contamination of crops with human pathogens will largely depend on the birds' exposure to environmental sources of pathogens (Figure 7.1), their capacity to physically transport the pathogen, and perhaps, but not necessarily limited by, their ability to act as a reservoir of the pathogen (Kruse et al., 2004). However, in the end, it is the likelihood and magnitude for fecal contamination of soils, substrates, and water that is the most direct link for risk to human health (Mohapatra et al., 2008; Jokinen et al., 2011).

Pigeons

Pigeons (*Columba livia*) live in close association with humans (Johnston and Janiga, 1995). They are a gregarious species that feed in flocks, form large roosts, and visit habitats that have a high likelihood of harboring human pathogens, e.g., dairies and

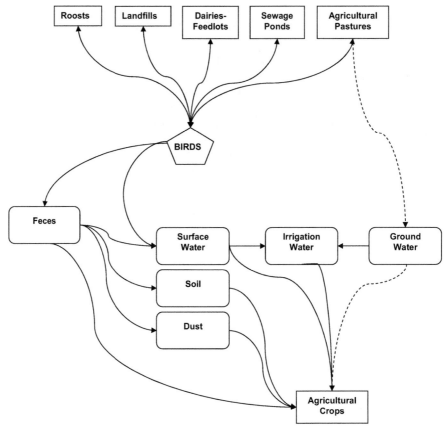

FIGURE 7.1

Possible routes of contamination of field or orchard crops by wild birds. Rectangles indicate environmental sources/habitats of pathogens. Pentagon represents various species of wild birds likely to visit both the source and agricultural landscapes. Rounded rectangles represent media pathogens may reside and be transported in.

feedlots. Pigeons will often flock to agricultural fields to pick up grit to aid in digestion, or to consume spilled grains. As with most commensal bird species considered to be high risk for transmission of human pathogens, pigeons have a prodigious capacity to produce feces. When occurring in large numbers, the fecal load for contamination of surface water and soils also can be large. Perhaps more importantly, is the propensity of pigeons to use architectural structures as day and night roosting sites. Open crop storage or processing sheds present an ideal circumstance for attracting pigeons, and in the absence of bird exclusionary mitigation measures, provide an opportunity for fecal accumulation. As with most species of birds, humans are not likely to come in direct contact with pigeons; rather, humans are likely to come in contact with feces or fecally contaminated substrates. Accumulation of feces presents greater opportunities for contamination of produce or exposure *via* contaminated soils (Costa et al., 2010), dusts (Gage et al., 1970), and water sources (Graczyk, 2007b) (Figure 7.1).

Over 60 human pathogens have been isolated from pigeons; however, only five pathogens have been documented to routinely be transmitted to humans (Haag-Wackernagel and Moch, 2004): *Chlamydophila psittaci, Histoplasma capsulatum, Aspergillus spp., Candida parapsilosis,* and *Cryptococcus neoformans.* Most cases of disease transmission are related to inhalation of dusts and aerosols (Anon, 2002), emphasizing the risk of accumulated feces at building structures to human health. Only one case of foodborne illness has been directly linked to pigeons, and that was attributed to lemon pudding made from the eggs of domestic pigeons (Clarenburg and Dornickx, 1933).

Gulls

A variety of species of gulls (Laridae) have been implicated in the carriage of human pathogens and contamination of watersheds, surface waters, and structures (Wither et al., 2005; Kinzelman et al., 2008). For the most part the number of gulls is increasing, largely owing to the availability of landfills, which provide a source of abundant food. Gulls are attracted to agricultural fields to forage on rodents, insects, and at times the crops themselves (O'Connor, 1992). Gulls can occur in large numbers, produce prodigious quantities of fecal material, and thus act as a source for contamination of soils and substrates. Gulls frequent landfills, dairies and feedlots, sewage ponds, other waste facilities, and agricultural pastures, all sources of human pathogens. Thus, they should be considered a major source of contamination risk to crops, produce, water easements, irrigation sources, and processing facilities (Coulson et al., 1983; Fricker, 1984; Yorio and Caille, 2004; Nelson et al., 2008; Ramos et al., 2010; Converse et al., 2012).

Waterfowl

Ducks and geese frequent agricultural fields to feed on crops, spilled grains, acquire grit, or loaf. Waterfowl are most likely to be a source of contamination of soils and surface waters, and indirectly, ground water (Somarelli et al., 2007; Graczyk et al., 2008). Large flocks are likely to begin use of agricultural fields in the fall and leave in the early spring (McKay et al., 2006; Amano et al., 2007). During this period the flocks may move across the landscape on a local, regional, or continental scale, thus increasing the opportunities for transporting pathogens from one site to another. Waterfowl are likely to use surface water areas, agricultural pastures, and sewage ponds, all of which can act as environmental sources of human pathogens (Conn et al., 2007).

Passerines

The number of passerine species that use agricultural fields is large, but those that travel in large flocks are more limited. Blackbirds (Ictaridae) and starlings (*Sturnus vulgaris*) will frequently use feedlots, dairies, and agricultural pastures, and thus, are likely to encounter human pathogens at those sites (Odermatt et al., 1998; Kaleta, 2002; Nielsen et al., 2004). Both species are also likely to use agricultural crop fields as sources of food and grit, thereby providing an opportunity to contaminate soils and crops with their feces. Starlings may also use crop storage and processing facilities as roost sites, providing additional opportunities to contaminate produce. House sparrows (*Passer domesticus*) have a more restricted range. However, this species is commonly associated with poultry houses, feedlots, and dairies and are common roost and nesting inhabitants of crop storage and processing facilities (Craven et al., 2000, Kirk et al., 2002). Other species of passerines are more likely to use only the agricultural fields; thus the risk of their contaminating field crops will be based on their exposure to pathogens in the general environment (Boutin et al., 1999; Laiolo, 2005).

Bacterial diseases

Campylobacter

Campylobacter is widespread in wild birds (Luechtefeld et al., 1980). High isolation rates have been obtained in gulls (*Larus* spp., 15–50%), crows (*Corvus corone cornix, Corvus levaillanti and Corvus corone,* 34–89%), blue magpies (*Cyanopica cyanus,* 20%), gray starlings (*Sturnus cineraceus,* 14%), pigeons (4–26%), (Ito et al., 1988; Quessy and Messier, 1992; Casanovas et al., 1995; Ramos et al., 2010), and Canada geese (*Branta canadensis,* 5–40%) (Rutledge et al., 2013). In cases where stomach contents have been analyzed, it is evident that the birds have visited landfills, thus indicating the importance of food habits as a primary factor in the varying prevalence of *C. jejuni* (Kapperud and Rosef, 1983). *Campylobacters* are also widely distributed in aquatic environments and in sewage effluents and agricultural runoff, environments conducive to exposing waterfowl to the pathogen (Brown et al., 2004). *Campylobacter* species may survive, and remain potentially pathogenic, for long periods in aquatic environments, but less so on terrestrial substrates (Krampitz, and Hollander, 1999). *Campylobacter* has been found in water sources for human consumption and irrigation, and bird fecal material has been implicated in the contamination (Denis et al., 2011).

Campylobacters are one of the most significant causes of human gastrointestinal infections worldwide, and the role that terrestrial birds and waterfowl have in the spread of disease is beginning to be elucidated. For example, Starlings shed *Campylobacter* at high rates, suggesting that they may be a source of human and farm animal infection. However, based on genetic analysis Colles et al. (2008a) concluded that these bacteria were distinct from poultry or human disease isolates with the ST-177 and ST-682 clonal complexes, possibly representing starling-adapted genotypes. Thus, these authors concluded that there was no evidence that wild starlings represent a major source of *Campylobacter* infections of food animals or humans. Similarly, Colles et al. (2008b) investigated wild geese as a potential source of *Campylobacter* infection for humans and farm animals in waterborne disease outbreaks. The authors found that large numbers of wild geese carry *Campylobacter*; however, there was limited mixing of *Campylobacter* populations among the different sources examined. Thus, they concluded that genotypes of *C. jejuni* isolated from geese are highly host specific, and geese are unlikely to be the source of the human disease outbreaks. In contrast, French et al. (2008) identified members of *Campylobacter* clonal complexes ST-45, ST-682, and ST-177 recovered from starling feces as being indistinguishable from those observed in human cases and concluded that wild birds could contribute to the burden of campylobacteriosis in preschool children at playgrounds.

Chlamydia

Avian chlamydiosis was originally called "parrot fever." However, recent studies have shown that "parrot fever" and ornithosis are the same disease manifested in different species and are all caused by the bacterium, *Chlamydophila psittaci* (Andersen and Vanrompay, 2000). Chlamydial infections have been identified in over 150 species of wild birds (Brand, 1989). Generally, these wild birds are asymptomatic. Bacteria are shed sporadically in nasal secretions and feces. Although the natural host reservoir systems are unknown, its wide occurrence in wild bird populations and the intermittent infections of farm stock are consistent with exposure to wild birds. Sporadic shedding was seen in experimentally inoculated great-tailed grackles (*Cassidix mexicanus*) and cowbirds (*Molothrus ater*), indicating their potential as host-reservoir systems (Roberts and Grimes, 1978). The most probable risk to farm stock and poultry is when wild birds gain access to feed bins and contaminate the bins with their feces. Infection usually occurs through exposure to contaminated aerosol dusts (Page, 1959). Turkeys can become infected by exposure to starlings, common grackles (*Quiscalus quiscula*), and brown-headed cowbirds (Roberts and Grimes, 1978; Grimes et al., 1979). Serovars D and E can result in 50 to 80% morbidity and 5 to 30% mortality in turkeys (Andersen, 1997). In ducks the economic impact is also significant, with morbidity and mortality ranging from 10 to 80% and 0 to 30%, respectively (Andersen et al., 1997). The high morbidity and mortality in domestic fowl and waterfowl suggest that mortality events in the wild species occurring near agroecosytems may be a useful warning system for triggering enhanced field produce monitoring. Wild avian strains also can infect mammals, including humans, and can cause severe disease or death (Andersen and Vanrompay, 2000).

Escherichia coli

The most important reservoir for verocytotoxin-producing *Escherichia coli* (VTEC) is considered to be ruminants, particularly cattle, though VTEC can be isolated from many mammals and birds (Wallace et al., 1997; Rice et al., 2003). Infection of humans by STEC may result in combinations of watery diarrhea, bloody diarrhea, and hemolytic uremic syndrome. Severe disease in the form of bloody diarrhea and the hemolytic uremic syndrome is attributable to Shiga toxin (Stx), which exists as two major types, Stx1 and Stx2 (Gyles, 2007). Humans can become infected through contamination of food and water, as well as through direct contact. Given the propensity of certain species of birds to frequent facilities or pastures used by ruminants, there remains a distinct possibility that wild birds may play a role in the transmission of VTEC to field crops and fruits.

Several studies have implicated wild birds in the transmission of VTEC, based on the similar characterizations between avian and human isolates (Asakura et al., 2001; Kobayashi et al., 2002). Numerous studies have documented the presence of *Stx*-producing *E. coli* (STEC) in pigeons (Morabito et al., 2001), gulls (Makino et al., 2000), waterfowl (Feare et al., 1999; Kullas et al., 2002), and passerines (Nielsen et al., 2004; Kobayashi et al., 2008). Other human virulence factors have also been identified in these species, including *eae, cldt,* CNF-1, CNF2, K1, LT, *hly*A, SLT-I, SLT-II, STa, and STb, (Morabito et al., 2001; Kullas et al., 2002; Fukuyama et al., 2003; Pedersen et al., 2006). In addition, phylogenic studies have illustrated the similarity of VTEC based on a number of measures (Makino et al., 2000). Though less common, a few studies have shown direct linkages and phylogenetic relatedness between avian VTEC and isolates causing human illness (Sonntage et al., 2005; Ejidokun et al., 2006).

Though VTEC are widely reported in the species groups discussed above, Canada geese have been implicated as the most likely source of non-point source pollution of inland waters. Molecular fingerprints of *E. coli* isolated from regional populations showed an unexpectedly high percentage of isolates identified as having a wildlife origin (geese and deer). Geese were the dominant source of *E. coli* (44.7–73.7% of the total sources) in four sub-watersheds followed by cows (10.5–21.1%), deer (10.5–18.4%), humans (5.3–12.9%), and unidentifiable sources (0.0–11.8%) (Somarelli et al., 2007).

Public pressure is mounting to reduce or eliminate antimicrobials as ingredients of feed for poultry and other agricultural animals, primarily due to the fear of multidrug-resistant bacteria in clinical infections in both animals and humans. Wild birds have been implicated as reservoirs and as vectors for the spread of antibiotic resistant strains of *E. coli*. Gibbs et al. (2007) found drug-resistant strains of *E. coli* in feces of yellow-headed blackbird (*Xanthocephalus xanthocephalus*). Dolejska et al. (2008) reported *E. coli* resistant to 12 antimicrobials in 9% (n=54) of isolates from house sparrows (*Passer domesticus*). Greater than 95% of *E. coli* isolates from Canada geese in agricultural environments were resistant to penicillin G, ampicillin, cephalothin, erythromycin, lincomycin, sulfathiazole, and vancomycin; no *E. coli* were resistant to bacitracin or ciprofloxacin (Fallacara et al., 2001; Cole et al., 2005; Middleton and Ambrose, 2005).

Most studies have only focused on the serotype O157:H7/H–, however, there are suggestions that wild birds may be involved in the transmission of other pathogenic serotypes (Kullas et al., 2002). Morabito et al. (2001) described widespread, clonally-related isolates of *E. coli* O45, O18ab, and O75 serotypes in several pigeon flocks. The overall prevalence was similar between three flocks (10.8%), with evidence of *Stx*-producing *E. coli* (STEC).

Listeria

Listeriosis in humans is caused by infection by *Listeria monocytogenes*. While all serovars of *L. monoctyogenes* are considered human pathogens, the most potentially virulent are 4b, 1/2b, and 1/2a. *L. monocytogenes* is commonly associated with soils and feces in the environment and may be one of the most ubiquitous microorganisms in the soil. Human exposure is most likely through ingestion of contaminated food, but hand–oral contact or inhalation may also occur. Approximately 1 to 10% of the human population is thought to carry *L. monocytogenes* asymptomatically in the intestines. Healthy people rarely become ill after exposure. Serious cases almost always occur in the elderly, pregnant women, newborns, and those who are debilitated or immunocompromised (Acha and Szyfres, 2003).

Listeria spp. are commonly found in birds (Harkin et al., 1986; Fenlon, 1985). Overall, the prevalence detected across species lies within the same range. On the lower end Quessy and Messier (1992) found a prevalence of 9.5% in fecal samples of ring-billed gulls (*Larus delawarensis*). Clark and Sullivan (unpublished data) found prevalence ranging from 8 to 12% from Canada goose (*Branta canadensis*) fecal samples from five states (Colorado, New York, Pennsylvania, Washington, and Wisconsin) collected throughout the year. On the higher end, investigators have found prevalence of 43.2% in crows and 36% in gulls (Helstrom et al., 2008). Wild birds may pose a risk to human health. The magnitude of prevalence may reflect local environmental conditions that birds may visit, e.g., landfills, sewage treatment facilities, or livestock facilities. Visitation of such environments may also impact exposure and carriage of virulent strains of *Listeria*. The serovars 1/2a and 4b predominated in the eight serotyped *L. monocytogenes* isolates in the Yoshida et al. (2000) study. Similar genotypes have been found in wild birds and local fresh food markets as well (Zhang et al., 2007; Mosupye and von Holy, 2000; Hellstrom et al., 2007).

Salmonella

Various *Salmonella* strains have been isolated from a range of wild bird species. Given the ubiquitous nature of the host range, and the pathogenicity of the organism to humans, caution should be exercised anytime there is exposure to fecally contaminated surfaces or soils. When sapro-zoonotic infections do occur, 99% of the documented cases appear to have resulted from aerosol exposure (Haag-Wackernagel and Moch, 2004).

Gulls appear to pose the greatest risk of infection and carriage. This is perhaps owing to their greater propensity to visit sewage treatment ponds and thus acquire pathogens. *Salmonella* prevalence in gulls ranges between 1 to 55% (Butterfield et al., 1983; Fenlon, 1981, 1983; Sixl et al., 1997; Casanova, 1995). Several genetic and epidemiological studies have linked transport of pathogens from the site of acquisition to distant sites, including food-processing facilities and stockyards (Coulson et al., 1983; Nesse et al., 2005). Persistence in marked free-ranging gulls was shown to be limited to about 10 days (Snoeyenbos et al., 1967; Palmgren et al., 2006). Other genetic studies have shown that strains carried by gulls are similar to human pathogens.

Pigeons are generally characterized by low prevalence (3–4%) of *Salmonella* (Pasmans et al., 2004; Tanaka et al., 2005; González–Acuña et al., 2007). Despite the low prevalence, studies have implicated pigeons and sparrows in the maintenance of the pathogen at feedlots and dairies (Quevedo et al., 1973; Connolly et al., 2006; Pedersen et al., 2006).

Fungal diseases

Aspergillus

Aspergillus spp. are rapidly growing molds most commonly associated with decaying matter and the feces of waterfowl and raptors, though a variety of wild birds and domestic poultry are also known to become infected with *Aspergillus* spp. (Buxton and Sommer, 1980; Friend, 2006). In animals, greater than 90% of infections are caused by *A. fumigatus* (Quinn, 1994). The highest prevalence of *A. fumigatus* in waterfowl is in winter. The most susceptible people in populations for the respiratory and allergic complications of infection are those who are immunocompromised or on extended regimens of medication (Latge, 2001).

Cryptococcus

Cryptococcus neoformans is a fungus typically associated with bird feces (Blaschke-Hellmessen, 2000). *C. neoformans* typically only affects the immunocomprimised. Cryptococcal infection may cause a pneumonia-like illness, with shortness of breath, coughing, and fever. Skin lesions may also occur. Another common form of cryptococcosis is central nervous system infection, such as meningoencephalitis. The primary risk for infection is inhaling dusts containing contaminated feces. Pigeons appear to be the primary wild bird involved in transmission to humans, though *C. neoformans* has been detected in a variety of other species (Pollock, 2003; Cafarchia et al., 2006; Rosario et al., 2008). Prevalence in pigeons may range between 9–19% (Weber and Shafer, 1991; Soogarun et al., 2006). Unlike *Histoplasma*, *C. neoformans* viability in the environment is limited. Ruiz et al. (1982) showed that viability of *C. neoformans* decreased from 50 to 86% over the course of a year, once pigeons were excluded from a roost. They attributed this decrease in viability to desiccation.

Histoplasma

Histoplasma capsultatum is a zoonotic fungal pathogen, commonly found in soils and bird feces, that affects the respiratory system (Ajello, 1964). In endemic areas of the United States as much as 80 to 90% of the human population is infected (Rubin et al., 1959). Fewer than 10% of those who inhale airborne spores develop a pulmonary infection. However, the pulmonary form can disseminate and is potentially fatal if not treated. Acute pulmonary histoplasmosis is the most dramatic form of the disease and occurs in people who have inhaled massive doses of spores. Chronic infection in humans can result in permanent lung damage. People with HIV are most susceptible to the disseminated form of the illness. People at highest risk are those working in agriculture, particularly poultry operations, or those people coming in contact with bird feces associated with bird roosts (Dodge et al., 1965; Tosh et al., 1966). Such roosts are likely to be associated with dense vegetation (i.e., agricultural wind breaks), trees, or storage or processing sheds. Because transmission is through breathing dust particles containing spores, any disturbance of contaminated soil can cause infection (Storch et al., 1980; Stobierski et al., 1996). Soil studies have shown that the viable spores persist in contaminated soils over many years (9–13+ years, DiSalvo and Johnson, 1979), at times long after bird activity at a site has ceased (Gustafson et al., 1981). Moreover, residents downwind from disturbed contaminated soils can become infected (Latham et al., 1980; Chick et al., 1981). Formalin has been used to sterilize soils contaminated with *Histoplasma* and deemed to be a high risk of further infection to local human populations (Smith et al., 1964; Tosh et al., 1967; Bartlett et al., 1982). Though *Histoplasma* has been detected in the feces of many bird species, pigeons, blackbirds, and starlings are the most likely wild birds to be a source of soil contamination, and the species of wild birds most likely to be associated with roosts near agriculture production or processing (Schwarz et al., 1957; Pollack, 2003; Cermeno et al., 2006).

Parasitic diseases

Cryptosporidia

Cryptosporidium parvum is an important gastrointestinal parasite of humans and other animals that can be transmitted via contamination of food and water (Mackenzie et al., 1994; Millar et al., 2002). Symptoms may be long lasting and include diarrhea, loose or watery stools, stomach cramps, upset stomach, and a slight fever (Fayer et al., 1998). Some people have no symptoms. In persons with average immune systems, symptoms usually last about 2 weeks. Waterfowl in general, but Canada geese in particular, have been implicated in the contamination of water (Hatch, 1996; Smith et al., 1993; Graczyk et al., 1997; Fallacara et al., 2004), and oocytes recovered from feces have been demonstrated to be infectious Graczyk et al. (1998). Starlings and other birds have also been implicated as carriers of *Cryptosporidium parvum* near farms (McCarthy et al., 2008; Lord et al., 2010).

Microsporidia

Microsporidians (*Encephalitozoon intestinalis, E. hellem, E. cuniculi, E. bieneusi*) are obligate intracellular parasites that increasingly are involved in opportunistic infections of immunocompromised and immunocompetent people (Weber et al., 1994). *E. hellem* has been the species most commonly associated with avian hosts (Slodkowicz-Kowalska et al., 2006). Epidemiological evidence strongly supports contaminated water, including water used for crop production, as a significant risk factor for human disease (Dowd et al., 1998; Fournier et al., 2000; Mathis et al., 2005; Thurston-Enriquez et al., 2002), and avian contamination of surface water via defecation as an important contributory risk factor for pathogen transmission (Slodkowicz-Kowalska et al., 2006). Although waterborne transmission is the most likely avenue for opportunistic infection of humans, a study by Haro et al. (2005) showed that pigeon feces was positive for *E. Bieneusi* (9.7% prevalence)*, E. intestinalis* (4% prevalence), *and E. hellem*

(1% prevalence) and 4.8% of pigeons were co-infected. Bart et al. (2008) found similar prevalence. The authors concluded that there was no barrier to transmission from pigeons to humans.

Toxoplasma

Toxoplasma gondii is a common single-celled parasite responsible for infection of more than 60 million people in the United States each year (Dubey, 2002). Infection can be acquired via hand to mouth contact with feces, contaminated soil, water, or raw meat. In most cases, the hosts' natural immune system clears the disease, and most healthy humans are rarely aware that they are infected. Symptoms are flu-like and include swollen joints and fatigue. However, people with impaired immune systems, embryos, and neonates are particularly vulnerable to severe consequences of infection, e.g., eye and brain damage. Birds are included in the extensive list of wildlife species implicated as carriers of this parasite (Coutelen et al., 1953; Drobeck et al., 1953; Siim et al., 1963; Dubey, 2002). The prevalence of *T. gondii* in wild birds likely to be associated with agricultural landscapes is moderately high (Table 7.1). This is of some concern because *T. gondii* is readily transmitted through the fecal-water route and represents a risk for contaminating crops or water sources used by humans for consumption or food processing (Bahiea-Oliverira et al., 2003). Finally, human populations can be affected by exposure to feces associated with roosts (Peach et al., 1989).

Table 7.1 Prevalance of *Toxoplasma gondii* in Selected Species of Birds Commonly Associated with Agricultural Production or Processing

Order	Common Group Name	Prevalence Range (%)	Reference
Charadriiformes	gulls	6–16	Dubey, 2002; Burridge et al., 1979
Anserifmores	duck, geese	1–28	Dubey, 2002; Pak, 1976
Galliformes	pheasants, quail, turkey	2–19	Dubey, 2002
Columbiformes	pigeons, doves	5–12	Dubey, 2002; Gibson and Eyles, 1957; Jacobs et al., 1952; Catar, 1974
Passeriformes	house sparrow *Passer domesticus*	1–18	Hejlicek et al., 1981; Ruiz and Frenkel, 1980; Dubey, 2002; Pak, 1976
	starling *Sturnus vulgaris*	1	Dubey, 2002; Pak, 1976
	crow *Corvus brachyrhynchos*	1	Finlay and Manwell, 1956

Determinating risk and control points

The first step in developing cost-effective contamination management is to develop robust risk analyses. Carefully constructed, such analyses will identify the important control points. For example, Duffy and Schaffner (2002) developed a Monte Carlo simulation that showed contamination with *E. coli* was higher in dropped apples owing to higher probability of fecal contamination by gulls, and increased site risks when orchards were near sewage ponds or landfills. Because gulls have a high association with landfills and sewage ponds, and those associations can result in higher carriage rates for pathogens (Ferns and Mudge, 2000; Nelson et al., 2008; Ramos et al., 2010), the risk to the orchard becomes higher either because it is an attractive foraging site, or merely by being en route when gulls fly over and drop feces. In this case the practical solution is to avoid using dropped apples in production of unpasteurized foods and drinks. Alternatively, management of contamination sources for *E. coli* may also be effective, though beyond the control of the orchardist. Such land-use patterns, as part of the risk assessment, are emerging as part of a developing field for aviation safety as it relates to birds and could equally be applied to agricultural landscapes.

The current emphasis of the agricultural and microbiological community is in the refinement of the risk analysis process. Improved genetic and biochemical methodologies are resulting in better quantification in drawing the linkage between wildlife as vectors and carriers, wildlife use patterns of agricultural production and processing facilities, and the contamination of agricultural products and processes (Table 7.2). The next step in developing cost-effective contamination management is to better integrate existing bird control and management technologies and methodologies that have been used in mitigating crop damage towards a focus on preventing pathogen and parasite contamination (see below). In doing so, incorporation of bioeconomic modeling into the management simulations for the evaluation of risk mitigation strategies will be a critical step for managers. Such fully integrated simulation models generally have not been applied. The discipline is still largely in the empirical descriptive and univariate management mode. As food safety issues become increasingly important and more diverse technical collaborative teams are assembled, this constraint on management will dissipate.

Mitigation options

Several excellent reviews exist on the general practices of excluding and repelling birds from preharvest agricultural landscapes (Hyngstom et al., 1994; Mason, 1997; Conover, 2002; Linz, 2003). Other venues for research and methods for animal damage control can be found in the *Proceedings of the Vertebrate Pest Conference*, *Proceedings of Animal Damage Management Conference*, and the following journals: *Human-Wildlife Conflict*, *Journal of Wildlife Management*, and *Wildlife Society Bulletin*, and *Wildlife Research*.

Table 7.2 Selected Examples of Microbiological Risks Associated with Wild Birds and Agricultural Production and Processing

Bird Species	Pathogen	Contamination Circumstance	Risk	Reference
Gulls	Fecal coliforms, *Salmonella* spp., *Aeromonas* spp.	Beach water quality degraded as a function of visitation of site by gulls.	A	Levesque et al., 1993
birds	*Salmonella*	Mango processing	A	Sivapalasingam et al., 2003
birds	*E. coli*	Pecked apples	A	Riordan et al., 2010
birds	*E. coli*	Watershed	I	Jiang et al., 2007
Canada goose	*Cryptosporidium parva*	Detected in feces of geese that had rested in agricultural fields.	CC	Fayer et al., 2000
birds	*Salmonella*	Droppings associated with orange juice production	CC	Cook et al., 1998
Magpies	*Campylobacter jejuni*	Magpies pecking milk bottle tops contaminated milk. Children drinking milk from compromised containers became ill.	SC	Riordani et al., 1993
birds	*Mycobacterium* spp., *Salmonella* spp., *Cryptosporidia* spp.	Livestock food storage contamination. Model developed showing that fecal contamination was good predictor of disease outbreak in livestock.	SC	Daniels et al., 2003
birds	*E. coli*	*E. coli* strains were common in sweet cherry orchard environment and birds. *E. coli* was not commonly found on fruit still on the tree. However, *E. coli* was found on pickers' and processors' hands and could be related to environmental and bird sources. Cherries increasingly became contaminated from picking to processing. The implication is that hand-picking cherries also provided opportunity to pick up environmental contaminants thus transferring contamination.	SC, DE	Bach and Delaquis, 2009
Sandhill Cranes	*Campylobacter jejuni*	Cranes feeding on peas at farm-side contaminated peas with feces and a human outbreak ensued from consuming raw peas. Pathogens linked by PFGE.	SC, DE	Gardner et al., 2011

Table 7.2 Selected Examples of Microbiological Risks Associated with Wild Birds and Agricultural Production and Processing—cont'd

Bird Species	Pathogen	Contamination Circumstance	Risk	Reference
Gulls, pigeons, sparrows, crows	*Listeria monocytogenes*	A variety of birds using municipal landfills also visited food processing plants (fish, meat). *L. monocytogenes* strains at contaminated food processing sites and birds were similar by PFGE.	SC, DE	Hellstrom et al., 2007 Neese et al., 2010
European Starling	*E. coli* O157:H7	Dairy	DE	LeJeune et al., 2008
birds	*E. coli*: (stx1, stx2, hlyA, ehxA, LT1, ST1 cdtB, east 1, cnfl, cvaC)	Contamination of rainwater storage tanks from birds showed biochemical homology to contaminated water samples.	DE	Ahmed et al., 2013
birds	fungi	Damage to fruit allows fungal infection of wound, leading to rot of stored fruit.	E	Creemers, 1989
birds	*Aspergillus* spp.	Damage to pistachio allowing *Aspergillus* contamination	E	Doster and Michailides, 1994

(I) Implicated. Genetic evidence of pathogen being same strain as found on produce or human infection.
(A) Associated. Concordant evidence of animal feces and pathogen contamination. Animal known carrier of pathogen.
(CC) Casual circumstantial evidence, avian fecal contamination found along with other risk factors.
(SC) Strong circumstantial evidence, avian fecal contamination identified as most likely risk factor.
(DE) Direct genetic evidence linking animal, pathogen, and product/substrate.
(E) Enabling. Physical activity of bird allows opportunistic infection or contamination.

Lethal control generally is not an option for bird control, owing to federal and state statues and prohibitions, except as allowed under permit. It is also not practical or economical to exclude birds from large agricultural fields. The exception is for bird netting for high value fruits such as grapes (Tracey and Saunders, 2010; Baldwin et al., 2013). Non-lethal methods to scare birds off fields are the only recourse available to growers. However, in reality, it is not practical to exclude all wild birds from large agro-ecosystems, though it is feasible using integrated pest management approaches to limit populations for smaller areas (Avery, 1989; Mason and Clark, 1992, 1996; Clark, 1998). Methods can include the use of frightening devices, such as visual deterrents, pyrotechniques, propane exploders, or alarm calls, but it is important to avoid presenting cues on a systematic or regular basis so as to avoid habituation. A review of each of these methods and the successes and failures can be found in the handbook, *The Prevention and Control of Wildlife Damage* (Hyngstrom et al., 1994). The difficulty with the approaches outlined in the handbook is one of expense and human vigilance, both of which are at the root cause of failure. Stevens et al. (2000) explored the use of demand performance systems using radar technologies to activate a variety of bird-scaring devices. The method was successful at keeping migratory birds off hazardous waste ponds of 180 and 90 acres for over a year. The principal drawback of the system was expense. Most agricultural operations could not afford such protection.

Chemical repellents have been tried, but not all crops are amenable to their use owing to expense and regulatory restrictions (Clark, 1997). Nonetheless, several products have been developed, e.g., methyl anthranilate and anthraquinone-based products, and several other investigatory repellents have been evaluated (Avery and Mason, 1997; Cummings et al., 2002; Avery et al., 2005; Werner et al., 2008).

As mentioned, preharvest exclusion of birds from agriculture production may not be realistic either from an efficacy point of view, or where feasible, from an economic point of view. The scale of the areas to be protected in agricultural landscapes are simply too large, and the attractiveness of the resource to birds (i.e., food) is too compelling for repellents and other non-lethal technologies to work towards a goal of total exclusion. The solutions at the preharvest level need to be more carefully thought out. The first assessment should be one of risk. The literature appears to indicate that fecal contamination of water sources and not direct contamination of crop is the control point that should be addressed (Bach and Delaquis, 2009; Slifko et al., 2000; Riordan et al., 2010; Steele and Odumeru, 2004; Ijabadeniyi et al., 2011; Table 7.2). If this is the case, then irrigation practices and water storage and access can be addressed from a resource protection perspective toward bird control, using management strategies that are more temporally and spatially focused, hence practical and economically feasible using existing technologies.

It is becoming increasingly evident that an important control point is prevention of irrigation water contamination (McCarthy et al., 2008; Jokinen et al., 2010; Ijabadeniyi et al., 2011). While there is not much that can be done at the watershed scale, changes in irrigation pattern, water treatment prior to processing, and protection of farm-side water sources may be manageable. For example, drip-irrigation reduces splashing, which can distribute contaminated water on plant surfaces (Forslund et al., 2012). Passing water through a central treated and protected holding facility prior to drip irrigation or processing may further reduce risk of contamination from birds. Postharvest risk of contamination from birds can be addressed by physical exclusion of birds from processing areas, e.g., use of netting. The goal is to prevent fecal contamination of produce, grains, or water sources. Several investigators have studied the efficacy of water treatments and other chemical treatments to make sure that water used for washing becomes decontaminated.

SUMMARY

Wild birds are capable of pathogen carriage, acting as reservoirs, and becoming infected with a wide variety of pathogens, some of which are zoonoses. This review did not attempt to provide an exhaustive list. Rather its intent was to focus on the avian species and pathogens that represented the greatest likelihood to be of concern at the agricultural production and processing level. Unfortunately, little direct evidence bears on this issue; hence risk was assessed using information derived from urban human health, water quality, poultry, livestock production, wildlife health, and the veterinary literature. Despite the lack of direct evidence available relating to birds and the risks to farm-side production and processing of produce, it appears there is ample evidence to support the notion that birds can pose a human health risk by serving as a source of contamination of produce and crops. Nonetheless, more detailed empirical and risk modeling studies are needed. Moreover, such studies should be integrated into a larger ecological perspective of the values of birds to agro-ecosystems balanced against the risks they pose. Finally, studies and analyses should also incorporate assessments of mitigation management strategies in the context of economic, ecological, and public health valuations. These approaches are clearly beyond the scope of this review, but should be seriously considered over a simplistic interpretation of disease risk posed by wild birds and measures needed to eliminate them from agro-ecosystems.

References

Acha, P.N., Szyfres, B., 2003. (Pan American Health Organization [PAHO]). Zoonoses and communicable diseases common to man and animals. Volume 1. Bacterioses and mycoses, third ed. PAHO, Washington, DC Scientific and Technical Publication No. 580. Listeriosis; pp. 168–179.

Ahmed, W., Sidhu, J.P.S., Toze, S., 2013. An attempt to identify the likely sources of *Escherichia coli* harboring toxin genes in rainwater tanks. Environ. Sci. Technol. 46, 5193–5197.

Ajello, L., 1964. Relationship of *Histoplasma capsulatum* to avian habitats. Public Health Rep. 79, 266–270.

Amano, T., Ushiyama, K. Fujita, G., and Higuchi, H., 2007. Predicting grazing damage by white-fronted geese under different regimes of agricultural management and the physiological consequences for the geese. J. Appl. Ecol. 44, 506–515.

Andersen, A.A., 1997. Two new serovars of *Chlamydia psittaci* from North American birds. J. Vet. Diagn. Invest. 9, 159–164.

Andersen, A.A., Vanrompay, D., 2000. Avian chlamydiosis. Rev. Sci. Tech. 19, 396–404.

Anon. 2002. Compendium of measures to control *Chlamydophila psittaci* (Formerly *Clamydia psittaci*) infection among humans (Psittacosis and pet birds) In: www.avma.org/pubhlth/psittacosis.asp, vol. 2003. National Association of State Public Health Veterinarians.

Asakura, H., Makino, S., Kobori, H., Watarai, M., Shirahata, T., Ikeda, T., et al., 2001. Phylogenetic diversity and similarity of active sites of Shiga toxin (stx) in Shiga toxin-producing *Escherichia coli* (STEC) isolates from humans and animals. Epidemiol. Infect. 127, 27–36.

Avery, M.L., 1989. Experimental evaluation of partial repellent treatment for reducing bird damage to crops. J. Appl. Ecol. 26, 433–439.

Avery, M.L., Mason, J.R., 1997. Feeding responses of red-winged blackbirds to multisensory repellents. Crop Protect. 16, 159–164.

Avery, M.L., Werner, S.J., Cummings, J.L., Humphrey, J.S., Milleson, M.P., Carlson, J.C., et al., 2005. Caffeine for reducing bird damage to newly seeded rice. Crop Protect. 24, 651–657.

Bach, S., Delaquis, P., 2009. The origin and spread of human pathogens in fruit production systems. In: Fan, X., Niemira, B.A., Doona, C.J., Feeherry, F.E., Gravani, R.B. (Eds.), Microbial Safety of Fresh Produce. Wiley-Blackwell, Ames, IA.

Bahia-Oliveira, L.M.G., Jones, J.L., Azevedo-Silva, J., Alves, C.C.F., Orefiice, F., Addiss, D.G., 2003. Highly endemic, waterborne toxoplasmosis in North Rio de Janeiro State, Brazil. Emerg. Infect. Dis. 9, 55–62.

Baldwin, R.A., Salmon, T., Schmidt, R.H., Timm, R.M., 2013. Wildlife pests of California agriculture: regional variability and subsequent impacts on management. Crop Protect. 46, 29–37.

Bart, A., Wentink-Bonnema, M., Heddema, E.R., Buijs, J., van Gool, T., 2008. Frequent occurrence of human-associated *microsporidia* in fecal droppings of urban pigeons in Amerstdam, The Netherlands. Appl. Environ. Microbiol. 74, 7056–7058.

Bartlett, P.C., Weeks, R.J., Ajello, L., 1982. Decontamination of a *Histoplasma capsulatum*-infested bird roost in Illinois. Arch. Environ. Health 37, 221–223.

Benayas, J.M.R., Bullock, J.M., 2012. Restoration of biodiversity and ecosystem services on agricultural land. Ecosystems 15, 883–899.

Benskin, C.McW.H., Wilson, K., Jones, K., Hartley, I.R., 2009. Bacterial pathogens in wild birds: a review of the frequency and effects of infection. Biol. Rev. 84, 349–373.

Berges, S.A., Schulte Moore, L.A., Isenhart, T.M., Schultz, R.C., 2010. Bird species diversity in riparian buffers, row crop fields, and grazed pastures within agriculturally dominated watersheds. Agroforest Syst. 79, 97–110.

Beuchat, L.R., Ryu, J.H., 1997. Produce handling and processing practices. Emerg. Infect. Dis. 3, 459–465.

Blaschke-Hellmessen, R., 2000. *Cryptococcus* species–etiological agents of zoonoses or sapronosis? (in German). Mycoses 43 (Suppl. 1), 48–60.

Brand, C.J., 1989. Chlamydial infections in free-living birds. J. Am. Vet. Med. Assoc. 195, 1531–1535.

Brown, P.E., Christensen, O.F., Clough, H.E., Diggle, P.J., Hart, C.A., Hazel, S., et al., 2004. Frequency and spatial distribution of environmental *Campylobacter* spp. Appl. Environ. Microbiol. 70, 6501–6511.

Boutin, C., Freemark, K.E., Kirk, D.A., 1999. Farmland birds in southern Ontario: field use, activity patterns, and vunerability to pesticide use. Agric. Ecosystems Environ. 72, 239–254.

Burridge, M.J., Bigler, W.J., Forrester, D.J., Hennemann, J.M., 1979. Serologic survey for *Toxoplasma gondii* in wild animals in Florida. J. Am. Vet. Med. Assoc. 175, 964–967.

Butterfield, J., Coulson, J.C., Kearsey, S.V., Monaghan, P., McCoy, J.H., Spain, G.E., 1983. The herring gull *Larus argentatus* as a carrier of *Salmonella*. J. Hyg. (Lond.) 91, 429–436.

Buxton, I., Sommer, C.V., 1980. Serodiagnosis of *Aspergillus fumigatus* antibody in migratory ducks. Avian Dis. 24, 446–454.

Cafarchia, C., Romito, D., Iatta, R., Camarda, A., Montagna, M.T., Otranto, D., 2006. Role of birds of prey as carriers and spreaders of *Cryptococcus neoformans* and other zoonotic yeasts. Med. Mycol. 44, 485–492.

Casanovas, L., de Simón, M., Ferrer, M.D., Arqués, J., Monzón, G., 1995. Intestinal carriage of campylobacters, salmonellas, yersinias and listerias in pigeons in the city of Barcelona. J. Appl. Bacteriol. 78, 11–13.

Catar, G., 1974. Toxoplazmoza v ekologickych podmienkach na Slovensku (in Slovakian). Biologicke Prace (Bratislava) 20, 1–38.

Cermeno, J.R., Hernandez, I., Cabello, I., Orillan, Y., Cermeno, J.J., Albornoz, R., et al., 2006. *Cryptococcus neoformans* and *Histoplasma capsulatum* in dove's (*Columba livia*) excreta in Bolivar state, Venezuela. Rev. Latinoam. Microbiol. 48, 6–9.

Chick, E.W., Compton, S.B., Pass 3rd, T., Mackey, B., Hernandez, C., Austin Jr., E., et al., 1981. Hitchcock's birds, or the increased rate of exposure to *histoplasma* from blackbird roost sites. Chest 80, 434–438.

Clarenburg, A., Dornickx, C.G.J., 1933. Nahrungmittelvergiftung bei Menschen in Zusammendgang mit Tauben paratyphose. Zscher. Hyg. Inf. Krankh. 114, 31–41.

Clark, L., 1997. A review of the bird repellent effects of 117 carbocyclic compounds. In: Mason, J.R. (Ed.), Repellents in wildlife management. (August 8–10, 1995, Denver, CO). Colorado State University, Fort Collins, Colorado, pp. 343–352.

Clark, L., 1998. Review of bird repellents. In: Barker, R.O., Crabb, A.C. (Eds.), Eighteenth Vertebrate Pest Conference (March 2–5, 1998, Costa Mesa, California), pp. 330–337.

Cole, D., Drum, D.J., Stalknecht, D.E., White, D.G., Lee, M.D., Ayers, S., et al., 2005. Free-living Canada geese and antimicrobial resistance. Emerg. Infect. Dis. 11, 935–938.

Colles, F.M., Dingle, K.E., Cody, A.J., Maiden, M.C., 2008a. Comparison of *Campylobacter* populations in wild geese with those in starlings and free-range poultry on the same farm. Appl. Environ. Microbiol. J74, 3583–3590.

Colles, F.M., McCarthy, N.D., Howe, J.C., Devereux, C.L., Gosler, A.G., Maiden, M.C., 2008b. Dynamics of *Campylobacter* colonization of a natural host, *Sturnus vulgaris* (European Starling). Environ. Microbiol. 11, 258–267.

Conn, D.B., Weaver, J., Tamang, L., Graczyk, T.K., 2007. Synanthropic flies as vectors of *Cryptosporidium* and *Giardia* among livestock and wildlife in a multispecies agricultural complex. Vector-Borne Zoonotic Dis. 7, 643–651.

Connolly, J.H., Alley, M.R., Dutton, G.J., Rogers, L.E., 2006. Infectivity and persistence of an outbreak strain of *Salmonella enterica* serotype *Typhimurium* DT160 for house sparrows (*Passer domesticus*) in New Zealand. N. Z. Vet. J. 2006 (54), 329–332.

Conover, M.R., 2002. Resolving Human-wildlife Conflicts: The Science of Wildlife Damage Management. CRC Press, New York, pp. 418.

Converse, R.R., Kinzelman, J.L., Sams, E.A., Hudgens, E., Dufour, A.P., Ryu, H., et al., 2012. Dramatic improvements in beach water quality following gull removal. Environ. Sci. Technol. 46, 10206–10213.

Cook, K.A., Dobbs, T.E., Hlady, W.G., Wells, J.G., Barrett, T.J., Puhr, N.D., et al., 1998. Outbreak of *Salmonella* serotype Hartford infections associated with unpasteurized orange juice. JAMA 280, 1504–1509.

Costa, A.K., Sidrim, J.J.C., Cordeiro, R.A., Brilhante, R.S.N., Monteiro, A.J., rocha, M.F.G., 2010. Urban pigeons (*Columba livia*) as a potential source of pathogenic yeasts: a focus on antifungal susceptibility of *Cryptococcus* strains in northeast Brazil. Mycopathologia 169, 207–213.

Coulson, J.C., Butterfield, J., Thomas, C., 1983. The herring gull *Larus argentatus* as a likely transmitting agent of *Salmonella montevideo* to sheep and cattle. J. Hyg. (Lond.) 91, 437–443.

Coutelen, F., Biguet, J., Doby, J.M., Deblock, S., 1953. Le probleme de toxoplamoses aviares. Receptivite variable de vuelques oiseaux a une souche humaine de toxoplasmes. Ann. Parasitol. Hum. Comp. 28, 129–156.

Craven, S.E., Stern, N.J., Line, E., Bailey, J.S., Cox, N.A., Fedorka-Cray, P., 2000. Determination of the incidence of *Salmonella* spp., *Campylobacter jejuni*, and *Clostridium perfringens* in wild birds near broiler chicken houses by sampling intestinal droppings. Avian Dis. 44, 715–720.

Creemers, P., 1989. Chemical control of parasitic storage diseases on apple and pear. Acta Hort. (ISHS) 258, 645–654.

Crowder, D.W., Dykstra, E.A., Brauner, J.M., Duffy, A., Reed, C., Martin, E., et al., 2013. West nile virus prevalence across landscapes is mediated by local effects of agriculture on vector and host communities. PLoS ONE 8 (1), e55006. http://dx.doi.org/10.1371/journal.pone.0055006.

Cummings, J.L., Avery, M.L., Mathre, O., Wilson, E.A., York, D.L., Engeman, R.M., et al., 2002. Field evaluation of Flight Control™ to reduce blackbird damage to newly planted rice. Wildl. Soc. Bull. 30 (3), 816–820.

Daniels, M.J., Hutchings, M.R., Greig, A., 2003. The risk of disease transmission to livestock posed by contamination of farm stored feed by wildlife excreta. Epidemiol. Infect. 130, 561–568.

Davies, J.W., Anderson, R.C., Karstad, L., Trainer, D.O., 1971. Infectious and parasitic diseases of wild birds. Iows State University Press, Ames, Iowa, pp. 344.

Denis, M., Tanguy, M., Chidaine, B., Laisney, M.-J., Megraud, F., Fravalo, P., 2011. Description and sources of contamination by *Campylobacter* spp. of river water destined for human consumption in Brittany, France. Pathol. Biol. 59, 256–263.

Disalvo, A.F., Johnson, W.M., 1979. Histoplasmosis in South Carolina: support for the microfocus concept. Am. J. Epidemiol. 109, 480–492.

Dodge, H.J., Ajello, L., Engelke, O.K., 1965. The association of a bird-roosting site with infection of school children by *Histoplasma capsulatum*. Am. J. Public Health Nations Health 55, 1203–1211.

Dolejská, M., Senk, D., Cízek, A., Rybaríková, J., Sychra, O., Literák, I., 2008. Antimicrobial resistant *Escherichia coli* isolates in cattle and house sparrows on two Czech dairy farms. Res. Vet. Sci. 85, 491–494.

Doster, Michailides, T.J., 1994. *Aspergillus* molds and aflatoxins in pistachio nuts in California. Phytopathology 84, 583–590.

Dowd, S.E., Gerba, C.P., Pepper, I.L., 1998. Confirmation of the human pathogenic *microsporidia Enterocytozoon bieneusi, Encehphalitozoon intestinalis,* and *Vittaforma corneae* in water. Appl. Environ. Microbiol. 64, 3332–3335.

Drobeck, H.P., Manwell, R.D., Bernstein, E. Dillon, R.D., 1953. Further studies of toxomplasmosis in birds. Am. J. Hyg. 58, 329–339.

Dubey, J.P., 2002. A review of toxoplasmosis in wild birds. Vet. Parasitol. 106, 121–153.

Duffy, S., Schaffner, D.W., 2002. Monte Carlo simulation of the risk of contamination of apples with *Escherichia coli* O157:H7. Int. J. Food. Microbiol. 78, 245–255.

Ejidokun, O.O., Walsh, A., Barnett, J., Hope, Y., Ellis, S., Sharp, M.W., et al., 2006. Human Vero cytotoxigenic *Escherichia coli* (VTEC) O157 infection linked to birds. Epidemiol. Infect. 134, 421–423.

Fallacara, D.M., Monahan, C.M., Morishita, T.Y., Wack, R.F., 2001. Fecal shedding and antimicrobial susceptibility of selected bacterial pathogens and a survey of intestinal parasites in free-living waterfowl. Avian Dis. 45, 128–135.

Fallacara, D.M., Monahan, C.M., Morishita, T.Y., Bremer, C.A., Wack, R.F., 2004. Survey of parasites and bacterial pathogens from free-living waterfowl in zoological settings. Avian Dis. 48, 759–767.

Fayer, R., Graczyk, T.K., Lewis, E.J., Trout, J.M., Farley, C.A., 1998. Survival of infectious *Cryptosporidium parvum* oocysts in seawater and Easter oysters *(Crassostrea viginica)* in the Chesapeake Bay. Appl. Environ. Microbiol. 64, 1070–1074.

Fayer, R., Morgan, U., Upton, S.J., 2000. Epidemiology of *Cryptosporidium*: transmission, detection and identification. Int. J. Parasitol. 30, 1305–1322.

Feare, C.J., Sanders, M.F., Blasco, R., Bishop, J.D., 1999. Canada goose *(Branta canadensis)* droppings as a potential source of pathogenic bacteria. J. R. Soc. Health 119, 146–155.

Fenlon, D.R., 1981. Seagulls *(Larus* spp.) as vectors of salmonellae: an investigation into the range of serotypes and numbers of salmonellae in gull faeces. J. Hyg. (Lond.) 86, 195–202.

Fenlon, D.R., 1983. A comparison of *Salmonella* serotypes found in the faeces of gulls feeding at a sewage works with serotypes present in the sewage. J. Hyg. (Lond.) 91, 47–52.

Fenlon, D.R., 1985. Wild birds and silage as reservoirs of *Listeria* in the agricultural environment. J. Appl. Bacteriol. 59, 537–543.

Ferns, P., Mudge, G.P., 2000. Abundance, diet and *Salmonella* contamination of gulls feeding at sewage outfalls. Water Res. 34, 2653–2660.

Forslund, A., Ensink, J.H.J., Markussen, B., Battilani, A., Psarras, G., Gola, L., et al., 2012. *Escherichia coli* contamination and health aspects of soil and tomatoes (Solanum lycopersicum L.) subsurface drip irrigated with on-site treated domestic wastewater. Water Res. 46, 5917–5934.

Fournier, S., Liguory, O., Santillana-Hayat, M., Guillot, E., Sarfari, C., Dumoutier, N., et al., 2000. Detection of *microsporidia* in surface water: a one-year follow-up study. FEMS Immunol. Med. Microbiol. 29, 95–100.

Francesca, N., Canale, D.E., Settanni, L., Moschetti, G., 2012. Dissemination of wine-related yeasts by migratory birds. Environ. Microbiol. Rep. 4, 105–112.

French, N.P., Midwinter, A., Holland, B., Collins-Emerson, J., Pattison, R., Colles, F., et al., 2008. Molecular epidemiology of *Campylobacter jejuni* isolated from wild bird faecal material in children's playgrounds. Appl. Environ. Microbiol. 75, 779–783.

Fricker, C.R., 1984. A note on *Salmonella* excretion in the black headed gull *(Larus ribibundus)* feeding at sewage treatment works. J. Appl. Bacteriol. 56, 499–502.

Friend, M., 2006. Disease Emergence and Resurgence: The Wildlife-Human Connection. Circular 1285, U.S. Department of Interior and U.S. Geological Survey, Reston, VA, pp. 388.

Fukuyama, M., Furuhata, K., Oonaka, K., Sakata, S., Hara, M., Kakuno, Y., et al., 2003. Isolation and serotypes of Vero toxin-producing *Escherichia coli* (VTEC) from pigeons and crows. Kansenshogaku Zasshi 77, 5–9.

Gage, A.A., Dean, D.C., Schimert, G., Minsley, N., 1970. *Aspergillus* infection after cardiac surgery. JAMA Surg. 101, 384–387.

Gardner, T.J., Fitzgerald, C., Xavier, C., Klein, R., Pruckler, J., Stroika, S., et al., 2011. Outbreak of campylobacteriosis associated with consumption of raw peas. Clin. Infect. Dis. 53, 26–32.

Gebhardt, K., Anderson, A.M., Kirkpatrick, K.N., Shwiff, S.A., 2011. A review and synthesis of bird and rodent damage estimates to select California crops. Crop Protect. 30, 1109–1116.

Gibbs, P.S., Kasa, R., Newbrey, J.L., Petermann, S.R., Wooley, R.E., Vinson, H.M., et al., 2007. Identification, antimicrobial resistance profiles, and virulence of members from the family Enterobacteriaceae from the feces of yellow-headed blackbirds *(Xanthocephalus xanthocephalus)* in North Dakota. Avian Dis. 51, 649–655.

Gibson, C.L., Eyles, D.E., 1957. *Toxoplasma* infections in animals associated with a case of human congenital toxoplasmosis. Am. J. Trop. Med. Hyg. 6, 990–1000.

González-Acuña, D., Silva, G.F., Moreno, S.L., Cerda, L.F., Donoso, E.S., Cabello, C.J., et al., 2007. Detection of some zoonotic agents in the domestic pigeon (*Columba livia*) in the city of Chillán, Chile. Rev. Chil. Infectol. 2007 (24), 199–203.

Graczyk, T.K., Cranfield, M.R., Fayer, R., Trout, J., Goodale, H.J., 1997. Infectivity of *Cryptosporidium parvum* oocysts is retained upon intestinal passage through a migratory waterfowl species (Canada goose, *Branta canadensis*). Trop. Med. Int. Health 2, 341–347.

Graczyk, T.K., Fayer, R., Trout, J.M., Lewis, E.J., Farley, C.A., Sulaiman, I., et al., 1998. *Giardia* sp. Cysts and infectious *Cryptosporidium parvum* oocysts in the feces of migratory Canada geese (*Brant canadensis*). Appl. Environ. Microbiol. 64, 2736–2738.

Graczyk, T.K., Sunderland, D., Rule, A.M., da Silva, A.J., Moura, I.N.S., Tanang, L., et al., 2007b. Urban feral pigeons (*Columba livia*) as a source of air- and waterborne contamination with Enterocytozoon bieneusi spores. Appl. Environ. Mibrobiol. 73, 4357–4358.

Graczyk, T.K., Majewska, A.C., Schwab, K.J., 2008. The role of birds in dissemination of human waterborne enteropathogens. Trends Parasitol. 24, 55–59.

Grimes, J.E., Owens, K.J., Singer, J.R., 1979. Experimental transmission of *Chlamydia psittaci* to turkeys from wild birds. Avian Dis. 23, 915–926.

Gustafson, T.L., Kaufman, L., Weeks, R., Ajello, L., Hutcheson Jr., R.H., Wiener, L.L., et al., 1981. Outbreak of acute pulmonary hisotplasmosis in members of a wagon train. Am. J. Med. 71, 759–765.

Gyles, C.L., 2007. Shiga toxin-producing *Escherichia coli*: an overview. J. Anim. Sci. 85 (13 Suppl), E45–E62.

Haag-Wackernagel, D., Moch, H., 2004. Health hazards posed by feral pigeons. J. Infect. 48 (4), 307–313.

Harkin, J.M., Phillips Jr., W.E., 1986. Isolation of *Listeria monocytogenes* from an eastern wild turkey. J. Wildl. Dis. 22, 110–112.

Haro, M., Izquierdo, F., Henriques-Gil, N., Andres, I., Alonso, F., Fenoy, S., et al., 2005. First detection and genotyping of human-associated nicrosporidia in pigeons from urban parks. Appl. Environ. Microbiol. 71, 3153–3137.

Hatch, J.J., 1996. Threats to public health from gulls (*Laridae*). Int. J. Environ. Health. Res. 6, 5–16.

Hejlicek, K., Prosek, F., Treml, F., 1981. Isolation of *Toxoplasma gondii* in free-living small mammals and birds. Acta. Vet. Brno. 50, 233–236.

Hellström, S., Kiviniemi, K., Autio, T., Korkeala, H., 2007. *Listeria monocytogenes* is common in wild birds in Helsinki region and genotypes are frequently similar with those found along the food chain. J. Appl. Microbiol. 104, 883–888.

Hubálek, Z., 2004. An annotated checklist of pathogenic microorganisms associated with migratory birds. J. Wildl. Dis. 40, 639–659.

Hughes, L.A., Bennett, M., Coffey, P., Elliott, J., Jones, T.R., Jones, R.C., et al., 2009. Risk factors for the occurrence of *Escherichia coli* virulence genes ease, stx1 and stx2 in wild bird populations. Epidemiol. Infect. 137, 1574–1582.

Hyngstrom, S.E., Timm, R.M., Larson, G.E., 1994. Prevention and Control of Wildlife Damage. University of Nebraska Cooperative Extension. US Department of Agriculture-Animal and Plant Health Inspection Service-Animal Damage Control. Great Plains Agricultural Council - Wildlife Committee, Lincoln, Nebraska.

Ijabadeniyi, O.A., Debusho, L.K., Vanderlinde, M., Buys, E.M., 2011. Irrigation water as a potential preharvest source of bacterial contamination of vegetables. J. Food Safety 31, 452–461.

Ito, K., Kubokura, Y., Kaneko, K., Totake, Y., Ogawa, M., 1988. Occurrence of *Campylobacter jejuni* in free-living wild birds from Japan. J. Wildl. Dis. 24, 467–470.

Jacobs, L., Melton, M.L., Jones, F.E., 1952. The prevalence of toxoplasmosis in pigeons. Exp. Parasitol. 2, 403–416.

Jiang, S.C., Chu, W., olson, B.H., He, J.W., Choi, S., Zhang, J., et al., 2007. Microbial source tracking in a small southern California urban watershed indicates wild animals and growth as the source of fecal bacteria. Appl. Microbiol. Biotechnol. 76, 927–934.

Johnston, R.R., Janiga, M., 1995. Feral Pigeons. Oxford Univ. Press, Oxford, pp. 320.

Jokinen, C., Edge, T.A., Ho, S., Koning, W., Laing, C., Mauro, W., et al., 2011. Molecular subtypes of *Campylobacter* spp., *Salmonella enterica*, and *Escherichia coli* O157:H7 isolated from faecal and surface water samples in the Oldman River watershed, Alberta, Canada. Water Res. 45, 1247–1257.

Kale, M., Balfors, B., Mortberg, U., Bhattacharya, P., Chakane, S., 2012. Damage to agricultural yield due to farmland birds, present repelling techniques and its impacts: an insight from the Indian perspective. J. Agricul. Technol. 8, 49–62.

Kaleta, E.F., 2002. Foot-and-mouth disease: susceptibility of domestic poultry and free-living birds to infection and to disease-A review of the historical and current literature concerning the role of birds in spread of foot-and-mouth disease viruses. Dtsch. Tierarztl. Wochenschr. 109, 391–399.

Kapperud, G., Rosef, O., 1983. Avian wildlife reservoir of *Campylobacter fetus* subsp. jejuni, Yersinia spp., and *Salmonella* spp. in Norway. Appl. Environ. Microbiol. 45, 375–380.

Kinzelman, J., McLellan, S.L., Amick, A., Preedit, J., Scopel, C.O., Olapade, O., et al., 2008. Identification of human enteric pathogens in gull feces at Southwestern Lake Michigan bathing beaches. Can. J. Microbiol. 54, 1006–1015.

Kirk, J.H., Holmberg, C.A., Jeffrey, J.S., 2002. Prevalence of *Salmonella* spp in selected birds captured on California dairies. J. Am. Vet. Med. Assoc. 220, 359–362.

Kobayashi, H., Pohjanvirta, T., Pelkonen, S., 2002. Prevalence and characteristics of intimin- and Shiga toxin-producing *Escherichia coli* from gulls, pigeons and broilers in Finland. J. Vet. Med. Sci. 64, 1071–1073.

Kobayashi, H., Kanazaki, M., Hata, E., Kubo, M., 2008. Prevalence and characteristics of eae- and stx-Positive *Escherichia coli* from wild birds in the immediate environment of Tokyo Bay. Appl. Environ. Microbiol. 75, 292–295.

Krampitz, E.S., Holländer, R., 1999. Longevity of pathogenic bacteria especially *Salmonella* in cistern water. Zentralbl. Hyg. Umweltmed. 202, 389–397.

Kruse, H., Kirkemo, A.M., Handeland, K., 2004. Wildlife as source of zoonotic infections. Emerg. Infect. Dis. 10, 2067–2072.

Kullas, H., Coles, M., Rhyan, J., Clark, L., 2002. Prevalence of *Escherichia coli* serogroups and human virulence factors in faeces of urban Canada geese *(Branta canadensis)*. Int. J. Environ. Health Res. 12, 153–162.

Laiolo, P., 2005. Spatial and seasonal patterns of bird communities in Italian agroecosystems. Conservation Biol. 19, 1547–1556.

LeJeune, J., Homan, J., Linz, G., Pearl, D.L., 2008. Role of the European starling in the transmission of E. coli O157 on dairy farms. Proc. Vert. Pest Conf. 23, 31–38.

Latgé, J.P., 2001. The pathobiology of *Aspergillus fumigatus*. Trends Microbiol. 9, 382–389.

Latham, R.H., Kaiser, A.B., Dupont, W.D., Dan, B.B., 1980. Chronic pulmonary histoplasmosis following the excavation of a bird roost. Am. J. Epidemiol. 68, 504–508.

Levesque, B., Brousseau, P., Simard, P., Dewailly, E., Meisels, M., Ramsay, D., et al., 1993. Impact of the ring-billed gull *(Larus delawarensis)* on the microbiological quality of recreational water. Appl. Environ. Microbiol. 59, 1228–1230.

Luechtefeld, N.A., Blaser, M.J., Reller, L.B., Wang, W.L., 1980. Isolation of *Campylobacter fetus* subsp. *jejuni* from migratory waterfowl. J. Clin. Microbiol., 12406–12408.

Linz, G.M. (Ed.), 2003. Management of North American blackbirds. National Wildlife Research Center, Fort Collins, Colorado, USA.

Lord, A.T.K., Mohandas, K., Somanath, S., Ambu, S., 2010. Multidrug resistant yeasts in synanthropic wild birds. Annal. Clin. Microbiol. Antimibiotics 9, 1–5.

Mackenzie, W.R., Hoxie, N.J., Proctor, M.E., Gradus, M.S., Blair, K.A., Peterson, D.E., et al., 1994. A massive outbreak in Milwaukee of Cryptosporidium infection transmitted through the public water supply. N. Engl. J. Med. 331, 161–167.

Makino, S., Kobori, H., Asakura, H., Watarai, M., Shirahata, T., Ikeda, T., et al., 2000. Detection and characterization of Shiga toxin-producing *Escherichia coli* from seagulls. Epidemiol. Infect. 125, 55–61.

Mason, J.R. (Ed.), 1997. Repellents in wildlife management: proceedings of a symposium. Proceedings of the Second DWRC Special Symposium (August 8–10, 1995, Denver, Colorado). National Wildlife Research Center, Fort Collins, Colorado, USA.

Mason, J.R., Clark, L., 1992. Nonlethal repellents: the development of cost-effective, practical solutions to agricultural and industrial problems. Proc. Vertebrate Pest Con. 15, 115–129.

Mason, J.R., Clark, L., 1996. Grazing repellency of methyl anthranilate to snow geese is enhanced by a visual cue. Crop Protect. 15, 97–100.

Mathis, A., Weber, B., Deplazes, P., 2005. Zoonotic potential of the *microsporidia*. Clin. Microbiol. Rev. 18, 423–445.

McCarthy, S., Ng, J., Gordon, C., Miller, R., Wyber, A., Ryan, U.M., 2008. Prevalence of Cryptosoridium and Giardia species in animals in irrigation catchments in southwest Australia. Exp. Parasitol. 118, 596–599.

McDiarmid, A., 1969. Diseases in free-living wild animals. Academic Press, London, UK, pp. 332.

McKay, H., Watola, G.V., Langton, S.D., Langton, S.A., 2006. The use of agricultural fields by re-established greylag geese (*Anser anser*) in England: a risk assessment. Crop Protect. 25, 996–1003.

Merkens, M., Bradbeer, D.R., Bishop, C.A., 2012. Landscape and field characteristics affecting winter waterfowl grazing damage to agricultural perennial forage crops on the lower Fraser River delta, BC, Canada. Crop Protect. 37, 51–58.

Middleton, J.H., Ambrose, A., 2005. Enumeration and antibiotic resistance patterns of fecal indicator organisms isolated from migratory Canada geese (*Branta canadensis*). J. Wildl. Dis. 41, 334–341.

Millar, B.C., Finn, M., Xiao, L., Lowery, J.C., Dooley, J.S.G., Moore, J.E., 2002. Cryptosporidium in food-stuffs—an emerging aetiological route of human foodborne illness. Trends Food Sci. Technol. 13, 168–187.

Mohapartra, B.R., Broersma, K., Mazumder, A., 2008. Differentiation of fecal *Escherichia coli* from poultry and free-iving birds by (GTG)5-PCR genomic fingerprinting. Int. J. Med. Microbiol. 298, 245–252.

Morabito, S., Dell'Omo, G., Agrimi, U., Schmidt, H., Karch, H., Cheasty, T., Caprioli, A., 2001. Detection and characterization of Shiga toxin-producing *Escherichia coli* in feral pigeons. Vet. Microbiol. 82, 275–283.

Mosupye, F.M., von Holy, A., 2000. Microbiological hazard identification and exposure assessment of street food vending in Johannesburg, South Africa. Int. J. Food Microbiol. 61 (2–3), 137–145.

Mouysset, L., Doyen, L., Jiguet, F., Allaire, G., Leger, F., 2011. Bio economic modeling for a sustainable management of biodiversity in agricultural lands. Ecol. Econ. 70, 617–626.

Nelson, M., Jones, S.H., Edwards, C., Ellis, J.C., 2008. Characterization of *Escherichia coli* populations from gulls, landfill trash, and wastewater using ribotyping. Dis. Aquat. Org. 81, 53–63.

Nesse, L.L., Refsum, T., Heir, E., Nordby, K., Vardund, T., Holstad, G., 2005. Molecular epidemiology of *Salmonella* spp. isolates from gulls, fish-meal factories, feed factories, animals and humans in Norway based on pulsed-field gel electrophoresis. Epidemiol. Infect. 133, 53–58.

Nielsen, E.M., Skov, M.N., Madsen, J.J., Lodal, J., Jespersen, J.B., Baggesen, D.L., 2004. Verocytotoxin-producing and in wild birds and rodents in close proximity to farms. Appl. Environ. Microbiol. 70, 6944–6947.

O'Connor, K., 1992. The Herring Gull. Dillon Press, Toronto, Canada, pp. 65.

Odermatt, P., Gautsch, S., Rechsteiner, D., Ewald, R., Haag-Wackernagel, D., Mühlemann, R., et al., 1998. Swarms of starlings in Basel: a natural phenomenon, a nuisance or a health risk? Gesundheitswesen 60, 749–754.

Pak, S.M., 1976. Toxoplasmosis of birds in Kazakhstan (in Russian). Contrib. Nat. Nidality Dis. 5, 116–125.

Palmgren, H., Aspán, A., Broman, T., Bengtsson, K., Blomquist, L., Bergström, S., et al., 2006. *Salmonella* in Black-headed gulls (*Larus ridibundus*); prevalence, genotypes and influence on *Salmonella* epidemiology. Epidemiol. Infect. 134, 635–644.

Pasmans, F., Van Immerseel, F., Hermans, K., Heyndrickx, M., Collard, J.M., Ducatelle, R., et al., 2004. Assessment of virulence of pigeon isolates of *Salmonella enterica* subsp. *enterica* serovar *typhimurium* variant *copenhagen* for humans. J. Clin. Microbiol. 42, 2000–2002.

Pedersen, K., Clark, L., Andelt, W.F., Salman, M.D., 2006. Prevalence of Shiga toxin-producing *Escherichia coli* and *Salmonella enterica* in rock pigeons captured in Fort Collins, Colorado. J. Wildl. Dis 42, 46–55.

Pollock, C., 2003. Fungal diseases of columbiformes and anseriformes. Vet. Clin. North. Am. Exot. Anim. Pract. 6, 351–361.

Power, A., 2011. Ecosystem services and agriculture: tradeoffs and synergies. Phil. Trans. R. Soc. B. 365 2989–2971.

Quessy, S., Messie, R.S., 1992. Prevalence of *Salmonella* spp., *Campylobacter* spp. and *Listeria* spp. in ring-billed gulls (*Larus delawarensis*). J. Wildl. Dis. 28, 526–531.

Quevedo, F., Lord, R.D., Dobosch, D., Granier, I., Michanie, S.C., 1973. Isolation of *Salmonella* from sparrows captured in horse corrals. Am. J. Trop. Med. Hyg. 22, 672–674.

Quinn, P.J., 1994. Clinical Veterinary Microbiology. Elsevier Health Sciences, New York, pp. 648.

Ramos, R., Cerda-Cuellar, M., Ramirez, F., Jover, L., Ruiz, X., 2010. Influence of refuse sites on the prevalence of *Campylobacter* spp. and *Salmonella* serovars in seagulls. Appl. Environ. Microbiol. 76, 3052–3056.

Rice, D.H., Sheng, H.Q., Wynia, S.A., Hovde, C.J., 2003. Rectoanal mucosal swab culture is more sensitive than fecal culture and distinguishes *Escherichia coli* O157:H7-colonized cattle and those transiently shedding the same organism. J. Clin. Microbiol. 41, 4924–4929.

Roberts, J.P., Grimes, J.E., 1978. *Chlamydia* shedding by four species of wild birds. Avian Dis. 22, 698–706.

Riordan, T., Humhrey, T.J., Fowles, A., 1993. A point source outbreak of *Campylobacter* infection related to bird-pecked milk. Epidemiol. Infect. 110, 261–265.

Rosario, I., Acosta, B., Colom, M.F., 2008. Pigeons and other birds as a reservoir for *Cryptococcus* spp. Rev. Iberoam. Micol. 25, S13–S18.

Rubin, H., Furcolow, M.L., Yates, J.L., Brasher, C.A., 1959. The course and prognosis of histoplasmosis. Am. J. Med. 27, 278–288.

Ruiz, A., Neilson, J.B., Bulmer, G.S., 1982. A one year study on the viability of *Cryptococcus neoformans* in nature. Mycopathologia 77, 117–122.

Rutledge, M.E., Siletzky, R.M., Gu, W., Degernes, L.A., Moorman, C.E., DePerno, C.S., et al., 2013. Characterization of *Campylobacter* from resident Canada geese in an urban environment. J. Wildl. Dis. 40, 1–9.

Schwarz, J., Baum, G.L., Wang, C.J., Bingham, E.L., Rubel, H., 1957. Successful infection of pigeons and chickens with *Histoplasma capsulatum*. Mycopathol. Mycol. Appl. S8, 189–193.

Siim, J.C., Biering=Sorensen, U., Moller, T., 1963. Toxoplasmosis in domestic animals. Adv. Vet. Sci. 8, 335–429.

Sivapalasingam, S., Barrett, E., Kimura, A., Van duyne, S., De Witt, W., Ying, M., et al., 2003. A multistate outbreak of *Salmonella* enteric serotype Newport infection linked to mango consumtion: impact of water-tip disinfestations technology. Clin. Infect. Dis. 37, 1585–1590.

Sixl, W., Karpísková, R., Hubálek, Z., Halouzka, J., Mikulásková, M., Salava, J., 1997. *Campylobacter* spp. and *Salmonella* spp. in black-headed gulls (*Larus ridibundus*). Cent. Eur. J. Public Health 5, 24–26.

Slodkowicz-Kowalska, A., Graczyk, T.K., Tamang, L., Jedrzejewski, S., Nowosad, A., Zduniak, P., et al., 2006. Microsporidian species know to infect humans are present in aquatic birds: implications for transmission via water? Appl. Environ. Microbiol. 72, 4540–4544.

Smith, C.D., Furcolow, M.L., Tosh, F.E., 1964. Attempts to eliminate *Histoplasma capsulatum* from soil. Am. J. Hyg. 79, 170–180.

Smith, H.V., Brown, J., Soulson, J.C., Morris, G.P., Dirdwood, R.W.A., 1993. Occurrence of oocysts of *Cryptosporidium* sp. In *Larus* spp. Gulls. Epidemiol. Infect. 110, 135–143.

Snoeyenbos, G.H., Morin, E.W., Wetherbee, D.K., 1967. Naturally occurrine *Salmonella* in "blackbirds" and gulls. Avian Dis. 11, 642–646.

Sonntag, A.K., Zenner, E., Karch, H., Bielaszewska, M., 2005. Pigeons as a possible reservoir of Shiga toxin 2f-producing *Escherichia coli* pathogenic to humans. Berl. Munch. Tierarztl. Wochenschr 118, 464–470.

Somarelli, J.A., Makarewicz, J.C., Sia, R., Simon, R., 2007. Wildlife identified as major source of *Escherichia coli* in agriculturally dominated watersheds by BOX A1R-derived genetic fingerprints. J. Environ. Manag. 82, 60–65.

Soogarun, S., Wiwanitkit, V., Palasuwan, A., Pradniwat, P., Suwansaksri, J., Lertlum, T., et al., 2006. Detection of *Cryptococcus neoformans* in bird excreta. Southeast Asian J. Trop. Med. Public Health 37, 768–770.

Steele, M., Odumeru, J., 2004. Irrigation water as a source of foodborne pathogens on fruit and vegetables. J. Food. Prot. 67, 2819–2849.

Stevens, G.R., Rogue, J., Weber, R., Clark, L., 2000. Evaluation of a radar-activated, demand-performance bird hazing system. Int. Biodeterioration Biodegradation 45, 129–137.

Stobierski, M.G., Hospedales, C.J., Hall, W.N., Robinson-Dunn, B., Hoch, D., Sheill, D.A., 1996. Outbreak of histoplasmosis among employees in a paper factory–Michigan, 1993. J. Clin. Microbiol. 34, 1220–1223.

Storch, G., Burford, J.G., George, R.B., Kaufman, L., Ajello, L., 1980. Acute histoplasmosis. Description of an outbreak in northern Louisiana. Chest 77, 38–42.

Tanaka, C., Miyazawa, T., Watarai, M., Ishiguro, N., 2005. Bacteriological survey of feces from feral pigeons in Japan. J. Vet. Med. Sci. 67, 951–953.

Taylor, L.H., Latham, S.M., Woolhouse, M.E., 2001. Risk factors for human disease emeregence. Philos. Trans. R. Soc. Lond. B. Biol. Sci. 356, 983–989.

Thurston-Enriquez, J.A., Watt, P., Dowd, S.E., Enriquez, R., Pepper, I.L., Gerba, C.P., 2002. Detection of protozoan parasites and *microsporidia* in irrigation waters used for crop production. J. Food. Prot. 65, 378–382.

Tosh, F.E., Weeks, R.J., Pfeiffer, F.R., Hendricks, S.L., Greer, D.L., Chin, T.D., 1967. The use of formalin to kill *Histoplasma capsulatum* at an epidemic site. Am. J. Epidemiol. 85, 259–265.

Tracey, J.P., Saunders, G.R., 2010. A technique to estimate bird damage in wine grapes. Crop Protect. 29, 435–439.

Tsiodras, S., Kelesidis, T., Kelesidis, I., Bauchinger, U., Falagas, M.E., 2008. Human infections associated with wild birds. J. Infect. 56, 83–98.

Verkery, J.A., Feber, R.E., Fuller, R.J., 2009. Arable field margins managed for biodiversity conservation: A review resource provision for farmland birds. Agricul. Ecosyst. Environ. 133, 1–13.

Wallace, J.S., Cheasty, T., Jones, K., 1997. Isolation of vero cytotoxin-producing *Escherichia coli* O157 from wild birds. J. Appl. Microbiol. 82, 399–404.

Weber, A., Schäfer, R., 1991. The occurrence of *Cryptococcus neoformans* in fecal samples from birds kept in human living areas. Berl. Munch. Tierarztl. Wochenschr. 104, 419–421.

Weber, R., Bryan, R.T., Schwartz, D.A., Owen, R.L., 1994. Human microsporidial infections. Clin. Microbiol. Rev. 7, 426–461.

Wenny, D.G., DeVault, T.L., Johnson, M.D., Kelly, D., Sekercioglu, C.H., Tomback, D.F., et al., 2011. The need to quantify ecosystem services provided by birds. Auk. 128, 1–14.

Werner, S.J., Cummings, J.L., Tupper, S.K., Goldade, D.A., Beighley, D., 2008. Blackbird repellency of selected registered pesticides. J. Wildlife Manag. 72, 1007–1011.

Wither, A., Rehfisch, M., Austin, G., 2005. The impact of bird populations on the microbiological quality of bathing waters. Water Sci. Technol. 51, 199–207.

Yorio, P., Caille, G., 2004. Fish waste as an alternative resource for gulls along the Patagonian coast: availability, use, and potential consequences. Mar. Pollut. Bull. 48, 778–783.

Yoshida, T., Sugimoto, T., Sato, M., Hirai, K., 2000. Incidence of *Listeria monocytogenes* in wild animals in Japan. J. Vet. Med. Sci. 62, 673–675.

Zhang, Y., Yeh, E., Hall, G., Cripe, J., Bhagwat, A.A., Meng, J., 2007. Characterization of *Listeria monocytogenes* isolated from retail foods. Int. J. Food Microbiol. 113 (1), 47–53.

Produce Contamination by Other Wildlife

Daniel H. Rice

Food Laboratory Division, New York State Department of Agriculture and Markets, Albany, NY

CHAPTER OUTLINE

Introduction

Fresh produce is well established as an important source of foodborne illness in humans, as evidenced by its being a vehicle of transmission in several large, multistate outbreaks of disease involving a variety of pathogens and fresh produce. The 2008 FAO/WHO microbiological hazards in fresh fruits and vegetables meeting report lists leafy green vegetables as level 1 priority commodities for contamination with enterohemorrhagic *E. coli, S. enterica, Campylobacter* spp., *Shigella* spp., hepatitis A virus, noroviruses, *Cyclospora cayatenensis, Cryptosporidium, Yersinia pseudotuberculosis,* and *L. monocytogenes,* most of which are zoonotic pathogens (FAO/WHO, 2008). In this report wildlife are singled out as being potential sources of contamination of fresh produce with human pathogens. The report lists a number of questions to be addressed related to what the potential roles are of wildlife in contaminating produce, either directly or by environmental contamination routes. The lack of information on wildlife reservoirs of human diseases that can be transmitted by contaminated fresh produce is evidenced by this report. Some of the gaps regarding this issue will be addressed in this chapter.

The majority of agricultural production takes place in rural areas that also provide habitat to a wide variety of feral domestic and wild animals. Fresh produce is utilized as a food source by a variety of wild animals. Farmers lose an estimated $4.5 billion annually in crop production losses due to wild animals consuming unharvested

The Produce Contamination Problem. http://dx.doi.org/10.1016/B978-0-12-404611-5.00008-7

produce or damaging crops prior to harvest (Conover, 2002). Although there are some measures that can be taken to minimize the opportunity for produce in the field to come in contact with wild animals, the options are limited, and in most instances it is impossible to keep wild animals out of farm land. Very often wildlife control programs to mitigate crop damage are at odds with recreational activities such as hunting. It is for this reason and others that wildlife will continue to be directly associated with most agricultural production practices. Control measures to protect produce postharvest are more readily available and apparent. These are important because, as described in other chapters of this book, there are important considerations in the processing, packaging, and transport sectors that relate to the potential for wildlife to contaminate produce once harvested.

Up to one-third of the population of developed countries experiences a foodborne illness each year (Schlundt et al., 2004). Wildlife have been implicated as possible sources of contamination in produce-borne illness including several outbreaks of disease (Besser et al., 1993; Beuchat and Ryu, 1997; Brackett, 1999; Burnett and Beuchat, 2001; Cody et al., 1999; Greene et al., 2008; Jay et al., 2007), and although an outbreak of human illness associated with produce directly contaminated by wildlife has never been reported, there are a number of reasons to believe these have occurred. First, each year foodborne pathogens are associated with an estimated 76 million illnesses, 325,000 hospitalizations, and 5200 deaths in the United States (Mead et al., 1999). The CDC and others estimate that only a fraction of foodborne illnesses are actually detected, and of those that are detected, a food vehicle for infection is rarely determined. Second, wildlife are reported to carry many infectious agents that have been associated with foodborne illness in humans. Consequently, the probability that wildlife have been and will continue to be sources of contamination of produce that subsequently cause human illnesses is high; however, the extent to which wildlife contribute to the contamination of produce resulting in human illness is not known. The most compelling evidence for a wildlife role in contaminating produce is that of feral swine being implicated in a 2006 multistate outbreak of *E. coli* O157:H7 linked to packaged fresh baby spinach (Jay et al., 2007). Deer have been implicated as a possible source of *E. coli* O157:H7 contamination of apples used to make both unpasteurized juice and cider that were linked to outbreaks of illness in humans, although the links in these outbreaks were weak at best (Besser et al., 1993; Cody et al., 1999).

For the purpose of discussion in this chapter, the focus will be primarily on the species typically thought of as wildlife, primarily mammals, including feral domestic mammals. Other types of animals will be discussed where relevant. The potential role of birds in contaminating produce is significant and is discussed in another chapter. In addition, this chapter will discuss the potential role that insects, mollusks, and helminths play in transmitting pathogens to produce. This chapter will discuss the potential for occurrence of direct contamination of produce by wildlife. Indirect contamination, for example, via irrigation water, is covered in another chapter. A comprehensive review of the peer reviewed published literature indicates that many zoonotic pathogens have been reported to be carried by one or more species

Table 8.1 Zoonotic Pathogens Isolated from both Fresh Produce and Wild Animals

	Zoonotic Pathogen	**Known Wildlife Hosts and Relevant Intermediate Hosts**
Virus	Hepatitis E	Feral swine, wild boar, rodents, primates, and a wide variety of other mammals
Bacteria	*Campylobacter* spp. *S. enterica* Pathogenic *E. coli* *L. monocytogenes* and *Yersinia enterocolitica* *Arcobacter* spp.	Birds, mammals, marsupials, reptiles, amphibians Mammals Mammals, birds Not known
Protozoa	*C. parvum* *Giardia* spp. *Toxoplasma gondii*	Mammals Mammals Wild felids
Helminths— nematodes	*Ascaris suum* *Toxocara canis* *Toxocara cati* *Toxascaris leonine* *Lagochilascaris minor* *Angiostrongylus cantonensis* *Angiostrongylus costaricensis*	Feral swine, wild boar Wild canids Wild felids Wild canids Rodents, felids, and raccoons Rodents via gastropod intermediate hosts Rodents via gastropod intermediate hosts
Helminths— trematodes	*Fasciola hepatica* and *gigantica* *Fasciolopsis buski*	Wild ruminants, equids, and lagomorphs via a lymnaeid intermediate host Feral swine via a lymnaeid intermediate host

of wildlife and have the potential to be transmitted to humans through fresh produce (Table 8.1). As pointed out by Schlundt et al. (2004), the top five emerging foodborne diseases are caused by *S. enterica*, *Campylobacter* spp., enterohaemorrhagic *E. coli*, *Toxoplasma gondii*, and *Cryptosporidium* spp., all of which are zoonotic, and all having the potential of being transmitted to humans through fresh produce.

Viral pathogens

It is estimated that 67.2% of foodborne illnesses are viral (Mead et al., 1999); however, viral pathogens are almost exclusively host-specific. Hence the vast majority of these cannot be transmitted to humans by wildlife. The only apparent exception is hepatitis E, which is carried by a variety of wild, domestic, and feral mammals (Goens and Perdue, 2004; Smith, 2001). The primary exposure to humans in industrialized countries is through direct contact with infected animals, primarily swine, or eating undercooked meat and organs from infected animals (Teo, 2006). Seropositive rates for hepatitis E in blood donors from developed countries range from 0.4 to 3.3% positive (Smith, 2001). In non-industrialized countries, waterborne exposure is the primary route of infection (Seymour and Appleton, 2001), and seropositive rates range from 9.5 to 54.8% (Smith, 2001). Hepatitis E is transmitted mainly via the fecal–oral route, so in theory, wild animals shedding hepatitis E in their feces could contaminate produce that is consumed by humans, resulting in transmission of disease. Viruses are very difficult to detect in food samples, with detection techniques historically based on scanning electron microscopy. More recently, molecular-based technologies have been employed and should provide better information on the incidence of hepatitis E and other viral agents in food, including produce (Seymour and Appleton, 2001). Hepatitis E has not been reported to have been detected in fresh produce.

Bacterial pathogens

Bacterial agents cause an estimated 30.2% of all foodborne illnesses (Mead et al., 1999). The zoonotic bacterial organisms most commonly associated with foodborne disease are *C. coli* and *jejuni, S. enterica, E. coli* O157:H7, and *L. monocytogenes*, all of which have been isolated from fresh produce (Brackett, 1999; Mead et al., 1999; Schlundt et al., 2004; Tauxe, 2002). *Yersinia enterocolitica,* also a zoonotic foodborne pathogen, has been isolated from fresh produce, although never linked to a produce-borne outbreak of illness in humans (Beuchat, 1995; Brackett, 1999; Burnett and Beuchat, 2001). In addition, *Arcobacter* spp. appear to be involved in a limited number of foodborne illnesses each year, but the data on this potential zoonotic pathogen are very limited (Ho et al., 2006). Additional serotypes of enterohemorrhagic *E. coli* including O26, O103, O111, O118, and O145 in the future may be identified as important zoonotic pathogens associated with fresh produce. These have all been associated with human illness (Schlundt et al., 2004), many are associated with livestock (Cobbold et al., 2008; Frank et al., 2008), and at least one (O111) has been implicated in an outbreak of illness associated with unpasteurized apple juice (Cobbold et al., 2008; Frank et al., 2008; Vojdani et al., 2008).

Many of the zoonotic pathogens in humans that have been linked to a produce vehicle of infection, including *S. enterica, E. coli* O157:H7, *Camplyobacter* spp., and *L. monocytogenes*, have also been isolated from wildlife (Brown et al., 2004; Jijon et al., 2007; Lyautey et al., 2007; Parish, 1997; Renter et al., 2001, 2006; Rice et al., 2003; Scaife et al., 2006; Wahlstrom et al., 2003). However, wildlife have yet to be demonstrated as being a direct source of produce contamination that resulted in an outbreak of illness in humans. There is compelling evidence that feral swine may have contributed to the contamination of spinach with *E. coli* O157:H7 in a large multistate outbreak in 2006 (Jay et al., 2007). In the course of investigating this outbreak, it was found that a number of environmental origin isolates of *E. coli* O157:H7, including those from feral swine, genetically matched the human outbreak strain. Interestingly, over 33% of cattle feces, 23% of feral swine feces, 4% of water samples, and 8% of soil samples, collected from close proximity to cropland implicated as source fields for contaminated spinach, were *E. coli* O157:H7 positive, and most matched the outbreak strain (Jay et al., 2007). Swine feces were commonly found in fields used to grow spinach. This indicates that a long-term reservoir for *E. coli* O157:H7 must have existed in close proximity to these fields; whether or not this reservoir was maintained in animals or the environment was not determined. Feral swine have previously been identified as carrying *E. coli* O157:H7 in a survey of wildlife in Sweden (Wahlstrom et al., 2003). In this study several species of wild animals were surveyed for zoonotic pathogens. Only one of 68 feral swine samples tested were *E. coli* O157:H7 positive, suggesting that wild pigs were transiently colonized rather than a reservoir of this organism.

Other outbreaks of produce-borne illness in humans with epidemiological links to wildlife include apple juice and cider contaminated with *E. coli* O157:H7, attributed to use of apples that had been picked off the ground (Besser et al., 1993; Cody et al., 1999). The authors hypothesized that these apples may have been contaminated with deer feces positive for *E. coli* O157:H7 since there was evidence that deer frequented the orchard where the apples came from. Enteric illness of unknown etiology subsequent to consuming freshly pressed apple cider that was made using dropped apples, obtained from orchards where presence of deer was evident, has also been reported (Vojdani et al., 2008).

Environmental reservoirs may play an important role in the colonization of wild animals by *E. coli* O157:H7 and other zoonotic pathogens. *E. coli* O157:H7 has been demonstrated to remain in farm environments for extended periods of time, providing an apparent source for farm animal colonization over time (LeJeune et al., 2001; Rice et al., 1999; Van Donkersgoed et al., 2001). LeJeune et al. (2001) demonstrated that *E. coli* O157:H7 can remain viable in cattle water trough sediments for up to 245 days and that contaminated water from troughs with no animal contact for over six months were capable of infecting cattle. Cattle water troughs have been identified as potentially important reservoirs of *E. coli* O157:H7 in additional studies (Hancock et al., 1998, Van Donkersgoed et al., 2001; Wetzel and LeJeune, 2006), and persistence of strains within herds for up to 24 months indicates that environmental reservoirs are important in maintaining this organism in cattle herds (Rice et al., 1999).

An extensive longitudinal survey of range cattle environments by Renter et al. (2003) demonstrated that 0.51% of cattle water tanks, 0.25% of lakes and ponds, and 0.41% of free-flowing rivers and streams were positive for a diverse number of *E. coli* O157:H7 strains. In this study an identical strain of *E. coli* O157:H7 was shared by cattle, water, and wildlife. Since environmental reservoirs have been demonstrated to be important in maintaining *E. coli* O157:H7 in cattle production facilities, by implication, environmental reservoirs may be important in maintaining this and other pathogens in wildlife.

Human illnesses with *C. coli* and *jejuni*, *S. enterica*, *E. coli* O157:H7, and *L. monocytogenes* are often associated with foodborne exposure, and each of these organisms has been isolated from produce and from wild animals. Wildlife have been implicated as sources of foodborne pathogens for both outbreaks and individual cases of illness in humans. Only a limited number of surveys of wild animals for zoonotic pathogens exist. In general, these reports identify species of animals that tested positive for specific pathogens, based on a limited number of samples within a limited geographical area, and most are not capable of providing true prevalence estimates in selected populations of animals. Testing of wild animal feces indicates that deer, moose, rabbits, opossums, and wild boar/feral swine can carry *E. coli* O157:H7 (Jay et al., 2007; Renter et al., 2001, 2003, 2006; Rice et al., 2003; Scaife et al., 2006; Wahlstrom et al., 2003). Renter et al. (2001) demonstrated that *E. coli* O157:H7 prevalence in Nebraska white-tailed deer during the fall hunting season was 0.25%, indicating a generally low prevalence in this population of deer. Other surveys have demonstrated a similarly low prevalence in wildlife. However, a survey of wild rabbits in the UK demonstrated that over 8% of rabbit fecal samples tested during the summer were *E. coli* O157:H7 positive (Scaife et al., 2006).

Surveys of wildlife for *S. enterica* indicate that the majority of animals that test positive are birds; however, this organism has also been found in foxes, opossums, gray squirrels, woodchucks, and toads (Jijon et al., 2007; Parish, 1997; Wahlstrom et al., 2003). Eight of 71 (11%) wild animals tested at a rehabilitation center were positive for four serovars of *S. enterica*, and five (7%) of these were non-avian wildlife (opossum, gray squirrel, and woodchuck), indicating a relatively high prevalence of this organism in this population of nonavian wildlife (Jijon et al., 2007). One survey of Swedish wildlife for *Campylobacter* spp. indicated that hares, moose, and feral swine can carry thermophilic *Campylobacter* with prevalences ranging between 1 and 12%, depending upon species tested (Wahlstrom et al., 2003). Another survey of wild animal feces reported that 11% of non-avian feces tested were *Campylobacter* positive; this study did not report feces by species of animal (Brown et al., 2004).

Given its nearly ubiquitous nature and the fact that it is the most common cause of enteric bacterial illness, it is likely that many if not most wild animals have the potential to carry pathogenic species of *Campylobacter*. Unlike most foodborne bacterial pathogens, *Campylobacter* is very sensitive to a wide variety of environmental stressors and unable to multiply outside of an animal host (Park, 2002). In spite of these limitations, *Campylobacter* is the leading cause of bacterial foodborne illness (Mead et al., 1999) possessing a variety of mechanisms that allow it to persist in the environment and on food once contaminated (Cook and Bolster, 2007; Karenlampi and Hanninen, 2004; Murphy et al., 2006). A study that compared *L. monocytogenes* isolates from livestock, wildlife, and humans in Ontario, Canada demonstrated that this organism could be found in 6% of deer, 5% of moose, and 50% of both otter and raccoon feces, indicating that this organism is readily carried by a variety of wildlife (Lyautey et al., 2007). Identical Pulsed Field Gel Electrophoresis (PFGE) patterns of *L. monocytogenes* were found in deer, moose, and cattle, indicating either exposure to a common source or direct contact-associated serial infections.

None of the reported surveys for zoonotic pathogens in wildlife are comprehensive enough to allow estimates of true geographical prevalence among any species of wildlife or to comprehensively identify which species of animals do and do not carry specific zoonotic pathogens. The value of these reports lies in the fact that a wide range of wild animals have been reported to carry all the major foodborne zoonotic bacterial pathogens.

Yersinia enterocolitica has been isolated from fresh produce, but an outbreak of illness in humans has never been associated with produce (Beuchat, 1995). *Y. enterocolitica* is most commonly associated with domestic pigs but has been isolated from a wide variety of wild animals (Shayegani et al., 1986). This study reported that 10% of wild mammals tested from New York State had detectable *Y. enterocolitica* in their feces. It is not known if *Y. enterocolitica* is responsible for produce-borne illness in humans and whether or not wildlife are sources of produce contamination; however, the potential for both exists.

Arcobacter is a relatively new genus of organisms that has been associated with human illness, is zoonotic, has been isolated from food, and is frequently isolated from the environment (Ho et al., 2006). At this time nothing is known about what role, if any, *Arcobacter* plays in produce-borne illness in humans and whether or not there is a wildlife reservoir.

In addition to wildlife being sources of contamination of fresh produce, insects can be a mechanical vector for many foodborne bacterial pathogens. Fruit flies have been demonstrated under experimental conditions to transport *E. coli* to uncontaminated fruit (Janisiewicz et al., 1999; Sela et al., 2005). Fruit flies exposed to composting apples, inoculated with a genetically distinct strain of *E. coli*, readily transported *E. coli* to uncontaminated apples. Both generic *E. coli* and *E. coli* O157:H7 were shown to grow exponentially in apple wounds, demonstrating that fruit flies are viable vectors for contaminating fresh fruit with pathogens, and that once contaminated, fruit wounds are an excellent site for these pathogens to multiply (Janisiewicz et al., 1999). Fruit flies, exposed to feeding stations and feces artificially contaminated with a genetically unique strain of *E. coli*, were subsequently demonstrated to carry *E. coli* for up to seven days and to introduce *E. coli* to apple wounds (Sela et al., 2005). House flies and filth flies can carry a variety of zoonotic foodborne pathogens including *E. coli* O157:H7, *S. enterica*, and *C. jejuni* (Hancock et al., 1998; Iwasa et al., 1999; Olsen, 1998). These flies are frequently found in crop production areas that are adjacent to livestock facilities or other environments conducive to their reproduction. Additionally, the aquatic midge *Chironomus tentans* has been demonstrated to carry *S. enterica* from the larval stage to the adult fly stage in experimental settings using contaminated aquatic sediments and fresh water as the pathogen source (Moore et al., 2003). The larval stages of this organism inhabit a wide variety of aquatic environments including riparian habitats, irrigation canals, and cattle watering troughs; consequently, this insect may participate in the ecology of *S. enterica* in the environment. It is clear that in order for insects to play a role in contaminating fresh fruit or produce, a nearby source of contamination is needed since these vectors do not generally travel very far, have short life spans, and infections are apparently not passed on from generation to generation. Many orchards and produce farms are located in livestock-intensive areas and areas with abundant wildlife. Consequently, fecal sources of pathogens for contaminating insects exist, and insect vectors are potentially important sources of contamination.

There is an interesting body of literature on the soil nematode *Caenorhabditis elegans*. This nematode can be colonized with zoonotic foodborne pathogens including *S. enterica, E. coli* O157:H7, and *L. monocytogenes* when exposed to contaminated soil (Caldwell et al., 2003a, 2003b; Kenney et al., 2005, 2006; Labrousse et al., 2000). These small soil nematodes (1.5 mm long) feed on bacteria, and it has been hypothesized that they are candidates for mechanically contaminating certain crops that come into direct contact with soil where a pathogen source exists. It has been shown experimentally that once *S. enterica* is internalized in *C. elegans*, it is apparently protected from many commercial sanitizers, indicating that produce containing *S. enterica*-infected *C. elegans* may remain a source of human infection in spite of cleaning and sanitizing (Caldwell et al., 2003a).

Parasitic pathogens

Determining the prevalence of parasitic foodborne diseases in humans is difficult mainly because detecting the infective stages of these organisms in food presents several challenges. Epidemiological evidence indicates that 2.6% of foodborne diseases are parasitic (Mead et al., 1999). The vast majority of parasitic foodborne infections are from protozoa; however, infections with nematodes and trematodes have been documented. The primary foodborne zoonotic protozoa are *Cryptosporidium parvum, Giardia duodenalis*, and *Toxoplasma gondii*. All three have been associated with outbreaks of illness in humans, but never with a fresh produce vehicle of infection. Additionally, *Balantidium coli* and *Blastocystis* spp. have been isolated from fresh produce and are known to be carried by animals (Slifko et al., 2000), although infection in humans with these organisms never has been reported to be associated with fresh produce. The only zoonotic helminths linked to fresh produce are the trematodes *Fasciola hepatica* and *gigantica, Fasciolopsis buski*, and the nematodes *Ascaris suum, Toxocara canis* and *cati, Toxascaris leonine, Lagochilascaris minor*, and both *Angiostrongylus cantonensis* and *costaricensis* (Polley, 2005; Slifko et al., 2000). Only *Fasciola hepatica* has been responsible for outbreaks of produce-borne infections in humans (Macpherson, 2005; Mas-Coma, 2005; Rondelaud et al., 2005).

Protozoa

Most instances of foodborne protozoal disease in humans are linked to poor sanitation by food handlers who are themselves infected or through waterborne exposure. It is estimated that only 1 out of 10 cases of foodborne protozoal infections are reported (Casemore, 1990), and of those reported, an etiological agent is often not identified. Detection of protozoal agents in produce is difficult, and even if foodborne transmission were common, confirmation through detection of the agent in produce would be rare. Foodborne transmission has been documented for *C. parvum,* and *G. duodenalis*, both of which also occur in wild animals (Appelbee et al., 2005; Polley,

2005; Rose and Slifko, 1999). These agents are the most common enteric parasites in humans and are widespread in the environment from a number of sources, including discharges from food animal production facilities and treated and untreated waste-water effluent (Dawson, 2005; Hunter and Thompson, 2005). These organisms are also reported to be widespread in irrigation water used for crop production in the United States and Central America (Thurston-Enriquez et al., 2002). Presumably, environmental contamination is common because these organisms have a wide range of hosts including domestic and wild animals. No outbreaks of foodborne giardiasis have been reported in industrially manufactured or processed food. A few outbreaks of foodborne cryptosporidiosis from fresh produce including green onions (CDC, 1997) and unpasteurized apple juice (Dawson, 2005; Smith et al., 2007) have been reported. A few studies have demonstrated the presence of *C. parvum* on a variety of fresh produce including cilantro, lettuce, blackberries, cabbage, basil, parsley, celery, leeks, green onions, green chilis, mung bean sprouts, and other seed sprouts (Calvo et al., 2004; Ortega et al., 1997; Robertson et al., 2002). Whether any of these were contaminated by wildlife could not be determined.

There have been a number of widespread outbreaks of produce-borne disease caused by *Cyclospora cayetanensis*. This protozoan parasite is apparently host-specific and thought to occur only in humans. Thus, it appears that outbreaks of *C. cayetanensis* reported in fresh berries and other produce were not from wild animals (Mansfield and Gajadhar, 2004).

There are many wild and feral animal reservoirs for *Cryptosporidium* spp. and *Giardia* spp., including most mammals, birds, and amphibians; both *C. parvum* and *G. duodenalis* have been reported in a wide variety of mammal species (Appelbee et al., 2005; Polley, 2005; Smith et al., 2007). Molecular-based analyses of *C. parvum* and *G. duodenalis* indicate that many of the variants isolated from wild animals are not apparently associated with human illness, indicating that these organisms may not be ubiquitously pathogenic to humans (Appelbee et al., 2005). Future studies may better define the roles and potential roles of both domestic and wild animals in the epidemiology of both *C. parvum* and *G. duodenalis* infections in humans. Wild and feral cats present the only wildlife sources of *Toxoplasma gondii* with relevance to produce contamination. The feces of wild felids can contain oocysts that are infectious to humans; however, direct transmission of this or any other protozoal agents from wild animals to produce that subsequently infects humans has never been documented. *Balantidium coli* and *Blastocystis* spp. are occasionally associated with human illness but never directly linked to fresh produce. Since these organisms also occur in animals, they also have the potential to contaminate produce, although a produce-borne outbreak of illness in humans from these organisms has not been reported.

In addition to the potential for direct contamination by wildlife, mechanical transport of *C. parvum* by house flies and wild filth flies has been demonstrated (Graczyk et al., 1999a, 1999b, 2000). This may play a role in produce contamination from farm to retail distribution. Similar to bacterial pathogens, this mechanism of transport requires a pathogen source in close proximity to produce.

Helminths

Foodborne outbreaks of human illness with zoonotic helminths are rare with the vast majority of infections being from contaminated meat. There are several species of zoonotic helminths that occur in wild animals and can theoretically infect humans through contaminated produce, including the nematodes *Ascaris suum, Toxocara canis* and *cati, Toxascaris leonine, Lagochilascaris minor, Angiostrongylus cantonensis* and *costaricensis*, and the trematodes *Fasciola hepatica, gigantica*, and *Fasciolopsis buski* (Macpherson, 2005; Polley, 2005; Slifko et al., 2000). The majority of zoonotic nematode infections in humans are attributed to direct contact with companion or farm animals (Polley, 1978); however, produce-borne infection is possible. Only the trematode *Fasciola hepatica* has been linked to human produce-borne infections with a wild animal source (Polley, 2005; Rondelaud et al., 2001). As with bacterial zoonotic foodborne pathogens, there is very limited information on the carriage rates of helminths in wildlife that are infective to humans.

Ascaris suum is carried by feral pigs and has a worldwide distribution. Infective ova from this parasite can contaminate produce through direct contact with the feces of wild pigs. One study in the United States reports the prevalence of *A. suum* in wild pigs from Kansas to be 20% (Gipson et al., 1999). Infections in humans are very rare in industrialized countries, and only limited reports exist of outbreaks of illnesses in humans in developing countries. There are no reports of foodborne *A. suum* infections in humans. Theoretically any fecal–oral route of exposure to infective ova is a possible route of infection in humans.

Toxocara canis and *Toxascaris leonina* are carried by wild and feral canines. *Toxocara cati* is carried by feral and wild felines, and *Lagochilascaris minor* is carried by wild felines, raccoons, and rodents (Polley, 2005; Slifko et al., 2000). The infective stages of these nematodes occur in the feces of these definitive hosts. Consequently, human exposure is fecal–oral, and produce is a potential vehicle for human infection. Human disease, including larval visceral migrans and ocular disease from *Toxocara canis* and *cati* and *Toxascaris leonine*, has been reported although never linked to food (Polley, 1978).

The infective larvae of *Angiostrongylus cantonensis* and *costaricensis* occur in gastropod intermediate hosts, presenting a route of human infection in raw vegetables grown in aquatic environments (Polley, 2005). The definitive hosts for these organisms are wild rodents; however, the host range appears to be expanding to include other mammals (Macpherson, 2005). Although there are no reports of foodborne infections of these nematodes in humans, infective ova could theoretically contaminate produce through direct contact with contaminated feces from wild hosts or their waterborne intermediate hosts. Several of these nematodes are common in domestic animals and are reported to occur in their wild counterparts. However, there are no reported surveys of wild animals for the presence of these nematodes, making it difficult to determine the extent to which wild animals are infected and what the potential is for wild animals to contribute to human illness.

Worldwide, there are an estimated 2.4 million cases of fascioliasis annually (Rim, 1992) resulting in significant morbidity and mortality in certain endemic areas (Garcia et al., 2007). In hyperendemic areas, human prevalence is as high as 72% and is coassociated with proximity to aquatic sources of appropriate lymnaeid intermediate hosts (Mas-Coma et al., 1999, 2005). The primary wildlife reservoirs of *Fasciola hepatica* and *gigantica* are wild ruminants and equids (Polley, 2005; Slifko et al., 2000); *F. hepatica* is also common in wild lagomorphs (Rondelaud et al., 2005). The definitive host range for *F. hepatica* is expanding, and depending on geographic location, can include a wide variety of mammals and marsupials (Mas-Coma, 2005; Mas-Coma et al., 2005).

Human infection with *F. hepatica* in developed countries is almost exclusively due to ingestion of contaminated aquatic vegetation; primarily watercress (Macpherson, 2005; Rondelaud et al., 2005), and a number of aquatic plant-borne infections of *F. hepatica* have been reported (Mas-Coma et al., 2005). The primary routes of exposure to humans in developing countries are contaminated drinking water and food. Wildlife sources of *Fasciolopsis buski* include wild pigs (Slifko et al., 2000), and both human and animal infection with this organism are apparently rare. These trematodes are transmitted to humans through the ingestion of infective metacerariae from a variety of gastropods; consequently, foodborne exposure to humans would result from contact with contaminated water or the gastropod intermediate host.

Mitigating wildlife–crop interactions

The Leafy Green Marketing Agreement of California published a food safety guidance document for lettuce and leafy greens in 2008 (LGMA, 2008). This document is tied closely to a previous guidance document from the industry, spearheaded by the International Fresh-Cut Produce Association (Gorny, 2006), but goes into greater detail on how to mitigate wild animal contact with lettuce and leafy greens than did the document published in 2006. The LGMA guidance document lists deer and wild pigs as "animals of significant risk" and recommends that lettuce and leafy green crops with evidence of heavy contact with these animals not be harvested. In addition, this document recommends that barriers be used whenever possible to mitigate contact of these crops with animals of significant risk through the removal of habitat and installation of fencing. To their credit the authors of this guidance document caution producers from removing habitat that may be important to beneficial insects, and that local regulations may prevent removal of habitat adjacent to croplands.

Limited information exists on management strategies that have been demonstrated to be successful in minimizing contact of produce with wild animals. Jay and Wiscomb (2008) discuss mitigation strategies for controlling the interaction of feral swine with crop lands. Hunting has been identified as a potentially viable tool for controlling feral swine populations under certain conditions; however, the very high reproductive rate of feral swine, combined with a lack of access to private land, makes hunting alone an inadequate tool for controlling feral swine populations.

Fencing appears to be the mitigation strategy with the greatest potential for success but requires a significant investment of resources to build and maintain, making it an unattractive option for many producers. Removal of habitat is considered to be of no use in controlling feral swine since these animals are highly mobile and have large home ranges.

Physical barriers such as greenhouses are protective to a limited number of crops; however, the vast majority of fresh produce is grown outdoors with unrestricted access to wild animals. The utility of currently available control interventions to minimize wild animal interactions with the majority of growing crops appears weak at best.

SUMMARY

The list of zoonotic pathogens that have the potential to infect humans through a fresh produce vehicle is extensive. Many of these organisms are rarely associated with human illness; however, several are among the current list of most significant emerging foodborne pathogens. Establishing the role that wildlife and feral domestic animals play in produce-borne human illness is very difficult, as evidenced by the number of outbreak investigations that imply wildlife involvement in spite of a lack of conclusive evidence that this has indeed occurred. Almost certainly, some human illnesses with zoonotic pathogens are directly linked to wild animals. Environmental reservoirs are critical components to the ecology and epidemiology of produce-borne illness in humans, and programs aimed at reducing human illness associated with consuming fresh produce must account for a variety of real and potential environmental reservoirs. Wildlife and feral domestic animals are components of these environmental reservoirs, and control measures aimed at the environmental level could very well reduce the incidence of certain zoonotic pathogens in wildlife.

Mitigating wildlife–produce production interactions is a challenge. In most cases there are no economically feasible mechanisms to prevent wildlife from coming into direct contact with produce while being grown. At best barriers can be installed to prevent some species of wildlife from entering production areas, but these are not all exclusive. Once harvested, the opportunities to keep produce and wildlife separate are greatly expanded, and this stage of the production process presents several critical intervention points that HACCP programs should consider. Historically, the incidence of foodborne illness associated with produce was linked to growing season in the Northern Hemisphere; however, the proportion of fresh produce that is imported versus domestically produced is expanding rapidly. In the last 10 years the amount of fresh fruits and vegetables imported to the United States has nearly doubled and is approaching half the produce that is consumed in this country (USDA:ERS, 2007). The phenomenon of year-round access to fresh produce from many parts of the world, including several developing countries in the Southern Hemisphere, has resulted in increased risks of widespread exposure to zoonotic pathogens on produce and an expanded repertoire of pathogens to consider.

References

Appelbee, A.J., Thompson, R.C., Olson, M.E., 2005. *Giardia* and *Cryptosporidium* in mammalian wildlife—Current status and future needs. Trends Parasitol. 21, 370–376.

Besser, R.E., Lett, S.M., Weber, J.T., Doyle, M.P., Barrett, T.J., Wells, J.G., et al., 1993. An outbreak of diarrhea and hemolytic uremic syndrome from *Escherichia coli* O157:H7 in fresh-pressed apple cider. J. Am. Med. Assoc. 269, 2217–2220.

Beuchat, L.R., 1995. Pathogenic microorganisms associated with fresh produce. J. Food Prot. 59, 204–216.

Beuchat, L.R., Ryu, J.H., 1997. Produce handling and processing practices. Emerg. Infect. Dis. 3, 459–465.

Brackett, R., 1999. Incidence, contributing factors, and control of bacterial pathogens in produce. Postharvest. Biol. Technol. 15, 305–311.

Brown, P.E., Christensen, O.F., Clough, H.E., Diggle, P.J., Hart, C.A., Hazel, S., et al., 2004. Frequency and spatial distribution of environmental *Campylobacter* spp. Appl. Environ. Microbiol. 70, 6501–6511.

Burnett, S.L., Beuchat, L.R., 2001. Human pathogens associated with raw produce and unpasteurized juices, and difficulties in decontamination. J. Ind. Microbiol. Biotechnol. 27, 104–110.

Caldwell, K.N., Adler, B.B., Anderson, G.L., Williams, P.L., Beuchat, L.R., 2003. Ingestion of *Salmonella enterica* serotype Poona by a free-living nematode, *Caenorhabditis elegans*, and protection against inactivation by produce sanitizers. Appl. Environ. Microbiol. 69, 4103–4110.

Caldwell, K.N., Anderson, G.L., Williams, P.L., Beuchat, L.R., 2003. Attraction of a free-living nematode, *Caenorhabditis elegans*, to foodborne pathogenic bacteria and its potential as a vector of *Salmonella* poona for preharvest contamination of cantaloupe. J. Food Prot. 66, 1964–1971.

Calvo, M., Carazo, M., Arias, M.L., Chaves, C., Monge, R., Chinchilla, M., 2004. Prevalence of *Cyclospora sp., Cryptosporidium sp.*, microsporidia and fecal coliform determination in fresh fruit and vegetables consumed in Costa Rica. Arch. Latinoam. Nutr. 54, 428–432.

Casemore, D.P., 1990. Foodborne protozoal infection. Lancet 336, 1427–1432.

CDC, 1997. Foodborne outbreak of cryptosporidiosis Spokane Washington. Morb. Mort. Wkly. Rpt. 73, 353–355.

Cobbold, R.N., Davis, M.A., Rice, D.H., Szymanski, M., Tarr, P.I., Besser, T.E., Hancock, D.D., 2008. Associations between bovine, human, and raw milk, and beef isolates of non-O157 Shiga toxigenic *Escherichia coli* within a restricted geographic area of the United States. J. Food Prot. 71, 1023–1027.

Cody, S.H., Glynn, M.K., Farrar, J.A., Cairns, K.L., Griffin, P.M., Kobayashi, J., et al., 1999. An outbreak of *Escherichia coli* O157:H7 infection from unpasteurized commercial apple juice. Ann. Intern. Med. 130, 202–209.

Conover, M.R., 2002. Resolving human-wildlife conflicts: The science of wildlife damage management. Lewis Publishers, Boca Raton: FL.

Cook, K.L., Bolster, C.H., 2007. Survival of *Campylobacter jejuni* and *Escherichia coli* in groundwater during prolonged starvation at low temperatures. J. Appl. Microbiol. 103, 573–583.

Dawson, D., 2005. Foodborne protozoan parasites. Int. J. Food. Microbiol. 103, 207–227.

FAO/WHO, 2008. Microbiological hazards in fresh fruits and vegetables Meeting report. WHO Publications.

Frank, C., Kapfhammer, S., Werber, D., Stark, K., Held, L., 2008. Cattle density and Shiga toxin-producing *Escherichia coli* infection in Germany: Increased risk for most but not all serogroups. Vector Borne. Zoonotic. Dis. 8, 635–643.

Garcia, H.H., Moro, P.L., Schantz, P.M., 2007. Zoonotic helminth infections of humans: echinococcosis, cysticercosis and fascioliasis. Curr. Opin. Infect. Dis. 20, 489–494.

Gipson, P.S., Veatch, J.K., Matlack, R.S., Jones, D.P., 1999. Health status of a recently discovered population of feral swine in Kansas. J. Wildl. Dis. 35, 624–627.

Goens, S.D., Perdue, M.L., 2004. Hepatitis E viruses in humans and animals. Anim. Health. Res. Rev. 5, 145–156.

Gorny, J.R., Gilcas, H., Gombas, D., Means, K., 2006. Commodity specific food safety guidelines for the lettuce and leafy greens supply chain (1st ed.). International Fresh-Cut Produce Association, Produce Marketing Association, United Fresh Fruit and Vegetable Association and Western Growers.

Graczyk, T.K., Cranfield, M.R., Fayer, R., Bixler, H., 1999. House flies (*Musca domestica*) as transport hosts of *Cryptosporidium parvum*. Am. J. Trop. Med. Hyg. 61, 500–504.

Graczyk, T.K., Fayer, R., Cranfield, M.R., Mhangami-Ruwende, B., Knight, R., Trout, J.M., Bixler, H., 1999. Filth flies are transport hosts of *Cryptosporidium parvum*. Emerg. Infect. Dis. 5, 726–727.

Graczyk, T.K., Fayer, R., Knight, R., Mhangami-Ruwende, B., Trout, J.M., Da Silva, A.J., Pieniazek, N.J., 2000. Mechanical transport and transmission of *Cryptosporidium parvum* oocysts by wild filth flies. Am. J. Trop. Med. Hyg. 63, 178–183.

Greene, S.K., Daly, E.R., Talbot, E.A., Demma, L.J., Holzbauer, S., Patel, N.J., et al., 2008. Recurrent multistate outbreak of *Salmonella* Newport associated with tomatoes from contaminated fields, 2005. Epidemiol. Infect. 136, 157–165.

Hancock, D.D., Besser, T.E., Rice, D.H., Ebel, E.D., Herriott, D.E., Carpenter, L.V., 1998. Multiple sources of *Escherichia coli* O157 in feedlots and dairy farms in the northwestern USA. Prev. Vet. Med. 35, 11–19.

Ho, H.T., Lipman, L.J., Gaastra, W., 2006. Arcobacter, what is known and unknown about a potential foodborne zoonotic agent. Vet. Microbiol. 115, 1–13.

Hunter, P.R., Thompson, R.C., 2005. The zoonotic transmission of *Giardia* and *Cryptosporidium*. Int. J. Parasitol. 35, 1181–1190.

Iwasa, M., Makino, S., Asakura, H., Kobori, H., Morimoto, Y., 1999. Detection of *Escherichia coli* O157:H7 from *Musca domestica* (Diptera: Muscidae) at a cattle farm in Japan. J. Med. Entomol. 36, 108–112.

Janisiewicz, W.J., Conway, W.S., Brown, M.W., Sapers, G.M., Fratamico, P., Buchanan, R.L., 1999. Fate of *Escherichia coli* O157:H7 on fresh-cut apple tissue and its potential for transmission by fruit flies. Appl. Environ. Microbiol. 65, 1–5.

Jay, M.T., Cooley, M., Carychao, D., Wiscomb, G.W., Sweitzer, R.A., Crawford-Miksza, L., et al., 2007. *Escherichia coli* O157:H7 in feral swine near spinach fields and cattle, central California coast. Emerg. Infect. Dis. 13, 1908–1911.

Jijon, S., Wetzel, A., LeJeune, J., 2007. *Salmonella enterica* isolated from wildlife at two Ohio rehabilitation centers. J. Zoo. Wildl. Med. 38, 409–413.

Karenlampi, R., Hanninen, M.L., 2004. Survival of *Campylobacter jejuni* on various fresh produce. Int. J. Food. Microbiol. 97, 187–195.

Kenney, S.J., Anderson, G.L., Williams, P.L., Millner, P.D., Beuchat, L.R., 2005. Persistence of *Escherichia coli* O157:H7, *Salmonella* Newport, and *Salmonella* Poona in the gut of a free-living nematode, *Caenorhabditis elegans*, and transmission to progeny and uninfected nematodes. Int. J. Food. Microbiol. 101, 227–236.

Kenney, S.J., Anderson, G.L., Williams, P.L., Millner, P.D., Beuchat, L.R., 2006. Migration of *Caenorhabditis elegans* to manure and manure compost and potential for transport of *Salmonella* newport to fruits and vegetables. Int. J. Food. Microbiol. 106, 61–68.

Labrousse, A., Chauvet, S., Couillault, C., Kurz, C.L., Ewbank, J.J., 2000. *Caenorhabditis elegans* is a model host for *Salmonella typhimurium*. Curr. Biol. 10, 1543–1545.

LeJeune, J.T., Besser, T.E., Hancock, D.D., 2001. Cattle water troughs as reservoirs of *Escherichia coli* O157. Appl. Environ. Microbiol. 67, 3053–3057.

LGMA, 2008. Commodity specific food safety guidelines for the production and harvest of lettuce and leafy greens June 13, 2008. www.caleafygreens.ca.gov/members/documents/LGMAAcceptedGAPs06.13.08.pdf.

Lyautey, E., Hartmann, A., Pagotto, F., Tyler, K., Lapen, D.R., Wilkes, G., et al., 2007. Characteristics and frequency of detection of fecal *Listeria monocytogenes* shed by livestock, wildlife, and humans. Can. J. Microbiol. 53, 1158–1167.

Macpherson, C.N., 2005. Human behaviour and the epidemiology of parasitic zoonoses. Int. J. Parasitol. 35, 1319–1331.

Mansfield, L.S., Gajadhar, A.A., 2004. *Cyclospora* cayetanensis, a food- and waterborne coccidian parasite. Vet. Parasitol. 126, 73–90.

Mas-Coma, S., 2005. Epidemiology of fascioliasis in human endemic areas. J. Helminthol. 79, 207–216.

Mas-Coma, S., Angles, R., Esteban, J.G., Bargues, M.D., Buchon, P., Franken, M., Strauss, W., 1999. The Northern Bolivian Altiplano: A region highly endemic for human fascioliasis. Trop. Med. Int. Health 4, 454–467.

Mas-Coma, S., Bargues, M.D., Valero, M.A., 2005. Fascioliasis and other plant-borne trematode zoonoses. Int. J. Parasitol. 35, 1255–1278.

Mead, P.S., Slutsker, L., Dietz, V., McCaig, L.F., Bresee, J.S., Shapiro, C., et al., 1999. Food-related illness and death in the United States. Emerg. Infect. Dis. 5, 607–625.

Moore, B.C., Martinez, E., Gay, J.M., Rice, D.H., 2003. Survival of *Salmonella enterica* in freshwater and sediments and transmission by the aquatic midge *Chironomus tentans* (Chironomidae: Diptera). Appl. Environ. Microbiol. 69, 4556–4560.

Murphy, C., Carroll, C., Jordan, K.N., 2006. Environmental survival mechanisms of the foodborne pathogen *Campylobacter jejuni*. J. Appl. Microbiol. 100, 623–632.

Olsen, A.R., 1998. Regulatory action criteria for filth and other extraneous materials. III. Review of flies and foodborne enteric disease. Regul. Toxicol. Pharmacol. 28, 199–211.

Ortega, Y.R., Roxas, C.R., Gilman, R.H., Miller, N.J., Cabrera, L., Taquiri, C., Sterling, C.R., 1997. Isolation of *Cryptosporidium parvum* and *Cyclospora cayetanensis* from vegetables collected in markets of an endemic region in Peru. Am. J. Trop. Med. Hyg. 57, 683–686.

Parish, M.E., 1997. Public health and nonpasteurized fruit juices. Crit. Rev. Microbiol. 23, 109–119.

Park, S.F., 2002. The physiology of *Campylobacter* species and its relevance to their role as foodborne pathogens. Int. J. Food. Microbiol. 74, 177–188.

Polley, L., 1978. Visceral larva migrans and alveolar hydatid disease. Dangers real or imagined. Vet. Clin. North. Am. 8, 353–378.

Polley, L., 2005. Navigating parasite webs and parasite flow: Emerging and re-emerging parasitic zoonoses of wildlife origin. Int. J. Parasitol. 35, 1279–1294.

Renter, D.G., Gnad, D.P., Sargeant, J.M., Hygnstrom, S.E., 2006. Prevalence and serovars of *Salmonella* in the feces of free-ranging white-tailed deer *(Odocoileus virginianus)* in Nebraska. J. Wildl. Dis. 42, 699–703.

Renter, D.G., Sargeant, J.M., Hygnstorm, S.E., Hoffman, J.D., Gillespie, J.R., 2001. *Escherichia coli* O157:H7 in free-ranging deer in Nebraska. J. Wildl. Dis. 37, 755–760.

Renter, D.G., Sargeant, J.M., Oberst, R.D., Samadpour, M., 2003. Diversity, frequency, and persistence of *Escherichia coli* O157 strains from range cattle environments. Appl. Environ. Microbiol. 69, 542–547.

Rice, D.H., Hancock, D.D., Besser, T.E., 2003. Faecal culture of wild animals for *Escherichia coli* O157:H7. Vet. Rec. 152, 82–83.

Rice, D.H., McMenamin, K.M., Pritchett, L.C., Hancock, D.D., Besser, T.E., 1999. Genetic subtyping of *Escherichia coli* O157 isolates from 41 Pacific Northwest USA cattle farms. Epidemiol. Infect. 122, 479–484.

Rim, H.-J., Farag, H.F., Sornmani, S., Cross, J.H., 1992. Food-borne trematodes: Ignored or emerging. Parasitol. Today 10, 207–209.

Robertson, L.J., Johannessen, G.S., Gjerde, B.K., Loncarevic, S., 2002. Microbiological analysis of seed sprouts in Norway. Int. J. Food Microbiol. 75, 119–126.

Rondelaud, D., Hourdin, P., Vignoles, P., Dreyfuss, G., 2005. The contamination of wild watercress with Fasciola hepatica in central France depends on the ability of several lymnaeid snails to migrate upstream towards the beds. Parasitol. Res. 95, 305–309.

Rondelaud, D., Vignoles, P., Abrous, M., Dreyfuss, G., 2001. The definitive and intermediate hosts of *Fasciola hepatica* in the natural watercress beds in central France. Parasitol. Res. 87, 475–478.

Rose, J.B., Slifko, T.R., 1999. *Giardia, Cryptosporidium*, and *Cyclospora* and their impact on foods: A review. J. Food Prot. 62, 1059–1070.

Scaife, H.R., Cowan, D., Finney, J., Kinghorn-Perry, S.F., Crook, B., 2006. Wild rabbits (*Oryctolagus cuniculus*) as potential carriers of verocytotoxin-producing *Escherichia coli*. Vet. Rec. 159, 175–178.

Schlundt, T.H., Jansen, J., Herbst, S.A., 2004. Emerging food-borne zoonoses. Rev. Sci. Technol. Office Int. Epizootes. 23, 513–533.

Sela, S., Nestel, D., Pinto, R., Nemny-Lavy, E., Bar-Joseph, M., 2005. Mediterranean fruit fly as a potential vector of bacterial pathogens. Appl. Environ. Microbiol. 71, 4052–4056.

Seymour, I.J., Appleton, H., 2001. Foodborne viruses and fresh produce. J. Appl. Microbiol. 91, 759–773.

Shayegani, M., Stone, W.B., DeForge, I., Root, T., Parsons, L.M., Maupin, P., 1986. *Yersinia enterocolitica* and related species isolated from wildlife in New York State. Appl. Environ. Microbiol. 52, 420–424.

Slifko, T.R., Smith, H.V., Rose, J.B., 2000. Emerging parasite zoonoses associated with water and food. Int. J. Parasitol. 30, 1379–1393.

Smith, H.V., Caccio, S.M., Cook, N., Nichols, R.A., Tait, A., 2007. *Cryptosporidium* and *Giardia* as foodborne zoonoses. Vet. Parasitol. 149, 29–40.

Smith, J.L., 2001. A review of hepatitis E virus. J. Food Prot. 64, 572–586.

Tauxe, R.V., 2002. Emerging foodborne pathogens. Int. J. Food Microbiol. 78, 31–41.

Teo, C.G., 2006. Hepatitis E indigenous to economically developed countries: To what extent a zoonosis? Curr. Opin. Infect. Dis. 19, 460–466.

Thurston-Enriquez, J.A., Watt, P., Dowd, S.E., Enriquez, R., Pepper, I.L., Gerba, C.P., 2002. Detection of protozoan parasites and microsporidia in irrigation waters used for crop production. J. Food Prot. 65, 378–382.

USDA: ERS, 2007. Increased U.S. imports of fresh fruit and vegetables. Outlook Rep FTS-328–01 1–20.

Van Donkersgoed, J., Berg, J., Potter, A., Hancock, D., Besser, T., Rice, D., et al., 2001. Environmental sources and transmission of *Escherichia coli* O157 in feedlot cattle. Can. Vet. J. 42, 714–720.

Vojdani, J.D., Beuchat, L.R., Tauxe, R.V., 2008. Juice-associated outbreaks of human illness in the United States, 1995 through 2005. J. Food Prot. 71, 356–364.

Wahlstrom, H., Tysen, E., Olsson Engvall, E., Brandstrom, B., Eriksson, E., Morner, T., Vagsholm, I., 2003. Survey of *Campylobacter* species, VTEC O157 and *Salmonella* species in Swedish wildlife. Vet. Rec. 153, 74–80.

Wetzel, A.N., LeJeune, J.T., 2006. Clonal dissemination of *Escherichia coli* O157:H7 subtypes among dairy farms in northeast Ohio. Appl. Environ. Microbiol. 72, 2621–2626.

Commodities Associated with Major Outbreaks and Recalls

Leafy Vegetables

Karl R. Matthews

Department of Food Science, School of Environmental and Biological Sciences, Rutgers,
The State University of New Jersey, New Brunswick, NJ

CHAPTER OUTLINE

Introduction

The microbial safety of leafy vegetables is a continual cause for concern in the United States and throughout the world. In the United States, analysis of data from 1998 to 2008 indicates that about 22% of foodborne illness were associated with consumption of contaminated leafy greens (Painter et al., 2013). Leafy greens were the second most frequent cause of hospitalizations. The majority of illness linked to leafy greens were caused by norovirus; *Salmonella* and Shiga-toxin producing *Escherichia coli* were the major bacterial causes of illness. Outbreaks associated with *Salmonella* and *E. coli* O157:H7 tend to receive the most attention due to the severity of the illness

The Produce Contamination Problem. http://dx.doi.org/10.1016/B978-0-12-404611-5.00009-9

and occurrence of deaths. Foodborne outbreaks under the heading "leafy greens" encompass more than just those linked to lettuce and spinach. Leafy green products include romaine lettuce, green leaf lettuce, red leaf lettuce, butter lettuce, baby leaf lettuce, escarole, endive, spring mix, spinach, cabbage, kale, arugula, and chard. However, the majority of produce-related outbreaks are associated with lettuce.

Outbreaks associated with leafy greens

The major produce outbreaks that have occurred in the United States have been associated with bagged leafy greens. Large outbreaks receive the most attention by the public as was the case with the 2006 outbreak involving spinach contaminated with *E. coli* O157:H7, which resulted in 205 confirmed illnesses (FDA, 2007). Small-scale (< 50 cases) outbreaks involving *E. coli* O157:H7-contaminated lettuce continue to occur in the United States despite improved production and handling practices. The diversity of products on the market has now seen outbreaks linked to products containing a wide diversity of leafy greens. An outbreak involving 33 cases (13 hospitalizations, no deaths) was linked to a spring mix (blend of lettuces and other leafy greens) and spinach blend that was contaminated with *E. coli* O157:H7 (CDC, 2012). Voluntary recalls of bagged spinach, lettuce, and other leafy greens products have occurred frequently in recent years, perhaps the result of improved routine testing practices by the industry and government agencies.

In Europe, outbreaks have been attributed to both locally produced and imported leafy greens. In 2004, an outbreak of *Salmonella* Thompson infections was reported in Norway, Sweden, and England. These cases were likely all linked to the consumption of contaminated rucola lettuce imported from Italy (Nygard et al., 2008). In Sweden, a total of 135 cases, including 11 cases of hemolytic uremic syndrome (HUS), were linked to the consumption of locally produced lettuce that was contaminated with *E. coli* O157. Water samples from a stream used for irrigation were positive for the outbreak strain, as were cattle at a farm upstream of the irrigation point (Soderstrom et al., 2008).

Foodborne illness outbreaks associated with escarole, endive, kale, arugula, and chard are rare, if they have ever occurred. In Canada and the United States, outbreaks of foodborne illness have been traced back to the consumption of cabbage. The causative agents were *E. coli* and *Listeria*. In each outbreak the cabbage was used in coleslaw. Improper washing of the cabbage prior to use and the use of raw sheep manure in production fields were indicated as causes (Sewell and Farber, 2001).

Surveys designed to evaluate the microbial safety of fresh fruits and vegetables indicate that few samples test positive for foodborne pathogens (Abadia et al., 2008; Arthur et al., 2007; Smith DeWaal and Glassman, 2013). No *E. coli* O157:H7 was isolated from lettuce purchased in retail establishments in Canada, Spain, and the United States (Abadia et al., 2008; Arthur et al., 2007; FDA/CFSAN, 2004). In each of the studies, *Salmonella* was associated with only one lettuce sample. Two lettuce samples from the study conducted in Spain tested positive for

Listeria monocytogenes. No samples of parsley and cilantro from Canada yielded *Salmonella, E. coli* O157:H7, or *Shigella*. These types of survey studies would suggest that the incidence of contamination with any foodborne pathogen is extremely low, and particularly so for *E. coli* O157:H7. Although not often discussed, parasites may also contaminate produce. A study of leafy vegetables in southwestern Saudi Arabia found that 17% of watercress, 17% of lettuce, and 13% of leek were positive for parasites (Al-Binali et al., 2006).

The widespread global contamination of leafy greens with *E. coli* O157 and subsequent outbreaks of foodborne illness are difficult to explain. No specific genes are attributed to *E. coli* O157:H7 that would explain why the pathogen is so intricately associated with leafy greens outbreaks. Research demonstrates the ability of the pathogen to survive for extended periods in water, soil, and manure. The microbe is also capable of surviving shifts in temperature, exposure to sunlight (ultraviolet), moisture, and nutrients. Cattle are considered the primary reservoir of *E. coli* O157:H7. Fecal shedding of the pathogen by domestic and feral pigs, wild birds, deer, and by other domestic livestock and wildlife has been described (Cooley et al., 2007; Jay et al., 2007). Other foodborne pathogens exhibit similar survival characteristics, but for reasons yet unknown, are not associated with outbreaks of foodborne illness linked to the consumption of leafy greens. Intensive research is being conducted to better understand the interaction of *E. coli* O157:H7 and other pathogens with leafy greens. This research should fill data gaps and ultimately aid in protecting the consumer from human illness linked to the consumption of contaminated leafy greens.

The consumer enjoys having foods that are convenient and require minimal preparation prior to use. The sales of bagged leafy greens have exploded since their introduction nearly 30 years ago. In recent years a number of large outbreaks associated with bagged leafy greens have made consumers question the safety of such products. When an outbreak occurs associated with leafy greens, most consumers first learn of the outbreak through television broadcasts. The large spinach outbreak that occurred in 2006 in the United States serves as an excellent case study on consumer knowledge and attitudes following a large outbreak. The outbreak involved bagged spinach contaminated with *E. coli* O157:H7. A total of 205 cases in 26 states with 103 hospitalizations, 31 with HUS, and 3 deaths were reported (CFERT, 2007).

A survey of public response to the recall was conducted by the Food Policy institute at Rutgers University (Cuite et al., 2007). Prior to the outbreak most consumers perceived bagged leafy greens to be safe to eat. Most consumers (~80%) were aware of the recall, but not certain of the types of spinach involved in the recall. Nearly half of the respondents thought that washing contaminated produce would make it safe to eat. Surprisingly, about 13% of Americans who were aware of the spinach recall continued to eat fresh spinach during the recall. This was despite the efforts of the U.S. FDA during the recall to promote its key message that no fresh spinach was considered safe to eat. Consumption of bagged spinach declined dramatically for months following the outbreak. The impact of the outbreak was felt across the produce industry since retail sales of bagged salad without spinach also declined

(Calvin, 2007). This change in purchasing habits of fresh leafy greens was associated in part with the information that consumers received in the weeks and months following the outbreak. Consumers may have found it difficult to make informed decisions concerning the safety of leafy greens given the broad information provided by government agencies and experts to the news agencies (Todd et al., 2007).

Growing conditions by geographical region: link to outbreaks

In the United States leafy vegetable production moves from California to Arizona and Mexico as the seasons change, ensuring a constant supply of product throughout the year. Spain produces approximately half of all commercially grown lettuce in Europe. Differences in soil, climate, and cultivars independently or collectively impact microbial quality of the crops grown in those regions. Rainfall in these regions averages 5 to 15 cm per year. Irrigation is essential to grow crops under such limited rainfall conditions. Contamination can occur in the field by exposure to contaminated irrigation water or floodwaters. A study designed to track *E. coli* O157:H7 in a major produce production region of California suggests that the pathogen, when found in water, is generally close to a point source (Cooley et al., 2007). The authors do point out that in periods of high water-flow, often associated with flooding, the pathogen may be transported over 30 km.

The handling of water used for irrigation can also result in contamination. Water from wells may be pumped into retaining ditches prior to use. Wildlife and feral animals can contaminate the water, and under the proper environmental conditions, bacterial populations may increase, placing any crops irrigated with the water at risk of being contaminated. Unregulated release of untreated sewage into rivers and streams can result in the contamination of irrigated crops with a range of microorganisms (Okafo et al., 2003). Growers must be aware of Good Agricultural Practices (GAPs) and be willing to implement GAPs to reduce the risk to human health associated with consumption of contaminated leafy greens (Jackson et al., 2007). Specific food safety guidelines have been developed for the production of lettuce and leafy greens (Western Growers, 2008). An important component of those guidelines is the use of nonsynthetic soil amendments.

Crops may be irrigated using overhead sprinkler, subsurface drip, or surface furrow; the method used can have a direct impact on risk of contamination of the edible portion of a crop. The risk of lettuce crop contamination increased with sprinkle irrigation compared to other methods evaluated (Fonseca et al., 2011). A minimum 6-day lag between irrigation and harvest was suggested following spray irrigation with *E. coli*-contaminated water since the pathogen survived that long on spinach leaves (Wood et al., 2010). Oliveira et al. (2011) reported that lettuce leaves were positive for *Listeria* following surface or spray irrigation using water contaminated with that organism. *Listeria* could be detected for four weeks post-exposure, although initial populations on leaves decreased rapidly.

Soil amendments are commonly used to add organic and inorganic nutrients to the soil. Human pathogens may potentially be present in animals' manures, and if not composted properly or thermally processed, may provide a source of leafy vegetable contamination. Many large organic and conventional leafy vegetable operations use chicken pellets. Studies addressing the microbial safety of chicken pellets are extremely limited. A recent reported indicated that drying the fresh chicken litter/manure at 250°C eliminated *Salmonella*, fecal streptococci, and enterobacteria (Lopez–Mosquera et al., 2008). Bacterial population may be reduced but not completely eliminated depending on the pellet production method used (Hammed, 2013). Greater attention must be focused on the microbial safety of chicken or other manure source pellets and other types of soil amendments. Assumptions cannot be made that these products are safe to use simply because they have been through a thermal process. Standard practices with defined time/temperature parameters must be followed to ensure that human pathogens, if present, will be inactivated.

Harvesting practices: influence on contamination

The handling of lettuce during and immediately postharvest can have a dramatic effect on the microbial safety of the product. Depending on the market and intended utilization for processing, lettuces may be harvested by hand or mechanically. Head lettuces are usually harvested by hand, cored, trimmed (removal of outer leaves), sprayed with a sanitizing wash, bagged, and boxed in the field. Although moving processing to the field likely has distinct economic advantages, the impact on microbial safety has not been examined adequately. Cross-contamination of lettuce through contact with workers' hands (or gloves), knives, automated equipment (conveyor belt), and wash water may occur. The cut end of the lettuce is laden with nutrients that support bacterial growth (Brandl, 2008). Baby lettuce and young lettuce destined for use in bagged salads are machine harvested, dispensed into bins, placed into a refrigerated truck trailer, and then transported to the packing facility. The lettuce may be used immediately or stored, depending on processing practices. The greater the handling and processing of a product, the more pronounced physiological changes will be in that product. Those changes will shorten the product shelf-life and enhance the growth and survival of microbes associated with the product (Aruscavage et al., 2006). Valentin–Bon et al. (2008) reported that the mean total bacterial count of bagged lettuce and spinach samples was 7.0 log cfu/g. Bacterial counts were similar for conventional and organically grown spinach and lettuce mixes. No *E. coli* count exceeded 10 MPN/g; presently, there are no *E. coli* limits for bagged produce in the United States.

Leafy greens other than lettuce and spinach have received little attention. Specialty crops are more likely to be harvested by hand, but are subjected to steps similar to those used for baby lettuce. The behavior of *E. coli* O157:H7 in association with leafy greens and lettuce has received the greatest attention. The leaf age was shown to be a factor in the growth and subsequent population of *Salmonella* and *E. coli*

O157:H7 associated with Romaine lettuce (Brandl and Amundson, 2008). Populations of the pathogens were greater on young inner leaves compared to middle and outer leaves (older). A few studies have investigated the fate of *L. monocytogenes* and *Salmonella* on leafy greens (Brandl and Mandrell, 2002; Jablasone et al., 2005; Lapidot et al., 2006). Research demonstrated that the population of *L. monocytogenes* on the surface of parsley, grown under field conditions, declines rapidly within 2 days (Dreux et al., 2007). The researchers suggested that the risk is minimal unless contamination of aerial surfaces occurs very shortly before harvest. In 2013 the FDA released a proposed rule that seeks to establish science-based minimum standards for the safe growing, harvesting, packing, and holding of produce grown for human consumption (FDA, 2013). The proposed rule will likely have a positive impact on the microbial safety of all types of leafy greens.

Processing practices and product contamination
Handling prior to processing

The improper handling of product immediately after harvest can compromise the safety of leafy greens. This is particularly true for baby greens that are harvested into bins for transport to the processing facility. Placing the bins directly onto the soil could result in contaminants contacting the bottom of the bin. The bins are often stacked one on top of the other, permitting the transfer of contaminants from the bottom of one bin to the contents of the bin below. The bins should be placed into a refrigerated truck trailer as rapidly as possible to cool down the product and limit the ability of microorganisms to grow. A large amount of latex is released from cut stems, providing nutrients for the growth of microbes. *E. coli* O157:H7, artificially inoculated onto cut lettuce stems, increased 11-fold over four hours of incubation at 28°C (Brandl, 2008). Proper refrigeration is imperative to cool the product, thereby limiting or preventing growth of the pathogen.

Cooling crops to 4°C or less will slow or prevent the growth of pathogens including *Salmonella*, *E. coli* O157:H7, and *L. monocytogenes*. Leafy greens are generally cooled under forced air, but passive storage under refrigeration is still a widely used method. Vacuum cooling is a common practice in the leafy greens industry. However, research suggests that the process can promote the infiltration of *E. coli* O157:H7 into lettuce (Li et al., 2008). Therefore, the cooling process could have a significant impact on the microbial safety of leafy greens. Most studies addressing the influence of storage temperature on growth and survival of pathogens in association with leafy greens focus more on retail product rather than bulk loose greens. The populations of *E. coli* O157:H7 and *L. monocytogenes* on iceberg lettuce increased approximately 1.5 to 2.5 log during 12 days' storage at 8°C (Francis and O'Beirne, 2001). Reducing the storage temperature to 4°C limited growth of the pathogens; however, viable populations remained at the end of the storage period. Others have also demonstrated the growth of *E. coli* O157:H7, *Salmonella*, and *L. monocytogenes* on iceberg lettuce held at refrigeration temperatures (Koseki and Isobe, 2005). The population of *E. coli*

O157:H7 and *Salmonella* on cilantro, oregano, basil, chive, parsley, and rosemary declined by 1.5 log or less after storage at 4°C for 19 days (Hsu et al., 2006). These results underscore the ability of both bacteria to persist under common storage conditions for leafy greens.

The ability to track a product during its journey from the field to the table is critical to the prevention of human illness. The traceability of a product to its origin will limit the number of cases of illness during an outbreak and decrease the amount of product that must be recalled. Until recently, the produce industry has not had in place a suitable system for tracking of produce. Difficulties in traceability are associated with the commingling of product from different fields into the same bag. Most systems use barcodes to trace product as it travels from the field to the consumer. There are many disadvantages to the use of barcodes since they are affixed to the outside and are subject to wear and scratching. A direct line-of-sight is required to read the code, and once on the package, the code cannot be changed to reflect changes in product profile or handling. The use of a radio frequency identification device (RFID) would permit the tracking of leafy greens from the field to the retail level. The RFID tags can be continuously written to with information, and read remotely. The RFID system can be monitored via the Internet, and the technology can be used by large and small operations alike.

There exists the possibility that in the future, consumers would be able to access the information about a particular product at the retail level. Some producers are using RFID technology to trace product from the field to retail markets. The technology has been evaluated in several pilot studies involving produce shipped from California or Taiwan to Hawaii (Swedberg, 2012). The pilot study provided a wealth of new information including that product at the top of a pallet may be exposed to warmer temperatures ultimately deceasing shelf-life. Ultimately, alerts can be made throughout the supply chain when and where a food product is not being handled appropriately. A seal will be available to fresh produce buyers, providing assurance that the product purchased is directly traceable to a particular farm or processor. The technology exists so that temperature and humidity information can be tracked to determine if food products are transported and stored under appropriate conditions. The RFID system can be integrated into Good Manufacturing Practices (GMPs) and Hazard Analysis Critical Control Point (HACCP) systems. In Sweden, a large supplier of pallets to the produce industry uses RFID to track pallets. Proper washing and sanitizing of the pallets can be monitored using RFID, ensuring that pallets potentially contaminated with human pathogens are not used in the harvest, storage, or transport of fresh produce.

The microbial safety of leafy greens is paramount to human health. Microbiological testing of water, soil amendments, equipment (field and processing), and commodity must be conducted to comply with GMPs, HACCP programs, buyer agreements, and other food safety guidelines. In some instances, quantitative microbiological analysis is required for water that is intended for irrigation (Western Growers, 2008). A decision tree was developed whereby crops directly contacted with water exceeding accepted microbiological criteria (most probable number testing for generic *E. coli*) would be sampled and tested for *E. coli* O157:H7 and

Salmonella. Microbiological screening is also being implemented prior to product processing. Many large grower/packers are now practicing test-and-hold programs. In general, palletized harvested product is identified at delivery to the processor. The product is sampled and held in a designated area of the warehouse until microbiological testing can be completed. Typically, product is tested for *E. coli* O157:H7 and *Salmonella*; more recently, processors are testing for *L. monocytogenes*. The testing methods used must be completed within 12 hours; a longer period would require large holding structures and limit production capabilities. PCR-based methods are best suited for this purpose since they are rapid, specific, and cost effective. A fairly comprehensive list of rapid methods for use with water and food samples has been compiled (www.wga.com).

Washing and sanitizing
Equipment design

New approaches to the washing/sanitizing of leafy greens are continually being proposed. Unfortunately, most are not sufficiently efficacious or practical for industry implementation. Chemical sanitizers generally provide a 1- to 2-log reduction in viable bacteria associated with leafy greens. Changes in equipment and processing strategies are required to enhance microbial reduction. Prior to bagging, leafy greens are dumped into water flumes containing sanitizing water. The total exposure time may range from 60 to 120 seconds. Leaves may float on the surface, a phenomenon often referred to as "lily padding," which limits exposure of the entire leaf to the antimicrobial contained in the water. Equipment that effectively submerges all product upon entry into the flume system is now available. Other systems use bends in the flume to redirect product, causing it to become submersed, or air jets that agitate the water and effectively cause leaves floating on the surface to become submerged. Although every processor may have different procedures, leafy greens often go through a triple-wash process. The first wash occurs when product goes into an agitating tank containing weakly chlorinated water. This step is intended to remove gross physical debris (bugs, soil, stones). The major cleaning occurs in a second tank or flume containing chlorinated water. Although bacteria associated with the surface of a product may be killed, the chlorine in the water is intended to control bacterial numbers in the wash water and associated with equipment. The final wash is actually a rinse step that is intended to remove residual chlorine from the product.

Equipment used for conveying product, removing excess water, and for packaging must be user friendly and facilitate ease of cleaning. Cleaning of equipment during shift breaks will minimize the build-up of organic matter on equipment and cross-contamination of product. Microorganisms can rapidly build up on equipment and effectively "inoculate" product during processing. This may result in a shortened shelf-life and increased risk to consumer safety.

The present methods used to wash and sanitize leafy greens will inevitably result in cross-contamination should a few leaves be contaminated initially. Research clearly shows that levels of foodborne pathogens on leafy greens may be only minimally reduced following washing in water containing sanitizing agents such as chlorine (Gil et al., 2009). These studies demonstrate the potential for even a few contaminated leaves to disseminate a pathogen to a large mass of leafy greens, exacerbating the magnitude of an outbreak. As discussed, bins of a given commodity (spinach, red lettuce, arugula) are dumped into a hopper and conveyed to a wash tank or flume, ideal conditions for direct contact of leaves and equipment.

Sanitizing agents

At present chlorine is the primary postharvest sanitizing agent in use by the fresh produce industry. This was the situation more than 10 years ago (Parish et al., 2003), and remains true today (Gile et al., 2009). Indeed, leafy vegetable processors recognize that water disinfectants are used to prevent cross-contamination, and not to surface-sanitize produce. However, such processes do impact microbial populations on the product. Sanitizing/disinfecting agents other than chlorine must be more aggressively investigated.

New technologies being applied to pathogen reduction of leafy greens include electrolytic oxidizing water and ultrasound alone or in combination with a chemical sanitizer. Acid electrolyzed water (AEW) and neutral electrolyzed water (NEW) have been studied as alternative sanitizers. Electrolyzed water is generated through the electrolysis of a dilute NaCl solution. AEW only passes through the anode chamber and has a strong bactericidal effect due to its low pH (2–4), high oxidation-reduction potential (ORP > 1000 mV), and content of hypochlorous acid. NEW is generated by passing the NaCl solution through the anode and cathode chambers of a membrane electrolyzer. NEW is a near-neutral (pH 8 ± 0.5) solution in which the main bactericidal agents are HOCl, ClO$^-$, and HO$_2$ radical. The potential acceptance of NEW by the leafy greens industry is greater since it would be less corrosive to equipment and cause less irritation to hands. Its effectiveness in reducing *E. coli* O157:H7, *Salmonella*, and *Listeria* populations associated with lettuce and other leafy greens was similar to that of chlorine, with reductions of 1 to 2 log units (Abadias et al., 2008). The advantage was that NEW, containing about 50 ppm free chlorine, was as effective as applying chlorinated water at 120 ppm free chlorine.

Research associated with the application of ultrasound for the microbial decontamination of produce has been revived. The action of ultrasound is associated with the formation of cavitation bubbles. The movement of these bubbles generates the mechanical energy to remove microbes from a surface. The use of ultrasound in combination with chlorinated water reduced levels of *Salmonella* associated with iceberg lettuce by 2.7 log (Seymour et al., 2002). More recent research also suggests that utilization of ultrasound in combination with a sanitizing agent enhanced the reduction in *E. coli* O157:H7 associated with spinach leaves (Zhou et al., 2009). However, the high capital cost and expensive process of optimization and water treatment has to date precluded its use by the fresh produce industry. The development of new equipment that can be retrofitted on existing flume and wash tanks may make the method more appealing from a cost standpoint.

A novel control method that has been receiving interest in recent years is the use of bacteriophage. Researchers have investigated the use of bacteriophages to reduce *S.* Enteritidis and *L. monocytogenes* associated with honeydew melons (Leverentz et al., 2001, 2003). The application of bacteriophages for the control of foodborne pathogens is an emerging area of research (Hagens and Loessner, 2007). The technology could be applied to the control of foodborne pathogens associated with leafy greens. Bacteriophages are usually very host-specific. Many can infect only one bacterial genus, and others infect only one serotype within a bacterial species. Therefore, bacteriophages that specifically attack *E. coli* O157:H7 could be applied during postharvest processing. Public acceptance of the application of bacteriophage to fresh leafy greens as a biocontrol agent may be met with resistance. Consumers may perceive bacteriophage as dangerous, potentially capable of causing human illness.

Packaging

A multitude of measures must be employed to ensure the microbial safety of leafy greens. Prolonged exposure of leafy greens to biocidal compounds during transport and retail sale may aid in killing or preventing the growth of foodborne pathogens. The use of novel packaging materials, packaging design, and package inserts may provide an added dimension to strategies for the control of microbes associated with leafy vegetables.

Treatment of inoculated lettuce leaves with ClO_2 gas, generated from dry chemical sachets, for 30 minutes resulted in log reductions of 3.4, 4.3, and 5.0 for *E. coli* O157:H7, *S.* Typhimurium, and *L. monocytogenes*, respectively; treatment for one hour resulted in log reductions of 4.4, 5.3, and 5.2, respectively (Lee et al., 2004). No treatment-induced quality defects were reported. However, the large population reductions reported in this study may reflect the brief time interval (30 minutes) between inoculation and treatment, which might have been insufficient to allow for aggregate formation and internalization of the targeted pathogens, conditions expected to enhance their survival.

Researchers evaluated the efficacy of gaseous chlorine dioxide in killing *E. coli* O157:H7, *Salmonella*, and *L. monocytogenes* on various fresh-cut commodities including cabbage and lettuce. They reported log reductions of 3.13 to 4.42 for cabbage but only 1.53 to 1.58 for lettuce (Sy et al., 2005). It is not clear why population reductions with lettuce were relatively small. The most likely explanations are internalization at cut surfaces and interference from a film of condensate on stomata and other leaf surfaces under the high humidity conditions of the treatment cabinet (70–80%).

It may be possible to develop packaging that permits extended uniform exposure of leafy greens to ClO_2 gas while in the package. This would mitigate or prevent the development of adverse quality issues such as browning, often associated with high concentration-short period exposure. Treatment of lettuce and cabbage with ClO_2 gas prior to bagging may actually result in high bacteria populations during storage (Gomez–Lopez et. al., 2008). The ClO_2 gas treatment would reduce populations of all bacteria but might allow for certain types of bacteria to rapidly increase without the competition for nutrients. Another potential concern with the use of ClO_2 gas is browning.

Within the past 5 years new natural, biodegradable, and edible packaging films have been studied that inhibit the growth of foodborne pathogens. An antimicrobial film effective against *E. coli* and *Salmonella* was shown to extend the shelf-life of iceberg lettuce held at 10°C by 5 days (Kang et al., 2007). The researchers conducted additional studies that extended by an additional 2 days the shelf-life of fresh-cut iceberg lettuce packaged in the antimicrobial film under modified atmosphere (MAP) conditions (95% N_2, 2.6% O_2, and 2.4% CO_2; Kang et al., 2008). The atmosphere conditions can independently inhibit spoilage and foodborne pathogenic bacteria, thereby prolonging shelf-life and increasing microbial food safety (Fonseca, 2006). The possibility exists to combine in-package ClO_2 gas treatment with an antimicrobial packaging film to injure or kill foodborne pathogens present on the product and prevent recovery or growth of survivors.

Novel types of packaging and edible films are being tested that will potentially extend product shelf-life, reduce risk of pathogen growth on contaminated food surfaces, and have consumer acceptance. Carvacrol, a major constituent of oregano with antimicrobial activity, has been incorporated into edible tomato and apple films (Du et al., 2008a, 2008b). The antimicrobial films were produced by combining carvacrol with tomato or apple purees. Both films exhibited antibacterial activity against *E. coli* O157:H7. Antimicrobial films made from plants may have wider acceptance than those developed using chemical compounds. Such products could be used in bagged salad mixes where potential flavor notes may be considered acceptable, and add to the appeal of the product.

The antimicrobial agent nisin has been incorporated into nonedible films. Nisin is a broad-spectrum bacteriocin, produced by lactic acid bacteria, which binds to and forms pores in the cell membranes of Gram-positive bacteria. Nisin-coated plastic films were shown to reduce levels of *L. monocytogenes* on vacuum-packaged, cold-smoked salmon (Neetoo et al., 2008). Redesign of packaging to include inserts or bridges increasing the contact area of product with the antimicrobial packaging material would enhance utility of the method.

In the future, these technologies have the potential to be applied in the control of foodborne pathogens associated with produce. Nonconventional technologies must be considered since there exists no single method that will effectively eliminate foodborne pathogens that may be associated with leafy greens.

Interaction of microbes with leafy greens

The interaction of human enteric pathogens with plants has been extensively reviewed (Brandl, 2006; Solomon et al., 2006; Aruscavage et al., 2006; Delaquis et al., 2007; Heaton and Jones, 2008). The conclusions that can be drawn from those reviews suggest that researchers have a limited understanding of the behavior of enteric pathogens in the phylosphere. Greater effort must be made to understand plant microbe interactions and the interaction between enteric pathogens and epiphytic microbes (e.g., yeast, molds, bacteria).

Plant leaf characteristics

The plant leafy surface is not an ideal environment for the survival of enteric pathogens. The leaf surface is exposed to ultraviolet light, shifts in temperature and relative humidity, and presence of available moisture (presence or absence of rain/irrigation). A range of epiphytic microbes is also present, impeding the ability of enteric bacteria to colonize the leaf surface. The availability of carbon and nitrogen sources leached from the plant will also influence survival and growth. The most common carbon sources available on the surface of plants are glucose, sucrose, and fructose (Mercier and Lindow, 2000). Leakage of nutrients from the plant can occur following damage to the cuticle. The leaf environment is not uniform with respect to conditions that would support survival and growth of bacteria. Epiphytic bacteria preferentially colonize the base of trichomes, around the stomata, and along veins in the leaves. Foodborne pathogens were shown to localize near leaf veins. The reason for this may be related to the greater wettability of the area, increasing nutrient leaching and water availability (Brandl and Mandrell, 2002). Leaf age may be an important factor in the growth and survival of epiphytic and opportunistic bacteria. The exudate from young lettuce leaves was reported to be 2.9 and 1.5 times richer in total nitrogen and carbon, respectively, than exudates from middle leaves (older leaves) (Brandl and Amundson, 2008).

Native flora of leafy greens

The community of microbes on the plant leaf surface is complex and includes many species of bacteria, yeasts, and molds. Populations of aerobic bacteria on leafy greens may average 10^5 to 10^6 cfu/g of leaf tissue (Aruscavage et al., 2006). The Gram-negative bacteria are the most predominant group of epiphytic microorganisms found in the plant phylosphere. Microbes on the plant leaf surface form biofilms, whereby cells are encased in an exopolysaccharide matrix that provides protection against adverse environmental conditions. Early research showed that bacteria in biofilms comprise between 10 and 40% of the total bacterial population on leafy greens (parsley and endive) (Morris et al., 1998). Microbial biofilms on leaves may influence the attachment, growth, and survival of enteric pathogens on the leaf surface. Researchers demonstrated that a biofilm-deficient mutant of *S.* Typhimurium persisted on parsley at elevated levels. They suggested that the microbe may have been able to penetrate the preexisting biofilm and therefore persist at levels similar to the parent strain (Lapidot at al., 2006). A more comprehensive discussion of biofilms can be found in Chapter 2.

Interaction of enteric pathogens with the leaf surface may also benefit from the action of phytopathogenic bacteria. Phytopathogens such as *Erwinia* can degrade plant tissue, providing enteric pathogens with a broadened spectrum of nutrients for growth (Agrios, 1997). Whether an enteric pathogen can colonize a leaf surface may be influenced by the native flora found on the leaf and characteristics of the enteric microbe.

Microbe characteristics

Enteric bacteria have a vast array of cell surface moieties that may influence the ability of the cell to interact with plant tissue. Relatively few reports on the intimate interaction of foodborne pathogens with leafy greens have been published. Specific moieties or factors involved in attachment/interaction include exocellular polysaccharide, cell surface charge, presence/absence of fimbriae, and hydrophobicity. The presence of curli appeared to have no influence on attachment of *E. coli* O157:H7 to lettuce (Boyer et al., 2007). Curli are coiled extracellular structures on the cell surface that bind fibronectin and other proteins facilitating adherence to and invasion of epithelial cells (Chapman et al., 2002). An exhaustive study involving the screening of a mutant library of *L. monocytogenes* and identification of three mutants that had reduced adherence to radish tissue suggested that temperature may affect expression of attachment factors used by *L. monocytogenes* (Gorski et al., 2003). A series of experiments were conducted to determine the influence of cell surface structures of *E. coli* O157:H7, such as flagella, curli, lipopolysaccharide, and exopolysaccharide on survival and colonization of the pathogen on plant leaf tissue (Seo and Matthews, 2012). The research demonstrated that flagella contributed to the induction of plant defense response. Curli were detected by the plant defense system, consequently limiting survival of the pathogen on the plant.

Solomon and Matthews (2006) reported that gene expression and bacterial processes, such as motility or production of extracellular compounds, were not required for initial attachment of *E. coli* O157:H7 to lettuce. In the study, live and glutaraldehyde-killed *E. coli* O157:H7 and fluorescent polystyrene microspheres were used in experiments to investigate interactions with lettuce. The microspheres are comparable in size to single bacterial cells, yet free of any surface moieties or appendages that have been hypothesized to be involved in attachment. The role of gene expression and cell surface moieties are likely to be important in further colonization and survival on the leaf. The role of exopolysaccharide production by enteric pathogens to facilitate intimate adherence to plants has not been sufficiently explored. Research suggests that the ability to produce biofilm does not play a significant role in initial adhesion of *Salmonella* to lettuce and survival after disinfection (Lapidot et al., 2006). Notwithstanding the role in adherence, exopolysaccharides protect bacteria from the action of antimicrobial agents and desiccation, and may aid in the integration of the enteric bacteria into the epiphytic biofilm.

A comparison of the genomes of the plant pathogen *Erwinia cartovora* subsp. *atroseptica* and the human pathogen *S. enterica* Typhi suggests the genomes are similar, but that genes associated with plant interactions are absent in *S.* Typhi (Toth et al., 2006). However, gene expression by enteric pathogens associated with plants may influence survival on the plant and enhance virulence within a human host. The plant pathogen *Erwinia chrysanthemi* induces the *sap* operon to provide resistance to plant antimicrobial compounds and enhance its virulence. The *sap* operon may also be induced in *S. enterica*, perhaps providing resistance to plant antimicrobials and providing enhanced virulence in a human host (Taylor, 1998). Expression of the type III secretion system (TTSS) by *Salmonella* when in association with plants may actually be detrimental to survival of the microbe in association with the plant (Iniguez et al., 2005). The type III secretion system enables delivery of pathogenicity proteins to host cells; the system is found in Gram-negative bacteria and animal pathogens (Hueck, 1998). Colonization studies of *Arabidopsis thaliana* suggest that the plant recognizes components of the TTSS as part of its defense system (Iniguez et al., 2005).

Internalization

The localization of enteric pathogens at subsurface sites on leafy green plant tissue impedes their removal during washing and the ability of sanitizers to inactivate the microbe. Bacteria may gain access to interior regions of a leaf through stomata, abrasions or cuts, action of plant pathogens, and through the root system (Bernstein et al., 2007). Internationalization may not occur in all food crops may. Golberg et al. (2011) reported that the highest incidence of internationalization occurred in iceberg lettuce and arugula and the lowest incidence in parsley. Internalization varied significantly within a commodity, from 0 to 100% in iceberg lettuce. The studies were not designed to determine uptake through the root system. Takeuchi and Frank (2000) reported that *E. coli* O157:H7 localized within stomata of lettuce were protected from sanitation with chlorine. *E. coli* were found within the roots, hypocotyls, and cotyledons of soil-grown plants inoculated with contaminated irrigation water (Wachtel et al., 2002). Solomon et al. (2002) reported the internalization of *E. coli* O157:H7 into edible tissue of lettuce, detected by laser scanning confocal microscopy, through root-associated uptake of the pathogen. Internalized cells were detected in plants exposed to 10^8 cfu *E. coli* O157:H7, but not to 10^4 cfu *E. coli* O157:H7. These results are supported by a subsequent study that showed exposure of lettuce to 10^4 or less cfu *E. coli* O157:H7 in irrigation water, soil, or manure resulted in only a limited number of plants having internalized target bacteria (Mootian et al., 2009). The entry point(s) (roots, stomata, wound) for the pathogen was not determined.

In the environment, the levels of *E. coli* O157:H7 and other enteric pathogens are likely to be extremely low (Cooley et al., 2007). Therefore, mechanistically, enteric pathogens may be internalized via the root system and transported to edible tissue, but the risk of contamination by this route is likely low. The ability of *E. coli* O157:H7 to become internalized in tissues of leafy greens is supported (Franz et al., 2007) and contradicted (Hora et al., 2005; Warriner et al., 2003) in other published reports. A comprehensive review of the topic is available (Deering et al., 2012). Chapter 2 contains a more in-depth review of internalization of enteric foodborne pathogens in leafy greens and other fruits and vegetables.

Influence of cutting on microbial populations

The action of harvesting and processing of leafy greens inherently results in the release of plant exudate along cut edges. *E. coli* O157:H7 were found to attach preferentially to the cut edges of lettuce leaves (Takeuchi and Frank, 2000). *Salmonella* and *Shigella* were reported to grow more rapidly and reach higher populations on chopped leaves of cilantro and parsley, respectively, compared to whole leaves (Campbell et al., 2001; Wu et al., 2000). Brandl et al. (2004) reported that despite the presence of suitable substrates on lettuce and spinach leaves, *Campylobacter jejuni* was unable to grow. Following large foodborne illness outbreaks associated with the consumption of *E. coli* O157:H7-contaminated lettuce and spinach, greater focus has been placed on the growth of the pathogen on intact and damaged leaves. *E. coli* O157:H7 populations increased 4, 4.5, and 11-fold within four hours on romaine lettuce that received mechanical, physiological, and disease-induced lesions, respectively (Brandl, 2008). A 2-fold increase in *E. coli* O157:H7 populations occurred on leaves that were left intact. The influence of leaf age was also investigated; the *E. coli* O157:H7 population on young leaves was 27-fold greater than on middle-aged leaves. The study underscores the potential for low numbers of a pathogen to rapidly increase to levels that present a significant human health concern.

The growth of native microflora and foodborne pathogens during storage can dramatically impact shelf-life and safety of a product. Microbial populations on leafy greens increased significantly on product held at temperatures considered abusive for chilled foods. Studies investigating the behavior of *E. coli* O157:H7 on lettuce held at 5 to 22°C showed that at temperatures at or greater than 8°C, the pathogen population increased during storage (Delaquis et al., 2007). In bagged lettuce held at or less than 5°C, the population of *E. coli* O157:H7 remained unchanged or decreased. Maintaining the cold chain from the packer through the consumer is essential to prevent the growth of foodborne pathogens that may be present in low numbers.

Conclusion

Significant advances have been made in understanding interaction of enteric foodborne pathogens with leafy green plant tissue. In broad terms this knowledge will aid in the development of methods for detection and identification of target microbes, facilitate the development of novel strategies for reducing or eliminating target pathogens, and permit integration of Good Agricultural Practices, Good Manufacturing Practices, and Hazard Analysis Critical Control Point programs to ensure that leafy green commodities are handled appropriately from the field to retail establishment to consumers' homes. Localization of bacteria at subsurface locations is supported by research. Regardless of the route by which the pathogen became internationalized, the location of the organisms greatly reduces the likelihood of removal using conventional methods.

New technologies must be sought that will effectively inactivate internalized bacteria. Novel approaches and methods must not be dismissed since existing methods are not sufficient to improve or ensure the safety of leafy greens. Experiments need to be conducted that better reflect conditions encountered in the real world and that recognize restrictions associated with manufacturing operations. Basic surface sanitizing strategies fail to consistently reduce levels of target pathogens by greater than

2 log. New technologies using bacteriophage, package inserts for delivery of antimicrobial agents, and novel natural antimicrobials must be considered. Greater emphasis should be given to the utilization of existing technologies and the development of new technologies for tracking product from the field to retail establishments. This would facilitate more rapid and precise recalls, potentially reducing the number of human illnesses and limiting monetary loss to an entire industry, the result of blanket recalls of a given commodity.

References

Abadias, M., Usall, J., Anguera, M., Solsona, C., Vinas, I., 2008. Efficacy of neutral electrolyzed water (NEW) for reducing microbial contamination on minimally-processed vegetables. Int. J. Food Microbiol. 123, 151–158.

Agrios, G.N., 1997. Plant pathology, fourth ed. Academic Press, San Diego, CA.

Al-Binali, A.M., Bello, C.S., El-Shewy, K., Abdulla, S.E., 2006. The prevalence of parasites in commonly used leafy vegetables in South Western, Saudi Arabia. Saudi. Med. J. 27, 613–616.

Arthur, L., Jones, S., Fabri, M., Odumeru, J., 2007. Microbiological survey of selected Ontario-grown fresh fruits and vegetables. J. Food Prot. 70, 2864–2867.

Aruscavage, D., Lee, K., Miller, S., LeJune, J.T., 2006. Interactions affecting the proliferation and control of human pathogens on edible plants. J. Food Sci. 71, 89–99.

Bernstein, N., Sela, S., Pinto, R., Ioffe, M., 2007. Evidence for internalization of *Escherichia coli* into the aerial parts of maize via the root system. J. Food Prot. 70, 471–475.

Boyer, R.R., Sumner, S.S., Williams, R.C., Pierson, M.D., Popham, D.L., Kniel, K.E., 2007. Influence of curli expression by *Escherichia coli* O157:h7 on the cell's overall hydrophobicity, charge, and ability to attach to lettuce. J. Food Prot. 70, 1339–1345.

Brandl, M.T., 2006. Fitness of human enteric pathogens on plants and implication for food safety. Annu. Rev. Phytopathol. 44, 367–392.

Brandl, M.T., 2008. Plant lesions promote the rapid multiplication of *Escherichia coli* O157:H7 on postharvest lettuce. Appl. Environ. Microbiol. 74, 5285–5289.

Brandl, M.T., Haxo, A.F., Bates, A.H., Mandrell, R.E., 2004. Comparison of survival of *Campylobacter jejuni* in the phylosphere with that in the rhizosphere of spinach and radish plants. Appl. Environ. Microbiol. 70, 1182–1189.

Brandl, M.T., Amundson, R., 2008. Leaf age as a risk factor in contamination of lettuce with *Escherichia coli* O157:H7 and *Salmonella enterica*. Appl. Environ. Microbiol. 74, 2298–2306.

Brandl, M.T., Mandrell, R.E., 2002. Fitness of *Salmonella enterica* serovar Thompson in the cilantro phyllosphere. Appl. Environ. Microbiol. 68, 3614–3621.

CDC, 2012. Multistate outbreak of shiga toxin-producing *Escherichia coli* O157:H7 infections linked to organic spinach and spring mix blend (final update). http://www.cdc.gov/ecoli/2012/O157H7-11-12/index.html.

Calvin, L., 2007. Outbreak linked to spinach forces reassessment of food safety practices. Economic Research Service USDA. Amber Waves. 5, 24–31. www.ers.usda.gov/amberwaves.

Campbell, V.J., Mohle-Boetani, J., Reporter, R., Abbott, S., Farrar, J., Brandl, M.T., et al., 2001. An outbreak of *Salmonella* serotype Thompson associated with fresh cilantro. J. Infect. Dis. 183, 984–987.

Chapman, M.R., Robinson, L.S., Pinkner, J.S., Roth, R., Heuser, J., Hammer, M., et al., 2002. Role of *Escherichia coli* curli operons in directing amyloid fiber formation. Science 295, 851–855.

Cooley, M., Carychao, D., Crawford-Miksza, L., Jay, M.T., Myers, C., Rose, C., et al., 2007. Incidence and tracking of *Escherichia coli* O157:H7 in a major produce production region in California. Plosone. 11, e1159. www.plosone.org.

Cuite, C.L., Condry, S.C., Nucci, M.L., Hallman, W.K., 2007. Public response to the contaminated spinach recall of 2006 Publication number RR-0107–013. www.foodpoli cyinstitute.org.

Deering, A., Mauer, L.J., Pruitt, R.E., 2012. Internalization of *E. coli* O157:H7 and *Salmonella* spp. in plants: A review. Food Res. Int. 45, 567–575.

Delaquis, P., Bach, S., Dinu, L.-D., 2007. Behavior of *Escherichia coli* O157:H7 in leafy vegetables. J. Food Prot. 70, 1966–1974.

Dreux, N., Albagnac, C., Carlin, F., Morris, C.E., Nguyen-The, C., 2007. Fate of *Listeria* spp. on parsley leaves grown in laboratory and field cultures. J. Appl. Microbiol. 103, 1821–1827.

Du, W.X., Olsen, C.W., Avena-Bustillos, R.J., McHugh, T.H., Levin, E.E., Friedman, M., 2008. Antibacterial activity against *E. coli* O157:H7, physical properties, and storage stability of novel carvacrol-containing edible tomato films. J. Food Sci. 73, 378–383.

Du, W.X., Olsen, C.W., Avena-Bustillos, R.J., McHugh, T.H., Levin, E.E., Friedman, M., 2008b. Storage stability and antibacterial activity against *Escherichia coli* O157:H7 of carvacrol in edible apple films made by two different casting methods. J. Agric. Food Chem. 56, 3082–3088.

FDA/CFSAN, 2004. Produce safety from production to consumption: A proposed action plan to minimize foodborne illness associated with fresh produce consumption - June 2004. http://www.fda.gov/Food/FoodborneIllnessContaminants/BuyStoreServeSafeFood/ucm056859.htm.

FDA, 2007. FDA finalizes report on 2006 Spinach Outbreak. P07–51. http://www.fda.gov/NewsEvents/Newsroom/PressAnnouncements/2007/ucm108873.htm.

FDA, 2013. Standards for the growing, harvesting, packing, and holding of produce for human consumption. Fed. Regist. /Vol. 78, No. 11 /Wednesday, January 16, 2013 / Proposed Rules 3504–3646.

Francis, G.A., O'Briene, D., 2001. Effects of vegetable type, package atmosphere and storage temperature on growth and survival of *Escherichia coli* O157:H7 and *Listeria monocytogenes*. J. Ind. Microbiol. Biotechnol. 27, 111–116.

Franz, E., Visser, A.A., Van Diepeningen, A.D., Klerks, M.M., Termorshuizen, A.J., van Bruggen, A.H.C., 2007. Quantification of contamination of lettuce by GFP-expressing *Escherichia coli* O157:H7 and *Salmonella enterica* server Typhimurium. Food Microbiol. 24, 106–112.

Fonseca, J.M., 2006. Postharvest handling and processing: Sources of microorganisms and impact of sanitizing procedures. In: Matthews, K.R. (Ed.), Microbiology of fresh produce. Amer. Soc. Microbiol., Washington, DC, pp. 85–120.

Fonseca, J.M., Fallon, S.D., Sanchez, C.A., Nolte, K.D., 2011. *Escherichia coli* survival in lettuce fields following its introduction through different irrigation systems. J. Appl. Microbiol., 110, 893–902.

Gil, M.I., Selma, M.V., Lopez-Galvez, F., Allende, A., 2009. Fresh-cut product sanitation and wash water disinfection: problems and solutions. Int. J. Food Microbiol. 134, 37–45.

Golberg, D., Kroupitski, Y., Belausov, E., Pinto, R., Sela, S., 2011. *Salmonella* Typhimurium internalization is variable in leafy vegetables and fresh herbs. Int. J. Food Microbiol. 145, 250–257.

Gomez-Lopez, V.M., Ragaert, P., Jeyachchandran, V., Debevere, J., Devlieghere, F., 2008. Shelf-life of minimally processed lettuce and cabbage treated with gaseous chlorine dioxide and cysteine. Int. J. Food Microbiol. 121, 74–83.

Gorski, L., Palumbo, J.D., Mandrell, R.E., 2003. Attachment of *Listeria monocytogenes* to radish tissue is dependent upon temperature and flagellar motility. Appl. Environ. Microbiol. 69, 258–266.

Hagens, S., Loessner, M.J., 2007. Application of bacteriophages for detection and control of foodborne pathogens. Appl. Microbiol. Biotechnol. 76, 513–519.

Hammed, T.B., 2013. The effect of locally fabricated pelletizing machine on the chemical and microbial composition of organic fertilizer. Brit. Biotech. J. 3, 29–38.

Heaton, J.C., Jones, K., 2008. Microbial contamination of fruit and vegetables and the behaviour of enteropathogens in the phylosphere: a review. J. Appl. Microbiol. 104, 613–626.

Hsu, W.-Y., Simonne, A., Jitareerat, P., 2006. Fates of seeded *Escherichia coli* O157:H7 and *Salmonella* on selected fresh culinary herbs during refrigerated storage. J. Food Prot. 69, 1997–2001.

Hueck, C.L., 1998. Type III secretions systems in bacterial pathogens of animals and plants. Microbiol. Rev. 62, 379–433.

Jablasone, J., Warrnier, K., Griffiths, M., 2005. Interactions of *Escherichia coli* O157:H7, *Salmonella* Typhimurium and *L. monocytogenes* plants cultivated in a gnotobiotic system. Int. J. Food Microbiol. 99, 7–18.

Jackson, C., Archer, D., Goodrich-Schneider, R., Gravani, R.B., Bihn, E.A., Schneider, K.R., 2007. Determining the effect of Good Agricultural Practices awareness on implementation: A multi-state survey. Food Prot. Trends. 27, 684–693.

Jay, M.T., Cooley, M., Carychao, D., Wiscomb, G.W., Sweitzer, R.A., Crawford-Miksza, L., et al., 2007. *Escherichia coli* O157:H7 in feral swine near spinach fields and cattle, central California coast. Emerg. Infect. Dis. 13, 1908–1911.

Kang, S.-C., Kim, M.-J., Choi, U.-K., 2007. Shelf-life extension of fresh-cut iceberg lettuce (*Lactuca sativa* L) by different antimicrobial films. J. Microbiol. Biotechnol. 17, 1284–1290.

Kang, S.-C., Kim, M.-J., Park, I-S., Choi, U.-K., 2008. Antimicrobial (BN/PE) film combined with modified atmosphere packaging extends the shelf life of minimal processed fresh-cut iceberg lettuce. J. Microbiol. Biotechnol. 18, 568–572.

Koseki, S., Isobe, S., 2005. Prediction of pathogen growth on iceberg lettuce under real temperature history during distribution from farm to table. Int. J. Food Microbiol. 104, 239–248.

Lapidot, A., Romling, U., Yaron, S., 2006. Biofilm formation and the survival of *Salmonella* Typhimurium on parsley. Int. J. Food Microbiol. 109, 229–233.

Lee, S.-Y., Costello, M., Kang, D.-H., 2004. Efficacy of chlorine dioxide gas as a sanitizer of lettuce leaves. J. Food Prot. 67, 1371.

Leverentz, B., Conway, W.S., Camp, M.J., Janisiewicz, W.J., Abuladze, T., Yang, M., et al., 2003. Biocontrol of *Listeria monocytogenes* on fresh-cut produce by treatment with lytic bacteriophages and a bacterocin. Appl. Environ. Microbiol. 23, 4519–4526.

Leverentz, B., Conway, W.S., Alavidze, Z., Janisiewicz, W.J., Fuchs, Y., Camp, M.J., et al., 2001. Examination of bacteriophage as a biocontrol method for *Salmonella* on fresh-cut fruit: A model study. J. Food Prot. 64, 1116–1121.

Li, H., Tajkarimi, M., Osburn, B.I., 2008. Impact of vacuum cooling on *Escherichia coli* O157:H7 infiltration into lettuce tissue. Appl. Environ. Microbiol. 74, 3138–3142.

Lopez-Mosquera, M.E., Cabaleiro, F., Sainz, M.J., Lopez-Fabel, A., Carral, E., 2008. Fertilizing value of broiler litter: Effects of drying and palletizing. Bioresour. Technol. 99, 5626–5633.

Mercier, J., Lindow, S.E., 2000. Role of leaf surface sugars in colonization of plants by bacterial epiphytes. Appl. Environ. Microbiol. 66, 369–374.

Mootian, G., Wu, W.-H., Pang, H.-J., Matthews, K.R., 2009. Transfer prevalence of *Escherichia coli* O157:H7 from soil, water, and manure contaminated with low numbers of the pathogen to lettuce plants of varying age. J. Food Prot. 72, 2308–2312.

Neetoo, H., Ye, M., Chen, H., Joerger, R.D., Hicks, D.T., Hoover, D.G., 2008. Use of nisin-coated plastic films to control *Listeria monocytogenes* on vacuum-packaged cold-smoked salmon. Int. J. Food Microbiol. 122, 8–15.

Nygard, K., Lassen, J., Vold, L., Andersson, Y., Fisher, I., Lofdahl, S., et al., 2008. Outbreak of *Salmonella* Thompson infections linked to imported rucola lettuce. Foodborne. Pathog. Dis. 5, 165–173.

Okafo, C.N., Umoh, V.J., Galadima, M., 2003. Occurrence of pathogens on vegetables harvested from soils irrigated with contaminated streams. Sci. Total. Environ. 311, 49–56.

Painter, J.A., Hoestra, R.M., Ayers, T., Tauxe, R.V., Braden, C.R., Angulo, F.J., Griffin, P.M., 2013. Attribution of foodborne illnesses, hospitalizations, and deaths to food commodities by using outbreak data, United States, 1998-2008. Emerg. Infect. Dis. 19, 407–415.

Parish, M.E., Beuchat, L.R., Suslow, T.V., Harris, L.J., Garret, E.H., Farber, J.N., Busta, F.F., 2003. Methods to reduce/eliminate pathogens from fresh and fresh-cut produce. Comp. Rev. Food Sci. Food Safety. Vol. 2 (Suppl), 161–173.

Seo, S., Matthews, K.R., 2012. Influence of the plant defense response to *Escherichia coli* O157:H7 cell surface structures on survival of that pathogen on plant surfaces. Appl. Environ. Microbiol. 78, 5882–5889.

Sewell, A.M., Farber, J.M., 2001. Foodborne outbreaks in Canada linked to produce. J. Food Prot. 64, 1863–1877.

Seymour, I.J., Burfoot, D., Smith, R.L., Cox, L.A., Lockwood, A., 2002. Ultrasound decontamination of minimally processed fruits and vegetables. Int. J. Food Sci. Tech. 37, 547–557.

Smith DeWaal, C., Glassman, M., 2013. Outbreak Alert 2013. Center for Science in the Public Interest cspinet.org/new/pdf/outbreak_alert_2013_final.pdf.

Soderstrom, A., Osterberg, P., Linquist, A., Johnson, B., Lindberg, A., Blide Ulander, S., et al., 2008. A large *Escherichia coli* O157 outbreak in Sweden associated with locally produced lettuce. Foodborne. Pathog. Dis. 5, 339–349.

Solomon, E.B., Brandl, M.T., Mandrell, R.E., 2006. Biology of foodborne pathogens on produce. In: Matthews, K.R. (Ed.), Microbiology of Fresh Produce, first ed. American Society for Microbiology Press, Washington, DC 55.

Solomon, E.B., Yaron, S., Matthews, K.R., 2002. Transmission and internalization of *Escherichia coli* O157:H7 from contaminated manure and irrigation water into lettuce plant tissue. Appl. Environ. Microbiol. 68, 397–400.

Swedberg, C., 2012. International group tests RFID for Food Safety to Hawaii. RFID. J. http://www.rfidjournal.com/articles/view?9216/.

Sy, K.V., Murray, M.B., Harrison, M.D., Buechat, L.R., 2005. Evaluation of gaseous chlorine dioxide as a sanitizer for killing *Salmonella, Escherichia coli* O157:H7, *Listeria monocytogenes*, and yeasts and molds on fresh and fresh-cut produce. J. Food Prot. 68, 1176–1187.

Takeuchi, K., Frank, J.F., 2000. Penetration of *Escherichia coli* O157:H7 into lettuce tissues as affected by inoculum size and temperature and the effect of chlorine treatment on cell viability. J. Food Prot. 63, 434–440.

Taylor, C.B., 1998. Defense responses in plants and animals—More of the same. Planta 10, 873–876.

Todd, E.C.D., Harris, C.K., Knight, A.J., Worosz, M.R., 2007. Spinach and the media: How we learn about a major outbreak. Food Prot. Trends 27, 314–321.

Toth, I.K., Pritchard, L., Birch, P.R.J., 2006. Comparative genomics reveals what makes an enterobacterial plant pathogen. Annu. Rev. Phytopathol. 44, 305–306.

Valentin-Bon, I., Jacobson, A., Monday, S.R., Feng, P.C.H., 2008. Microbiological quality of bagged cut spinach and lettuce mixes. Appl. Environ. Microbiol. 74, 1240–1242.

Wachtel, M.R., Whitehand, L.C., Mandrell, R.E., 2002. Association of *Escherichia coli* O157:H7 with preharvest leaf lettuce upon exposure to contaminated irrigation water. J. Food Prot. 65, 18–25.

Warriner, K., Ibrahim, F., Dickinson, M., Wright, C., Waites, W.M., 2003. Interaction of *Escherichia coli* with growing salad spinach plants. J. Food Prot. 66, 1790–1797.

Western Growers, 2008. Commodity Specific food safety guidelines for the production and harvest of lettuce and leafy greens. www2.wga.com/popups/bestpracticesdraft.html.

Wood, J.D., Bezanson, G.S., Gordon, R.J., Jamieson, R., 2010. Population dynamics of *Escherichia coli* inoculated by irrigation into the phyllosphere of spinach grown under commercial production conditions. Int. J. Food Microbiol. 143, 198–204.

Wu, F.M., Doyle, M.P., Beuchat, L.R., Wells, J.G., Mintz, E.D., Swaminathan, B., 2000. Fate of *Shigella sonnei* on parsley and methods of disinfection. J. Food Prot. 63, 568–572.

Zhou, B., Feng, H., Luo, Y., 2009. Ultrasound enhanced sanitizer efficacy in redcution of *Escherichia coli* O157:H7 population in spinach leaves. J. Food Sci. 74, 308–313.

Melons

10

Alejandro Castillo[1], Miguel A. Martínez-Téllez[2], M. Ofelia Rodríguez-García[3]

[1]*GDepartment of Animal Science Center for Food Safety, Texas A&M University College Station, TX,* [2]*Dirección de Tecnología de Alimentos de Origen Vegetal Centro de Investigación en Alimentación y Desarrollo, A.C. Hermosillo, Sonora, Mexico,* [3]*Departamento de Farmacobiología Centro Universitario de Ciencias Exactas e Ingenierías Universidad de Guadalajara, Guadalajara, Jalisco, Mexico*

CHAPTER OUTLINE

The Produce Contamination Problem. http://dx.doi.org/10.1016/B978-0-12-404611-5.00010-5

Introduction

The nutritional characteristics of melons and the benefits of consuming these cucurbits in the diet is well documented (Eitenmiller et al., 1985; Hakerlerler et al., 1999). This makes melons a widely consumed commodity worldwide, which may make the growing of these commodities a long-term guarantee for profitable agriculture (Karchi, 2000).

In contrast, melons, especially cantaloupes (netted), have become a recurrent source of pathogens causing outbreaks of foodborne disease, especially *Salmonella* infection and more recently, *Listeria monocytogenes* infection resulting in listeriosis, a disease well known to have a high fatality rate (Gellin and Broome, 1989). Cantaloupes grow at ground level, thus increasing the potential for fruit-surface contamination. Factors such as the potential for pathogens to attach to the porous rind of the melon and internalize (Asta, 1999; Fan et al., 2006) and to form biofilms (Annous et al., 2005) may promote the occurrence of melon-linked outbreaks of foodborne disease. Although the contamination may be restricted to the rind, it can be transferred to the flesh during cutting. Cut cantaloupe is considered to be a potentially hazardous food according to the FDA food code because it is capable of supporting the growth of pathogens due to mild acidity (pH 5.2–6.7) and high water activity (0.97–0.99) (Bhagwat, 2006). In this chapter, information will be discussed with regards to foodborne disease outbreaks linked to melons, contamination sources and mechanisms of melon contamination, and possible mitigating strategies to reduce the risk of illness associated with consumption of melons.

Prevalence of human pathogens in and on melons

Melons have been reported as vehicles of pathogens causing outbreaks. However, relatively little definitive information on sources of human pathogens in melons is available (Ukuku and Sapers, 2006). In a binational study conducted by Texas A&M University and the University of Guadalajara, Mexico, eight cantaloupe farms and packing sheds from the U.S. and Mexico were sampled to evaluate cantaloupe contamination with *Salmonella* and *E. coli*. Samples collected from external surfaces of cantaloupes and water environments of packing sheds of cantaloupes farms were examined. Of a total of 1,735 samples collected, 31 (1.8%) tested positive for *Salmonella*; this pathogen was isolated from 5 (0.5%) of 950 samples in the south of Texas and from 1 (0.3%) of 300 samples in the State of Colima, Mexico. Fifteen *Salmonella* serotypes were isolated from samples collected in the U.S., and 9 from samples collected in Mexico (Castillo et al., 2004). Cantaloupes may be especially prone to accumulating microorganisms and harboring bacterial pathogens. In a study of the prevalence of *Salmonella* in the growing and processing environment of oranges, parsley, and cantaloupes in South Texas, Duffy et al. (2005) recovered *Salmonella* only from cantaloupes. In another study involving more than 900 field-collected melons produced in different regions of California during 1999 to 2001, no *Salmonella* were ever recovered (Suslow, 2004).

In Brazil, Penteado and Leitão (2009) did not find *Salmonella* or *L. monocytogenes* on the surface of 120 samples of cantaloupe, watermelon, and papaya, although other *Listeria* spp. were found in 2.4 to 7.5% of these samples.

In contrast, Materon et al. (2007) found that most of the cantaloupe samples collected from South Texas contained *Salmonella* and *Listeria* spp., with populations on the cantaloupe rind ranging between 0.2 and 3.1 log cfu/cm^2 for *Salmonella* and between 0.3 and 3.1 log cfu/cm^2 for *Listeria* spp. However, these results seemed to be affected by the use of unconventional methods for confirmatory identification of these bacteria. In Mexico, Espinoza-Medina et al. (2006) reported the differences in *Salmonella*-positive samples of cantaloupe when using a PCR method vs. the conventional enrichment-plating method, stressing the importance of confirming by conventional method all samples that tested positive by the rapid method.

In two independent surveys conducted by the U.S. Food and Drug Administration (FDA), which included Mexican cantaloupes, *Salmonella* was isolated from 0.78 and 1.08%, with 8 and 12 serotypes identified in each survey. In another survey *Salmonella* was isolated from 8 (5.3%) and *Shigella* from 3 (2.0%) of 151 cantaloupes samples collected from nine countries exporting to the U.S. (Bhagwat, 2006). In 1999 the FDA carried out a study to estimate the frequency of *Salmonella*, *Shigella*, and *E. coli* O157: H7 in 1003 samples (FDA, 2001). Ninety-six percent of the samples tested negative, 35 (3.5%) were positive for *Salmonella*, 9 (0.9%) for *Shigella*, and none were positive for *E. coli* O157: H7. In another study intended to determine *Salmonella* spp. and *Shigella* spp. in domestic and imported melons, these pathogens were isolated from 2.4% and 0.5% of the domestic melons and from 5.3% and 1.2% of the imported fruit, respectively (FDA, 2001, 2006).

As expected, the microbiota of the fresh produce reflects the environment in which they grow. Figueroa-Aguilar (2005) collected cantaloupe and environmental samples at the cantaloupe farms where the melons involved with the *Salmonella* Poona outbreaks of 2000 and 2001 (CDC, 2002) were produced. *Salmonella* spp. were isolated from four samples of river water, two of water from a hydrocooler, four fecal samples from iguanas, and one fecal sample of an unidentified animal. No *Salmonella* was isolated from melons or well water. In this study, the serotypes identified were *S.* Poona, *S.* Infantis, and *S.* Anatum.

Outbreaks of foodborne disease linked to melons
Characteristics of outbreaks

Melons have been associated with foodborne disease outbreaks caused by several serovars of *Salmonella, E. coli* O157:H7, *Campylobacter*, and Norovirus (Chapman, 2005). In the period of 1973 to 1997, 13 outbreaks were recorded with 341 cases of disease involving cantaloupes (6), watermelons (6), and one with a combination of musk and honeydew melons. The etiologic agents of melon-associated disease were *Salmonella* (6), *E. coli* O157:H7 (1), and *Campylobacter jejuni* (1) (Sivapalasingam et al., 2004). Between 1998 and 2011, 35 outbreaks involving melons as vehicles of various pathogens were reported to the CDC. The melons involved were cantaloupes (11), watermelons (7), honeydew melons (3), and a combination of cantaloupe with watermelons or honeydew melons (14). The etiologic agents for these outbreaks were Norovirus (14), *Salmonella* (14) *E. coli* O157:H7 (2), *C. jejuni* (1), *L. monocytogenes* (1), and *Shigella* (1). In two outbreaks the etiologic agent was not known (CDC, 2013).

Salmonella is the pathogen most frequently related with cases and outbreaks linked to melon consumption. The first report of *Salmonella* infection associated with melon consumption was in 1955; 17 cases of disease caused by *S.* Miami were linked to sliced watermelon purchased from a local store (Gaylor et al., 1955). Another outbreak linked to precut watermelon was reported in 1979, caused by *S.* Oranienburg. Six people in two families were affected, and each family acquired a watermelon half by the same vendor by the side of the highway (CDC, 1979).

In 1990 in the U.S., a large multi-state salmonellosis outbreak was documented by the CDC to be associated with melon consumption. It was reported that *Salmonella* Chester caused 245 cases (two fatal) in 30 states of the U.S. (Ries et al., 1990), although the real number of cases was estimated to exceed 25,000 (CDC, 1991; Tamplin, 1997). Apparently, transmission of pathogens to the interior of the cantaloupe may have occurred while cutting the unwashed melons (Beuchat, 1996). In a survey of melons from the same area where the melons involved with the outbreak originated, *Salmonella* was found in 24 of the 2,200 melons tested. Twelve different serovars were identified, but none of the isolates was *S.* Chester, the serovar that caused the outbreak (Madden, 1992). In another outbreak in 1991, more than 400 cases were reported from 23 states in the U.S. and 4 Canadian provinces. The causative agent of the illness in this outbreak was *S.* Poona, and the outbreak was linked to the consumption of contaminated cantaloupes produced in Texas (CDC, 1991).

In June of 1991, an outbreak of *S.* Javiana linked to watermelon affected 39 children. The microorganism was isolated from the feces of some of the victims as well as from leftovers of the sliced watermelon. Analysis of the plasmid profile and DNA chromosomal restriction led to the conclusion that it was the same strain. Apparently, the causative factor was the cutting of slices without previously washing the fruit and the consumption of the leftover melon, which had been stored at room temperature. The contamination probably took place during transportation of the fruit.

After these cases the FDA defined melons as potentially hazardous to health (Blostein, 1993). After these outbreaks of salmonellosis, linked mainly to imported melons, the melon industry considered as a prudent measure the implementation of the "Melon Safety Plan," which was focused on the chlorination of the water used to wash melons, as well as on the ice used in cooling or the refrigerated containers used for transporting the melons (Tauxe, 1997).

In 1997, an outbreak of salmonellosis caused by a strain of *S.* Saphra affected 25 California residents. Pulsed field gel electrophoresis (PFGE) was used to identify the outbreak strain, and identical PFGE patterns were found for 24 *S.* Saphra isolates from infected patients. This PFGE pattern was different from that of the 5 *S.* Saphra strains isolated in prior years in California. Most of the patients in this outbreak were young children whose parents recalled feeding them cantaloupe. The epidemiological and trace-back study in this outbreak supported the conclusion that the cantaloupes implicated in this outbreak were imported from a small region of Mexico. Only 17% of the patients washed the cantaloupes before cutting them (Mohle-Boetani et al., 1999). This outbreak is another example of gastrointestinal disease in the United States related to contaminated, imported melons.

An additional outbreak of salmonellosis involving the consumption of contaminated cantaloupes occurred in Ontario in 1998. Twenty-two cases were linked to the consumption of cantaloupes contaminated with *S.* Oranienburg (Deeks et al., 1998). During the time of the outbreak, cantaloupes were imported into Ontario from many sources, including the U.S., Mexico, and Central America. In 1999, melons were involved in an outbreak where *S.* Enteritidis caused 82 cases of disease in the U.S. (CDC, 2003a).

In 2000, 46 cases of salmonellosis were reported during an outbreak that occurred in over 6 states. At least 26 of the cases in California were attributed to *S.* Poona and linked to the consumption of contaminated cantaloupe. Three successive multistate outbreaks occurred in the spring of 2000 (47 cases), 2001 (50 cases including 2 deaths), and 2002 (58 cases), each of which was linked to eating cantaloupes from Mexican farms (CDC, 2002). The PFGE patterns of the 2000 and 2002 outbreak strains were indistinguishable, whereas the outbreak of 2001 differed (Tauxe et al., 2008). The FDA evaluated the farms in 2000 and 2001, identifying several possible sources of contamination during field operations, such as river water used for irrigation and fecal matter from the iguanas found to thrive in the fields, and during washing and packing, finding that the measures being applied were insufficient to minimize microbial contamination. After the 2002 outbreak, the FDA issued an import alert on all cantaloupes from Mexico (Tauxe et al., 2008).

Between January 18 and March 5, 2008, state health departments identified 50 ill persons in 16 states of the U.S. who were infected with *S.* Litchfield with the same genetic fingerprint. In addition, 9 ill persons with the outbreak strain were reported in Canada. The CDC collaborated with public health officials in multiple states across the U.S. and with the FDA to investigate this multi-state outbreak of *S.* Litchfield infections. Traceback studies pointed to cantaloupes from Honduras as the likely source of infections (CDC, 2008). The Honduran company that produced these melons had to dismiss 1,800 employees, and experienced a $13 million loss due to the *Salmonella* contamination. In 2011, 20 persons infected with a single strain of *Salmonella* Panama were reported in 10 states. This outbreak was epidemiologically linked to consumption of cantaloupes that were imported from Guatemala (CDC, 2011). In the following year, a large multistate outbreak of salmonellosis was reported to the CDC (CDC, 2012). This outbreak was linked to cantaloupes grown at a single farm in the state of Indiana. This outbreak was unique because most of the previous incidents of cantaloupe-related salmonellosis had been linked to imported cantaloupes. In this outbreak, a total of 261 persons were infected, and the infection was caused by two different serotypes of *Salmonella* (*S.* Typhimurium, affecting 228 people, and *S.* Newport, affecting 33 people). Ninety-four patients were hospitalized, and 3 people died. This outbreak, along with an outbreak of foodborne listeriosis occurring in 2011 (CDC, 2011a), was linked to cantaloupes growing in the U.S. and renewed interest in the safety of cantaloupes by the U.S. government and researchers (Chen et al., 2012; FDA, 2013a; Mahmoud, 2012; Vadlamudi et al., 2012).

Several outbreaks linked to melons have been caused by pathogens different from *Salmonella.* In 1993, 27 cases of *E. coli* O157: H7 were linked to melon consumption (Del Rosario et al., 1995). In 1997, cantaloupes were again considered to be the vehicle in 9 cases of hemorrhagic enteritis caused by *E. coli* O157: H7 (CDC, 2003). In 2004, a new outbreak of *E. coli* O157: H7 occurred in Montana, which was associated with cantaloupe consumption and involved six cases of HUS and PTT (ISID, 2004). Additionally, 48 cases of *Campylobacter* enteritis were linked to melons in 1993 (CDC, 2003). Melon-associated outbreaks have also been attributed to norovirus disease. In 1999 three outbreaks involving 23, 61, and 5 cases were reported in the months of May, June, and September, respectively. For these norovirus outbreaks, cantaloupes were identified as the vehicle (CDC, 2003). In 2011, a multi-state outbreak of listeriosis was linked to cantaloupes grown and packed at a single operation in Colorado. This outbreak was caused by four strains of *L. monocytogenes* and, compared to previous listeriosis outbreaks (Cartwright et al., 2013), is the most extensive outbreak of listeriosis recorded in the U.S. In addition, *L. monocytogenes* infection had not been linked to cantaloupes before. During this incident, a total of 146 people were infected over 28 states. Of these, 142 (95%) were hospitalized and 30 (20%) died. The implicated farm recalled cantaloupes that had been distributed in 24 states, and companies that used these cantaloupes for producing fresh-cut products also had to recall their products (FDA, 2012). After the FDA conducted a study to determine the factors that might have led to the outbreak, they determined that the packing facility and cold storage areas seemed to be the major points of contamination. Although *L. monocytogenes* was never found in field environments, the FDA could not rule out the preharvest areas as a possible source of sporadic contamination with *L. monocytogenes* in the cantaloupes. After this, the FDA proposed that *L. monocytogenes* may have created niches in the processing environment, where this pathogen can grow and then be transferred to the product via contaminated contact surfaces (FDA, 2011).

Contributing factors

Melon is a type of produce with unique characteristics. Keeping the cut melons for several hours at ambient temperature was a factor that appeared multiple times during outbreak investigations. Cut melons placed in salad bars, at group dinners, etc., can support the growth of *Salmonella* if the storage temperature is inadequate (Golden et al., 1993). A common theme in the outbreaks was that the melons were cut and then subjected to temperature abuse. In some cases, melons were cross-contaminated through inadvertent contact with raw meat, or a human handler (Iversen et al., 1987). In most instances, contamination was thought to have originated in the field from contact with contaminated soil (Bhagwat, 2006), and in other cases the contamination was thought to have come from contact of the melon rind with soil during harvesting and packing. The impact of such contamination is subsequently increased by displaying the precut melons at the store, where appropriate refrigeration cannot be guaranteed (Mohle-Boetani et al., 1999). Transfer of bacteria from the rind to edible melon flesh during cutting has been demonstrated in laboratory conditions (Ukuku et al., 2004). Contamination of produce can arise from a variety of sources including soil, water, equipment, and humans (Beuchat, 1996). Cantaloupes can become contaminated during growth, postharvest handling and packing, transportation, distribution, or during final preparation at food service or in the home. During postharvest handling and packing, melons can become contaminated by equipment used to hold, transport, clean, grade, sort, or pack melons, or from unsanitary washing (immersion of gondolas/trailers in dump tank water) or use of contaminated cooling water or ice (Parnell et al., 2005).

Impact of regulatory actions

The FDA is responsible for the safety of U.S. produce imports. Generally, the FDA carries out random sample collection at the border and prevents the entry of products that fail to pass laboratory inspection. The FDA can also detain a product without physical examination, based on past history of a shipping firm or other information indicating that the product might violate standards. When the source of a public health concern can be identified, only those firms with a problem have their shipments blocked (Calvin, 2003). Integration of imports into previously domestic fruit and vegetable markets also requires that exporters meet the food safety standards of other countries. Over the past several years, U.S. and Mexican authorities have worked to establish a framework that would broaden the allowance for cantaloupe exports from Mexico. In November 2002, the FDA issued an Import Alert on all Mexican cantaloupes, effectively banning all such imports from Mexico. This action followed three successive years (2000, 2001, and 2002) where outbreaks of *Salmonella*, associated with the consumption of contaminated cantaloupe from Mexico, occurred in Canada and the United States. Strong epidemiological evidence linked the outbreaks of 2001 to two deaths in California (Green et al., 2005).

Between 1999 and 2005, Mexican cantaloupe production declined 24% as a result of export restrictions. Between 1999 and 2006, cantaloupe imports from Mexico declined 92% and in 2006 accounted for just 3% of U.S. imports (SIAP-SAGARPA, 2002; Zahniser, 2006).

Since the imposition of the countrywide import alert, FDA has exempted several growers from this ban, and on October 26, 2005, the FDA and SENASICA (Mexican National Service for Animal and Plant Health, Food Safety and Quality) signed a memorandum of understanding (MOU) that allows for the differential treatment of prospective Mexican cantaloupe exporters, based on their past food safety performance (Anon, 2005). Efforts by Mexican growers to respond to the growing demand of Mexican consumers for safer food products are likely to further integrate the U.S. and Mexican fruit and vegetable sectors. However, as of June 2013, the Import

Alert #22-0, Detention Without Physical Examination (DWPE) of Cantaloupes from Mexico, is still valid (FDA, 2013), and there are still differences of opinion about whether the point of contamination has been controlled (Tauxe et al., 2008). Nevertheless, cantaloupe exports from Mexico are allowed for some firms. The MOU between the FDA and SENASICA defines three categories of exporting firms: Category 1, firms exempt from DWPE; Category 2, firms that have been directly implicated in an outbreak, or that have shipments from Mexico testing positive for *Salmonella* or other pathogens; and Category 3, firms that have not been implicated in any outbreak, or have had shipments testing positive for *Salmonella* or other pathogens. Under the terms of this MOU, firms in Categories 2 and 3 must be certified by SENASICA to comply with stringent requirements that include, but are not limited to, compliance with good agricultural practices. The FDA will accompany SENASICA on up to the first 12 inspections, and these inspections must be successful before these firms are accepted as certified and allowed to ship cantaloupes to the U.S. However, shipments from Category 2 firms will, in addition to being required to be certified, be subjected to testing for *Salmonella* and will be allowed entrance only after the shipment has tested negative for this pathogen. When a certified firm has had five consecutive shipments testing negative for *Salmonella*, it is classified as Category 1. The firms that fall into Category 1 are placed on a list called the Green List by the FDA, whereas firms in Category 2 are placed on a list called the Yellow List. As of June 2013, only 12 firms are included on the Green List and 5 on the Yellow List (FDA, 2013).

As a result of the outbreaks of listeriosis and salmonellosis that occurred in 2011 and 2012, respectively, which were both linked to U.S.-grown cantaloupes, the FDA released a letter to the cantaloupe industry on produce safety. In this letter, the FDA indicated that since these outbreaks were traced back to the packinghouses, they intended to start inspections, including sampling and testing, of cantaloupe packinghouses. The purpose of these inspections, as per the letter, is **"to assess the current practices by this segment of the produce industry and to identify insanitary conditions that may affect the safety of cantaloupe destined for distribution to consumers."** In addition the letter explained that they would take actions in case of adverse findings, although the type of actions were not described (FDA, 2013a). The activities described in this letter are related to some of the activities related to the proposed rules for fruits and vegetables under the Food Safety Modernization Act (FSMA) (Federal Register, 2013, 2013a).

Potential sources and mechanisms of contamination and measures recommended to prevent contamination

In the last decade cantaloupe and other fruits and vegetables have been involved in numerous outbreaks or recalls in the United States and Canada because of *Salmonella* contamination. Investigations conducted by researchers and or government agencies of the involved countries indicate that the main sources of contamination in the cantaloupe melon could be 1) water for irrigation or preharvest practices and postharvest management, 2) worker activities, 3) organic fertilizer, 4) animal or human feces, and

5) equipment and installations. The contamination of products can occur through direct contact, internalization, or cross-contamination. The application of Good Agricultural Practices in the full chain of production is the best tool for preventing biological contamination of cantaloupe melon.

Harvesting and packaging of cantaloupe melon are carried out in different ways. In California melons are harvested and packed in the field, but in Georgia, the harvested melons are transported to a shed for washing and packing (Akins et al., 2008). In Mexico, cantaloupes are handled in a similar way as in Georgia, according to guidelines for the voluntary implementation of good agricultural practices and good management practices in the production and packing of fruits and vegetables for fresh consumption by humans, released by the Mexican Government (SENASICA, 2008) published after the MOU between the FDA and the Mexican government (Anon, 2005).

The U.S. melon industry recognizes that once a melon is contaminated, it is difficult to remove or to inactivate a pathogen. Therefore, prevention of microbial contamination at all steps from production to distribution is widely favored over treatments to eliminate contamination after it has occurred (PMA and UFFVA, 2005). The proposed FSMA rules on produce safety include requirements for packing facilities to develop and implement a food safety plan that includes written hazard analysis, a list of preventive controls and their application details, corrective actions, verification procedures, and recall plans (Federal Register, 2013a). These requirements are for all commodities that will not receive further processing such as cooking or canning, including melons, and are expected to reduce the risk of foodborne illness.

Preharvest

Water for irrigation and production practices. Water is an essential element used in several production activities, including irrigation, pesticide, and liquid fertilizer applications, among others. Some consider water the main contamination vehicle for fruits and fresh vegetables. When water is in contact with fruits or vegetables, there is a high risk of contamination (Martinez-Tellez et al., 2007), which is why it is important to monitor the microbiological quality of water supplies to avoid pathogen spread among agricultural products. Water can be a transmission source for a large number of microorganisms such as the pathogen species of *Escherichia coli, Salmonella, Vibrio cholera, Shigella*, and also *Cryptosporidium parvum, Giardia lamblia, Cyclospora cayetanensis, Toxoplasma gondii* and norovirus, and hepatitis A viruses (FDA, 1998b).

Before considering the microbiological analysis used to evaluate water quality, we must recognize the fact that water sources have a high risk of contamination with human pathogens, and establish necessary measures to avoid such contamination. These measures must apply to both surface and ground water sources, such as ponds, rivers/streams, or lakes. Water sources close to cattle areas or other potential sources of contamination must be considered for pathogen testing. Additionally, it is necessary to consider water as a potential vehicle for spreading of pathogens from one product to another (Zawel, 1999). For example, irrigation water may carry contamination originating from animal feces in the crop field to fruits attached to the irrigated plants. The presence of pathogens in different sources of water has been reported (Falcão et al., 1993). If water, contaminated with a human pathogen, is used for applying pesticides, the presence of the pesticide may favor survival of the pathogen and even support their growth (Ng et al., 2005). Although the growth of pathogens in water containing pesticides has been reported in laboratory studies, this potential source of pathogen contamination is not generally considered to be important, and the use of contaminated water for pesticide application is a recurrent practice in production of fresh agricultural products. Exposure to contaminated water during production and handling has been identified as a major factor promoting the presence of viable human pathogens on cantaloupe melon surfaces (CDC, 2002). Once the pathogens adhere to the cantaloupe rind, the rough characteristics of the rind surface promote irreversible attachment and subsequent biofilm formation (Annous et al., 2005). Contaminated water should not be used for such purposes.

Worker activities. During the production of cantaloupe, fields workers have minimal contact with the developing fruit; in some cases, they turn the fruit when it is attached to the plant to avoid the formation of soil spots. Due to deficient hygienic practices of workers during fruit handling, human pathogens and/or parasites may be transmitted from workers to the fruits by direct contact. The use of wool or cotton gloves during fruit handling has been identified as an important contamination source (Martínez–Téllez, 2007). Personnel training and education about biologic hazards in agriculture are a priority in order to prevent contamination during primary production. If field personnel do not know, and hence, do not apply hygienic practices, workers may involuntarily contaminate crops, the water supply, equipment, and other workers, (FAO, 2003a). Portable, clean, and sanitary facilities (toilets and washing stations) with enough supplies (clean water, bactericidal soap, toilet paper, and disposable towels) must be installed in the field for workers' hygiene so that they are no more than 2 to 3 minutes' walk from the work location to favor their utilization.

Biological hazards also can be reduced during the production of cantaloupe by establishing preventive health programs for field workers, and avoiding the use of those workers who are sick or have symptoms of a certain disease, wounds on their hands or open sores, in activities involving direct contact with the product.

Organic fertilizer. Application of organic fertilizers in melon production is a matter of great concern due to the fact that compost, produced under deficient conditions or recontaminated with human pathogens as a consequence of incorrect handling, can present a high risk of biologic contamination during production. Materials that are traditionally used for the production of organic fertilizers are animal, crustacean, and vegetable wastes. The first two represent an important source of human pathogens that must be inactivated during the production process of organic fertilizer. The origin of the materials can define the type of microorganisms present in the compost if it is not produced correctly. Wastes of birds may contribute mainly *Salmonella* contamination, whereas enterotoxigenic *E. coli* can be introduced from waste of pigs or *E. coli* O157:H7 from cattle. Aerobic and anaerobic processes in organic fertilizer production reach adequate temperatures for the elimination of human pathogens, but it is necessary to verify the elimination of human pathogens in the compost by microbiological analysis of each produced batch. The U.S. Natural Resources Conservation Service (NRCS) has a standard for operation of composting facilities (Code 317) to reduce the pollution potential of organic agricultural wastes to surface and ground water. The requirements include, but are not limited to, ensuring that an operating temperature of 130 to 170°F (54–77°C) be achieved within 7 days and remain at these temperatures up to 14 days to ensure efficient composting (NRCS, 2005). The current regulation in the U.S. requires that if raw animal manure is used as organic fertilizer, it must be composted unless it meets specific conditions for spontaneous composting, such as allowing at least 120 days between raw manure application and harvesting of a commodity that has been in contact with the soil so fertilized. The farmer must assure the safety of his composts, whether self-produced or acquired from specialized suppliers. In general, application of fresh manure or waste of marine origin must never be authorized for any reason in the production of fruits and vegetables.

Animal or human feces. Fecal matter, whether human or animal, constitutes an important source of human pathogens, such as *E. coli* O157:H7, *Salmonella*, and *Campylobacter*, among others. These microorganisms may be present in soil and manure for more than three months, depending on temperature and soil conditions (Craigmill, 2000). For this reason, the presence of domestic animals or pets, and their defecation outdoors in or near fields, factors that may favor the spread of contamination to crop areas, must be forbidden. Contamination may be carried involuntarily by workers on their shoe soles or on their work tools, increasing the risk of product contamination. Human waste collected in sanitary facilities placed in the crop fields must be disposed of correctly to avoid contamination.

Wild fauna is also considered a contamination source in crop fields. The water sources and fresh food available in crop fields are the main attractions for wildlife, and it is necessary to establish barriers to prevent their access to crop areas. Incidents with other commodities, such as the outbreak of *E. coli* O157:H7 that occurred in 2006 linked to spinach, permitted investigators to establish the role of feral pigs and other fauna in spreading contamination (Jay et al., 2007). Maintaining ecological equilibrium and conservation are also priorities in modern agriculture, so it is necessary to consider the establishment of available water sources for wild fauna, away from the production zones, to reduce the access of wild animals to crop areas.

Farmers must know if the neighboring terrains and facilities are used for animal husbandry, and should take the necessary measures to assure that animal fecal matter will not be transported from contaminated land to their crops during rainy seasons (FDA, 1998). This can be avoided by the construction of physical barriers (canals) to divert runoffs to non-cultivated zones.

There is a great need for information on mechanisms for transmission of pathogens from feces to the melons, especially whether animal species that may have intimate contact with the melons, such as insects or nematodes, can serve as vectors of the pathogen to the melon surface. Caldwell et al. (2003) reported the ability of

S. Poona to survive in the digestive tract of *Caenorhabditis elegans*, a free-living microbivorous nematode, which later can shed this pathogen. This study proved the concept that free-living worms can serve as a source of bacterial pathogens in melons and other produce growing on the ground. In a more recent study, Gibbs et al. (2005) demonstrated the ability of *Diploscapter* sp., another free-living nematode commonly found in the rhizosphere of crop soils, to survive in manure, ingest *S.* Poona and other pathogens, and then to shed these pathogens 24 h after exposure to the inoculated manure. These studies indicate that the primary point of contamination of melons may be the field via environmental vectors.

Facilities and equipment. Buildings for the storage of field-packing materials, crop-production tools, pesticides and fertilizers, common facilities in agriculture fields, must be kept in order and waste-free to avoid the nesting of pests and mice. This kind of fauna may contaminate the materials and equipment for field packing that will be used later on during harvest. The exterior of these installations should be examined to assure that the surrounding areas (2–3 m) are free of undergrowth. In this zone it is also necessary to install traps for rodents, aimed at preventing their access to the interior of the storage rooms. All the tools used during production must undergo a cleaning and disinfection process each time they are going to be used or when they are going to be utilized for a different activity; for example, in different crop fields, which would favor cross-contamination. This kind of contamination may occur if a plow or a fertilizer spreader is contaminated in one locality and is used again in another crop field without prior cleaning and disinfection. This practice has the added advantage of reducing the presence of phytopathogens, since many practices focused on the reduction or elimination of human pathogens also accomplish the goal of eliminating microorganisms that damage the crops.

The proposed FDA rule, "Standards for the Growing, Harvesting, Packing, and Holding of Produce For Human Consumption" (Federal Register, 2013), includes specific requirements for preventing contamination addressing the above areas, as well as others that, when applied, should result in the reduction in the risk of introduction of biological hazards during preharvest operations.

Postharvest

The melon-packing process is the main point where fecal contamination might be introduced postharvest and where the most opportunities are found for spreading and increasing levels of contamination with human pathogens that originated in the field (Castillo et al., 2004; Johnston et al,. 2005). Therefore, growers must pay special attention to personnel training and management of sanitary facilities. The main sources of contamination in the postharvest handling of cantaloupes are 1) water for washing and sanitation, 2) worker activities, and 3) equipment and installations. Contamination of the products can occur through direct contact or cross-contamination. The application of sanitation practices, throughout the full chain of packing, is the best tool for preventing contamination of cantaloupe (FDA 2008; Martinez-Tellez et al., 2007).

Water for washing and sanitizing. In operations where melons are not packed in the field, postharvest procedures include cooling, rinsing, washing, and disinfecting the melons. In such steps water has direct contact with the product, and it is used in great amounts. As mentioned before, water can be a major vehicle of biological contamination of fruits and vegetables (Falcão et al., 1993; Gerba and Choi, 2006), and only potable water should be used in these practices. Once the melons arrive at the packing facilities, the melon rinsing process takes place; this is to eliminate residual soil or organic matter stuck to the melons. Some washing systems use water recirculation in this step, which involves the risk of a massive spread of contamination. That is why recycling of water is not recommendable in this part of the process. Washing may be carried out by dipping the melons in a dump tank with chlorinated water, or by spraying chlorinated water, and is intended to eliminate dirt from melons rather than remove microorganisms. The FDA discourages the use of dump tanks due to the potential for cross-contamination and internalization of pathogens (Michelle Smith, FDA, 2004; personal communication). Disinfection is the next step after washing. The concentration of antimicrobial chemicals used must be monitored and documented routinely to assure that disinfectant is kept at adequate levels. A number of investigations have pointed out the effectiveness of various disinfectants (Barak et al., 2003; Kozempel et al., 2002; Materon, 2003). Nevertheless, the different processes and operation conditions of the packinghouse cause variations in the disinfection results; therefore, validation of the disinfection process in each packing unit is recommended. A detailed description of disinfection procedures is provided later in this chapter.

Worker activities. As a consequence of deficient hygienic practices of workers, human pathogens and parasites may be transmitted directly to the fruits during packing, by direct contact with contaminated workers. Poor hygiene of workers has been identified as an important cause of pathogen contamination. Packing facilities with a high degree of mechanization in the processes of selection and classification of the melons reduce the risk of biological contamination by reducing the manipulation of melons by personnel. The correct application of personal hygienic practices by workers, and applied to their handling of tools, with constant supervision, would help to reduce the incidence of human pathogens in cantaloupe melon.

It is important to ensure that all the personnel, including those that do not participate directly in the manipulation of the fruits, including visitors, adhere to the hygiene practices established by management. In this context a training program in good hygienic practices and product handling must be in place and be offered for all the employees including supervisors, temporary, part-time and full time personnel, as well as subcontracted personnel (FDA, 1998). Training personnel in good hygienic practices is not an easy task, given the diversity in cultural backgrounds and languages among workers in this kind of industry. This difficulty is reflected when trying to teach them new ideas due to cultural and language barriers. Nevertheless, a great effort must be made to reduce the risks of contamination, in part through motivating employees by treating them as an important part of the company and showing appreciation of their value and contributions (Hurst and Shuler, 1992).

Equipment and facilities. Human pathogen sources in the centralized packing, cooling room, and storage room are mainly contributed by animal vectors such as rodents and birds that intrude into the installations. Such installations must be designed to avoid the introduction of these vectors and to counter such contamination with monitoring and documentation programs of hygiene practices and maintenance of installations and equipment. To reduce the possibility of spreading contamination in the sorting and packing areas, it is recommended that a program of insect and rodent control outside and inside the buildings be established and maintained. The design of buildings and selection of washing machinery should favor ease of cleaning and disinfection, and their mechanical parts should be resistant to corrosion to resist detergents and antimicrobial agents. The product conveyor belts must always be kept clean, sanitized, and free of cracks that may damage the melons and make the washing and sanitization process difficult.

The refrigerated containers used to transport the product, either by sea or land, must be inspected before loading the product to assure their cleanliness, absence of odors and visible residues, and that the temperature is adequate to preserve the quality of cantaloupe melon. The loading zone must maintain hygienic conditions so the loading equipment will not introduce contamination to the interior of the transport container. Supervision and documentation of these operations is necessary for the maintenance of the food safety program.

Cutting practices

Potential sources of pathogens. Unsatisfactory postharvest handling and kitchen practices can increase the risk of contamination of the edible portion of cantaloupes. The pathogen is usually located on the rind and is transferred to the flesh during cutting (Beuchat, 1996). In addition, inappropriate kitchen practices such as use of knives or cutting boards for processing of different commodities (e.g., meats) without suitable washing and sanitizing, or poor personal hygiene during preparation of the product for consumption may result in cross-contamination. An outbreak of cholera in the U.S. was associated with the consumption of cut cantaloupes. The cantaloupe was sliced and handled by an asymptomatic, infected person, who was suspected to have contaminated the cut melon during preparation and handling (Ackers et al., 1997).

Mechanisms for contamination. The most likely mechanism for contamination of melon flesh during cutting is the transfer of pathogens from contaminated rind to the flesh via the knife blade. In particular, cantaloupes are prone to harboring large bacterial populations, and the netting material that covers the rind increases the area for bacterial attachment. In a survey of melons collected from grocery stores in Texas, the aerobic plate count and total coliform counts were found to be 1.0 to 1.4 log cycles higher on the surface of cantaloupes than on the surface of honeydew melons (Cabrera–Diaz and Castillo, unpublished data, 2003). Similar information has been published by others (Ukuku, 2004; Ukuku et al., 2005). The heavier bioburden of the cantaloupe rind may increase the probability for pathogen transfer to the flesh during cutting, which when coupled with the ability of pathogens to grow in melon flesh at abusive temperatures and survive for long periods under refrigeration, may increase the risk of infection (Golden et al., 1993; Vadlamudi et al., 2012).

Although there is scarcity of information about mechanisms for contamination of melons during peeling, slicing, and serving, multiple outbreaks have been linked to cut melons in salad bars, or mixtures of cut melons (CDC, 2013). This may indicate that, in addition to a passive moving of pathogens from the rind to the flesh during cutting, other mechanisms may operate in transferring pathogens to cut melons. Cross-contamination from beef to watermelon was the mechanism in an outbreak of *E. coli* O157:H7 disease that was linked to cut watermelon and other salad-bar items in the summer of 2000, affecting at least 62 people in Wisconsin (Beers, 2000). During the outbreak investigation, the outbreak strain was also isolated from ground beef that was being made from sirloin chunks in close proximity to where the salads were being prepared, and from a sample of sirloin collected at the meat-packing plant that supplied the restaurant. Besides the cross-contamination mechanism that was likely to have contributed to this outbreak, worker's hygiene may have played a role as well. One of the cooks and another worker were suffering from diarrhea before the outbreak, and it is possible that they were still infectious and contaminated the salad-bar items by inappropriate handling. Another possible mechanism was the recycling of leftovers from the salad bar, which were refrigerated overnight and placed in the salad bar on successive days. This may have resulted in continued contamination of the salad bar over several days (Beers, 2000).

Preventing and minimizing contamination. The transfer of pathogens to the flesh of cantaloupes during cutting has been well established (Suslow and Cantwell, 2001; Ukuku, 2004; Ukuku and Fett, 2002; Ukuku and Sapers, 2002). According to Suslow and Cantwell (2001), the particular characteristics of the cantaloupe rind determines that the presence of as few as 150 bacteria per cm^2 of rind may result in contamination during cutting. Although these investigators did not specify whether this number was estimated from studies with native microbiota of cantaloupe or with inoculated pathogens, other studies involving both epiphytic bacteria and inoculated pathogens prove that counts similar or slightly higher than what was indicated by Suslow and Cantwell (2001) have been sufficient to allow transfer of organisms to the fresh-cut pieces (Ukuku and Fett, 2002a; Ukuku, 2004).

Under the premise that reducing the load of pathogens on the surface of melons will ultimately reduce the probability of conveying these pathogens to the melon flesh, a series of studies have been conducted at the USDA's Agricultural Research Service. Ukuku and Fett (2002) studied the effect of sanitizing the surface of cantaloupes, inoculated with *L. monocytogenes*, on the presence of this organism on the cubes of fresh-cut cantaloupe. The melons were inoculated with *L. monocytogenes* at a level of 3.5 log cfu/cm^2, allowed to dry, washed with water, chlorine (1,000 mg/L), or hydrogen peroxide (H_2O_2, 5%), and then cut. *L. monocytogenes* was consistently found on the cantaloupe cubes obtained from the unwashed controls and from the melons that were subjected to a water wash, but was consistently absent on the cubes obtained from melons subjected to chlorine or H_2O_2 wash. These sanitizers also were tested on cantaloupe inoculated with *S.* Stanley, and the counts of *S.* Stanley on cantaloupe cubes obtained from unwashed or water-washed melons were between 0.20 and 0.22 log cfu/g, whereas the organism was not detectable on cubes obtained from melons washed with chlorine or H_2O_2 unless the melons were stored at 4 or 20°C for 3 to 5 days before cutting (Ukuku and Sapers, 2002). This may indicate that cutting the cantaloupes promptly after surface disinfection can help to minimize *Salmonella* contamination of the flesh. A delay in cutting after disinfection may allow sufficient growth of surviving pathogens to yet again increase the chances for transfer to the flesh. Therefore, determining the maximum time period between sanitizing and cutting that still prevents microbial growth is paramount in establishing best practices for fresh-cut processing.

Treating inoculated cantaloupe and honeydew melons with a mixture of H_2O_2 (1%), nisin (25 µg/ml), sodium lactate (1%), and citric acid (0.5%) reduced *L. monocytogenes* and *E. coli* O157:H7 to undetectable levels on melon surfaces. Except for one instance, these pathogens could not be recovered, even by enrichment, from melon cubes obtained from whole melons treated with this mixture of sanitizers (Ukuku et al., 2005). Submerging cantaloupes inoculated with a mixture of five strains of *Salmonella* for 60 s in hot H_2O_2 (70°C) or water (97°C) reduced the transfer of this pathogen to the cut cantaloupes to undetectable levels, whereas cubes cut from untreated melons had *Salmonella* counts of 2.9 log cfu/g (Ukuku et al., 2004).

The cutting practices may also play a role in the transfer of *Salmonella* from contaminated rind to the cantaloupe flesh. Vadlamudi et al. (2012) reported that cutting inoculated cantaloupes after first peeling the rind resulted in decreased transfer of *Salmonella* Poona into the tissue in comparison with cutting of melons and removing the rind later. For both cutting methods, this investigator used a new, sterile knife between steps. Handwashing, or handwashing combined with the use of hand sanitizers and gloves, has been studied as measures that can reduce the risk of foodborne disease considerably (Paulson, 2000; Taylor, 2000).

Structural characteristics of melons promoting microbial survival and growth

Current knowledge about growth and survival of pathogens in melons

Early reports of the ability of melon edible flesh to sustain the survival and/or growth of pathogens were published in the 1980s. Fredlund et al. (1987) reported an outbreak of shigellosis linked to watermelon in Sweden, which was caused by *Shigella sonnei*. The watermelon had been purchased in Morocco and brought to Sweden, then cut and consumed 3 days after arrival to Sweden. The hypothesis was that the strain of *S. sonnei* had been internalized into the watermelon, perhaps by deceitful commercial practices including injecting water to increase the weight. To test this hypothesis, the investigators injected the outbreak strain into 17 watermelons, which then were stored at 20 or 30°C. At intervals during storage, core samples were obtained using a biopsy needle. *S. sonnei* grew rapidly inside the watermelons to reach counts of 8.0 to 9.0 log cfu/g within 3 days. In a study by Escartin et al. (1989), different strains of *Salmonella* and *Shigella* were inoculated onto watermelon, papaya, and jicama. Papaya is a tree fruit and jicama is a tuber. Although they are not melons (cucurbits), they are prepared similarly as melons, i.e., peeled and cut before consumption. Thus, should the rind be contaminated, pathogen transfer to the edible tissue also would be expected to occur. All of the pathogens tested were able to grow on inoculated cubes of the three commodities and in suspensions of papaya and watermelon flesh when stored at 22°C. Later, Castillo and Escartin (1994) reported the ability of *Campylobacter jejuni* to survive on watermelon and papaya cubes stored at 25 to 29°C. Golden et al. (1993) found the growth rate of a mixture of *Salmonella* strains on sliced cantaloupe, watermelon, and honeydew melons to be similar to the growth rate in tryptic soy broth at 23°C. At 5°C, *Salmonella* could not grow, but survived with no reduction in counts over 24 h of storage. Penteado and Leitão (2004) reported the ability of *Salmonella* Enteritidis to grow in homogenates of cantaloupe, watermelon, and papaya at different temperatures, including 10°C. In another study, they also observed growth of *L. monocytogenes* at 10°C (Penteado and Leitão, 2004a). This demonstrates the importance of correctly refrigerating cut melons to prevent growth. Pathogens such as *Salmonella* and *E. coli* O157:H7 can have a low infectious dose and their growth may not be required to cause illness. However, the higher the concentration of the pathogen, the higher the risk of disease and of the pathogen spreading throughout the entire batch of cut melons, thereby increasing the exposure if the contaminated melons are consumed. If appropriate conditions of temperature and humidity are provided, pathogens may even grow on the rind, as reported for *E. coli* O157:H7 (Del Rosario and Beuchat, 1995). Growth on cantaloupe rind was also observed for *S.* Poona when stored at 37°C, especially on wounded rinds (Beuchat and Scouten, 2004). Pathogen growth on the rind surface can increase the chances for transfer to the flesh during cutting.

Cantaloupe netting

The surface of cantaloupe includes a meshwork referred to as the netting (Webster and Craig, 1976). During preliminary work on characteristics of the cantaloupe surface, we inoculated pieces of cantaloupe and honeydew rinds with a suspension of *S.* Poona, and then the inoculated rind was stored at room temperature for 60 min. Superficial samples for *Salmonella* counts were collected with a sterile sponge, immediately after inoculating and after the 60 min storage. The counts on cantaloupe rind decreased by 1 log cycle after 60 min storage, whereas no count decrease was observed on honeydew rind (Cabrera-Diaz and Castillo, unpublished data, 2003). It seems unlikely that the cantaloupe surface had an antimicrobial activity. Instead, the netting material may have absorbed the inoculum providing sites for irreversible attachment of *Salmonella*, which resulted in the inability of our sampling device to recover the attached cells. Parnell et al. (2005) observed that *Salmonella* was detached to a greater extent from honeydew than from cantaloupe melons by a simple soaking in plain water for 60 s. When testing the effect of antimicrobial agents on the reduction of microorganisms on melons, Ukuku (2004), Ukuku and Fett (2002a), and Ukuku et al. (2005) found the overall magnitude of reduction to be similar for both types of melon; however, the microbial counts were consistently lower on honey-dew, which in turn resulted in a reduced transfer of microorganisms to the fresh-cut product.

The formation of the netting in cantaloupes is thought to be a response to cracking of the fruit surface (Meissner, 1952). This raised net tissue gives the surface of the cantaloupe an inherent roughness, and this surface roughness may favor bacterial attachment and hinder microbial detachment. Attachment of *S.* Poona to cantaloupe rind was reported by Barak et al. (2003). Annous et al. (2005) also reported that *Salmonella* can attach to the rind of cantaloupes, and suggested that the unique characteristics of the cantaloupe surface provide a large number of attachment sites for bacteria and impede contact between bacteria and aqueous sanitizers. Richards and Beuchat (2004) inoculated cantaloupes by dipping in a suspension of *S.* Poona, at 4 or 30°C, and then measured the water uptake from the inoculum suspension. The expected greater water uptake when dipping was carried out at 4°C was only observed for Eastern cantaloupes. Also, the water uptake was significantly greater for Western cantaloupes than for Eastern cantaloupes. These authors attributed these differences to, among other factors, a more dense netting on Western cantaloupes in comparison to the Eastern cantaloupes. Netting is a manifestation of cuticle disruption (Meissner, 1952). The cuticle is part of the dermal system, which governs the regulation of water loss. The cuticle is composed of surface waxes, cutin embedded in wax, and a layer of mixtures of cutin, wax, and polysaccharides; its thickness and structure vary greatly depending on the level of development of the plant (Kader, 2002). Greater disruption of the cuticle on Western cantaloupes may enable reten-tion of more water than the netting of Eastern cantaloupes allows (Richards and Beuchat, 2004). This water retention is indicative of the potential for microor-ganisms to internalize within the rind of cantaloupes if they are washed in dump tanks with contaminated water.

Biofilm formation

This section will focus on biofilm formation on melons (see Chapter 2 for more information concerning biofilms on produce). The attachment of *Salmonella* onto cantaloupe rind has been documented. If the attached microorganisms can grow, bio-film formation may occur. Annous et al. (2005) observed rapid biofilm formation of *Salmonella* spp. inoculated on cantaloupe rind. The biofilm provides a protective glycocalyx, which will make the organism recalcitrant to the antimicrobial activity of sanitizers (Frank, 2001). In addition, the biofilm structure may enhance the ability of the microorganism to spread to non-contaminated areas of the product and even to food preparation surfaces during cutting or peeling, which may result in cross-contamination from the melons to other foods.

Microbial infiltration and internalization

Any fissures, cuts, etc., of the melon will ultimately favor the entry of microorganisms to the flesh of the fruits. Bacterial internalization in melons has not been studied as extensively as for other commodities. Richards and Beuchat (2004) proposed that the adherence or infiltration of microorganisms into the melon may not be entirely promoted by temperature/pressure differentials but by the surface characteristics of the melons as well. Researchers with the FDA tested the potential for fluid infiltration during cantaloupe hydrocooling using brilliant blue as an indicator of water infiltration. They observed dye infiltration in 28% of 170 melons after immersion in iced water containing the dye, although dye buildup was observed in cankers (rind blemishes); intact melons also were infiltrated (Michelle Smith, FDA, personal communication, 2004). Research conducted at the University of California in Davis (Suslow, 2004a) showed that *S.* Typhimurium was able to internalize into cantaloupes through the ground spot and, secondarily, through the stem scar. After postharvest processing, the microorganism was found 5 mm under the rind. The ground spot is the area of the melon that is in contact with the ground during melon development. For cantaloupes, ground contact results in an area with a thin and underdeveloped rind, which is also poorly netted and more susceptible to fungal or bacterial growth. These characteristics make the ground spot an area of great potential for microbial internalization during postharvest practices. Soft rot has been reported to promote bacterial internalization. In a study on naturally contaminated fresh market produce, Wells and Butterfield (1997) found a higher incidence of bacteria that were biochemically similar to *Salmonella* on cantaloupes showing soft rots than on healthy cantaloupes. These authors also confirmed the presence of *Salmonella* in the decayed area of 2 out of 9 cantaloupes that were affected by fungal rots. When these authors tested the populations of *Salmonella* in other commodities that were co-inoculated with *Salmonella* and various strains of molds, the populations were approximately 1 log greater on produce that was damaged by the inoculated molds (Wells and Butterfield, 1999).

Use of antimicrobial treatments to decontaminate melons

Chapters 17 and 18 cover different aspects of the current technologies for reducing pathogens on produce. Therefore, this chapter will describe efforts to reduce pathogens on fresh and fresh-cut melons. Most studies have focused on cantaloupes and to a lesser extent, on honeydew melons. The effect of surface sanitizers on the transfer of pathogens from contaminated rind to the flesh during peeling or cutting has been discussed above.

Treatments tested on fresh melons

Chemical disinfectants. Although different researchers indicate that bacteria are recalcitrant to aqueous sanitizers when present in or on fruits and vegetables, mainly due to superficial as well as physicochemical characteristics of these food products of plant origin (Annous et al., 2005; Ukuku and Fett, 2004; Ukuku et al., 2006), cantaloupe washing is still common practice in the produce industry, except when melons are field-packed. A water wash is applied to remove soil, and other dirt from the melon surface, and usually is followed by a wash with a sanitizer. The water wash alone does not have a significant effect at removing bacterial pathogens, and only those that are loosely attached may be removed by the water wash, if not located in areas that are out of the reach of the water. The topography of the melon rind plays a role in the removal of microorganisms. Parnell et al. (2005) reported that soaking cantaloupes in water for 60 s resulted in a reduction of 0.7 log cycles in *Salmonella*, while soaking honeydew melons in plain water for 60 s resulted in a 2.8-log reduction. These authors also studied the rinds of both types of melons by scanning electron microscopy, and observed a large number of crevices and cracks on the cantaloupe rind, while honeydew rind was smooth. The natural roughness of the cantaloupe rind is thought to make removal of microorganisms more difficult than the smooth rind of honeydew melons, likely due to a larger number of protective areas on cantaloupe rind, which promotes bacterial attachment and allows the pathogens to evade contact with water.

Applying sanitizers during postharvest operations seems to be more meaningful when the purpose is to reduce microbial populations in the wash water, which can help prevent internalization and cross-contamination among product lots, rather than to sanitize the product. However, regardless of the limitations of produce disinfection procedures using aqueous sanitizers, such procedures can have some antimicrobial effect, reducing pathogen levels to some extent. Cantaloupe sanitizing may be applied as one more hurdle in a holistic approach to food safety, applicable not only to melons but to all fresh and fresh-cut produce, always keeping in mind that these treatments are not sufficient to stand alone as a kill step during processing. Rodgers et al. (2004) inoculated several fruits and vegetables including cantaloupes, with *E. coli* O157:H7 and *L. monocytogenes*, and then dipped the inoculated samples in solutions of chlorinated trisodium phosphate, chlorine dioxide, ozone, and peroxyacetic acid. According to these authors, exposure of cantaloupes for 5 min to these sanitizers resulted in a reduction of both pathogens from ca. 6.0 log cfu/g to undetectable levels on the rind, by all treatments, while the reduction observed on melons exposed to water alone was of 0.9 log cfu/g. Large reductions (6.7–7.3 log cycles) were also reported for *E. coli* O157:H7 on inoculated cantaloupes by dipping in solutions of lactic acid (1.5%), lactic acid + H_2O_2 (1.5% each), or lactic acid (1.5%) + tergitol (0.3%), and of 4.3 to 5.5 log cycles by using sodium hypochlorite at 200 mg/L free chlorine, all solutions applied at 20 or 30°C (Materon, 2003). For chlorine, the reduction of this pathogen was significantly smaller for a contact time of 1 min than for 10 min of contact. For all lactic-acid preparations the time of contact did not have an effect.

In contrast to these reports, which show large reductions of bacterial pathogens on melons by treatment with aqueous sanitizers, the majority of studies indicate that cantaloupes are specially difficult to sanitize (Alvarado-Casillas et al., 2007; Parnell et al., 2005; Ukuku and Sapers, 2001; Ukuku and Fett, 2002; Ukuku, 2004; Ukuku et al., 2005). Alvarado-Casillas et al. (2007) reported that treating cantaloupes with hypochlorite at 200, 600, and 1,000 mg/L resulted in reductions of 2.1 to 2.9 log cycles for *S.* Typhimurium and 1.5 to 2.1 log cycles for *E. coli* O157:H7 on the cantaloupe surface. When hot (55°C), 2% lactic acid solution was sprayed, the reductions obtained were 3.0 log cycles for *S.* Typhimurium, and 2.0 log cycles for *E. coli* O157:H7. Parnell et al. (2005) reduced populations of *S.* Typhimurium LT2 (a non-virulent strain) on cantaloupes by 1.8 log cycles by soaking the melons in a chlorine solution at 200 mg/L free chlorine, and obtained an additional 0.9 log reduction by scrubbing with chlorine solution, in comparison to soaking only. Ukuku (2004) observed a 2.3 to 2.5 log reduction in *Salmonella* populations on cantaloupes by immersion in 2.5% or 5.0% H_2O_2. According to Sapers et al. (2001), washing cantaloupes with 5% H_2O_2 at 50°C, alone or in combination with a commercial detergent formulation, was more effective than washing with water, surfactant solutions, 1000 ppm Cl_2, trisodium phosphate, or a commercial detergent formulation in reducing the microbial load on cantaloupe rind. The effectiveness of H_2O_2 over other sanitizers at reducing bacterial pathogens on the rind of melons has been extensively documented by a single research group (Ukuku and Sapers, 2001; Ukuku and Fett, 2002; Ukuku, 2004; Ukuku et al., 2005). In another study comparing bacterial reductions by various antimicrobials, Vadlamudi et al. (2012) tested the effectiveness of dipping in hypochlorite solution (200 mg/L free chlorine) for 3 min, spraying with warm (55°C) L-lactic acid solution (2%) for 2 min, or dipping in ozonized water (30 mg/L) for 5 min at reducing *Salmonella* Poona on the surface of whole cantaloupes. In this study, lactic acid and ozone treatments reduced *S.* Poona by 2.3 to 2.7 log cycles on both the rind and the stem scar, whereas dipping in chlorine was not different from untreated controls.

Hot water treatment. Hot water treatments are not recommended for all products because of the possibility of product damage; however, when the peeled product is to be further processed, such as in juice preparation, or when the rind is sufficiently strong, surface pasteurization seems to be a very effective alternative for reducing pathogens. The lethal effect of heat can even reach microorganisms located in places that are not reached by chemical sanitizers. The use of hot water dips was proposed by Pao and Davis (1999) for reducing pathogens on oranges that were further used for juice. Later, researchers with the U.S. Department of Agriculture's Agricultural Research Service developed a method for surface pasteurization of melons. Annous et al. (2004) demonstrated that a dip in hot water (76°C for 6 min) was able to thermally inactivate *S.* Poona regardless of attachment or biofilm formation, while maintaining the melon firmness. Ukuku et al. (2004) compared the effectiveness of a 60 s dip in hot water (70–90°C) or H_2O_2 at 70°C in reducing populations of *Salmonella* (5-serotype cocktail) on inoculated cantaloupes. Both water at 90°C and H_2O_2 at 70°C resulted in ca. 4 log reduction without affecting the stability of the flesh. They concluded that hot water pasteurization or hot H_2O_2 treatment can reduce the risk of enteric disease through the consumption of cantaloupes that have been surface-contaminated with *Salmonella*. This treatment also was found effective at extending the shelf-life of cantaloupes. Solomon et al. (2006) verified the lethal effect of surface pasteurization against pathogens on cantaloupe surfaces and conducted computer analysis to determine the heat penetration during treatment. They concluded that the edible flesh of the cantaloupe remained cool while the temperature of the outer surface of the rind increased rapidly.

Other treatments. Kozempel et al. (2002) developed the Vacuum/Steam/Vacuum system, consisting of rapidly applying a vacuum to eliminating air and humidity, which act as heat insulators on the surface to be treated by steam, then applying an antimicrobial steam treatment, and finally applying another vacuum step to cool the surface and prevent heat damage of the product. Using this system, these authors obtained a reduction of *Listeria innocua* (used as a surrogate for *L. monocytogenes*) of 4.0 to 4.7 log cycles.

Fresh-cut melons
Treatment with antimicrobial agents

The effects of a water wash, 50 mg/L chlorine, and 10 μg/ml nisin mixed with EDTA on populations of native mesophilic aerobes, *Pseudomonas* spp., lactic-acid bacteria and yeast, and molds were tested on whole and on fresh-cut cantaloupes and honey-dew melons (Ukuku and Fett, 2002a). Aerobic plate counts (APC) on untreated fresh-cut pieces of cantaloupe were found to be approx. 1 log cycle higher than on fresh-cut pieces of honeydew, which supports the idea that the cantaloupe surface harbors larger populations than other melons, and therefore a larger number of microorganisms may be transferred to the flesh during cutting. According to these authors, treatment with chlorine proved more effective than nisin + EDTA at reducing all microorganisms, and there were no differences in odor, appearance, and overall acceptability ratings for both melons regardless of the treatment applied. Mesophilic aerobes were reduced to undetectable levels on fresh-cut honeydew, and by ca. 2 log cycles on fresh-cut cantaloupes. All other organisms were reduced to undetectable levels on both types of melon, although these counts increased during refrigerated storage. Even though these authors did not indicate the level of detection in their counting methods, from the APC obtained from the controls, it can be estimated that the reductions obtained were > 3 log cycles. In Spain, Raybaudi-Massilia et al. (2008) developed an edible coating for fresh-cut melon, which includes antimicrobial molecules carried by the alginate-based coating. The antimicrobials used were essential oils, and their active ingredients were added to the edible coating during preparation. In addition, 2.5% malic acid and 2% calcium lactate were also dissolved and mixed with the alginate coating base. Cantaloupe (*piel de sapo* melon) pieces were inoculated with *S.* Enteritidis and covered with the coating linked to the antimicrobials. All antimicrobials were effective at reducing *Salmonella* and other native microbiota on fresh-cut cantaloupes, and also, the use of the edible coatings containing antimicrobials resulted in an increased shelf-life and improved microbiological quality.

Irradiation

A promising technology for destroying pathogens on fresh and fresh-cut produce is ionizing irradiation. Irradiation kills microorganisms by exposing the matrix to be treated to ionizing energy, which may be gamma rays, x-rays, or electron beams.

All these types of irradiation follow the same biocidal mechanism. When the product is exposed to irradiation, the gamma, x-rays, or electron beams collide with the microbial DNA, causing multiple breaks in the DNA chain, thus rendering the cells unable to grow. The organization that regulates irradiation internationally is the International Atomic Energy Agency (IAEA), a part of the United Nations Organization. In the U.S., food irradiation is approved by FDA on a case-by-case basis, and it is currently not approved for treatment of melons in the U.S. or any other country with regulations on irradiation of specific commodities. On August 22, 2008, the FDA amended the current regulation to approve irradiation (up to 4 kGy) for controlling foodborne pathogens and extending the shelf-life in fresh iceberg lettuce and fresh spinach (Federal Register, 2008), and it is likely that other commodities will be approved for irradiation in the future.

Studies on the use of electron beam (e-beam) irradiation for reducing pathogens in fresh-cut cantaloupes indicated that doses close to 1 kGy will result in a reduction of *Salmonella* between 2.2 and 3.6 log cycles (Palekar, 2004). Quality studies for irradiated fresh-cut cantaloupes indicated that irradiation at 1.4 kGy did not have any effect on the quality or sensory characteristics of the cut melons (Palekar et al., 2004a). Fan et al. (2006) reported the effect of gamma irradiation on fresh-cut cantaloupes obtained from fresh cantaloupes that were treated by hot water pasteurization. The cantaloupe cubes were treated at doses up to 0.5 kGy, which resulted in a reduction in aerobic plate counts of 0.5 to 1.4 log cycles. This low-dose irradiation treatment, applied to fresh-cut melon cubes, also extended the shelf-life of the product and did not have any adverse effect on sensory characteristics of the fresh-cut melons.

Although e-beams are unidirectional and therefore cannot be used to deliver a uniform dose of irradiation on products with irregular geometry such as of melons, this issue can be overcome using a novel device (Maxim Electron Scatter Chamber), which takes advantage of the scattering of electrons when hitting a surface (Maxim et al., 2011). Scattered electrons can then be directed to the target from multiple angles and irradiate irregular surfaces evenly. Using the Maxim Electron Scatter Chamber, Cuervo et al. (2009) were able to irradiate cantaloupes with a uniform dose of approx. 3 kGy, being able to reduce *Salmonella* Poona from initial counts of 4.0 to 4.7 log cfu/cm^2 to undetectable using an adequate plate count dilution scheme.

Conclusion

The relatively frequent occurrence of outbreaks of foodborne disease linked to melons, often imported, indicates the relevance of effective control measures for reducing the risk of contamination with human pathogens. Cantaloupes are particularly impervious to chemical disinfectants, most likely due to the unique composition and structure of their netted rind, which favors bacterial attachment and biofilm formation. Nevertheless, melon disinfection can still reduce pathogens to some extent, and this measure may be linked to food safety programs that include procedures that prevent contamination during growing, harvesting, and packing this unique commodity. These good agricultural practices, together with postharvest disinfection and introduction of further pathogen reduction strategies during packing, fresh-cut processing, and marketing, can be linked concurrently in a holistic approach to food safety. However, more research is needed to understand the sources and mechanisms for contamination in the field, how this contamination can proliferate and spread over many product units in a shipment, and whether novel technologies can be applied as additional hurdles to reduce pathogen levels and make safe melons available to consumers.

References

Ackers, M., Pagaduan, R., Hart, G., et al., 1997. Cholera and sliced fruit: Probable secondary transmission from an asymptomatic carrier in the United States. Int. J. Infect. Dis. 1, 212–214.

Akin, E.D., Harrison, M.A., Hurst, W., 2008. Washing practices on the microflora on Georgia-grown cantaloupes. J. Food Prot. 71, 46–51.

Alvarado-Casillas, S., Ibarra-Sánchez, S., Rodríguez García, O., et al., 2007. Comparison of rinsing and sanitizing procedures for reducing bacterial pathogens on fresh cantaloupes and bell peppers. J. Food Prot. 70, 655–660.

Annous, B.A., Burke, A., Sites, J.E., 2004. Surface pasteurization of whole fresh cantaloupes inoculated with *Salmonella* Poona or *Escherichia coli*. J. Food Prot. 67, 1876–1885.

Annous, B.A., Solomon, E.B., Cooke, P.H., Burke, A., 2005. Biofilm formation by *Salmonella* spp. on cantaloupe melons. J. Food Saf. 25, 276–287.

Anon, 2005. Memorandum of Understanding between the Food and Drug Administration Department of Health and Human Services of the United States of America and Servicio Nacional de Sanidad, Inocuidad y Calidad Agroalimentaria of the United Mexican States concerning Entry of Mexican Cantaloupes into the United States of America. Available at http://www.fda.gov/InternationalPrograms/Agreements/MemorandaofUnderstanding/ucm110200.htm (accessed 17.07.13.).

Asta, G., 1999. Fresh cut and prepared rockmelon – a discussion on risk. AFS 2, 8–9.

Barak, J.D., Chue, B., Mills, D.C., 2003. Recovery of surface bacteria from and surface sanitization of cantaloupes. J. Food Prot. 66, 1805–1810.

Beers, A., 2000. Outbreak at Milwaukee Sizzler provides lessons to industry (E. coli prevention). Food Chem. News 42 (41), 4.

Beuchat, L.R., 1996. Pathogenic microorganisms associated with fresh produce. J. Food Prot. 59, 2004–2216.

Beuchat, L.R., Scouten, A.J., 2004. Factors affecting survival, growth, and retrieval of Salmonella Poona on intact and wounded cantaloupe rind and in stem scar tissue. Food Microbiol. 21, 683–694.

Bhagwat, A., 2006. Microbiological safety of fresh-cut produce: Where are we now? In: "Microbiology of fresh produce" (K. Matthews), Emerging Issues in Food Safety. ASM. Press, Washington, D.C., pp. 121–165.

Blostein, J., 1993. An outbreak of *Salmonella* Javiana associated with consumption of watermelon. J. Environ. Health 56, 29–31.

Caldwell, K.N., Anderson, G.L., Williams, P.L., Beuchat, L.R., 2003. Attraction of a free-living nematode, *Caenorhabditis elegans*, to foodborne pathogenic bacteria and its potential as a vector of *Salmonella* Poona for preharvest contamination of cantaloupe. J. Food Prot. 66, 1964–1971.

Calvin, L., 2003. Produce, Food Safety, and International Trade: Response to U.S. Foodborne Illness Outbreaks Associated with Imported Produce. In: Buzby, J. (Ed.), International Trade and Food Safety. United States Department of Agriculture, USDA, ERS, AER Number 828, Nov. 2003, pp. 74–96. Available at http://www.ers.usda.gov/media/321547/aer828g_1_.pdf (accessed 17.07.13.).

Cartwright, E.J., Jackson, K.A., Johnson, S.D., Graves, L.M., Silk, B.J., Mahon, B.E., 2013. Listeriosis Outbreaks and Associated Food Vehicles, United States, 1998–2008. Emerg. Infect. Dis. 19, 1–9.

Castillo, A., Escartin, E.F., 1994. Survival of *Campylobacter jejuni* on sliced watermelon and papaya. J. Food Prot. 57, 166–168.

Castillo, A., Mercado, I., Lucia, L.M., et al., 2004. *Salmonella* contamination during production of cantaloupe: a bi-national study. J. Food Prot. 67, 713–720.

CDC, 1979. *Salmonella oranienburg* gastroenteritis associated with consumption of pre-cut watermelons-Illinois. MMWR 28, 522–523.

CDC, 1991. Centers for Disease Control and Prevention 1991. U.S. epidemiologic notes and reports: multistate outbreak of *Salmonella* Poona infections, United States and Canada, 1991. MMWR 40, 549–552.

CDC, 2002. U.S. Multiestate Outbreaks of *Salmonella* Poona Infections Associated with Eating Cantaloupe from Mexico, United States and Canada, Period 2000–2002, November 2002. MMWR 51 (46), 1044–1047.

CDC, 2003. Centers for Disease Control and Prevention 2003. U.S. Foodborne disease outbreaks line listings. 1990–2003. Available at http://www.cdc.gov/foodborneoutbreaks/outbreak_data.htm (accessed 12.11.08.).

CDC, 2003a. U.S. Outbreak of *Salmonella* serotype Javiana infections, Orlando, Florida, June 2002. MMWR 51, 683–684.

CDC, 2008. Multistate Outbreak of *Salmonella* Litchfield Infections Linked to Cantaloupe (Final Update). Avaliable at http://www.cdc.gov/salmonella/litchfield/ (accessed 17.07.13.).

CDC, 2011. Multistate Outbreak of *Salmonella* Panama Infections Linked to Cantaloupe. Available at http://www.cdc.gov/salmonella/panama0311/062311/ (accessed 18.07.13.).

CDC, 2011a. Multistate Outbreak of Listeriosis Linked to Whole Cantaloupes from Jensen Farms, Colorado. Available at http://www.cdc.gov/listeria/outbreaks/cantaloupes-jensen-farms/120811/index.html (accessed 18.07.13.).

CDC, 2012. Multistate Outbreak of *Salmonella* Typhimurium and *Salmonella* Newport Infections Linked to Cantaloupe (Final Update). Available at http://www.cdc.gov/salmonella/typhimurium-cantaloupe-08-12/index.html (accessed 18.07.13.).

CDC, 2013. Foodborne Outbreak Online Database (FOOD). Available at http://wwwn.cdc.gov/foodborneoutbreaks/# (accessed 17.07.13.).

Chapman, B.J., 2005. An Evaluation of an On-Farm Safety Program for Ontario Greenhouse Vegetable Producers; A Global Blueprint for Fruit and Vegetable Producers. M.Sc. Thesis, The University of Guelph, Master of Science. Appendix 2.1. Produce related outbreaks from 1990–2003, pp. 170–177.

Chen, W., Jin, T.Z., Gurtler, J.B., Geveke, D.J., Fan, X., 2012. Inactivation of *Salmonella* on whole cantaloupe by application of an antimicrobial coating containing chitosan and allyl isothiocyanate. Int. J. Food Microbiol. 155, 165–170.

Craigmill, A.L., 2000. Good agricultural practices are critical to stemming increase in produce outbreaks. Environ. Toxicol. Newsletter Vol. 20 (No. 2). Available at http://extoxnet.orst.edu/newsletters/ucd2000/nltrapr00.htm (accessed 17.07.13.).

Cuervo, M.P., Rodrigues-Silva, D., Maxim, J., Castillo, A., 2009. Use of a Novel Device to Enable Irradiation of Fresh Cantaloupes by Electron Beam Irradiation. IAFP's 5th. European Symposium on Food Safety. Berlin, Germany 7–9 October, 2009. Abstract P1–04. Food Prot. Trends. 30, 88.

Deeks, S., Ellis, A., Ciebin, B., et al., 1998. *Salmonella* Oranienburg, Ontario. Can. Commun. Dis. Rep. 24, 177–179.

Del Rosario, B.A., Beuchat, L.R., 1995. Survival and growth of enterohemorrhagic *Escherichia coli* O157:H7 in cantaloupe and watermelon. J. Food Prot. 58, 105–107.

Duffy, E.A., Lucia, L.M., Kells, J.M., et al., 2005. Concentrations of *Escherichia coli* and genetic diversity and antibiotic resistance profiling of *Salmonella* isolated from irrigation water, packing shed equipment, and fresh produce in Texas. J. Food Prot. 68, 70–79.

Falcão, D.P., Valentini, S.R., Leite, C.Q.F., 1993. Pathogenic or potentially pathogenic bacteria as contaminants of fresh water from different sources in Araraquara. Brazil. Wat. Res. 27, 1737–1741.

Fan, X., Annous, B.A., Sokorai, K.J.B., et al., 2006. Combination of hot-water surface pasteurization of whole fruit and low-dose gamma irradiation of fresh-cut cantaloupe. J. Food Prot. 69, 912–919.

FAO, 2003. Food and Agriculture Organization of the United Nations, Codex Alimentarius. Code Hygienic Pract. Fresh Fruits Vegetables. CAC/RCP 53 – 2003. Available at http://www.codexalimentarius.org/standards/list-of-standards/en/?no_cache=1 (accessed 17.07.13.).

Federal Register, 2008. Irradiation in the Production, Processing and Handling of Food. Final Rule, 21 CFR 179.26. FR Aug 22, 2008, 164:49603.

Federal Register, 2013. Standards for the Growing, Harvesting, Packing, and Holding of Produce for Human Consumption; Proposed Rule. FR Jan 16, 2013, 78:3504.

Federal Register, 2013a. Current Good Manufacturing Practice and Hazard Analysis and Risk-Based Preventive Controls for Human Food; Proposed Rule. FR Jan 16, 2013, 78:3646.

FDA, 1998. US Food and Drug Administration. Guidance for Industry. Guide to Minimize Microbial Food Safety Hazards for Fresh Fruits and Vegetables. Available at http://www.foodsafety.gov/~dms/prodguid.html (accessed 12.11.08.).

FDA, 2011. Environmental Assessment: Factors Potentially Contributing to the Contamination of Fresh Whole Cantaloupe Implicated in a Multi-State Outbreak of Listeriosis. Available at http://www.fda.gov/Food/RecallsOutbreaksEmergencies/Outbreaks/ucm276247.htm (accesed 25.07.13.).

FDA, 2012. Information on the Recalled Jensen Farms Whole Cantaloupes. Available at http://www.fda.gov/Food/RecallsOutbreaksEmergencies/Outbreaks/ucm272372.htm (accessed 25.07.13.).

FDA, 2013. Detention Without Physical Examination of Cantaloupes from Mexico, Import Alert 22–20. Available at http://www.accessdata.fda.gov/cms_ia/importalert_67.html (accessed 17.07.13.).

FDA, 2013a. Letter to Cantaloupe Industry on Produce Safety. Available at http://www.fda.gov/AboutFDA/CentersOffices/OfficeofFoods/CFSAN/CFSANFOIAElectronicReadingRoom/ucm341029.htm (accessed 25.07.13.).

FDA, 2008. US Food and Drug Administration Guide to minimize microbial food safety hazards of fresh-cut fruits and vegetables. Available at http://www.fda.gov/Food/GuidanceRegulation/GuidanceDocumentsRegulatoryInformation/ProducePlantProducts/ucm064458.htm (accessed 17.07.13.).

FDA (U.S. Food and Drug Administration), 2001. FDA Survey of Imported Fresh Produce. Available at http://www.fda.gov/Food/GuidanceRegulation/GuidanceDocumentsRegulatoryInformation/ProducePlantProducts/ucm118891.htm (accessed 17.07.13.).

Eitenmiller, R.R., Johnson, C.D., Bryan, W.D., et al., 1985. Nutrient composition of cantaloupe and honeydew melons. J. Food Sci. 50, 136–138.

Escartin, E.F., Castillo-Ayala, A., Saldaña-Lozano, J., 1989. Survival and growth of *Salmonella* and *Shigella* on sliced fresh fruit. J. Food Prot. 52, 471–472.

Espinoza-Medina, I.E., Rodríguez-Leyva, F.J., Vargas-Arispuro, I., Islas-Osuna, M.A., Acedo-Félix, E., Martínez-Téllez, M.A., 2006. PCR identification of *Salmonella*: potential contamination sources from production and postharvest handling of cantaloupes. J. Food Prot. 69, 1422–1425.

Figueroa-Aguilar, G.A., González-Ramírez, M., Molina-García, A.J., et al., 2005. Identificación de *Salmonella* spp en agua, melones cantaloupe y heces fecales de iguanas en una huerta melonera. Med. Int. Mex. 21 (4), 255–258.

Frank, J.F., 2001. Microbial attachment to food and food contact surfaces. Adv. Food. Nutr. Res. 43, 319–370.

Fredlund, H., Back, E., Sjoberg, L., Tornquist, E., 1987. Water-Melon as a vehicle of transmission of shigellosis. Scand. J. Infect. Dis. 19, 219–221.

Gaylor, G.E., MacCready, R.A., Reardon, J.P., McKernan, B.F., 1955. An outbreak of salmonellosis traced to watermelon. Public Health Rep. 70, 311–313.

Gerba, C.P., Choi, C.Y., 2006. Role of irrigation water in crop contamination by viruses. In: Goyal, S.M. (Ed.), Viruses in Foods. Springer Science, NY, pp. 257–263.

Gibbs, D.S., Anderson, G.L., Beuchat, L.R., et al., 2005. Potential role of *Diploscapter* sp. strain lkc25, a bacterivorous nematode from soil, as a vector of food-borne pathogenic bacteria to preharvest fruits and vegetables. Appl. Environ. Microbiol. 71, 2433–2437.

Gellin, B.G., Broome, C.V., 1989. Listeriosis. J. Am. Med. Assoc. 261, 1313–1320.

Green, T., Hanson, L., Lee, L., et al., 2005. North American Approaches to Regulatory Coordination. In: Huff, K., Meilke, K.D., Nutson, R.D., Ochoa, R.F., Rude, J. (Eds.), Agrifood Regulatory and Policy Integration Under Stress. Proc. 2nd. Workshop of the North American Agri-food Market Integration Consortium, San Antonio, Texas, pp. 9–47. May 5, 2005. Available at http://naamic.tamu.edu/sanantonio.htm (accessed 17.07.13.).

Golden, D.A., Rhodehamel, E.J., Kautter, D.A., 1993. Growth of *Salmonella* spp. in cantaloupe, watermelon and honeydew melons. J. Food Prot. 56, 194–1296.

Hakerlerler, H., Okur, B., Irget, E., Saatç, N., 1999. Carbonhydrate fractions and nutrient status of watermelon grown in the alluvial soils of Küçük Menderes Watershed, Turkey. In: Anac, D., Martin-Prével, P. (Eds.), Improved Crop Quality by Nutrient Management. Kluwer Acad. Pub, Boston, pp. 163–165.

Hurst, W.C., Shuler, G.A., 1992. Fresh produce processing an industry perspective. J. Food Prot. 55, 824–827.

International Society for Infectious Diseases, 2004. *E. coli* O157, cantaloupes – USA (Montana). Promed Mail, Archive Number: 20040722.1997. Published Date: 2004-07-22. Available at http://www.promedmail.org/ (accessed 17.07.13.).

Iversen, A.M., Gill, M., Bartlett, C.L., et al., 1987. Two outbreaks of foodborne gastroenteritis caused by a small round structured virus: evidence of prolonged infectivity in a food handler. Lancet 2, 556–558.

Jay, M.T., Cooley, M., Carychao, D.E., et al., 2007. *Escherichia coli* O157:H7 in feral swine near spinach fields and cattle, central California coast. Emerg. Infect. Dis. 13, 1908–1911.

Johnston, L.M., Jaykus, L.A., Moll, D., et al., 2005. A field study of the microbiological quality of fresh produce. J. Food Prot. 68, 1840–1847.

Kader, A.A., 2002. Postharvest biology and technology: An overview. In: Kader, A.A. (Ed.), Postharvest Technology of Horticultural Crops, third ed. University of California Agriculture and Natural Resources, pp. 39–47. Publication 3311.

Karchi, Z., 2000. Development of melon culture and breeding in Israel. Acta. Hortic. 510, 13–17.

Kozempel, M., Radewonuk, E.R., Scullen, O.J., Goldberg, N., 2002. Application of the vacuum/steam/vacuum surface intervention process to reduce bacteria on the surface of fruits and vegetables. Innov. Food Sci. Emerg. Technol. 3, 63–72.

Madden, J.M., 1992. Microbial pathogens in fresh produce the regulatory perspective. J. Food Prot. 55, 821–823.

Mahmoud, B.S.M., 2012. Effects of X-ray treatments on pathogenic bacteria, inherent microflora, color, and firmness on whole cantaloupe. Int. J. Food Microbiol. 156, 296–300.

Martínez-Téllez, M.A., Vargas-Arispuro, I.C., Silva-Bielenberg, H.K., Espinoza-Medina, I., Rodríguez-Leyva, F., González-Aguilar, G.A., 2007. Parte 2. Productos Agrícolas. Capítulo 7. Producción y Manejo Poscosecha de Hortalizas. In: Gardea Béjar, A.A., González Aguilar, G.A., Higuera Ciapara, I., Cuamea Navarro, F. (Eds.), Buenas Prácticas en la Producción de Alimentos. Editorial Trillas, Mexico, D.F, pp. 223–238. ISBN:978-968-24-8175-8.

Materon, L.A., 2003. Survival of *Escherichia coli* O157:H7 applied to cantaloupes and the effectiveness of chlorinated water and lactic acid as disinfectants. World. J. Microbiol.

Materon, L.A., Martinez-Garcia, M., McDonald, V., 2007. Identification of sources of microbial pathogens on cantaloupe rinds from pre-harvest to post-harvest operations. World. J. Microbiol. Biotechnol. 23, 1281–1287.

Maxim, J.E., Neal, J.A., Castillo, A., 2011. Maxim electron scatter chamber. U.S. Patent No. 8,008,640. U.S. Patent and Trademark Office, Washington, DC. Available at http://assignments.uspto.gov/assignments/q?db=pat&pat=8008640 (accessed 25.07.13.).

Meissner, F., 1952. Die korkbildung der fruchte von *Aesculus*-und *Cucumis*-arten. Osterr. Bot. Ztschr. 99, 606–623.

Mohle-Boetani, J.C., Reporter, R., Werner, S.B., et al., 1999. An outbreak of *Salmonella* serogroup Saphra due to cantaloupes from Mexico. J. Infect. Dis. 180, 1361–1364.

Ng, P.J., Fleet, G.H., Heard, G.M., 2005. Pesticides as a source of microbial contamination of salad vegetables. Int. J. Food Microbiol. 101, 237–250.

NRCS, 2005. Conservation Practice Standard. Composting Facility. Code 317. Available at http://efotg.nrcs.usda.gov/references/public/AL/tg317.pdf (accessed 17.07.13.).

Palekar, M.P., 2004. Attachment of Salmonella on Cantaloupe and Effect of Electron Beam Irradiation on Quality and Safety of Sliced Cantaloupe. Ph.D. Dissertation, Texas A&M University.

Palekar, M.P., Cabrera-Diaz, E., Kalbasi-Ashtari, A., et al., 2004a. Effect of electron beam irradiation on the bacterial load and sensorial quality of sliced cantaloupe. J. Food Sci. 69, M267–M273.

Pao, S., Davis, C.L., 1999. Enhancing microbiological safety of fresh orange juice by fruit immersion in hot water and chemical sanitizers. J. Food Prot. 62, 756–760.

Parnell, T., Harris, L.J., Suslow, T., 2005. Reducing *Salmonella* on cantaloupes and honeydew melons using wash practices applicable to postharvest handling, foodservice, and consumer preparation. Int. J. Food Microbiol. 99, 59–70.

Paulson, D.S., 2000. Handwashing, gloving, and disease transmission by the food preparer. Dairy. Food Environ. Sanit. 20, 838–845.

Penteado, A.L., Leitão, M.F.F., 2004. Growth of *Salmonella* Enteritidis in melon, watermelon and papaya pulps stored at different times and temperatures. Food Cont. 15, 369–373.

Penteado, A.L., Leitão, M.F.F., 2004a. Growth of *Listeria monocytogenes* in melon, watermelon and papaya pulps. Int. J. Food Microbiol. 92, 89–94.

Penteado, A.L., Leitão, M.F.F., 2009. Detection of *Listeria* spp. and *Salmonella* spp. on the surface of melon (*Cucumis melo*), watermelon (*Citrullus vulgaris*) and papaya (*Carica papaya*), by the TECRA Visual Immunoassay (TECRA-VIA) method and cultural procedures. Higiene Aliment. 23, 130–135.

PMA and UFFVA (Produce Marketing Association and United Fresh Fruit and Vegetable Association), 2005. Commodity Specific Food Safety Guidelines for the Melon Supply Chain. first ed. Availabe at http://www.fda.gov/downloads/Food/GuidanceRegulation/UCM168625.pdf (accessed 17.07.13.).

Raybaudi-Massilia, R.M., Mosqueda-Melgar, J., Martín-Belloso, O., 2008. Edible alginate-based coating as carrier of antimicrobials to improve shelf-life and safety of fresh-cut melon. Int. J. Food Microbiol. 121, 313–327.

Richards, G.M., Beuchat, L.R., 2004. Attachment of *Salmonella* Poona to cantaloupe rind and stem scar tissues as affected by temperature of fruit and inoculum. J. Food Prot. 67, 1359–1364.

Ries, A.A., Zaza, S., Langkop, C., 1990. A multistate outbreak of *Salmonella* Chester linked to imported cantaloupe [Abstract]. In: Programs and abstracts of the 30th Interscience Conference on Antimicrobial Agents and Chemotherapy. American Society for Microbiology, Washington, DC, p. 238. Abstract. 195.

Rodgers, S.L., Cash, J.N., Siddiq, M., Ryser, E.T., 2004. A comparison of different chemical sanitizers for inactivating *Escherichia coli* O157:H7 and *Listeria monocytogenes* in solution and on apples, lettuce, strawberries, and cantaloupe. J. Food Prot. 67, 721–731.

Sapers, G.M., Miller, R.L., Pilizota, V., Mattrazzo, A.M., 2001. Antimicrobial treatments for minimally processed cantaloupe melon. J. Food Sci. 66, 345–349.

SENASICA (Servicio Nacional de Sanidad, Inocuidad y Calidad Agroalimentaria), 2008. Lineamientos de Buenas Prácticas Agrícolas y Buenas Prácticas de Manejo en los Procesos de Producción de Frutas y Hortalizas Para Consumo Humano en Fresco (Guidelines for good agricultural practices and good management practices in the production and packing processes of fruits and vegetables for fresh consumption by humans). Available at http://www.senasica.gob.mx/?doc=3790 (accessed 17.07.13.).

SIAP-SAGARPA (Servicio de Información Estadística Agroalimentaria y Pesquera). 2002. SIACON. Mexico. Available at www.siap.sagarpa.gob.mx/ (accessed 17.07.13)

Sivapalasingam, S., Friedman, C.R., Cohen, L., Tauxe, R.V., 2004. Fresh produce: A growing cause of outbreaks of foodborne illness in the United States, 1973 through 1997. J. Food Prot. 67, 2342–2353.

Solomon, E.B., Huang, L., Sites, J.E., Annous, B.A., 2006. Thermal inactivation of *Salmonella* on cantaloupes using hot water. J. Food Sci. 71, M25–M30.

Suslow, T.V., 2004. Minimizing the risk of foodborne illness associated with cantaloupe production and handling in California. Available at http://ucfoodsafety.ucdavis.edu/files/26308.pdf (accessed 17.07.13.).

Suslow, T.V., Cantwell, M., 2001. Recent findings on fresh-cut cantaloupe and honeydew melon. Fresh Cut. April 2001. Available at http://freshcut.com/fc2001.htm#fc20 (accessed 17.07.13.).

Tamplin, M., 1997. *Salmonella* and cantaloupes. Dairy. Food Environ. Sanit. 17, 284–286.

Tauxe, R.V., 1997. Emerging Foodborne Diseases: An Evolving Public Health Challenge. Centers for Disease Control and Prevention. Atlanta, Georgia, USA, Special Issue. Vol 3. No. 4.

Tauxe, R., O'Brien, S.J., Kirk, M., 2008. Outbreak of food-borne diseases related to the International Food Trade. In: Doyle, M., Erickson, M.C. (Eds.), Imported Foods. Microbiological Issues and Challenges. ASM Press., Washington, DC, pp. 69–112.

Taylor, A.K., 2000. Food protection: new developments in handwashing. Dairy. Food Environ. Sanit. 20, 114–119.

Ukuku, D.O., 2004. Effect of hydrogen peroxide treatment on microbial quality and appearance of whole and fresh-cut melons contaminated with *Salmonella* spp. Int. J. Food Microbiol. 95, 137–146.

Ukuku, D.O., 2006. Effect of sanitizing treatments on removal of bacteria from cantaloupe surface, and re-contamination with *Salmonella*. Food Microbiol. 23, 289–293.

Ukuku, D.O., Fett, W., 2002. Behavior of *Listeria monocytogenes* inoculated on cantaloupe surfaces and efficacy of washing treatments to reduce transfer from rind to fresh-cut pieces. J. Food Prot. 65, 924–930.

Ukuku, D.O., Fett, W., 2002a. Effectiveness of chlorine and nisin-EDTA treatments of whole melons and fresh-cut pieces for reducing native microflora and extending shelf-life. J. Food Saf. 22, 231–253.

Ukuku, D.O., Fett, W.F., 2004. Method of applying sanitizers and sample preparation affects recovery of native microflora and *Salmonella* on whole cantaloupe surfaces. J. Food Prot. 67, 999–1004.

Ukuku, D.O., Sapers, G.M., 2001. Effect of sanitizer treatments on *Salmonella* Stanley attached to the surface of cantaloupe and cell transfer to fresh-cut tissues during cutting practices. J. Food Prot. 64, 1286–1291.

Ukuku, D.E., Bari, M.L., Kawamoto, S., Isshiki, K., 2005. Use of hydrogen peroxide in combination with nisin, sodium lactate and citric acid for reducing transfer of bacterial pathogens from whole melon surfaces to fresh-cut pieces. Int. J. Food Microbiol. 104, 225–233.

Ukuku, D.O., Pilizota, V., Sapers, G.M., 2004. Effect of hot water and hydrogen peroxide treatments on survival of *Salmonella* and microbial quality of whole and fresh-cut cantaloupe. J. Food Prot. 67, 432–437.

Ukuku, D.O., Sapers, G.M., 2006. Microbiological safety issues of fresh melons. In: Sapers, G.M., Gorny, J.R., Yousef, A.E. (Eds.), Microbiology of Fruits and Vegetables. CRC Press/Taylor & Francis Group, Boca Raton, FL, pp. 231–251.

Vadlamudi, S., Taylor, T.M., Blankenburg, C., Castillo, A., 2012. Effect of chemical sanitizers on *Salmonella enterica* serovar Poona on the surface of cantaloupe and pathogen contamination of internal tissues as a function of cutting procedure. J. Food Prot. 75, 1766–1773.

Webster, B.D., Craig, M.E., 1976. Net morphogenesis and characteristics of the surface of muskmelon fruit. J. Am. Soc. Hortic. Sci. 101, 412–415.

Wells, J.M., Butterfield, J.E., 1997. *Salmonella* contamination associated with bacterial soft rot of fresh fruits and vegetables in the marketplace. Plant Dis. 81, 867–872.

Wells, J.M., Butterfield, J.E., 1999. Incidence of *Salmonella* on Fresh Fruits and Vegetables Affected by Fungal Rots or Physical Injury. Plant Dis. 83, 722–726.

Zahniser, S., 2006. U.S.-Mexico agricultural trade during the NAFTA era. Proceedings of the Doha, NAFTA and California Agriculture Conference. Giannini Foundation, Sacramento, CA. Jan 13, 2006. Available at http://giannini.ucop.edu/US_Mexico_Zahniser_060123.pdf (accessed 17.07.13.).

Microbiological Safety of Sprouted Seeds: Interventions and Regulations

11

Keith Warriner, Barbara Smal

Department of Food Science, University of Guelph, Guelph, Ontario, Canada

CHAPTER OUTLINE

The Produce Contamination Problem. http://dx.doi.org/10.1016/B978-0-12-404611-5.00011-7

Introduction

The first recorded production of sprouted seeds was in China 2737 BC when Emperor Shennong promoted sprouts as an effective medicine. Indeed, the long history of Chinese medicine can be traced back to sprouted seeds. The Emperor also had economic motives for promoting sprouted seeds as the commodity provided a means of feeding the population with limited agricultural resources. Once established, sprouted seeds became a staple in Asia and became popular across the globe for the very reasons they were introduced. Specifically, sprouted seeds are considered a health food rich in protein, minerals, and anti-cholesterol constituents (Kim et al., 2012). Aside from the health benefits and cost of production, consumers also use sprouts to add texture, color, and flavor in culinary dishes. Yet, despite the benefits of sprouted seeds there have been on-going food safety issues linked to the product to the point that most food agencies issue warnings for high-risk groups (Yishan et al., 2013). Within the last decade there have been over 40 outbreaks linked to sprouted seeds, two of which are amongst largest foodborne illness incidents in history.

The pathogens commonly implicated in sprout-related outbreaks are *Salmonella* and Shiga-toxin producing *Escherichia coli*. With respect to the latter, the most common serotype implicated is O157:H7, but a more recent trend is the recovery of non-O157 Shiga-toxin *Escherichia coli* in spouted-seed foodborne illness outbreaks. Within the U.S. the number of cases of foodborne illnesses between 1995 and 2010 linked to sprouts directly was over 2000, which is more than caused by any other salad vegetable type (Yishan et al., 2013). As a result of ongoing food safety issues there has been a sustained effort to identify sources of contamination, pathogen screening strategies, seed disinfection, guidelines, and introduction of regulations. However, despite such efforts the incidence of foodborne illness associated with sprouts remains (Yishan et al., 2013). The following provides an overview of sources of contamination and description of pathogens of concern. A chronology of foodborne illness outbreaks linked to sprouts will be described and landmark cases that stimulated research into finding intervention strategies. The development of policy guidelines to aid sprout producers will be provided and why such polices had limited impact on increasing the microbiological safety of sprouted seeds. Finally, the prospect of introducing regulations into the sprout industry will be discussed and how government-led initiatives may have greater success compared to the previous industry-led programs.

Sprouted seed market structure

Sprouting is the process of germinating seeds to produce seedlings that can be consumed raw or cooked. In Asian countries sprouted seeds are regarded as a staple food in the daily diet and are cooked as opposed to eaten raw. In the West, sprouted seeds are primarily consumed as a health food or to enhance the sensory characteristics of foods.

The market for seeds destined for sprout production is almost insignificant when placed into the context of the global seed market. The global trade for seeds is approximately $22 billion, and the most popular sprouted seed (mung bean sprouts) accounts for approximately $100 million. Therefore, in relative terms the market for seeds destined for sprout production forms only a minor proportion of seed market with the majority being directed toward crop production. Mung beans represent an exception to other seeds given that 75% of the crop is exclusively for sprout production. In terms of food safety this is relevant given that the market size and production justifies implementing on-farm food safety practices. In contrast, the economics of implementing on-farm food safety practices for other seed types is less economically viable given only a small proportion is directed toward sprout production.

In global terms, the most mass-produced sprouted seeds are bean sprouts, followed by alfalfa (lucerne) and soy sprouts (FPC, 2012) (Table 11.1). However, in more recent years the variety of seed types sprouted has diversified due to the perception of being more microbiologically safe but also due to imparting novel (different) sensory characteristics. In the U.S., the most commonly sprouted seeds include alfalfa, adzuki, buckwheat, cabbage, clover, cress, broccoli, radish, sesame, mung bean, and onion. Mung bean sprouts are the most popular within Canada and Australia with market values of $20 million and $42 million, respectively. The top assortment of seeds used for sprouting in the European Union (EU) includes mung bean, alfalfa, radish, peas, and sunflower and mixtures of different varieties (EFSA, 2011). The EU sprout sector's market value at the consumer level has been estimated at $600 m (EFSA, 2011). By far the largest market is in Asia where sprouted seeds are the staple diet with mung beans again being the dominant type produced. Within

Table 11.1 Examples of Seed Types Commonly Used in Sprout Production

Sprouted Seed Type	Examples
Pulses	Alfalfa, clover, fengreek, lentil, chickpea, mung bean, soybean
Cereals	Oat, wheat, maize, rice, barley, rye, kamut, buckwheat
Oilseeds	Sesame, sunflower, almond, hazelnut, linseed, peanut
Brassica	Broccoli, cabbage, watercress, mustard, mizunam, radish, daikon
Microgreens	Celery, carrot, parsley, fennel
Allium	Onion, leek, green onion
Others	Lettuce, spinach, lemon grass

the West the industry is primarily composed of small to medium producers that serve local markets. Yet, in China it is more common to find large producers producing in excess of 100 tons of sprouts daily.

Seed production

It should be noted that seeds are produced, handled, and delivered under different conditions depending on the seed type. Although seeds used for sprouting are commonly grouped together the reality is that certain types are more susceptible to contamination than others; for example, mung bean vs. alfalfa.

In broad terms, seed production can be divided into annual or perennial production cycles. As the term infers, annual crops only survive for a single growing season while perennial remain productive over several years. This can have implications on the microbiological safety of sprouted seeds in terms of exposure to contamination.

Alfalfa seed production

Alfalfa is a perennial crop that remains productive for 6 to 10 years depending on climatic conditions. It is not uncommon for animals to graze directly on alfalfa crops between rotations in seed production. The majority of alfalfa production takes place in dry, arid, lands with the main production centers being Australia and the United States. Although hot-dry conditions are conducive to the growth of alfalfa there is a need to flood-irrigate the fields 1 to 6 times per year depending on natural rainfall. If the floodwater is contaminated with human pathogens then it will represent an obvious source of potential contamination.

The main use of alfalfa grass is as a feed source for ruminants with animals being permitted to graze on the crop from early spring to fall. If seed production is required for a designated season the crop will be left to flower with pollination occurring around July and seed harvesting occurring in the early fall. The harvesting of alfalfa seedpods is primarily by combine harvester following an initial cutting and drying phase on the ground. Because the seedpods are close to the ground there is an inevitable pick up of soil and other sources of contamination during the harvesting phase. The seeds are stored in bins before being taken to cleaning and packing. The alfalfa roots remaining in the field can be further grazed by animals prior to burning the crop in the late fall. The new shoots emerge in the spring where the crop could be used for seed production again or more likely rotated for animal grazing/hay production.

The majority of alfalfa seed produced goes to export with a relatively low proportion being directed toward sprouted seeds. For example, of the 7000 tons of alfalfa seed produced within Australia only 300 tons is used in sprout production. Within the U.S. the total yield per year is 70 million kg with only a small fraction going to sprout production.

As one can deduce there are several food safety risk factors associated with alfalfa production. The most obvious is the contact with animals but also the field-irrigation

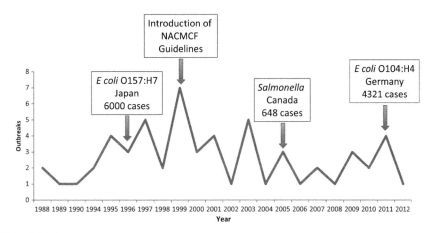

FIGURE 11.1

Flow diagram of sprouted seed production that includes souces of contamination and mitigation strategies at each point of the process.

water that if contaminated can transfer pathogens to the crop. The close contact of seeds with soil during harvesting and even the potential transfer of pathogens via bees can be considered additional risk factors (Figure 11.1).

Mung bean production

Mung bean is an annual crop that is cultivated in temperate climates such as central China and central Australia. There have been moves to produce mung beans within Canada although the climatic conditions are less favorable. As described, mung beans are primarily produced for sprouting, although use as whole beans or flour is also undertaken. Unlike alfalfa seeds, mung beans are graded depending on the intended use, with those for sprout production being premium (i.e., low percentage of defects) and those for flour production being of lower quality.

Mung bean plants have a growth period of 90 to 120 days with quality beans being produced by judicious choice of variety, soil type, and planting rates. It is key to ensure an adequate supply of water and hence irrigation is frequently applied throughout the season. Harvesting of the seed pods is critical and must be synchronized to ensure at least 95% is at the correct state of maturity and moisture content (14–16%). The seed pods can be harvested mechanically, although in China (the major producer) the operation is performed by hand.

Seed processing

The postharvest processing of seeds is similar regardless of type and essentially involves an initial cleaning step by passing through a series of sieves. The seed is then sorted based on weight via a gravity table that essentially comprises a tilted fluidized bed, with those of the highest bulk density traveling to the high point. The method is effective at removing damaged seeds and foreign objects such as stones. The seeds are tested for germination yield (>90%) prior to packing into 25 kg lots, then shipped to distributors. The seed distribution network has experienced extensive consolidation over the last decade. For example, within the U.S. there are six seed distributors that supply over 85% of the market.

Methods of sprout production

The generic approach to sprout production involves an initial quality assurance check on seeds followed by a washing and decontamination step (Yishan et al., 2013). The seeds are then soaked in water for 3 to 16 h to rehydrate and stimulate the germination process (Figure 11.2). There are different approaches to the actual sprouting process that includes drums, trays, punnets, or soil depending on the seed type (Thomas et al., 2003). At the end of the sprouting process the sprouts are harvested and can undergo a postharvest wash that depends on the sprout type. The sprouts are packed, chilled, and then distributed. The shelf-life of sprouts is relatively short and hence distribution is over a limited geographical area.

Alfalfa sprouts

Alfalfa seeds are sprouted in rotating drums, trays, or punnets (Yetim et al., 2010). The sprout room temperature is maintained between 18 to 21°C to stimulate the sprouting process, which takes 3 to 6 days. Over the course of the sprouting period irrigation water is applied frequently (every 3–5 h) to sustain sprout development and modulate temperature. Sprouting is also performed in the presence of natural or artificial light to stimulate chlorophyll development (i.e., green sprouts).

Mung bean sprouts

In contrast to alfalfa, mung beans are sprouted within bins (25–75 kg lots) within darkened rooms maintained within the range of 20 to 28°C (Warriner and Council, 2011). Irrigation water is applied via an overhead shower every 3 to 4 h to provide moisture to the developing seed and reduce bed temperature. The sprouting period is typically 5 days but is dependent on the sprouting temperature applied.

Postharvest handling

Sprouts produced in punnets can be directly distributed without any further processing. Those grown in soil are cut, whilst sprouts produce on trays, drums, or bins are subjected to postharvest washes. The main purpose of the postharvest wash is to remove the residual seed coat and other exogenous matter. The water is typically kept at 4°C to facilitate cooling of sprouts thereby reducing the respiration rates. The washed sprouts are drained subjected to a mild drying process prior to weighing and packaging.

Sprouts are highly perishable due to the high respiration rate, which can be controlled by reducing the storage temperature. Yet, certain sprouts such as mung beans are temperature sensitive and consequently need to be stored above 0°C with a relative humidity of 95 to 100%. Temperature control is critical in the distribution and handling of sprouts given even relatively brief exposure (30 min) to 20°C can reduce shelf-life by up to 50%.

Sprouts can be packed into plastic-lined boxes/crates or more common, perforated bags to facilitate gas exchange. Due to the short shelf-life of sprouts, the products are delivered directly to food service outlets or retail stores. Because of the ongoing food safety issues linked to sprouted seeds there are several high-profile retail chains within North America that have withdrawn the products from their inventory lists.

Pathogens linked to sprouted seeds

There have been a range of pathogens linked to outbreaks implicating sprouted seeds. For example, outbreaks have been sporadically associated with *Bacillus cereus, Staphylococcus aureus, Aeromonas hydrophilia,* and enteric protozoan (Beuchat, 1996; De Roever, 1999). There have been no outbreaks linked to sprouts contaminated with *Listeria monocytogenes,* although recalls have been issued when the pathogen was recovered during routine screening at retail (Kim et al., 2013). In the majority of foodborne illness cases the implicated pathogen is Shiga-toxin *Escherichia coli* (e.g., *E. coli* O157:H7) or more commonly *Salmonella* (Olaimat and Holley, 2012).

Shiga-toxin producing *Escherichia coli*

Although all pathogenic *E. coli* represent a significant health risk, those belonging to the Enterohemogenic *E. coli* (EHEC) group are of most concern, especially *E. coli* O157:H7 (Beutin and Martin, 2012). The cause for the high virulence of EHEC is the production of Shiga-like toxins (verotoxin or verocytotoxin). The genes for Shiga toxin are believed to have been horizontally transferred to *E. coli* from *Shigella* via bacteriophage. There are two types of toxins (encoded by Stx_1 and/or Stx_2) that act by cleaving a single adenine residue from 28S rRNA belonging to the ribosomal subunit resulting in the shutdown of protein synthesis (Ferens and Hovde, 2011). The kidney is rich in receptors for attachment of *E. coli* O157:H7 and consequently toxico-infection by the bacterium can be accompanied by renal failure (HUS syndrome).

Although *E. coli* O157:H7 is considered the most significant EHEC serotype it must be noted that other non-O157 Shiga-toxin producing types such as O111, O145, O113, O103, O91, O26 and O104 also exist (Grant et al., 2011; Pexara et al., 2012). Collectively all *E. coli*-possessing toxin genes are categorized as Shiga-toxin *Escherichia coli,* or STEC. However, the presence of *stx* genes is only one of several virulence factors required to cause illness (Grant et al., 2011). For this reason, even though there are 200 serotypes of STEC identified over 70% are of minor concern due to one or more missing virulence factors required to cause illness (Mathusa et al.,

2010). The main source of *E. coli* O157:H7 and other STEC is from the manure of ruminants (cattle, sheep) and sewage (Farrokh et al., 2013). The carriage of STEC by cattle has been estimated to range from 5 to 90% depending on which diagnostic method is applied (i.e., molecular vs. culture) (Warriner and Council, 2011). Yet, the majority of STEC recovered from cattle are non-O157 STEC with serotype O157:H7 carriage being 2 to 10% (Monaghan et al., 2012). *E coli* O157:H7 can persist for extended periods in the environment (over 100 days in soil amended soil) with non-O157 STEC persisting equally as long (Ma et al., 2012; Oliveira et al., 2012).

Entroaggregative *Escherichia coli* (EAggEC) is a subclass of pathogenic *E. coli* and considered to have evolved separately from STEC (Pabalan et al., 2013). EAggEC is commonly responsible for infant diarrhea primarily in developing countries and also travelers to developing countries (i.e., traveler's diarrhea) (Pabalan et al., 2013). There have been outbreaks within industrialized countries where EAggEC has been responsible for sporadic diarrhea, although it is rare. The symptoms of EAggEC infection comprises watery diarrhea, occasionally with blood and mucus lasting 7 to 14 days, but the condition is non-lethal provided the patient remains hydrated (Boll et al., 2013). The fimbria on the surface of EAggEC act to facilitate attachment between cells and the host mucosa. The bacterium forms a "brick-like" structure that essentially forms a biofilm on the lining of the gastrointestinal tract thereby providing firm attachment (Boll et al., 2013).

Traditionally, EAggEC has been primarily linked to person-to-person transmission and rarely associated with foods unless contaminated by the food handler. However, a strain (serotype O104:H4) of EAggEC that had acquired the Shiga-toxin gene was implicated in one of the largest outbreaks of foodborne illness linked to sprouted seeds. *E. coli* O104:H4 likely acquired the stx_2 gene though lysogenized bacteriophages as did the better-known O157:H7 serotype (Hauser et al., 2013).

Salmonella

The genus *Salmonella* includes over 2700 serovars 200 of which are commonly connected to human illness, with *S.* Typhimurium and *S.* Enteritidis being the most prevalent (Lee and Greig, 2013). *Salmonella* is carried within the gastrointestinal tract of wild animals, poultry, pigs, and humans. However, *Salmonella* recovered from vegetables typically belong to less common serotype groups; for example, Newport or Montevideo (Brandi et al., 2013).

There is concern with regard to the distribution of multi-drug resistant *Salmonella* within the food chain. Although drug resistance is commonly linked to healthcare settings and animal production there have been isolates recovered from sprouts exhibiting resistance to antibiotics (Snary et al., 2004). The main concern of multiple drug resistance is the potential fatal infections that can occur in the population and the limited number of treatment options.

Similar to *E. coli* O157:H7, the main transmission route of *Salmonella* to foods is through fecal contamination, cross-contamination, and food handling. However, unlike *E. coli* O157:H7, *Salmonella* has a broader range of carriers with poultry, pigs, cattle, and pets being most significant. Once present in the environment, *Salmonella* can persist over extended time periods and has evolved mechanisms to enhance survival through down-regulating metabolism, in addition to colonizing protozoan (Bradford et al., 2013). The adaption of *Salmonella* to persist on plants is also considered to extend the persistence of the pathogen in the field environment (Brandi et al., 2013).

Outbreaks linked to sprouted seeds

There have been over 40 foodborne illness outbreaks associated with sprouted seeds consumption (Table 11.2; Figure 11.2). The most well-documented outbreaks have occurred within the United States and have principally involved *Salmonella* and alfalfa (Ding et al., 2013). However, other major outbreaks have been observed across the globe and likely are underreported.

There have been notable foodborne illness outbreaks linked to sprouts, some of which are regarded as the largest in history. Although one can assume that outbreaks linked to sprouted seeds have occurred throughout history the first recorded cases occurred in the 1970s and more frequently in the 1980s. The most notable of the early outbreaks was linked to bean sprouts contaminated with *Salmonella* and occurred in Sweden and England in 1988 (Omahony et al., 1990) (Table 11.2). Five different *Salmonella* serotypes were recovered from mung beans sprouts and the source traced to seeds imported from Australia. Several other outbreaks followed, implicating *Salmonella* with cress and alfalfa (Omahony et al., 1990). A critical point was reached in 1996 with an incident involving radish sprouts contaminated with *E. coli* O157:H7 (HaraKudo et al., 1997; Itoh et al., 1998; Saegusa, 1998; Table 11.2). The seeds implicated in the outbreak were imported from the U.S. and sprouted within a single facility. The radish sprouts were distributed across Sakai to food-service outlets that included elementary schools. Within a week there were thousands of cases of *E. coli* O157:H7 infections reported, a proportion of which led to haemolytic uremic syndrome (HUS). The number of illnesses overwhelmed the Japanese health services and consequently there was a delay in the investigation. Based on the epidemiology studies it was evident that the radish sprouts were the most likely source of the pathogen, although sprout samples taken tested negative for the pathogen. As with many perishable foods, finding a product that tests positive for the implicated pathogen is rare given the short shelf-life and delay in confirming that an outbreak had occurred. Yet, based on epidemiology data the investigation concluded the likely originated from a batch of contaminated seed imported from the U.S. In total, there were at least 6000 cases confirmed, although it was estimated that over 9000 persons suffered illness as a result of the *E. coli* O157:H7 outbreak.

In the wake of the *E. coli* O157:H7 outbreak linked to radish sprouts there was a greater food safety focus placed on sprouted seeds. This could partly explain the increase in the number of recorded outbreaks that occurred post-1996 but also through advances in epidemiology and recognized food safety risks associated with sprouted

Table 11.2 Selected Foodborne Illness Outbreaks Linked to Contaminated Sprouts

Pathogen	Year	Cases	Sprout Type	Country
E. coli O26	2012	29	clover	U.S.
Salmonella Enteritidis	2011	10	alfalfa	U.S.
E. coli O104:H4	2011	3855	fenugreek	Germany, France
Salmonella Newport	2010	28	alfalfa	U.S.
Salmonella Bareilly	2010	190	bean sprouts	UK
Salmonella Saintpaul	2009	235	alfalfa	U.S.
Salmonella Cubana	2009	14	onion	Canada
Salmonella bovismorbifcans	2009	42	alfalfa	Finland
Salmonella Typhimurium	2008	24	alfalfa	U.S.
Salmonella Weltereden	2007	45	alfalfa	U.S.
Salmonella Stanley	2007	44	alfalfa	Sweden
Salmonella Oranineberg	2005	125	alfalfa	Australia
Salmonella Enteritidis	2005	648	bean sprouts	Canada
E. coli O157:H7	2005	1	alfalfa	U.S.
Salmonella	2004	12	alfalfa	U.S.
E. coli O157:H7	2003	7	alfalfa	U.S.
E. coli O157:H7	2002	7	alfalfa	U.S.
Salmonella Enteritidis	2001	84	bean sprouts	Canada
Salmonella Enteritidis	2000	12	bean sprouts	Holland
Salmonella Typhimurium	1999	119	alfalfa sprouts	U.S.
Salmonella Senftenberg	1997	60	alfalfa sprouts	U.S.
E. coli O157:H7	1997	126	radish sprouts	Japan
E. coli O157:H7	1996	6000	radish sprouts	Japan
Salmonella Goldcoast	1989	31	cress sprouts	UK
Salmonella	1988	143	mung bean	UK
Salmonella	1988	195	mung bean	Sweden
Bacillus cereus	1973	3	soy, mustard, and cress	U.S.

seeds (Figure 11.2). Two further outbreaks occurred in Japan in 1997 and again were traced to radish sprouts. In North America the number of cases of foodborne illnesses linked to sprouts was averaging 300 per year, primarily implicating alfalfa and clover sprouts (Yang et al., 2013). To address the increasing incidence of foodborne illness cases linked to sprouts the National Advisory Committee on Microbiological Criteria of Foods (NACMCF) published guidelines to improve food safety standards (NACMCF, 1999). The guide essentially provided a review on published research on food safety aspects of sprouted seeds and recommendations on how this could

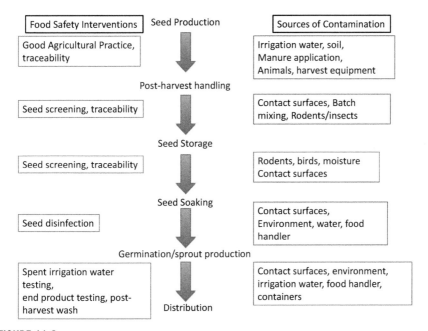

FIGURE 11.2

Number of foodborne illness outbreaks linked to sprouted seeds from 1988 to 2012.

be improved. Amongst the recommendations were to improve practices for producing seed and also initiatives that could be followed by sprout producers in terms of sanitary standards within the sprouting facility, in addition to seed disinfection and product testing. The publication of the guide led to a decline in outbreaks, although several still occurred each year (Figure 11.2).

In 2005, the largest foodborne illness outbreak linked to mung bean sprouts occurred in Canada. On November 2005 there was a spike in the number of *S.* Enteritidis cases within Southern Ontario (Mohle-Boetani et al., 2009). By November 22, 2005 the potential source of contamination was traced to a sprouting facility in Toronto. In the subsequent investigations no *Salmonella*-positive mung beans were detected, although there was evidence of pest infestation, in addition to equipment being transferred from a neighboring poultry processing facility. In total there were 648 confirmed cases of salmonellosis, and although the source was never identified it was likely the sprouting environment as opposed to the source seed. Further outbreaks linked to sprouted seeds continued, implicating mainly alfalfa but also the occasionally mung bean sprouts (Figure 11.1). From 2008 there was an increased incidence of *Listeria monocytogenes* linked to sprouts, although in the majority of cases the occurrence resulted in product recalls as opposed to actual outbreaks. The increase in the prevalence of *Listeria* was likely due to better surveillance given that the pathogen was routinely screened for in ready-to-eat foods such as sprouted seeds. Another interesting trend was the implication of different sprout types such as onion sprouts and the re-emergence of clover sprouts as a vehicle for foodborne pathogens (Yishan et al., 2013; Table 11.2).

With the ongoing food safety issues linked to sprouted seeds it was somewhat inevitable that a major outbreak would occur, and it did so in May 2011. The outbreak was centered in the German city of Hamburg and was on a similar scale to that of the Japanese incident in 1996 (Soon et al., 2013). Yet, both the pathogen (*E. coli* O104:H4) implicated and the sprout type (fenugreek) were unusual given neither had been associated with sprout-related outbreaks previously. Fenugeek is an annual plant with similarities to mung beans and is a commonly used herb or spice used in Asian cuisines. The outbreak started in early May, with the first cases of HUS being reported on May 22, 2011. By the time sprouted seeds had been identified as the likely source on June 9, 2013 over 3000 cases of infection had been reported along with 800 suffering from HUS and 33 deaths (Scharlach et al., 2013). Confirmation that sprouts were implicated was confirmed by a parallel outbreak in the Bordeaux region of France that affected 15 (Hauswaldt et al., 2013). The last reported case linked to the outbreak was reported on July 11, 2011, over two months following the first event. Clearly, the delay in detecting the outbreak resulted in a large number of cases with contaminated batches of sprouted seeds being continuously introduced into the population. Yet, the virulence of the pathogen undoubtedly contributed to the scale and impact of the outbreak. The pathogen recovered from infected persons was a STEC but not the more commonly encountered O157:H7 serotype. The strain was also unusual in that it lacked the intimin (eae) gene required for the attachment of EHEC to the host. Through advances in gene sequencing, it was possible to sequence the complete genome of causative pathogen within a week using a technique referred to as pyrosequencing (Mellmann et al., 2011). The results of the sequencing revealed that the pathogen implicated in the outbreak was not within the EHEC group but an EAggEC referred to as serotype O104:H4. EAggEC are more commonly non-lethal and linked to person-to-person transmission and mainly implicated in urinary tract infections. However, *E. coli* O104:H4 had acquired the stx_2 toxin, making the strain highly virulent as indicated by the large number of HUS and deaths. The same serotype had been recovered 10 years previously within a remote African village and a single case in Korea in the same year (Rasko et al., 2011). Before this time, a related *E. coli* strain, serotype O104:H21, was responsible for an outbreak linked to raw milk in 1994 within the state of Montana (Feng et al., 2001). As in the German outbreak, the outbreak in Montana involved a high proportion of young adult females, which accounted for 67% of cases reported. This compares to 68% of young adult females affected by the sprouted seed outbreak. A further unusual feature of *E. coli* O104:H4 is that unlike EHEC, which tends to affect the young and old, the EAggEC strain has a broader age range of susceptible hosts. Again, this likely contributed to the number of cases of foodborne illness resulting from the outbreak.

Although the outbreak was centered in Hamburg incidence of *E. coli* O104:H4 was reported in 17 countries (Soon et al., 2013). The majority of cases were from persons who visited Hamburg and returned to their home countries. However, it was reported that at least two cases in the U.S. were attributed to secondary transmission from infected persons. This was not totally unexpected given the ease by which EAggEC passes from person-to-person.

The identification of fenugeek sprouts as the source of *E coli* O104:H4 was performed by epidemiology data given that no seeds screened for the pathogen tested positive. Sprout samples taken from opened packs did test positive for *E. coli* O104:H4, although the contamination by the handler/consumer could not be discounted. The lack of the "smoking-gun" (i.e., positive samples from unopened packs) was a result of the delay in detecting the outbreak. Such a delay is also a reason for the inability to recover *E. coli* O1O4:H4 from environmental samples taken from the sprouting facility located on an organic farm outside Hamburg. There remains a theory that infected workers on the farm contaminated the sprouts during the production phase. However, the most compelling evidence to implicate seeds was the link between the French outbreak to the one in Germany. Through investigations the implicated seed was traced to a seed lot produced in Egypt. The seeds were imported into Germany and then subsequently supplied to seed distributors in Germany, Austria, Spain, and the UK. The seeds implicated in the French outbreak of O104:H4 were from a pack of seeds distributed by the UK supplier, but through traceability linked to the original Egyptian lot (Muniesa et al., 2012). How the seeds were originally contaminated remains open to speculation and will likely remain unknown.

There continue to be outbreaks of foodborne illness linked to sprouts with a notable example being a series of salmonellosis cases linked to alfalfa distributed by a national restaurant chain (CDC, 2012; Table 11.2). In response, the chain made the decision to switch to clover sprouts with the perception of being at less risk of being contaminated compared to alfalfa. Yet, in 2012 an outbreak of *E. coli* O26 across 5 states serving clover sprouts resulted in 29 confirmed cases. The outbreak further underlined the fact that pathogens can become associated with a diverse range of sprout types.

Interactions of pathogens with sprouting seeds

In the course of germination starch and protein sources within the seed are degraded by endogenous enzymes to provide nutrients for the developing seedling. In addition, the released nutrients function to attract and establish the rhizosphere microbial populations required to aid plant nutrition and provide protection against phytopathogens (Hassan and Mathesius, 2012). In the natural environment the seed would initiate germination in the soil where human pathogen would encounter difficulties in competing for sites in the rhizosphere due to the highly competitive, preadapted, background microflora and relatively low temperatures. However, seed germination in sprouted seed production represents a contrasting environment in terms of high temperatures (20–28°C) and humidity (>90%) that along with the nutrient rich exudates can support the growth of human pathogens. Indeed, it has been reported that levels of *Salmonella* and *E. coli* O157:H7 can reach in excess of 6 log cfu/g during the first 24 h of the sprouting process even when present on the seed at levels of

0.1 cfu/g (Stewart et al., 2001). Although the majority of studies have focused on the growth of *Salmonella* and *E. coli* O157:H7 on alfalfa seed or mung beans it should be noted that all seed types potentially can support the growth of the enteric pathogens (Jablasone et al., 2005; Kumar et al., 2007).

The most critical time period in defining the microflora of sprouted seeds is the first 24 h of seed germination. Here, microbes (bacteria and fungi) within the seed and immediate environment compete for the nutrient rich exudates (Howard and Hutcheson, 2003). Pathogens such as *Salmonella* introduced later in the sprouting period fail to become established due to a combination of endogenous competitive microflora and lower level of exudates released by the developing sprout (Howard et al., 2003).

The growth of *Salmonella* and *E. coli* O157:H7 is not restricted to the surface of sprouts as both can become established (internalized) within the vascular system (Hirneisen et al., 2012; Warriner et al., 2003). Once pathogens are internalized within the inner vascular system simply washing the sprouted seeds is ineffective, with the consequence that most interventions applied to enhance the microbiological safety of sprouts are focused on preharvest interventions.

The internalization of human pathogens into seedlings was previously considered to be a passive process. However, it has been demonstrated that bacteria need to be viable to be internalized within plants and the degree of interaction is dependent on the bacterial type (Solomon and Matthews, 2006). For example, *Listeria monocytogenes* fails to be internalized into sprouting seeds, in contrast to *E. coli* O157:H7 and *Salmonella,* which appear adapted to the process (Jablasone et al., 2005).

Sources of contamination

In the majority of outbreaks the seed used to prepare sprouts has been identified as the original source of pathogens. The main routes of how the seed is initially contaminated has not been studied to any great extent, although the causes are likely to be the same as for fresh produce. For example, it has been demonstrated that introducing *E. coli* O157:H7 or *Salmonella* onto the flowers of mung bean plants results in the pathogens being recovered on the exterior and interior of the subsequent seedpods (Hora et al., 2007).

From reviewing the seed production process it is evident that different types are more susceptible to being contaminated during production than others. For example, alfalfa seed is not specifically produced for sprout production, so manure management and irrigation water standards may not be a high priority. It should also be noted that animals are on occasion allowed to graze on the alfalfa crops therefore increasing the risk of introducing enteric pathogens such as *E. coli* O157:H7 and *Salmonella*. In addition, the nature of harvesting increases the risk of soil being mixed with the seed, which can be a potential cross-contamination point especially during postharvest seed handling (Figure 11.2). Alfalfa seed traceability is also relatively poor, and it is not uncommon for consignments to be composed of multiple seed lots derived from different geographical locations. Therefore, contamination of a large consignment of seed from one single contaminated lot is possible. Consequently, alfalfa seed can be considered high risk, which likely explains the large number of outbreaks linked to this sprout type.

In contrast to alfalfa, mung bean production carriers less risk given that the scale of production of beans specifically for sprout production can justify implementing Good Agricultural Practices (GAP). Farms operating under GAP take measures to exclude animals, restrict manure application, and perform microbiological testing of irrigation water, in addition to implementing traceability with minimal mixing of batches (Bremer et al., 2003). Yet, despite implementing GAP there always remains the risk of contamination from multiple sources given the open nature of seed production.

Seeds can be contaminated at postharvest via common contact surfaces and through pest infestation during transport/storage. Seeds can also undergo scarification prior to sprouting to enhance hydration and enhance germination rates (Holliday et al., 2001). However, seed scarification or natural damage enables pathogens to become imbedded within thereby decreasing the efficacy of seed disinfection techniques (Holliday et al., 2001).

The role of contamination derived from the sprouting facility during sprout production has not been considered to any great extent. Although it is commonly considered that the seed is the origin of contamination it is possible that the pathogens introduced at the early stages of germination could proliferate and become established on the subsequent sprouts (Howard and Hutcheson, 2003).

Postharvest contamination of sprouts has not been investigated, with most focus being placed on the early stages of sprout production. It is acknowledged that washing sprouts is of little value with regards to removing contamination and is essentially performed to remove debris such as the seed coat. However, cross-contamination between batches via the wash water could occur as observed with leafy green processing (Barrera et al., 2012).

Interventions to enhance the microbiological safety of sprouted seeds

The majority of interventions used in the production of sprouted seeds are focused on preventing the seed from being contaminated, detecting contaminated seed batches, and/or decontaminating seeds prior to sprouting (Figure 11.2). Research has also been performed on treating irrigation water and postharvest sprout disinfection, although to a lesser degree.

Seed screening

The major seed distributors have adopted seed-screening protocols in an attempt to capture contaminated batches thereby restricting distribution through the chain. In addition to assessing seed germination yield, damaged seed and foreign bodies, additional tests include *Salmonella* and *E. coli* O157:H7 screening. For pathogen screening a 25 g subsample is taken from each bag and then combined to form a composite sample up to 3 kg. Given that pathogens can be present in low numbers the seeds are germinated and the spent irrigation water collected for selective enrichment for *Salmonella* and *E. coli* O157:H7. The challenge in seed screening is to obtain a representative sample from a designated lot of seed. If, for example, a subsample of 20 g is taken from a consignment of 20 tons of seed the probability of detecting a contaminated batch when present at 4 cfu/g is only 7% (OMAF, 2002a). However, if 50 × 20 g samples are screened the probability of detection increases to 98%. To increase the probability to 99.99%, a 250 × 20 g sample is required with the assumption that contamination is distributed homogenously throughout the lot (OMAF, 2002b). Although seed screening is routinely applied by seed distributors there is little data available on how many contaminated batches have been intercepted before entering sprout production. There are undocumented reports that at least one seed batch that tested positive for *E. coli* O157:H7 was detected during routine screening (International Sprout Growers Association; personnel communication). Clearly, making data available on seed-screening results will underline the usefulness of testing.

Seed disinfection

The NACMCF guidelines published in 1999 recommend, amongst other measures, the implementation of seed disinfection prior to sprouting to remove field acquired contamination. Specifically, the guide recommends that "prior to sprouting, seed should be subjected to one or more treatments that could effectively reduce (achieve a 5-log reduction) or eliminate pathogenic bacteria, such as *Salmonella* spp. and *E. coli* O157:H7. Intervention strategies that achieve less than a 5-log reduction should be paired with microbial testing of sprouts or spent irrigation water." The only recommended seed disinfection provided was based on a 15 min soaking in 20,000 ppm hypochlorite solution followed by a potable water rinse (Fett, 2006b). Hypochlorite was likely selected given the historical use as a sanitizer and low cost. It may have been anticipated that seed disinfection using 20,000 ppm hypochlorite would have been a regulatory requirement similar to pasteurization of milk. However, it was acknowledged that hypochlorite at 20,000 ppm could reduce pathogen levels but could not ensure complete elimination (Montville and Schaffner, 2004). Indeed, based on modeling predictions it has been estimated that 20,000 ppm hypochlorite treatment could ensure complete elimination of pathogens on seeds only 9% of the time (Montville and Schaffner, 2004). In addition to the low efficacy, hypochlorite at high concentrations is hazardous and is also prohibited in some countries (e.g., Germany). Hypochlorite is also incompatible with organic production, although this varies between associations (Ding et al., 2013). As a consequence, sprout growers either did not apply the disinfection method or used lower concentrations than the recommended 20,000 ppm hypochlorite (Weissinger and Beuchat, 2000).

Despite the limitation of hypochlorite there is evidence that the seed disinfection method at least reduced the number of foodborne illness cases in outbreaks. For example, a *Salmonella* outbreak linked to clover sprouts was traced to two sprout producers who used the same seed lot. Sprout producer A applied the recommended 20,000 ppm hypochlorite, while Sprout Producer B did not. From the incidence rate it was found that a greater proportion of the 112 cases of salmonellosis cases reported could be attributed to Sprout Producer B (Brooks et al., 2001). Gill et al. (2003) reported an outbreak of *S.* Mbandaka linked to alfalfa sprouts linked to two sprout producers who did not apply a seed disinfection step. However, three other sprout producers who used the same contaminated seed lot applied a hypochlorite disinfection step with no cases of *Salmonella* linked to the produced sprouts. Based on such evidence it was concluded that applying hypochlorite was better than using no sanitizer. Yet, given the limitations of hypochlorite there is a sustained effort to seek more effective alternative seed disinfection treatments.

Alternative seed disinfection methods

Studies of seed disinfection methods have been dominated by attempts to achieve a 5 log cfu reduction of pathogens or attempting to show equivalency to hypochlorite (reviewed by Montville and Schaffner, 2004; Yang et al., 2013; Ding et al., 2013). However, given that even low levels of pathogens that survive seed disinfection can grow during sprout production one would question the relevance of using the 5-log reduction as a metric. Therefore, demonstrating complete elimination of pathogens on seeds is more appropriate even though this is rarely demonstrated in published works. As a consequence, despite the diverse selection of seed sanitation methods tested those demonstrating complete elimination of pathogens are relatively few in number.

Successful seed disinfection approaches

Chemical-based treatments

Stabilized sodium chlorite is commonly applied as a precursor for chlorine dioxide and activated using a suitable acid such as phosphoric or lactic acid. In the nonactivated (alkali-stabilized form) sodium chlorite exhibits moderate antimicrobial activity, although highly biocompatible so has found application in, for example, contact lens fluid. When applied in seed disinfection the stabilized sodium chlorite is supplemented into the soak water (200 ppm) used in the initial stages of seed production (Hora et al., 2007). To access the efficacy of the disinfection treatment, seeds/beans were inoculated with either *Salmonella* or *E. coli* O157:H7, soaked in the stabilized sodium chlorite treatment, and then sprouted over 5 days with periodic irrigation. The sprouts were harvested after 48 or 96 h and then screened for pathogens. The study demonstrated complete elimination of both *Salmonella* and *E. coli* O157:H7 inoculated onto mung beans, alfalfa seeds, and soybeans (Kumar et al., 2006). In comparison, seeds/beans treated with the recommended 20,000 ppm hypochlorite was ineffective treatment, with high numbers of pathogens being recovered on the subsequent sprouts. The critical parameters for the success of the stabilized sodium chlorite treatment were seed-sanitizer ratio (1:4 w/v), contact time (8–24 h), and sanitizer concentration (150–200 ppm) (Kumar et al., 2006). The treatment was effective at decontaminating naturally contaminated seed and had no negative effect on sprout development or germination yield (Hora et al., 2007).

A further successful seed decontamination treatment reported was based on a fatty acid-based sanitizer consisting of peroxyacid (250 ppm), caprylic-capric acids (1000 ppm), lactic acid (1000 ppm), and glycerol monolaurate (500 ppm) (Pierre and Ryser, 2006). Complete inactivation of *E. coli* O157:H7 and *Salmonella* introduced onto alfalfa seeds was reported using 15 times concentration of sanitizer with an exposure time of 3 mins. It should be noted that pathogen reduction was assessed by screening the seeds rather than the subsequent sprouts. Yet, no decrease in seed germination yield was observed (Pierre and Ryser, 2006).

Acetic acid in the vapor phase applied over a prolonged time period has proven effective at complete elimination of *Salmonella* and *E. coli* O157:H7 inoculated onto alfalfa or radish seeds (Nei et al., 2011). Treatment with 8.7% acetic acid at 55°C for 3 h supported a > 5 log cfu reduction of pathogens, although 24 h treatment was required to ensure complete elimination of *E. coli* O157:H7. *Salmonella* was recovered even when the treatment was extended to 48 h (Nei et al., 2011). Delaquis et al. (1999) applied gaseous acetic acid (242 ppm) at 22°C for 24 h and achieved complete elimination of *Salmonella* and *E. coli* O157:H7 inoculated onto mung beans (Delaquis et al., 1999).

Physical methods

Hot water treatment is one of the most common methods for seed disinfection by virtue of the fact that it is applied by the majority of Japanese sprout producers (Bari et al., 2010; Bari et al., 2011). Laboratory-based studies have demonstrated that complete elimination of *Salmonella* and *E. coli* O157:H7 can be achieved by treating seeds at 90°C for 90 s, although adverse effects on seed germination were reported (Bari et al., 2008). Consequently, in commercial practice the treatment regimes are in the order of 85°C for 10 s, which supports a 3 log cfu reduction in pathogen levels, although cannot ensure complete elimination. Yet, it should be noted that hot water treatment of seeds in Japan is primarily to inactivate phytopathogens as opposed to an intervention step against human pathogens given that the majority of sprouts are cooked (Bari et al., 2010).

Neetoo and Chen (2011) have shown that alfalfa seeds exposed to dry heat at 65°C for 10 days or 70°C for 24 h achieved a 5-log reduction of *Salmonella* and *E. coli* O157:H7 without affecting seed germination, but did reduce sprout yield by 21% (Neetoo and Chen, 2011). Bari et al. (2009) demonstrated that dry-heat treatment at 50°C for 17 or 24 h reduced *E. coli* O157:H7 numbers to below detectable levels in radish, broccoli, and alfalfa seeds, but was unable to reduce the pathogen numbers to below the detectable level in mung bean seeds (Bari et al., 2009).

High hydrostatic pressure (HHP) is an established technology for nonthermal pasteurization of a diverse range of foods including deli meats and juices. HPP treatment is a batch process whereby the vacuum-packed sample is placed within a chamber containing a non-compressible liquid (commonly water) and pressurized to 500 to 600 MPa. Neetoo et al. (2009) reported that pressure treatments at 600 MPa for 20 min at 20°C could reduce but not eliminate *E. coli* O157:H7 on alfalfa seeds (Neetoo et al., 2009). However, by using 550 MPa for 2 min and a treatment temperature of

40°C it was possible to completely inactivate *E. coli* O157:H7 as demonstrated by the lack of positive samples when treated seeds were enriched (Neetoo et al., 2009).

Although HHP is a promising technique there are potential barriers to commercial application as an alternative seed disinfection technique. Specifically, seeds HHP treated by emersion in water as opposed to vacuum packed, have a significant decrease in germination yield (Ariefdjohan et al., 2004). Yet, if seeds are presoaked in water and then placed into pouches for HHP treatment complete elimination of *E. coli* O157:H7 and *Salmonella* was achieved by applying 600 MPa for 5 min, although germination rates decreased by 10% (Neetoo and Chen, 2010). In addition, HHP technology is relatively expensive (11 cents per kg product), which may be a further limitation to commercial application.

Irradiation has a long history as a nonthermal intervention method, although it is mainly used for treating spices, insect inactivation, or to prevent sprouting of vegetables such as potatoes. Because irradiation targets DNA a balance needs to be made with respect to applying a sufficient dose to inactivate pathogens while preserving seed viability. Rajkowski and Thayer (2000) reported complete inactivation of *Salmonella* introduced onto alfalfa and treated with 0.5 kGy but sprout development was stunted (Rajkowski and Thayer, 2000).

Interventions during sprouting

Biological control can be classed under seed disinfection, although the main objective is to inhibit or reduce the growth of pathogens over the sprouting period. Interest has been placed on the application of probiotic lactic acid bacteria, which produce a range of antimicrobials including bacterocins, hydrogen peroxide, and organic acids (Fett, 2006b). The underlying basis of the approach is to ensure that the probiotic becomes established with sprouts thereby inhibiting pathogens through competitive inhibition and production of antimicrobials. Wilderdyke et al. (2004) isolated a diverse range of lactic acid bacteria from alfalfa, with a high proportion demonstrating inhibitory activity against *Salmonella*, *E. coli* O157:H7, and *L. monocytogenes in vitro* (Wilderdyke et al., 2004). However, no trials were performed to assess if the isolates could inhibit human pathogens during sprout production.

Pseudomonas fluorescens 2-79 has demonstrated strong antagonistic activity against *Salmonella in vitro*. When co-inoculated onto alfalfa seed the final population of *Salmonella* ranged from 1 to 2 log cfu/g compared to 3 to 4 log cfu/g of nontreated controls (Fett, 2006; Liao, 2008).

A further biological control approach was to inoculate seeds with a community of microbes recovered directly from mature sprouts. By "pitching" the undefined culture with alfalfa seeds inoculated with *Salmonella* the levels of the pathogen progressively decreased over the 7-day sprouting period reaching 2.4 log cfu/g at the time of harvest (Matos and Garland, 2005). In contrast, *Salmonella* in nontreated controls attained levels of 9 log cfu/g at the end of the sprouting period (Matos and Garland, 2005). The concept of using microbial communities to at least reduce pathogen levels shows promise, although how to recreate defined preparations will represent a challenge.

A further biological treatment of interest is the application of bacteriophages, which can selectively infect target pathogens without disrupting the endogenous sprout populations (Kocharunchitt et al., 2009). Kocharunchitt et al. (2009) co-inoculated bacteriophages with *Salmonella* onto alfalfa seed and reported only a 1 log cfu/g reduction of the pathogen on the subsequent sprouts (Kocharunchitt et al., 2009). Pao et al. (2004) also reported low log reductions of *Salmonella* when bacteriophages were co-inoculated onto mustard seeds. The limited efficacy of bacteriophages to suppress the growth of pathogens on sprouts is unclear but has been proposed to be through the natural equilibrium that is reached between host cell and phages, which ensures survival of both. In addition, the binding of phages to plant material, the presence of natural antimicrobial constituents, and the generation of resistant mutants has also been highlighted (Pao et al., 2004).

A successful bacteriophage treatment has been reported whereby the phages were co-inoculated with antagonistic bacteria (Ye et al., 2009). Here, virulent bacteriophages against *Salmonella* were isolated from pig farms with the antagonistic bacterial strain (*Enterobacter asburiae* strain JX1) being recovered from tomatoes. By using a combination of bacteriophage cocktail with *E. asburiae* no *Salmonella* were recovered from mung bean sprouts at the point of harvest, compared to controls, which had counts >7 log cfu/g. When bacteriophages and antagonistic bacteria were applied separately the residual *Salmonella* populations at the end of sprouting were reduced to 1 to 3 log cfu/g (Ye et al., 2009). Therefore, the combination of bacteriophages and antagonistic bacteria supported a synergistic anti-*Salmonella* effect, although the underlying mechanisms remain to be elucidated.

Irrigation water supplements

In the course of sprouting germinating seeds are frequently irrigated to provide moisture to the developing sprouts and to regulate temperature. This has led to testing different antimicrobial different antimicrobial agents to control pathogens during sprout production. Fett (2002) evaluated a range of antimicrobial agents supplemented into irrigation water on the total aerobic counts of sprouting alfalfa seeds. Acetic acid (88 mM, pH 3.5), acidified sodium chlorite (60 ppm, pH 5), hydrogen peroxide (1000 ppm, pH 7) peroxyacetic acid (80 ppm, pH 5.2), and phosphate buffer (0.3%, pH 11) all proved phytotoxic, stunting sprout development. Chlorinated water (3800–20,000 ppm) reduced the microflora of sprouts by 1 to 2 log cfu/g compared to controls irrigated with potable water (Fett, 2002). Taormina and Beuchat (1999) assessed the same range of sanitizers to control *E. coli* O157:H7 on sprouting alfalfa seed. From the range of irrigation water supplements tested only acidified sodium chlorite applied at 1200 ppm resulted in a significant decrease (2.92 log cfu/g compared to 5.00 log cfu/g for controls) in *E. coli* O157:H7 levels on the final sprouts (Taormina et al., 1999). Yet, it was concluded that antimicrobial supplements is an ineffective approach to control pathogens on sprouting seeds and more likely to negatively affect sprout development.

Postharvest treatment of sprouts

Postharvest washing of sprouts has been demonstrated to be relatively ineffective at removing pathogens, with log reductions of 2.5 log cfu being reported regardless of the sanitizer applied (Lee et al., 2002). It is likely that the limited log count reductions are due to biofilms on the surface of sprouts, sequestering of sanitizers by high organic content, and of course, the presence of internalized populations.

More success with respect to sprout decontamination has been met using physical methods. *L. monocytogenes* inoculated onto mung bean sprouts and then treated with HHP (400 MPa at 40°C) resulted in complete inactivation of the pathogen with no recovery during post-treatment storage (Munoz et al., 2006). Waje et al. (2009) evaluated irradiation to inactivate *E. coli* O157:H7, *Salmonella,* and *L. monocytogenes* inoculated onto different sprout types (Waje et al., 2009). The study estimated that a dose of 2.30 and 3.65 kGy was required to support a 5-log reduction using gamma ray and electron beams, respectively. Although the studies did not assess if irradiation can inactivate internalized populations the technique is being considered to enhance the microbiological safety of sprouts (Waje et al., 2009).

Spent irrigation water testing

In addition to seed sanitation it is also recommended to screen for pathogens in sprouting seed beds. Testing is performed 48 h into sprout production given that the pathogens, if present, would have attained high levels and also provides sufficient time to get the result of testing prior to releasing the sprouts to market. A key challenge in sampling sprouting seed beds regardless if performed in drums, trays, or bins is the heterogeneous distribution of contamination (Liu and Schaffner, 2007; McEgan et al., 2009). It has been demonstrated that contaminated seed introduced at a single point fails to become distributed throughout the bed (Liu and Schaffner, 2007; McEgan et al., 2008). Consequently, it is recommended to test spent irrigation water as opposed to sprouts directly. Through validation trials it has been found that microbial counts recovered in spent irrigation water that has permeated through the sprouting seed bed is sufficiently reflective of that associated with sprouts (Fu et al., 2001). Yet, testing spent irrigation water does not give total assurance due to the heterogeneous distribution of contamination within sprouting seed beds, especially with mung bean sprouts that are commonly sprouted in 25 to 75 kg batches. One option is to collect multiple samples from beds and then combine as a composite. Yet, studies performed comparing single vs. composite samples did not result in a greater probability of detecting contamination if present (McEgan et al., 2008). A further approach is to collect large volumes (>10 liters) of irrigation water and subsequently concentrate microbes using tangential flow filtration (McEgan et al., 2009). However, it is more common, if not universal, to collect a single spent irrigation water (300–1000 ml) per batch for *Salmonella* and *E. coli* O157:H7 screening. The subject of pooling spent irrigation water from different batches is subject to debate. By pooling samples the number of tests required to be performed are reduced. However, in the event of a positive all batches of sprouts used to prepare the composite must be discarded. There are also issues related to diluting pathogens by using composite samples that may result in a false-negative reaction.

Screening of spent irrigation water involves taking a 2×100 ml sample and enriching using the appropriate media followed by plating and then confirmation testing (Health Canada, 2006). One of the challenges encountered using plating techniques is derived from the high background microflora, which can make identification of typical *Salmonella* or *E. coli* O157:H7 problematic. Of more significance is the delay in confirming positive colonies. Even though sampling is performed early (48 h) into the sprouting period the time for analysis can be in excess of 4 days by which time the sprouts may well have been harvested and shipped. As a consequence, research in this area has focused on sampling the sprouts at 24 h rather than 48 h or applying rapid detection methods (Ding et al., 2013). The sampling of seed beds 24 h into sprout production has the risk of generating false negatives given that pathogens introduced at low levels may not become established. Therefore, rapid detection methods have emerged as a more reliable option. In this respect, technologies have been developed that can concentrate pathogens from large volumes (10 liters) using tangential flow filtration followed by detection using electrochemical immune sensors. By omitting the enrichment step it is possible to detect contaminated seedbeds within 4 h (Maks and Fu, 2013). Real-time PCR techniques have also been applied whereby spent irrigation water can be analyzed within 4 h following a 16 to 24 h enrichment step.

The value of spent irrigation water in terms of preventing contaminated batches of sprouts being distributed remains unclear. As with seed screening, the results of spent irrigation water testing have not been published. Yet, there have been foodborne illness outbreaks linked to sprouted seeds where the spent irrigation water tested positive. Although the batch of sprouts was discarded the producer continued to use the same batch of seeds, which was ultimately the origin of the outbreak. This raises the question of how to interpret results of spent irrigation water testing, which will likely become more significant as rapid techniques such as RT-PCR are adopted. Specifically, it was proposed that in the event of a presumptive positive the spent irrigation water should be re-tested. In addition, in the event of a positive RT-PCR result action should only be taken with a culture positive sample (NACMCF, 1999). The interpretation of results in spent irrigation water testing obviously requires clarification .

Guidelines to enhance the microbiological safety of sprouted seeds
United States

In the aftermath of the *E. coli* O157:H7 outbreak in Japan and the increasing number of other sprout-related outbreaks the NACMCF published guidelines on approaches to enhancing the microbiological safety of sprouted seeds (Figure 11.1) (NACMCF, 1999). The recommendations provided in the report formed the very foundation of subsequent guides published by other agencies. The 1999 NACMCF recommendations were more related to identifying gaps in the then current practices but within a short time became a guide to the industry. The noteworthy points of the guidelines are that it treated all seed types the same, with no designation for high or low risk. In addition, the recommendations only suggested 20,000 ppm hypochlorite as a method to disinfect seeds, despite evidence to the contrary. In some ways the guidelines provided a focus on the food safety issues related to sprouts but in another sense they also restricted the introduction of new approaches. For example, a common argument among sprout growers for not using alternative seed sanitation methods came down to the apparent lack of flexibility in the guide to apply anything else other than hypochlorite. The guide was also ambiguous in terms of how to meet the recommendations, which proved a major barrier to implementation.

The NACMCF issued a further guidance document in 1999 relating to sampling and microbial sampling of spent irrigation water in the course of sprout production (National American Committee for the Microbiological Criteria of Foods, 1999). Unlike its sister document, the spent irrigation water guide was very specific with respect to how and when to perform sampling. It is likely that publishing the two sprouted seed guidelines separately was intentional given that the testing of irrigation water had a stronger science base compared to the more generic recommendations on how to improve the microbiological safety of sprouted seeds.

In the two years following the publication of the two NACMCF guidelines in 1999 there was a decrease in the incidence of foodborne outbreaks linked to sprouts (Figure 11.1). However, form 2003 there was an upward trend in the number of outbreaks linked to sprouted seeds that were thought at that time due to sprout producers not adequately following the guidelines, especially in relation to seed disinfection and spent irrigation water testing. This was true up to a point in that the majority of sprout producers used a lower concentration of hypochlorite for seed disinfection than the recommended 20,000 ppm. Spent irrigation water was also performed sporadically given that it was a recommendation as opposed to a regulation.

In 2004 the FDA issued letters to the sprout industry to remind them to follow the NACMCF guidelines when producing sprouts, although it did fall short of imposing regulations. It is possible that the FDA pulled back from issuing regulations to the sprout industry due to the acknowledged limitations of the hypochlorite-based seed disinfection method. To address this gap the FDA held a public meeting on approaches to enhance the microbiological safety of sprouted seeds especially in relation to seed disinfection methods. The meeting included opinions from sprout associations that questioned the usefulness of seed disinfection and from academics who brought their alternative approaches to the floor. The net result of the public meeting was that no changes to the original 1999 guidelines were required and the status quo would prevail. However, the outbreaks linked to sprouted seeds continued and ultimately led to a further letter being issued by the FDA reminding sprout growers to follow the guidelines and review their food safety intervention methods, despite the guidelines providing ill-defined approaches apart from spent irrigation water testing. In effect, the sprout industry was being advised to formulate their own approaches to enhance the safety of sprouted seeds. This was re-enforced by the proposed rule for produce under FSMA, which devoted a special section specifically related to sprouted seed production. There were key points made in the proposed ruling that the grower of seeds or beans destined for sprout production have food safety systems in place (i.e., to work under GAP). A further requirement was for sprout producers to have a validated seed disinfection method in place and sanitary facilities, in addition to a *Listeria* environmental sampling plan. Spent irrigation water was also required to be performed to screen for *Listeria* spp, *Salmonella,* and *E. coli* O157:H7.

To support the implementation of the FSMA regulations the Sprout Safety Alliance (SSA) was formed by the FDA in collaboration with the Illinois Institute of Technology's Institute for Food Safety and Health (IIT IFSH). The SSA has worked to develop an auditor's checklist to help sprout producers upgrade their facilities and provides a list of potential alternative seed disinfection methods to hypochlorite.

Canada

Canada does not produce a significant amount of seeds or beans destined for sprouting, although the sprouted seed sector is significant. In 2002, the Ministry of Agriculture and Food (OMAF) published a risk assessment on sprouted seeds. The report highlighted that the seeds destined for sprout production represent the highest risk in the process. A Good Manufacturing Guide for sprout producers was published in 2007, which placed emphasis on seed and irrigation water testing, although did not provide recommendations on seed disinfection methods (OMAF, 2007).

On a federal scale, the Canadian Food Inspection Agency (CFIA) guidelines issued in 2007 essentially built on recommendations of the NACMCF 1999 guidelines. Specifically, recommendations were placed on sourcing seeds, general sanitary standards of sprouting facilities, and Good Manufacturing Practices. With respect to seed disinfection, the guidelines indicated that a process that can support a 3-log reduction in pathogen levels is acceptable, with examples given as 2000 ppm hypochlorite or 6 to 10% hydrogen peroxide being recommended as treatments. The guidelines did offer the option of other treatments, provided that initial approval was obtained from the Food Directorate of Health Canada. Spent irrigation water and sprout testing procedures were also outlined in the guidelines (Health Canada, 2007).

Europe

The *E. coli* O104:H4 outbreak prompted the European Food Safety Authority (EFSA) to establish a working committee to assess the food safety hazards and potential interventions in sprouted seed production (EFSA, 2011). The report was published in 2011 and provided an informative overview of the food safety risks associated with sprouts and potential sources of contamination. The recommendations made by the panel broadly followed those issued in other guides with respect to preventing the introduction of pathogens at the farm level. There was little detail provided with respect to seed decontamination but importantly it was identified that not one single treatment can treat all seed types. In addition, the report identified the need for harmonization across the EU with respect to testing the efficacy of different seed disinfection treatments – in essence, a standard protocol for testing seed disinfection technologies.

The EFSA report placed less focus on the significance of testing but more emphasis on the need for control at the primary production level and traceability. The implementation of HACCP was also proposed, although how to establish and verify the Critical Control Point was problematic. The need for standardized methods for testing seeds/beans and sprouts was also recommended.

In 2013 the European Commission issued Regulation 208/2013 based on the recommendations, which essentially stated that traceability of batches of spouts and seeds at all stages of production, processing, and distribution. Furthermore, it states that seeds used for sprouting whether produced within or outside the EU must be accompanied by a Certificate in Accordance that states that the seeds were produced under hygienic conditions (GAP). The regulations required traceability throughout the chain. That is, if a batch of sprouts was implicated in an outbreak it will be possible to trace back to the producer, seed distributor and seed producer by means of a reference number (barcode). Regulation 2073/2005 provided a new microbiological criteria for sprouted seeds that included screening for STEC (O157, O26, O11, O103, O145, and O104) that would be absent from 5 × 25g samples taken at least 48 h into the sprouting process. This relates to the screening of the Top 6 shiga toxin producing *Escherichia coli* in sprouts 48 h into the production period. The Top 6 were selected as these serotypes are commonly implicated in HUS. It should be noted that the EU regulation specifies sampling of sprouts with no mention of irrigation water despite the known limitations of the former sample regime. Notably, there is no mention of *Salmonella* screening despite the pathogen being the one most implicated in foodborne illness outbreaks linked to sprouts.

Australia and New Zealand

Australia can be considered more progressive compared to other sprout-producing nations in that the regulations provide more specific information within the published guidelines. In addition, Australia was the first to introduce regulations specific to the sprouted seed industry, albeit it within one state.

As in other countries, Australia and New Zealand have experience several outbreaks linked to sprouted seeds. In 2008 the Food Standards Agency of Australia and New Zealand published a set of guidelines for sprout producers (NZFSA, 2008). However, like in other nations, the adoption of the guidelines was low. Yet, retailers have been active in terms of imposing specifications that primarily focus on quality issues and end-product testing of sprouts and the pathogens of concern (*Salmonella*, *E. coli* O157:H7, and *L. monocytogenes*). Moreover, sprout producers supplying to the large retail chains must have in place a food safety system that is accredited and audited under the Global Food Safety Initiative (GFSI-SQF, BRC) standards. As a consequence, 60% of the industry is accredited, although those sprout producers supplying product to other markets do not run a food safety system.

In 2011 there was a proposed set of regulations published by the FZFSA. Interestingly, the request for regulations was promoted by the industry itself having recognized that guidelines were inadequate to enhance food safety standards across the industry. Currently, only sprout producers operating within New South Wales are required to follow regulations specifically related to sprouted seeds. The regulations introduced in 2005 state that the spout producer must implement a food safety program, undergo audits to assess compliance with the food safety program, and undertake regular spent irrigation water testing. The recommended food safety system to implement was a HACCP-based approach that was certified by state inspectors or auditors. An important part of the HACCP-based system is the inclusion of a Supplier Approval Program. Here, the seed supplier needs to provide evidence of a HACCP-based food safety program (i.e., on-farm HACCP) or provide evidence of working under GAP. Importantly, the seed supplier needs to provide a certificate of analysis and seed testing results for each batch produced. Additional elements of the HACCP-based system for sprout producers includes GMP, calibration of measuring equipment, and verification through fortnightly screening of 25 ml of spent irrigation water for *Salmonella*. Seed sanitation using 20,000 ppm hypochlorite or equivalent is also performed, in addition to postharvest washing and end-product screening. From surveys performed to assess the impact of the implemented regulations within NSW it was found that the microbiological quality of sprouts (predominantly mung bean and alfalfa) had gone from 10% marginal results recorded in 2005 to <5% in 2008 (NSW, 2008). The results of the survey suggest that the implementation of regulations within the sprouted seed industry has been a success, although it should be noted that NSW only accounts for 15% of the sprout production within Australia and the sample size used was limited to 122 batches of product. Nevertheless, it can be seen that ensuring control at the seed-production level and at processing will enhance food safety standards.

Conclusions and future directions

Sprouted seeds continue to be considered a high-risk food commodity, with most food safety agencies issuing health advisories to high-risk members of the population. From research dating back to the early 1990s it is clearly evident that virulent enteric pathogens can become established on sprouting seeds and become internalized into the inner vascular system thereby limiting the efficacy of postharvest washes. To address the ongoing food safety issues the initial approach was to provide generic-based recommendations, which had a limited impact on improving standards within the sector as evident by the continuing foodborne illness outbreaks. Throughout the 2000s the published guidelines placed an emphasis on seed screening and spent irrigation water testing. As history has demonstrated, reliance on testing to catch contamination is an unreliable intervention given the inherent difficulty in detecting low levels of contamination or pathogens heterogeneously distributed within a batch. In this respect, focusing on interventions such as seed disinfection would represent a more effective approach. Yet, despite several seed disinfection methods being available these have yet to be implemented on commercial scale.

A current trend to improve the microbiological quality of sprouts is to move toward implementation of regulations, previously resisted by both government and industry. Regulations specifically related to sprouts have proven an apparent success in New South Wales, but time will tell how the industry will be reshaped with the regulatory changes introduced in the EU and the U.S. A common theme within the regulations is the responsibility of the seed supplier to provide pathogen-free seeds. One can envision that such requirements can be met by mung bean producers, although may prove more challenging for other seed types where the sprouted seed market forms only a minor proportion of the sector. It is conceivable that a compromise may be met by classifying sprouts based on risk. For example, alfalfa seed has a higher probability of harboring pathogens than buckwheat or pea shoots. Yet, regardless of this fact the most effective approach to enhancing food safety will be the implementation of a seed disinfection step and/or an effective postharvest intervention such as irradiation of sprouts.

References

Ariefdjohan, M.W., Nelson, P.E., Singh, R.K., Bhunia, A.K., Balasubramaniam, V.M., Singh, N., 2004. Efficacy of high hydrostatic pressure treatment in reducing *Escherichia coli* O157 and *Listeria monocytogenes* in alfalfa seeds. J. Food Sci. 69, M117–M120.

Bari, M.L., Enomoto, K., Nei, D., Kawamoto, S., 2010. Practical Evaluation of Mung Bean Seed Pasteurization Method in Japan. J. Food Prot. 73, 752–757.

Bari, M.L., Enomoto, K., Nei, D., Kawamoto, S., 2011. Development of Effective Seed Decontamination Technology to Inactivate Pathogens on Mung Bean Seeds and Its Practical Application in Japan. Jarq-Japan. Agric. Res. Q. 45, 153–161.

Bari, M.L., Inatsu, Y., Isobe, S., Kawamoto, S., 2008. Hot water treatments to inactivate *Escherichia coli* O157: H7 and Salmonella in mung bean seeds. J. Food Prot. 71, 830–834.

Bari, M.L., Nei, D., Enomoto, K., Todoriki, S., Kawamoto, S., 2009. Combination Treatments for Killing *Escherichia coli* O157:H7 on Alfalfa, Radish, Broccoli, and Mung Bean Seeds. J. Food Prot. 72, 631–636.

Barrera, M.J., Blenkinsop, R., Warriner, K., 2012. The effect of different processing parameters on the efficacy of commercial post-harvest washing of minimally processed spinach and shredded lettuce. Food Control. 25, 745–751.

Beuchat, L.R., 1996. Pathogenic microorganisms associated with fresh produce. J. Food Prot. 59, 204–216.

Beutin, L., Martin, A., 2012. Outbreak of Shiga Toxin-Producing *Escherichia coli* (STEC) O104:H4 Infection in Germany Causes a Paradigm Shift with Regard to Human Pathogenicity of STEC Strains. J. Food Prot. 75, 408–418.

Boll, E.J., Struve, C., Boisen, N., Olesen, B., Stahlhut, S.G., Krogfelt, K.A., 2013. Role of Enteroaggregative *Escherichia coli* Virulence Factors in Uropathogenesis. Infect. Immun. 81, 1164–1171.

Bradford, S.A., Morales, V.L., Zhang, W., Harvey, R.W., Packman, A.I., Mohanram, A., et al., 2013. Transport and Fate of Microbial Pathogens in Agricultural Settings. Crit. Rev. Environ. Sci. Technol. 43, 775–893.

Brandi, M.T., Cox, C.E., Teplitski, M., 2013. *Salmonella* Interactions with Plants and Their Associated Microbiota. Phytopathology 103, 316–325.

Bremer, P.J., Fielding, T., Osborne, C.M., 2003. The safety of seeds and bean sprouts: risks and solutions. Food NZ. 3, 33–40.

Brooks, J.T., Rowe, S.Y., Shillam, P., Heltzel, D.M., Hunter, S.B., Slutsker, L., et al., 2001. *Salmonella typhimurium* infections transmitted by chlorine-pretreated clover sprout seeds. Am. J. Epidemiol. 154, 1020–1028.

Centers for Disease Control and Prevention (CDC), 2012. Multistate outbreak of Shiga toxin-producing *Escherichia coli* O26 infections linked to raw clover sprouts at Jimmy John's Restaurants. Available at http://www.cdc.gov/ecoli/2012/o26-02-12/index.html (accessed 04.13).

De Roever, C., 1999. Microbiological safety evaluations and recommendations on fresh produce (vol. 9, pg 321, 1998). Food Control 10, 117–143.

Delaquis, P.J., Sholberg, P.L., Stanich, K., 1999. Disinfection of mung bean seed with gaseous acetic acid. J. Food Prot. 62, 953–957.

Ding, H., Fu, T.-J., Smith, M.A., 2013. Microbial contamination in sprouts: how effective is seed disinfection treatment? J. Food Sci. 78, R495–R501.

European Food Safety, A, 2011. Shiga toxin-producing *E. coli* (STEC) O104:H4 2011 outbreaks in Europe: taking stock. EFSA. J. 9, 2390 [2322pp.]-2390 [2322pp.].

European Food Safety Authority (EFSA), 2011. Scientific opionion on the risk posed by Shiga toxin producing *Escherichia coli* (STEC) and other pathogenic bacteria in seeds and sprouted seeds. EFSA. J. 9, 1–100.

Farrokh, C., Jordan, K., Auvray, F., Glass, K., Oppegaard, H., Raynaud, S., et al., 2013. Review of Shiga-toxin-producing *Escherichia coli* (STEC) and their significance in dairy production. Int. J. Food Microbiol. 162, 190–212.

Feng, P., Weagant, S.D., Monday, S.R., 2001. Genetic analysis for virulence factors in *Escherichia coli* O104: H21 that was implicated in an outbreak of hemorrhagic colitis. J. Clin. Microbiol. 39, 24–28.

Ferens, W.A., Hovde, C.J., 2011. *Escherichia coli* O157:H7: Animal Reservoir and Sources of Human Infection. Foodborne Pathog. Dis. 8, 465–487.

Fett, W.F., 2002. Reduction of the native microflora on alfalfa sprouts during propagation by addition of antimicrobial compounds to the irrigation water. Int. J. Food Microbiol. 72, 13–18.

Fett, W.F., 2006a. Inhibition of *Salmonella enterica* by plant-associated pseudomonads in vitro and on sprouting alfalfa seed. J. Food Prot. 69, 719–728.

Fett, W.F., 2006. Interventions to ensure the microbiological safety of sprouts. In: Sapars, G.M., Gorny, J.R., Yousef, A.E. (Eds.), Microbiology of Fruits and Vegetables. Taylor & Francis, New York.

Food Standards agency of Australia and New Zealand (NZFSA), 2008. Primary production and processing standard for seed sprouts. Available at http://ris.finance.gov.au/files/2012/03/02-Seed-Sprouts-RIS.pdf (accessed 04.13).

Fresh Produce Consortium (FPC), 2012. Guidance for food business operators on the hygienic sourcing, production and safe handling of ready to eat sprouts. Accessabile at http://www.freshproduce.org.uk/index.php.

Fu, T., Stewart, D., Reineke, K., Ulaszek, J., Schlesser, J., Tortorello, M., 2001. Use of spent irrigation water for microbiological analysis of alfalfa sprouts. J. Food Prot. 64, 802–806.

Gill, C.J., Keene, W.E., Mohle-Boetani, J.C., Farrar, J.A., Waller, P.L., Hahn, C.G., et al., 2003. Alfalfa seed decontamination in a *Salmonella* outbreak. Emerg. Infect. Dis. 9, 474–479.

Grant, M.A., Hedberg, C., Johnson, R., Harris, J., Logue, C.M., Meng, J., et al., 2011. The significance of non-O157 Shiga toxin-producing *Escherichia coli* in food. Food Prot. Trends 31, 33–45.

HaraKudo, Y., Konuma, H., Iwaki, M., Kasuga, F., SugitaKonishi, Y., Ito, Y., et al., 1997. Potential hazard of radish sprouts as a vehicle of *Escherichia coli* O157:H7. J. Food Prot. 60, 1125–1127.

Hassan, S., Mathesius, U., 2012. The role of flavonoids in root-rhizosphere signalling: opportunities and challenges for improving plant-microbe interactions. J. Exp. Bot. 63, 3429–3444.

Hauser, E., Mellmann, A., Semmler, T., Stoeber, H., Wieler, L.H., Karch, H., et al., 2013. Phylogenetic and Molecular Analysis of Food-Borne Shiga Toxin-Producing *Escherichia coli*. Appl. Environ. Microbiol. 79, 2731–2740.

Hauswaldt, S., Nitschke, M., Sayk, F., Solbach, W., Knobloch, J.K.M., 2013. Lessons Learned From Outbreaks of Shiga Toxin Producing *Escherichia coli*. Curr. Infect. Dis. Rep. 15, 4–9.

Health Canada, 2006. Guidance for Industry: Sample collection and testing for sprouts and irrigation water. Available at http://www.hc-sc.gc.ca/fn-an/legislation/guide-ld/sprout_water_testing_analyse_pousses_eau-eng.php (accessed 04.13).

Health Canada, 2007. Code of practice for the hygienic production of sprouted seeds. Available at http://www.inspection.gc.ca/english/fssa/frefra/safsal/sprointe.shtml (accessed 04.13).

Hirneisen, K.A., Sharma, M., Kniel, K.E., 2012. Human Enteric Pathogen Internalization by Root Uptake into Food Crops. Foodborne Pathog. Dis. 9, 396–405.

Holliday, S.L., Scouten, A.J., Beuchat, L.R., 2001. Efficacy of chemical treatments in eliminating *Salmonella* and *Escherichia coli* O157: H7 on scarified and polished alfalfa seeds. J. Food Prot. 64, 1489–1495.

Hora, R., Kumar, M., Kostrzynska, M., Dixon, M.A., Warriner, K., 2007. Inactivation of *Escherichia coli* O157: H7 and *Salmonella* on artificially or naturally contaminated mung beans (*Vigna radiata* L) using a stabilized oxychloro-based sanitizer. Lett. Appl. Microbiol. 44, 188–193.

Howard, M.B., Hutcheson, S.W., 2003. Growth dynamics of *Salmonella enterica* strains on alfalfa sprouts and in waste seed irrigation water. Appl. Environ. Microbiol. 69, 548–553.

Itoh, Y., Sugita-Konishi, Y., Kasuga, F., Iwaki, M., Hara-Kudo, Y., Saito, N., et al., 1998. Enterohemorrhagic *Escherichia coli* O157: H7 present in radish sprouts. Appl. Environ. Microbiol. 64, 1532–1535.

Jablasone, J., Warriner, K., Griffiths, M., 2005. Interactions of *Escherichia coli* O157: 147, *Salmonella typhimurium* and *Listeria monocytogenes* plants cultivated in a gnotobiotic system. Int. J. Food Microbiol. 99, 7–18.

Kim, D.K., Jeong, S.C., Gorinstein, S., Chon, S.U., 2012. Total Polyphenols, Antioxidant and Antiproliferative Activities of Different Extracts in Mungbean Seeds and Sprouts. Plant Foods Hum. Nutr. 67, 71–75.

Kim, S.A., Kim, O.M., Rhee, M.S., 2013. Changes in microbial contamination levels and prevalence of foodborne pathogens in alfalfa *(Medicago sativa)* and rapeseed (*Brassica napus*) during sprout production in manufacturing plants. Lett. Appl. Microbiol. 56, 30–36.

Kocharunchitt, C., Ross, T., McNeil, D.L., 2009. Use of bacteriophages as biocontrol agents to control *Salmonella* associated with seed sprouts. Int. J. Food Microbiol. 128, 453–459.

Kumar, M., Hora, R., Kostrzynska, M., Waites, W.M., Warriner, K., 2006. Inactivation of *Escherichia coli* O157: H7 and *Salmonella* on mung beans, alfalfa, and other seed types destined for sprout production by using an oxychloro-based sanitizer. J. Food Prot. 69, 1571–1578.

Kumar, M., Hora, R., Kostrzynska, M., Warriner, K., 2007. Mode of *Salmonella* and *Escherichia coli* O157: H7 inactivation by a stabilized oxychloro-based sanitizer. J. Appl. Microbiol. 102, 1427–1436.

Lee, M.B., Greig, J.D., 2013. A review of nosocomial *Salmonella* outbreaks: infection control interventions found effective. Public Health 127, 199–206.

Lee, S.Y., Yun, K.M., Fellman, J., Kang, D.H., 2002. Inhibition of *Salmonella* Typhimurium and *Listeria monocytogenes* in Mung bean sprouts by chemical treatment. J. Food Prot. 65, 1088–1092.

Liao, C.H., 2008. Growth of *Salmonella* on sprouting alfalfa seeds as affected by the inoculum size, native microbial load and *Pseudomonas fluorescens* 2-79. Lett. Appl. Microbiol. 46, 232–236.

Liu, B., Schaffner, D.W., 2007. Mathematical modeling and assessment of microbial migration during the sprouting of alfalfa in trays in a nonuniformly contaminated seed batch using enterobacter aerogenes as a surrogate for *Salmonella* Stanley. J. Food Prot. 70, 2602–2605.

Ma, J., Ibekwe, A.M., Crowley, D.E., Yang, C.-H., 2012. Persistence of *Escherichia coli* O157:H7 in Major Leafy Green Producing Soils. Environ. Sci. Technol. 46, 12154–12161.

Maks, N., Fu, T.J., 2013. Evaluation of PCR detection of *Salmonella* in alfalfa sprouts and spent irrigation water collected during sprouting of naturally contaminated seed. J. Food Prot. 76, 314–317.

Mathusa, E.C., Yuhuan, C., Enache, E., Hontz, L., 2010. Non-O157 Shiga toxin-producing *Escherichia coli* in foods. J. Food Prot. 73, 1721–1736.

Matos, A., Garland, J.L., 2005. Effects of community versus single strain inoculants on the biocontrol of *Salmonella* and microbial community dynamics in alfalfa sprouts. J. Food Prot. 68, 40–48.

McEgan, R., Fu, T.J., Warriner, K., 2009. Concentration and Detection of *Salmonella* in Mung Bean Sprout Spent Irrigation Water by Use of Tangential Flow Filtration Coupled with an Amperometric Flowthrough Enzyme-Linked Immunosorbent Assay. J. Food Prot. 72, 591–600.

McEgan, R., Lee, S., Schumacher, B., Warriner, K., 2008. Composite versus single sampling of spent irrigation water to assess the microbiological status of sprouting mung bean beds. J. Sci. Food Agric. 88, 1549–1553.

Mellmann, A., Harmsen, D., Cummings, C.A., Zentz, E.B., Leopold, S.R., Rico, A., et al., 2011. Prospective Genomic Characterization of the German Enterohemorrhagic *Escherichia coli* O104:H4 Outbreak by Rapid Next Generation Sequencing Technology. Plos One 6, e22751.

Mohle-Boetani, J.C., Farrar, J., Bradley, P., Barak, J.D., Miller, M., Mandrell, R., et al., 2009. *Salmonella* infections associated with mung bean sprouts: epidemiological and environmental investigations. Epidemiol. Infect. 137, 357–366.

Monaghan, A., Byrne, B., Fanning, S., Sweeney, T., McDowell, D., Bolton, D.J., 2012. Serotypes and virulotypes of non-O157 Shiga-toxin producing *Escherichia coli* (STEC) on bovine hides and carcasses. Food Microbiol. 32 232–229.

Montville, R., Schaffner, D.W., 2004. Analysis of published sprout seed sanitization studies shows treatments are highly variable. J. Food Prot. 67, 758–765.

Muniesa, M., Hammerl, J.A., Hertwig, S., Appel, B., Bruessow, H., 2012. Shiga Toxin-Producing *Escherichia coli* O104:H4: a New Challenge for Microbiology. Appl. Environ. Microbiol. 78, 4065–4073.

Munoz, M., De Ancos, B., Sanchez-Moreno, C., Cano, M.P., 2006. Evaluation of chemical and physical (high-pressure and temperature) treatments to improve the safety of minimally processed mung bean sprouts during refrigerated storage. J. Food Prot. 69, 2395–2402.

National Advisory Committee on Microbiological Criteria of Foods NACMCF, 1999. Microbiological safety evaluations and recommendations on sprouted seeds. Int. J. Food Microbiol. 52, 123–153.

Neetoo, H., Chen, H., 2011. Individual and combined application of dry heat with high hydrostatic pressure to inactivate *Salmonella* and *Escherichia coli* O157:H7 on alfalfa seeds. Food Microbiol. 28, 119–127.

Neetoo, H., Chen, H.Q., 2010. Pre-soaking of seeds enhances pressure inactivation of *E. coli* O157:H7 and *Salmonella* spp. on crimson clover, red clover, radish and broccoli seeds. Int. J. Food Microbiol. 137, 274–280.

Neetoo, H., Ye, M., Chen, H., 2009. Factors affecting the efficacy of pressure inactivation of *Escherichia coli* O157:H7 on alfalfa seeds and seed viability. Int. J. Food Microbiol. 131, 218–223.

Nei, D., Latiful, B.M., Enomoto, K., Inatsu, Y., Kawamoto, S., 2011. Disinfection of Radish and Alfalfa Seeds Inoculated with *Escherichia coli* O157:H7 and *Salmonella* by a Gaseous Acetic Acid Treatment. Foodborne Pathog. Dis. 8, 1089–1094.

New South Wales Food Authority, 2008. Report on the microbiological quality of sprouts. Available at http://www.foodauthority.nsw.gov.au/_Documents/science/sprout_report.pdf (accessed 04.13).

Olaimat, A.N., Holley, R.A., 2012. Factors influencing the microbial safety of fresh produce: A review. Food Microbiol. 32, 1–19.

Oliveira, M., Vinas, I., Usall, J., Anguera, M., Abadias, M., 2012. Presence and survival of *Escherichia coli* O157:H7 on lettuce leaves and in soil treated with contaminated compost and irrigation water. Int. J. Food Microbiol. 156, 133–140.

Ontario Ministry of Agriculture and Food (OMAF), 2002. Sprouted seeds risk assessment. Available at http://www.omafra.gov.on.ca/english/food/inspection/fruitveg/risk_assessment_pdf/sproutedseeds/30ra.pdf (accessed 04.13).

Ontario Ministry of Agriculture and Food (OMAF), 2007. Sprouted seeds Good Manufacturing Practices guidebook. Available at http://www.sproutnet.com/pdfs/OMAFRA-Seed-Sampling.pdf (accessed 04.13).

Omahony, M., Cowden, J., Smyth, B., Lynch, D., Hall, M., Rowe, B., et al., 1990. An outbreak of *Salmonella* StPaul infection associated with bean sprouts. Epidemiol. Infect. 104, 229–235.

Pabalan, N., Singian, E., Jarjanazi, H., Steiner, T.S., 2013. Enteroaggregative *Escherichia coli* and acute diarrhea in children: a meta-analysis of South Asian populations. Eur. J. Clin. Microbiol. Infect. Dis. 32, 597–607.

Pao, S., Randolph, S.P., Westbrook, E.W., Shen, H., 2004. Use of bacteriophages to control *Salmonella* in experimentally contaminated sprout seeds. J. Food Sci. 69, M127–M130.

Pexara, A., Angelidis, D., Govaris, A., 2012. Shiga toxin-producing *Escherichia coli* (STEC) food-borne outbreaks. J. Hellenic. Vet. Med. Soc. 63, 45–53.

Pierre, P.M., Ryser, E.T., 2006. Inactivation of *Escherichia coli* O157: H7, *Salmonella typhimurium* DT104, and *Listeria monocytogenes* on inoculated alfalfa seeds with a fatty acid-based sanitizer. J. Food Prot. 69, 582–590.

Rajkowski, K.T., Thayer, D.W., 2000. Reduction of *Salmonella* spp. and strains of *Escherichia coli* O157: H7 by gamma radiation of inoculated sprouts. J. Food Prot. 63, 871–875.

Rasko, D.A., Webster, D.R., Sahl, J.W., Bashir, A., Boisen, N., Scheutz, F., et al., 2011. Origins of the *E. coli* Strain Causing an Outbreak of Hemolytic-Uremic Syndrome in Germany. N. Engl. J. Med. 365, 709–717.

Saegusa, A., 1998. U.S. and Japanese scientists in dispute over 'poisoned' radishes. Nature 392, 642–642.

Scharlach, M., Diercke, M., Dreesman, J., Jahn, N., Krieck, M., Beyrer, K., et al., 2013. Epidemiological analysis of a cluster within the outbreak of Shiga toxin-producing *Escherichia coli* serotype O104:H4 in Northern Germany, 2011. Int. J. Hyg. Environ. Health 216, 341–345.

Snary, E.L., Kelly, L.A., Davison, H.C., Teale, C.J., Wooldridge, M., 2004. Antimicrobial resistance: a microbial risk assessment perspective. J. Antimicrob. Chemother. 53, 906–917.

Solomon, E.B., Matthews, K.R., 2006. Interaction of live and dead *Escherichia coli* O157: H7 and fluorescent microspheres with lettuce tissue suggests bacterial processes do not mediate adherence. Lett. Appl. Microbiol. 42, 88–93.

Soon, J.M., Seaman, P., Baines, R.N., 2013. *Escherichia coli* O104:H4 outbreak from sprouted seeds. Int. J. Hyg. Environ. Health 216, 346–354.

Stewart, D.S., Reineke, K.F., Ulaszek, J.M., Tortorello, M.L., 2001. Growth of *Salmonella* during sprouting of alfalfa seeds associated with salmonellosis outbreaks. J. Food Prot. 64, 618–622.

Taormina, P.J., Beuchat, L.R., Slutsker, L., 1999. Infections associated with eating seed sprouts: An international concern. Emerg. Infect. Dis. 5, 626–634.

Thomas, J.L., Palumbo, M.S., Farrar, J.A., Farver, T.B., Cliver, D.O., 2003. Industry practices and compliance with U.S. food and drug administration guidelines among California sprout firms. J. Food Prot. 66, 1253–1259.

Waje, C.K., Jun, S.Y., Lee, Y.X., Kim, B.N., Han, D.H., Jo, C., et al., 2009. Microbial quality assessment and pathogen inactivation by electron beam and gamma irradiation of commercial seed sprouts. Food Control 20, 200–204.

Warriner, K., 2011. IUFoST Scientific Information Bulletin. Shiga toxin producing *Escherichia coli*: Germany 2011 *Escherichia coli* O104:H4 outbreak linked to sprouted seeds. IUFoST Scientific Information Bulletin. Shiga toxin producing *Escherichia coli*: Germany 2011 *Escherichia coli* O104:H4 outbreak linked to sprouted seeds. Available at http://iufost.org/sites/default/files/docs/IUF.SIB.E_coli.pdf (accessed 04.13).

Warriner, K., Spaniolas, S., Dickinson, M., Wright, C., Waites, W.M., 2003. Internalization of bioluminescent *Escherichia coli* and *Salmonella* Montevideo in growing bean sprouts. J. Appl. Microbiol. 95, 719–727.

Werssinger, W.R., Beuchat, L.R., 2000. Comparison of chemical treatments to eliminate *Salmonella* on alfalfa seeds. J. Food Prot. 63, 1475–1482.

Wilderdyke, M.R., Smith, D.A., Brashears, M.M., 2004. Isolation, identification, and selection of lactic acid bacteria from alfalfa sprouts for competitive inhibition of foodborne pathogens. J. Food Prot. 67, 947–951.

Yang, Y., Meier, F., Lo, J.A., Yuan, W., Sze, V.L.P., Chung, H.-J., et al., 2013. Overview of Recent Events in the Microbiological Safety of Sprouts and New Intervention Technologies. Compr. Rev. Food Sci. Food Saf. 12, 265–280.

Ye, J., Kostrzynska, M., Dunfield, K., Warriner, K., 2009. Evaluation of a Biocontrol Preparation Consisting of *Enterobacter asburiae* JX1 and a Lytic Bacteriophage Cocktail To Suppress the Growth of *Salmonella* Javiana Associated with Tomatoes. J. Food Prot. 72, 2284–2292.

Yetim, H., Tornuk, F., Ozturk, I., Sagdic, O., 2010. Microbial safety of edible seed sprouts. Akademik Gida 8, 18–23.

Yishan, Y., Meier, F., Lo, J.A., Wenqian, Y., Sze, V.L.P., Hyun-Jung, C., et al., 2013. Overview of recent events in the microbiological safety of sprouts and new intervention technologies. Compr. Rev. Food Sci. Food Saf. 12, 265–280.

Salmonella and Tomatoes 12

Jerry A. Bartz, Massimiliano Marvasi, Max Teplitski

Department of Plant Pathology, University of Florida, Gainesville, FL

CHAPTER OUTLINE

Introduction

Fresh tomatoes or their fresh-cut products have been linked to up to 19 outbreaks of salmonellosis since 1990. According to a search of http://wwwn.cdc.gov/foodborn eoutbreaks/Default.aspx., four outbreaks have been linked to tomatoes since 2008. Projected sources of contaminated tomato fruit associated with these outbreaks have varied. To date, there has been only one report of isolation of this pathogen from field-grown tomatoes and that was by personal communication (Barak and Schroeder, 2012). By contrast, serovars of *Salmonella enterica* have been isolated from wetlands and ditches beside or ponds near or in tomato fields (Green et al., 2008) as well as on many other types of crop plants (Barak and Schroeder, 2012). Surveys by the Federal Drug Administration (FDA) and U.S. Department of Agriculture (USDA) involving 2924 samples of tomatoes failed to detect *Salmonella* (Gorny, 2006). In a survey of supermarkets in two New Jersey counties between 1992 and 1995, *Salmonella* was detected in 10% of various types of tomatoes (Wells and Butterfield, 1997).

The Produce Contamination Problem. http://dx.doi.org/10.1016/B978-0-12-404611-5.00012-9

Fruit showing soft-rotted tissues were more likely to be positive. Orozco, et al. (2008) reported that 2.8% of tomatoes sampled from a protected agriculture farm in Mexico were contaminated with *Salmonella*. One fresh-market tomato production area has been linked with several different outbreaks, and outbreak strains have been isolated from ponds in that region. Over the period of July 2 through October 30, 2002, serotype Newport was isolated from 512 patients in 22 states (Greene et al., 2008). Standard round tomatoes packed on the eastern shore of Virginia (Delmarva Peninsula) were linked to that outbreak. Tomatoes from this same area were once again associated with an outbreak in 2005 that occurred over the period of July 7 through September 24. A third outbreak attributed to this production area, occurred between June 5 and October 20, 2006 and a fourth between June 13 and September 10, 2007. Each outbreak was caused by serotype Newport, and all strains had the same pulsed field gel electrophoresis (PFGE) pattern. This outbreak strain was recovered from an irrigation pond on one of the suspect farms. Water from a second contaminated pond in the area was used by its owner to mix pesticides, which is a clear violation of Good Agricultural Practices (DACS, 2008). A more recent survey of "mid-Atlantic" tomato farms detected numerous serotypes in soil and surface waters but not on leaves or fruit sampled from fields (Micallef et al., 2012).

The first two outbreaks of non-typhoidal salmonellosis during the period 1990 to 2008 were also linked to a single production area; this time in South Carolina (Hedberg, et al., 1999). Packer "A" was common to both outbreaks, although the first was caused by serotype Javiana and the second, Montevideo. Both occurred over the same multi-state area (Illinois, Wisconsin, Minnesota, and Michigan) over nearly identical calendar days (June 28 through August 2 and June 29 through August 2, in 1990 and 1993, respectively). The first involved both restaurant meals and retail grocers, whereas the second was restaurant meals alone. The marketing chain for the 1993 outbreak was investigated and reported in a memorandum by Dr. F. J. Angulo, the Epidemic Intelligence Service (EIS) officer responsible for the Centers for Disease Control (CDC) investigation. Typical fresh-market tomato marketing chains include several different handling steps between a production field and a consumers' fork, where contaminants could be introduced onto or proliferate in fruit. These include but are not limited to:

1. Field operations in producing crop. Workers pruning plants (removing side shoots) and/or tying them to stakes in the field can spread bacterial contaminants, both plant pathogenic as well as other types, particularly if plants are wet with rainfall, dew, or guttation.
2. Manual harvest of crop. Fields are picked up to three separate times with 7 to 10 days between each harvest. Crews may or may not wear gloves. Once again, growers are admonished to not allow harvest operations if plants are wet because of a risk of spreading contamination and of damaging fruit, which are more tender when congested with water (Bartz et al., 2012).
3. Transit to packinghouse. Flatbed trucks or gondolas are used to move freshly harvested tomatoes from field to packinghouse. Unimproved roads in fields connecting to normal roads can lead to injuries to fruit associated with transit vibration (literally bumps in the roads).

4. Packing fruit. At the packinghouse fruit are emptied from harvest bins or gondolas (4 to 18 h after harvest), usually into water. Fruit are subsequently washed, dried, waxed, sorted, sized, and packed in fiberboard boxes. Unloading procedures involve physical contact among fruit, water, and packingline surfaces. This contact can lead to dispersal of contamination, injury, and movement of water into fruit. Fruit dropped into water can be infiltrated due to hydrostatic forces (impact and submersion) or to temperature change-induced internal vacuums. Sanitizers in the water can reduce dispersal and will prevent most internal contamination due to water infiltration (Clement et al., 2000), but cannot completely sanitize the surface of fruit. For example, Felkey (2002) treated inoculated fruit in a simulated flume with 150 ppm free chlorine for up to 120 sec at pH 6.5 and 35°C. Populations of *Salmonella* spp. recovered from inoculated fruit were 6.4 logs lower when inoculum was placed on smooth, intact surfaces as compared with stem scars or wounds, where only an average of 1.0- and 0.7-log reduction was observed, respectively.

5. Ripening fruit. Green fruit are exposed to ethylene gas at 20°C for 6 to 8 days to induce uniform ripening and may later be stored (up to 8 days at 15.5°C).

6. Marketing fruit. "Gas-ripened fruit" may be shipped directly to supermarket warehouses or to "repack operations" where boxes are emptied onto belts for color sorting either by photo-sensors or by workers. The goal of these operations is to deliver a uniformly ripened product to end-users. Fruit at the desired stage of ripeness are packed into boxes and sold to a distributer who then sells them to individual markets. Fruit that have not ripened to the desired stage are repacked and then stored for an additional period of time. Fruit with sub-standard color or other defects may be sold from facility to "drive-up buyers."

7. Fruit in markets. Individual markets served by distributors include supermarkets, restaurants, as well as fresh-cut processors where fruit are sliced or diced depending on end-market. Total time between harvest and end-use ranges up to 4 weeks. Best quality fruit are those that become table ripe and are consumed within 10 to 14 days of harvest (Sargent et al., 2005; VanSickle, 2008).

The marketing chain for the 1993 outbreak, attributed to South Carolina tomatoes, includes the following: On June 21, a single semi-truck hauled 36,000 lbs. of tomatoes from Packer "A" to a Chicago repack operation, and a second semi took a similar load to Company "B" in Michigan. These two loads represented 5% of the tomatoes shipped from Packer "A" on that day and would represent a single harvest from about 4 acres, based on average tomato yields (Maynard et al., 2000) and packouts (packout = percentage of tomatoes entering packinghouse that are packed); the rest are discarded or picked out by small vendors. No evidence was presented that fruit handled by Company "B" in Michigan or the other 95% shipped by Packer "A" to various receivers were responsible for illnesses. However, one report noted that certain distributors linked with the outbreak received tomatoes directly from Packer "A" and not through the Chicago repack operation (Hedberg et al., 1999). A sanitation failure was believed responsible for the earlier 1990 outbreak, although there was no evidence of excessive postharvest decays, which often result from sanitation failures (Sargent et al., 2005). As noted above, such decays have been linked with a high incidence of *Salmonella* (Wells and Butterfield, 1997). The implicated strains were not found in or around Packer "A's" facilities in either 1990 or 1993.

Multiple outbreaks have also been linked to Florida-grown tomatoes, although certain aspects of these linkages are unclear. Most importantly, outbreak strains have not been detected in fields or packinghouses in Florida. Certain linkages of outbreaks to Florida tomatoes appear to have been inconsistent with fresh market tomato production patterns. For example, during the late fall/early winter of 1998 (outbreak ended February 2, 1999), 86 illnesses were linked to tomatoes produced and packed either in South Florida or in West Central Florida (Cummings et al., 2001). Three packinghouses were investigated. Neither the outbreak strain nor any other strain of *Salmonella* were detected. Up to eight farms supplying these three packinghouses were surveyed. None of the environmental samples contained *Salmonella*. Illnesses were not reported in Florida, whereas California had the most reported cases (44) appearing over the longest duration (46 days). Arizona reported 13 cases over a 43-day period. These two states are just north of a large winter crop, fresh market tomato production area in Mexico (Boyette et al., 2011).

In 2002, 141 cases of salmonellosis caused by serotype Javiana developed among summer visitors to a theme park near Orlando, Florida in what became known as the "Transplant Games Outbreak" (Srikantiah et al., 2005). Traceback led to diced Roma tomatoes supplied by Plant "X." Frozen samples of that product were found to contain 150 to 1000 cfu fecal coliforms/g, but *Salmonella* was not detected. Plant "X" purchased Roma tomatoes from a wholesaler, who bought most of his fruit from a single Florida grower/packer. It was not clear from the outbreak report if Plant "X" only serviced the amusement park or had other customers.

Fresh-cut processors such as Plant "X" have been prominent in the marketing chain of other outbreaks. For example, a multiple serotype, convenience store outbreak discussed below involved a fresh-cut, sliced-tomato operation where Roma tomatoes at 45 to 65°F were placed in a water bath at 33 to 35°F to firm them prior to the cutting operation. A culture-confirmed outbreak strain of serovar Anatum matching that isolated from four human cases was isolated from a package of sliced tomatoes. In 2005, an outbreak caused by serovar Braenderup was attributed to diced tomatoes produced by a processor in Kentucky. That processor was chilling fruit in a 32 to 35°F water bath before dicing them. Cooling submerged tomatoes is known to lead to an infiltration of stem-scar tissues with water (Bartz and Showalter, 1981; Vigneault et al., 2000). As such, if contamination were present on a few tomatoes it would likely be dispersed to many.

The first reported multiple serotype outbreak sickened 429 patients, who had consumed sandwiches at a chain of convenience stores during the summer (June 29 to July 12) of 2004 (CDC, 2005). Serovar Javiana was detected most often with 429 cases followed by sv. Typhimurium (27), Group D untypable (6), sv. Anatum (5), sv. Thompson (4), and sv. Muenchen (4). Fruit responsible for these cases could not be traced back to a single field or grower although two packinghouses and a field-pack operation in Florida were implicated. Tomatoes processed in those pack-inghouses were hauled from farms in Florida, Georgia, and South Carolina. Environmental samples around fields were positive for *Salmonella* but outbreak strains were not detected. Of note is that harvests from Florida production fields dwindle during the end of June as Georgia, South Carolina, and the eastern shore of Virginia begin their peak production periods (USDA-ERS, 2013). This rather tangled list of potential sources illustrates the difficulty in tracing back contamination to a source. Often, fresh-market tomato operations have packinghouses and fields in different states and sometimes in different countries. Harvest dates for fields in different states can overlap. For example, a 2004 sv. Braenderup-outbreak (June 15 to July 21) was traced to a grower/packer in South Florida that was receiving fruit from a farm in North Central Florida as well as from a farm in Coastal South Carolina. The latter farm was implicated in an outbreak in Canada caused by sv. Javiana over the same time period. Fruit from the two farms were being packed using the same equipment over the same 2-day period, just prior to the sv. Braenderup outbreak.

What is known about outbreaks and tomato production

The supermarket survey by Wells and Butterfield (1997) and the isolation of *Salmonella* spp. from fruit harvested from plants grown in a greenhouse (Orozco et al., 2008) suggest that consumers of tomatoes and other fresh produce are frequently exposed to this bacterium. Fortunately, illnesses are not common. A large minimum infective dose may be one reason for the lack of correlation between exposure and illness. One estimate for a minimum infectious dose was log 9.0 cells for 50% of a population of healthy individuals (Todar, 2005). However, for those with compromised immune systems, 15 to 20 cells could initiate an infection (FDA, 2003). Yet, outbreak investigation reports have not linked illness with immunity incompetent individuals. This implies that large populations of *Salmonella* are developing on tomatoes involved in outbreaks.

A problem linking field contamination with multistate outbreaks involves the volumes of fruit produced and handled as compared with outbreak dynamics. Maynard et al. (2000) noted that during a fall season, an average of 32,500 to 65,000 pounds of fruit were harvested per acre. The first of three harvests totaled 3450 to 23,300 pounds, depending on cultivar. If these fruit averaged 200 g in weight, the total number of fruit would equal 146,250 per acre. If packout averaged 90% and defects including decay among packed fruit averaged 5%, then over 124,000 fruits/acre were marketed. For a 100-acre field, this number would increase by 2-log units. With such numbers, widespread contamination in a field does not appear likely since thousands of tomato consumers do not become ill, and the pathogen is rarely detected in marketed tomatoes. The relatively few cases versus the number of tomatoes harvested suggests one of the following: 1) contamination by *Salmonella* is a rare event, 2) populations contaminating fruit in the field are unstable, or 3) postharvest populations mostly fail to proliferate to infectious levels. Plant-to-plant dispersal of *Salmonella* within a field would be uncommon because populations achieved in leaves or other canopy structures are far smaller than those produced by competent plant pathogens such as *Xanthomonas vesicatoria*, causal agent of bacterial spot, where lesions featuring ruptured plant surfaces or ooze of bacteria from natural openings enable efficient dispersal of bacteria by rainfall, wind, or mechanical means (Barak and Schroeder, 2012). Based on standard definitions, the introduction of *Salmonella* onto or into tomato plants or fruit would not qualify as inoculation, and subsequent proliferation of that bacterium would not qualify as infection, quite simply because *Salmonella* has not been proven to be parasitic on tomatoes.

An amplification of contamination, particularly surface contamination of just a few fruit, could easily result from handling operations, whether in field, greenhouse, or marketing systems. Fruit are harvested at 80 to 100% of full size and green color to fully ripe (Salveit, 2005). Harvest crews are usually paid by volume harvested and have little incentive to discard defective fruit. Fruit dropped on soil or bed surfaces are not supposed to be salvaged (DACS, 2008). Fruit with advanced decays would either be ignored or tossed to aisles between beds. The harvest crew effectively handles each individual fruit, and whether members are wearing gloves or not, contamination on one fruit or the foliage of one plant could be dispersed to many additional fruit, particularly if the crop is wet with rainfall, dew, or guttation.

How *Salmonella* contaminates tomato fruit

Salmonella is not known to survive well on surfaces of tomato fruit unless introduced as large populations or in a matrix that shields cells from external stresses (Wei et al., 1995). *Salmonella* seems particularly vulnerable to the UV from sunlight (Abulreesh, 2012). This may explain why there's only one report of detecting it on tomatoes growing in fields despite its presence in irrigation water, ponds, ditches, and surrounding fields (Barak and Schroeder, 2012). Based on the discovery of various bacteria within mature fruit (Samish and Etinger–Tulczynska, 1963), one could predict that strong storms would move *Salmonella* from nearby ponds and soils into fruit on plants in the field, just as such storms have been responsible for movement of various plant pathogenic bacteria into potential infection courts miles from a source (Gottwald et al., 1997). *Salmonella* has been found on tomatoes growing in protected agriculture (Orozco et al., 2008), most likely because greenhouse coverings largely filter out shorter UV wavelengths (UV-B), which would otherwise kill bacteria on the fruit surface (Sundlin, 2002). In field tomato production, exposure of fresh contaminants on fruit surfaces to UV as well as to desiccation may be reduced during harvest. Harvested fruit are quickly emptied into field bins or gondolas where an upper layer of fruit would not only shade those residing below, but also contribute to an overall high humidity within the fruit mass. Scuffs and punctures in fruit surfaces would provide surface contaminants access to moisture and metabolites from damaged epidermal cells as well as the fruit's apoplast. These surface injuries are of variable sizes, but have profound implications on harvests when just a few fruit are contaminated prior to harvest. Punctures that are the size of a grain of sand are unlikely to be detected and hence culled from the harvest at the packinghouse. Yet a wound that breaks through the cuticle is sufficient to enable development of bacterial soft rot (caused by a strict wound invader) as well as allow internalization of bacteria, such as *Salmonella* (Johnson, 1947).

Several lines of evidence suggest that stem depressions or stem scars are a likely area in which *Salmonella* cells could enter the apoplast of a tomato fruit, and once in the apoplast, they would be shielded from exposure to external stresses such as UV or dessication. Samish and Etinger–Tulcyzynska (1963) established that the highest population of bacteria in a tomato fruit was in the connective tissue at the stem end and decreased towards peripheral tissues such as the pericarp tissues and blossom scar. To elucidate how bacteria might enter that tissue prior to harvest, their research team used cultures of *Serratia marcesens*, a red-pigmented bacterium, and one which they had not found previously inside tomatoes. Cultures smeared on surfaces of young fruit could not be recovered later from internal sections. Cultures applied to wounds on young fruit remained limited to the wound. However, those placed on the outer surfaces of sepals on young, well-shaded fruit were subsequently detected in 22 of 40 fruit that were sampled between 6 and 40 days after inoculation. Additionally, tomatoes harvested from three farms using overhead irrigation consistently contained more bacteria than those from three farms that used furrow irrigation. Gas exchange for harvested tomato tissues occurs mainly through the stem depression (Brooks, 1937; Saltveit, 2005). Waxiness develops over fruit surfaces as they approach the mature green stage of ripeness. Corking around and partially beneath the pedicel would also develop as fruits mature. Waxy cuticles preclude direct gas exchange, whereas corking around attachment of pedicels to fruit, together with expansion of fruit leading to cracking of the cork, would open portals for gas exchange. If air is injected into tomato fruit with attached stems that are submerged in water, air bubbles will flow out from under the sepals (Bartz, unpublished observations). Bacteria can be dispersed miles by a strong storm (Gottwald et al., 1997). Together these observations provide strong evidence that *Salmonella* could be introduced under sepals by contaminated rain splash, overhead irrigation, or pesticide application. Once at the base of sepals or in corky areas adjacent to sepals, *Salmonella* would be protected from UV and could internalize via the process utilized by the naturally occurring endophytic microbes discovered by Samish and Etinger–Tulcyzynska (1963). Other ways *Salmonella* can internalize in tomato plants have been reviewed recently (Erickson, 2012). These pathways have been derived from laboratory, growth chamber, or greenhouse models. Validation of these models by showing similar movement by common epiphytes or other members of the microbial ecosystem on tomatoes has not been done.

Once *Salmonella* has become internalized, it cannot be removed. Dispersal of contaminants among fruit during harvest is especially acute if plants are wet. Wetness on fruit surfaces will likely persist until fruit are washed and processed at the packinghouse. Not only does wetness preclude desiccation of fruit surfaces but it also protects surface microbes and enables them to produce biofilms (Zottola, 1994; Mandrell et al., 2006; Morris et al., 2002). Free-water connections called "water channels"

(Johnson, 1947) may become established in wounds. These "water channels" enable particulate matter including bacteria on the surface to become internalized. Fruit temperatures at harvest are often ideal for proliferation of various bacteria, and those temperatures will persist until fruit have been cooled, which usually doesn't happen until after washing and packing.

Cut surfaces on fresh-cut tomato products are particularly vulnerable to internalized contamination. Surfaces of cut tomatoes feature an internal apoplast directly linked with its external environment by fluid channels. Fluids on and in surface apertures would contain contents of damaged cells which supports proliferation of bacteria. Therefore, fruit intended for fresh-cut processing must be completely free of *Salmonella* and must be handled according to time-temperature rules (CFSAN 2007).

Sources of *Salmonella* and other human pathogens in crop production environments

Most strains of *Salmonella* and pathogenic *E. coli* (EHEC) that infect domestic or wild animals are pathogenic to humans, although specificity for animal type exists among serovars (Albulreesh, 2012). Prevalence of these human pathogens in wild and domestic animals, and survival of such pathogens during movement between hosts and from host to sites where contact with humans is likely, are of extreme importance to designing an effective response. *Salmonella* and EHEC have been defined as "animal-associated" bacteria due to low detection rates (< 1%) of these pathogens in field samples of crop plants as well as small populations when detected. However, on average, only ~4% of wild animals test positive for *Salmonella*, with a great number of them (including certain birds) being free of either *Salmonella* or EHEC (Gorski et al., 2011; Wahlstrom et al., 2003). Abulreesh (2012) argued that there was "strong evidence to suggest that the organism is ubiquitous and widely distributed in the environment." There does not appear to be a consensus on which species of wildlife are consistent carriers of *Salmonella*: sparrows, towhee, and crows, as well as feral pigs, coyotes, deer, elk, opossum, and skunk, all have tested positive, while mice, rabbits, raccoons, squirrels, blackbirds, geese, mallard ducks, and starlings were negative for *Salmonella* (Gorski et al., 2011). Healthy, free-living birds appear to acquire *Salmonella* when exposed to contaminated environments such as landfills, manure applications, etc. (Abulreesh, 2012). Two animals commonly found in the wild and in crop production areas (feral pigs and white-tail deer) tend to test positive for these pathogens. Of the feral pigs tested in two surveys, 13 to 23% of them tested positive for *Salmonella* and/or pathogenic *E. coli* (Jay et al., 2007). Between 0.5% and 7% of white-tail deer tested positive for *Salmonella* and/or pathogenic *E. coli* in similarly conducted studies (Sargeant et al., 1999; Mandrell, 2009). Forty percent of all *Salmonella* serotypes have been predominantly cultured from reptiles, including small reptiles such as geckos (Mermin et al., 2004). In reptiles, *Salmonella* is generally considered to be a part of the normal flora of the gastrointestinal tract, and less frequently, as a cause of salmonellosis (Smith, 2012). Rodents also can play a role in the transmission of *Salmonella* contamination on farms (Umali et al., 2012).

Domestic animals raised in confinement tend to test positive for these pathogens more consistently: in three different surveys, 68%, 13 to 72%, and 100% of cattle were found to harbor either *E. coli* O157:H7 or *Salmonella* (Mandrell, 2009; Stephens et al., 2007). By contrast, in a recent survey of leafy green production areas in California only one in >700 samples from free-range cattle tested positive for these pathogens (Gorski et al., 2011).

Salmonella in surface waters in U.S. vegetable-producing regions

Water has the potential to contaminate fresh produce at various points in the crop production/marketing cycle. Therefore, microbiological quality of water at all points in this cycle must be ensured. Surveys of water quality in major vegetable-production regions suggest that well water generally is free of human pathogens (Jay et al., 2007). For example, only ~3% of surface water samples tested in California were positive for pathogenic *E. coli,* and over 7% were positive for *Salmonella* (Gorski et al., 2011). By contrast, in the Southeastern U.S. (North Carolina, Georgia, Florida), detection of *Salmonella* from surface water samples is nearly an order of magnitude higher than in California (Rajabi et al., 2011 and references therein). In either production area, use of untreated surface waters for crop irrigation, spray applications, or produce washes may pose a significant food safety risk.

Factors that influence the transfer of *Salmonella* from water to the crops are not well understood, especially under natural conditions where populations are relatively small. Parsley irrigated with water containing as little as ~300 CFU/ml resulted in the persistence of *S.* Typhimurium on the plants (Kisluk and Yaron, 2012; Brandl et al., 2013). The same study also showed that contamination of parsley from irrigation water during the winter season (when compared with other seasons) led to a higher level of *Salmonella* persistence in the parsley phyllosphere. Cooler temperatures appear to be more favorable for survival of *Salmonella* in aqueous environments (Albulreesh, 2012).

An additional factor may be related to seasonal changes in the quantity and diversity of phyllosphere microbial communities (Kisluk and Yaron, 2012). In general, non-typhoidal salmonellosis outbreaks are more common in warm than cool seasons. *Salmonella* populations recovered from surface water often appear to be related to storm runoff (Albulreesh, 2012). Higher populations were found in surface waters during rainy seasons and in fresh as compared with salt water. Additionally, *Salmonella* was more frequently detected in sediment than in the water column above sediment. Attachment to particulates provides protection against UV as well as predators (Abulreesh, 2012). Perhaps for the same reasons, *Salmonella* survives better in terrestrial environments as compared with aquatic ones. For example, *Salmonella* Typhimurium was found to be metabolically active in soil ranging from 54 and 231 days, depending on the soil characteristics and environmental conditions (Islam, 2004). Increases of *Salmonella* populations in sewage sludge have been recorded during low temperatures (Abulreesh, 2012). Irrigation of crops with waste water has been linked with a high incidence of contamination of vegetables. Patterns of human and animal behavior in warmer months (seasonal changes in diets, use of recreational waters, etc.) may also contribute to an increased exposure of individuals to the environmental reservoirs of the pathogens.

Salmonella ecology and its implications for produce safety

A link between survival and proliferation of *Salmonella* and pathogenic *E. coli* on plants under laboratory conditions was recently suggested to be part these pathogens' life cycles. In a three-step food-chain experiment Semenov et al. (2010) found that *Salmonella* and *E. coli* O157:H7 could move from seedlings that were sown into manure-amended soil to cows, mice, and snails that ate these seedlings. These animals became infected/colonized and then shed those pathogens back into the environment as feces. Rainfall and human activities would be important to the scope of this cycle since domestic livestock shun grazing around their own deposits of feces (Smith et al., 2009). Rainfall could disperse contaminants, and rainfall runoff could carry contaminants to nearby streams. Humans could apply manure slurries to fields used in the production of animal feed.

Surface waters have also been associated with movement of *Salmonella* and other pathogens from environment to new hosts. Passage through plants and/or survival in soil did not appear to affect the organisms' pathogenicity. Schikora et al. (2011) demonstrated that *Salmonella* Typhimurium inoculated into and recovered from *Arabidopsis* leaf homogenates was as virulent as the inoculum grown in a nutrient-rich culture medium. By contrast, in certain tests *Salmonella,* which had fallen into a viable but nonculturable state, were no longer pathogenic whether resuscitated or not (Abulreesh, 2012). These studies suggest that environmental persistence of *Salmonella* and other enteric pathogens is an important (perhaps, co-evolved) component of their life cycle. Mobile elements (especially those involved in virulence and stress responses) in the *Salmonella* genomes provide the necessary plasticity for the bacterium to easily adapt to new environments. The environment and available nutrition within the apoplastic space of a green or ripe tomato fruit could require an abrupt change in the metabolism of *Salmonella*. An abundance of mobile elements encoding pathogenic properties may facilitate the emergence of strains with novel combinations of pathogenic traits (Switt et al., 2012). Further studies will be required to clarify how these mobile elements support the persistence or proliferation of *Salmonella* outside of its animal hosts.

Under laboratory conditions, *Salmonella* is capable of forming colonies at various plant locations including wet leaves, roots, germinating seeds, or flowers (Brandl and Mandrell, 2002; Cooley et al., 2003; Guo et al., 2001). How these laboratory models relate to contamination of field or greenhouse tomato production is not clear. Microbes that rarely multiply on or in plants in the field are called casual (Leben, 1961). Such microbes usually succumb to desiccation, UV, and a lack of nutrition (Morris et al., 2002). However, under certain environments, particularly after harvest and handling, casual microbes can multiply in their hosts (Bartz, 2006). Water congestion of storage tissues appears particularly egregious for enabling bacterial proliferation. Young (1974) observed extensive population increases of various bacteria when intercellular spaces of plant tissues contained free water. Cell-membrane damage along with mineral and metabolite leakage appears to accompany tissue water congestion (King and Bolin, 1989). Most of the above pertains to internalized microorganisms. Those located on phyllosphere surfaces would be exposed to UV, starvation, and marginal nutrient availability (Hirano and Upper, 2000; Lindow and Brandl, 2003). In laboratory or growth chamber tests, *Salmonella* can multiply on the phyllosphere and form microcolonies on leaves, although not as robustly as bacterial species typically considered to be successful plant epiphytes (Brandl and Mandrell, 2002). Human pathogens on leaves preferentially move towards stomata and colonize the vein areas, the bases of trichomes, and lesions (resulting from disease or nutritional disorders) (Brandl and Mandrell, 2002; Kroupitski et al., 2009a; Barak et al., 2011; Brandl and Amundson, 2008; Brandl, 2008; Kroupitski et al., 2011; Kroupitski et al., 2009b; Aruscavage et al., 2008). These sites may provide shelter from environmental stresses or offer increased nutrient and water availability. However, such sites are also attractive to plant-associated microbes, and enteric pathogens must interact with the indigenous microbiota, which could lead to outcomes that are both positive and negative for survival of *Salmonella* (Teplitski et al., 2011).

Interactions between *Salmonella* and tomatoes: molecular insights

Under certain conditions *Salmonella,* enterovirulent *E. coli,* and other enteric microorganisms form colonies in plants, persist, and multiply for extended periods of time. Designing better sanitation programs is not likely to resolve this issue if the microbe is inside plant tissues. A better understanding of the molecular mechanisms of interactions between human enteric pathogens and plants will likely lead to approaches that target and disrupt specific undesirable behaviors in these pathogens, thereby increasing their vulnerability.

High throughput screens and transcriptomic analyses have identified dozens of genes involved in the ability of *Salmonella* and *E. coli* to attach to plant surfaces and utilize nutrients found in plant tissues (Barak et al., 2005; Noel et al., 2010a; Kroupitski et al., 2013; Kyle et al., 2010). Collectively, results of these studies suggest that formation of colonies on surfaces and tissues of tomato vegetative and reproductive organs may depend on specific *Salmonella* genes and may be a function of the plant genotype and the physiological state of the plant. Interestingly, horizontally acquired virulence genes located on the *Salmonella* pathogenicity islands (SPIs) do not appear to play a role in persistence within mature tomato fruits (Noel et al., 2010a). However, these virulence genes seem to have different roles during interactions with different plant species: in tomatoes, SPI mutants were as fit (able to proliferate) as the wild type (Noel et al., 2010a), whereas in alfalfa and lettuce, SPI mutants have phenotypes that are distinct from those of the wild-type strain (Schikora et al.,

2011; Dong et al., 2003; Iniguez et al., 2005). In *Arabidopsis,* for example, *Salmonella* virulence SspH1, SspH2 proteins may interact with pathogenicity related (PR) proteins produced by the plant in response to microbial activities, and both SPI-1- and SPI-2-encoded type III secretion systems are needed for successful proliferation in *Arabidopsis* (Schleker, 2012; Shikora, et al., 2011). When pathogenic *E. coli* cells were exposed to lettuce leaf exudates, the pathogen's horizontally acquired LEE virulence genes were up-regulated (Kyle et al., 2010). Collectively, these observations suggest that enteric pathogens' horizontally acquired virulence genes may have functions in microbial interactions with plants, but the differences in the phenotypes of the virulence mutants observed could be due to molecular strategies with which enteric pathogens interact with plant vegetative and reproductive organs involved in these studies.

When a collection of *Salmonella* plasmids carrying the promoter-gfp constructs were inoculated into tomatoes, ~50 unique *Salmonella* genes were up- or down-regulated within red ripe tomatoes (Noel et al., 2010a). While none of the corresponding single mutants were significantly reduced in competitive fitness within these fruit, a combination of five different single mutants within the same strain led to a reduction in ability of the bacterium to persist by approximately five-fold (Noel et al., 2010a). This is an important overall observation as it indicates that the search for chemicals capable of disrupting undesirable behaviors in *Salmonella* may need to focus on higher-level regulatory cascades (e.g., those suggested by the studies of Marvasi et al., 2013), or that a combination of multiple approaches will have to be implemented to successfully disrupt persistence of *Salmonella* and other enteric pathogens in the crop production environment.

The maturity stage of the tomato fruit affects *Salmonella* proliferation and its gene expressions

Even though *Salmonella* is able to proliferate in tomatoes at any ripening stage, mature tomatoes are more conducive to *Salmonella* growth when compared with immature tomatoes (Marvasi et al., 2013; Noel et al., 2010a). Red tomatoes have a lower pH (4.4) and more metabolites in their apoplastic fluid than green ones (pH 6.7) (Almeida and Huber, 1999). The pH decrease is related to a leakage of organic acids through cell membranes and exposure of carboxyl groups from the hydrolysis of pectin (Sakurai, 1998). In response to these changes in the apoplastic space, *Salmonella* genes may be regulated differently according to the maturity stage of the fruit. For example, expression of *Salmonella* Typhimurium genes involved in O-antigen capsule production has been shown to be activated mainly at the immature as compared with mature fruit tissues (Marvasi et al., 2013; Noel et al., 2010a). The ecological role of the O-antigen during proliferation in green tomatoes has to be clarified; it may be involved in response to the low levels of nutrients available, or it could involve an avoidance or protection of *Salmonella* from potential tomato defense responses. Changes in apoplastic pH associated with fruit ripening may contribute to differential roles of the O-antigen capsule in mature or immature tomatoes.

The role of plant genotype in interactions with *Salmonella*

Several research groups demonstrated that the outcomes of a crop's interactions with *Salmonella* and *E. coli* is affected by plant genotype (Klerks et al., 2007; Iniguez et al., 2005; Quilliam et al., 2012; Barak et al., 2008; Shi et al., 2007). Indeed, not only does the ability of *Salmonella* to internalize in plant tissues depend on plant species (Jablasone, 2005), but crop colonization also differs among cultivars of a given species (Barak et al., 2011; Barak et al., 2008; Klerks et al., 2007). There was an approximately 100-fold difference in the phyllosphere populations of *Salmonella* on four tomato varieties, with *Solanum pimpinellifolium* line WVa700 supporting the lowest number of bacteria (Barak et al., 2011). It is noteworthy that 'WVa700' is also less susceptible to bacterial speck caused by *P. syringae* pv tomato; however, it is not clear whether the same molecular mechanisms determine the resistance to both bacteria (Barak et al., 2011). Similarly, plant genotype has an important role in the proliferation of *E. coli* O157:H7 in the lettuce phyllosphere (Quilliam et al., 2012). During an infection, outer structures and/or exogenous substances on a plant pathogenic bacterium can elicit defense responses from the plant. For example, protein-protein interactions (PPIs) are the fundamental building blocks of communication or defense. A prediction platform developed by Schleker (2012) estimates up to 10,962 PPIs interact with between 33 *Salmonella* effectors and virulence factors and 4,676 *Arabidopsis* proteins; however, their functional interactions are yet to be demonstrated. These discoveries point to the potential of breeding for resistance to colonization by enteric pathogens, although the economic feasibility of breeding for resistance to these organisms is not yet clear (Teplitski et al., 2012). It remains unknown whether there is a correlation between plant basal immune responses to phytopathogens and to human pathogens or other casual microorganisms. Such a correlation would provide an opportunity to integrate breeding for increased basal resistance of crops to both plant and human enteric pathogens.

Interactions of *Salmonella* with plant-associated bacteria (including plant pathogens) and their implications in produce safety

The impetus for better understanding of the interactions between human enteric pathogens and phytopathogens stems from the supermarket surveys that demonstrated that 60% of produce showing symptoms of soft rot also harbored presumptive *Salmonella* (Wells and Butterfield, 1997). Follow-up laboratory studies including those with tomato fruits revealed that plant tissue macerated by pectinolytic pathogens such as *Dickeya dadantii* (*Erwinia chrysanthemi*) and *Pectobacterium carotovorum* promoted growth of *S.* Typhimurium and *E. coli* O157:H7 to population densities approximately 10 times greater than levels on healthy plants; sudden increases in proliferation of the human pathogens coincided with the appearance of soft-rot symptoms in leafy greens and tomatoes (Yamazaki et al., 2011; Goudeau et al., 2013; Noel et al., 2010b; Brandl, 2008). Transcriptomic studies by Goudeau et al. (2013) revealed that *Salmonella* cells colonizing lettuce and cilantro leaf soft-rot lesions caused by *D. dadantii* utilize a broad range of nutrients made available through the pectinolytic activity of the plant pathogen. When forming colonies within plant tissues, *Salmonella* and enterohemorrhagic *E. coli* also benefit from association with biotrophic phytopathogens (like *Pseudomonas syringae* and *Xanthomonas campestris*) (Aruscavage et al., 2008; Aruscavage et al., 2010; Barak and Liang, 2008). An increase in growth similar to that observed in response to the biotrophic phytopathogens was observed on lettuce leaves that were mechanically damaged or showed symptoms of tip burn (dry lesions on leaf margins resulting from a physiological disorder) (Aruscavage et al., 2008; Brandl, 2008). Further genetic studies are needed to clarify how plant pathogen activities promote *Salmonella* proliferation. While physical tissue degradation may release more nutrients, both *Salmonella* and phytobacterial pathogens initially compete for the same resources. It is likely that other factors, such as quorum sensing or messengers, may contribute to development of such high population densities.

Prevention of contamination of tomatoes by *Salmonella*

Sanitation measures during harvest and in a packinghouse are critical to marketing successfully. Until relatively recently, sanitation was designed mainly to prevent postharvest decays (Bartz, 1980, 1991). Today, the industry realizes that the threat of contamination by *Salmonella* can be reduced by the same steps that have been used to minimize postharvest decay. However, like postharvest decay pathogens, *Salmonella* cannot be eliminated if it has internalized or become buried in a matrix such as a biofilm. As such, preventative measures must be implemented starting at the very beginning of production.

Field selection should ensure the crop will not be exposed to contamination. Plant pathogenic bacteria have been blown as aerosols up to 5 miles from diseased citrus trees by a strong storm (Gottwald et al., 1997). While not locating fields within 5 miles of a confined animal operation or other potential source is a very restrictive measure, a careful analysis of potential movement of *Salmonella* contamination into a field should be undertaken. In particular, possible flood events should be considered, and a potential field should have a means to avoid introduction of flood waters from animal operations or from wetlands, rivers, etc. Sources of fertilization for a crop should never include manure containing *Salmonella*. *Salmonella* survives well in terrestrial environments, particularly if embedded in feces (Abulreesh, 2012). An additional concern about field location is the potential for dust or aerosol introduction from nearby sources. Strong rainfall can force particulates, including bacterial aerosols, into natural apertures on plant surfaces as well as into wounds caused by storm events (Clayton, 1936). Fattal et al. (1986) detected aerosolized enteric bacteria and viruses up to 730 m downwind of field plots that were being irrigated with wastewater: the 730-m detection was the outer limit of their experiment. A second consideration in preventing contamination is that water sources used to irrigate a crop or as finished pesticides or fertilizers applied to the crop should either be potable or tested to make sure *Salmonella* is not present. A third consideration is to isolate the crop from soil surfaces by use of mulches. This reduces chances of rain-splash contamination. Fourth, tomatoes grown by stake culture are less likely to be contacted by rain splash or temporary flooding due to heavy rainfall. Moreover, canopies of staked plants will dry more rapidly following rainfall or dew. Fifth, harvests should be scheduled when plant canopies are dry. Wet fruit are particularly vulnerable to contamination by decay pathogens and appear to be more prone to injuries. Sixth, freshly harvested fruit should be protected from direct exposure to sunlight, which can cause significant increases in fruit temperature and even surface damage. Seventh, water used to handle freshly harvested fruit should contain recommended levels of an approved sanitizer. The sanitizer won't remove all bacteria associated with fruit but will prevent transfers of bacteria among fruit in the tank. Eighth, fruit should not be allowed to float in water for more than 2 minutes, and if in water for that length of time, only a single layer of fruit should be permitted. Multiple layers of fruit lead to significant hydrostatic pressure on the lower layers. By contrast, most fruit, particularly green tomatoes, float with their stem scars above the water surface.

Research needs

Based on molecular insights into *Salmonella*'s interaction with tomato fruit, workers have called for an investigation into breeding for resistance to that human pathogen. However, the evidence to date does not include proof of Koch's postulates, which is the standard bar for proving pathogenicity (Barak and Schroeder, 2012). Moreover, no evidence has been presented for *Salmonella* moving from plant to plant in a field as would a typical plant pathogenic bacterium. Standard growth chamber/greenhouse models have enabled discovery of exciting facets of *Salmonella*'s life outside its animal host. Now it is time to begin evaluating these models in terms of their applicability to field conditions by including added UV exposure and standard soil microbial ecosystems. This organism does appear to be very plastic and is able to adapt to a wide range of environments including those within plants. However, based on current evidence, it appears to be a saprophyte while associated with plants. Better control measures are needed. Simply adding chlorine to wash water or dump tank water at a packinghouse doesn't appear sufficient to prevent outbreaks. An evaluation of potential interactions with plant-associated bacteria leading to biofilms appears warranted (Morris et al., 2002; Mandrell et al., 2006). Biofilms have long been known to protect embedded microbes from exposure to external stress factors, which would include sanitizing agents (Zottla, 1994).

Situations that appear likely to give rise to hazardous contamination need evaluation. If a field becomes temporarily flooded from nearby surface water, must the field be abandoned, or will introduced contaminants disappear after a reasonable period of time, and what would be "reasonable"? If a field is struck by a strong rainstorm, how long after that event must growers wait before starting or resuming harvest operations? Evidence suggests *Salmonella* on fruit during harvest will proliferate and persist through the postharvest lifetime of such fruit, but will it? Population dynamics, other than detection after various treatments, have not been clarified.

References

Abulreesh, H.H., 2012. Samonellae in the environment. In: Annous, B.A., Gurtler, J.B. (Eds.), Salmonella—Distribution, Adaptation, Control Measures and Molecular Technologies. pp. 19–50. InTech.

Almeida, D.P.F., Huber, D.J., 1999. Apoplastic pH and inorganic ion levels in tomato fruit: A potential means for regulation of cell wall metabolism during ripening. Physiol. Plant 105, 506–512.

Aruscavage, D., Miller, S.A., Ivey, M.L., Lee, K., LeJune, J.T., 2008. Survival and dissemination of *Escherichia coli* O157:H7 on physically and biologically damaged lettuce plants. J. Food Prot. 71, 2384–2388.

Aruscavage, D., Phelan, P.L., Lee, K., LeJeune, J.T., 2010. Impact of changes in sugar exudate created by biological damage to tomato plants on the persistence of *Escherichia coli* O157:H7. J. Food. Sci. 75, M187–M192.

Barak, J.D., Gorski, L., Naraghi-Arani, P., Charkowski, A.O., 2005. *Salmonella enterica* virulence genes are required for bacterial attachment to plant tissue. Appl. Environ. Microbiol. 71, 5685–5691.

Barak, J.D., Kramer, L.C., Hao, L.Y., 2011. Colonization of tomato plants by *Salmonella enterica* is cultivar dependent, and type 1 trichomes are preferred colonization sites. Appl. Environ. Microbiol. 77, 498–504.

Barak, J.D., Liang, A., Narm, K.E., 2008. Differential attachment to and subsequent contamination of agricultural crops by *Salmonella enterica*. Appl. Environ. Microbiol. 74, 5568–5570.

Barak, J.D., Liang, A.S., 2008. Role of soil, crop debris, and a plant pathogen in *Salmonella enterica* contamination of tomato plants. Plos One 3, e1657.

Barak, J.D., Schroeder, B.K., 2012. Interrelationships of Food Safety and Plant Pathology: the life cycle of human pathogens on plants. Annu. Rev. Phytopathol. 50, 241–266.

Bartz, J.A., 2009. Raw tomatoes and Salmonella. In: Sapers, G.M., Solomon, E.B., Mathews, K.R. (Eds.), The Produce Contamination Problem: Causes and Solutions. Academic Press, Elsevier, Inc, Burlington, MA, pp. 223–248.

Bartz, J.A., 2006. Internalization and infiltration. In: Sapers, G.M., Gorny, J.R., Yousef, A.E. (Eds.), Microbiology of fruits and vegetables. CRC Press, Taylor and Francis, Boca Raton, FL, pp. 75–94.

Bartz, J.A., 1991. Postharvest diseases and disorders in tomato fruit. In: Jones, J.B., Jones, J.P., Stall, R.I., Zitter, T.A. (Eds.), Compendium of tomato diseases. Am. Phytopath. Soc, St. Paul, MN, pp. 44–49.

Bartz, J.A., 1980. Causes of postharvest losses in Florida tomato shipments. Plant Dis. 64, 934–937.

Bartz, J.A., Showalter, R.K., 1981. Infiltration of tomatoes by aqueous bacterial suspensions. Phytopathology 71, 515–518.

Bartz, J.A., Sargent, S.A., Scott, J.W., 2012. Postharvest Quality and Decay Incidence among Tomato Fruit as Affected by Weather and Cultural Practices. U. F. IFAS Extension. http://edis.ifas.ufl.edu/pp294.

Boyette, M.D., Sanders, D.C., Estes, E.A., 2011. Postharvest cooling and handling of field- and greenhouse-grown tomatoes. AG-414-9 North Carolina State Univ. http://www.bae.ncsu.edu/programs/extension/publicat/postharv/tomatoes/tomat.html.

Brandl, M.T., 2008. Plant lesions promote the rapid multiplication of Escherichia coli O157:H7 on postharvest lettuce. Appl. Environ. Microbiol. 74, 5285–5289.

Brandl, M.T., Cox, C.E., Teplitski, M., 2013. *Salmonella* interactions with plants and their associated microbiota. Phytopathology 103, 316–325.

Brandl, M.T., Amundson, R., 2008. Leaf age as a risk factor in contamination of lettuce with *Escherichia coli* O157:H7 and *Salmonella enterica*. Appl. Environ. Microbiol. 74, 2298–2306.

Brandl, M.T., Mandrell, R.E., 2002. Fitness *of Salmonella enterica* serovar Thompson in the cilantro phyllosphere. Appl. Environ. Microbiol. 68, 3614–3621.

Brooks, C., 1937. Some effects of waxing tomatoes. Proc. Am. Soc. Hort. Sci. 35, 720.

CDC., 2005. Outbreaks of *Salmonella* infections associated with eating Roma tomatoes—United States and Canada, 2004. MMWR 54, 325–328.

CFSAN., 2007. Program information manual, retail food protection, storage and handling of tomatoes. www.cfsan.fda.gov/ear/pimtomat.html.

Clayton, E.E., 1936. Water soaking of leaves in relation to development of the wildfire disease of tobacco. J. Agric. Res. 52, 239–269.

Clement, V., Bartz, J.A., Sargent, S.A., 2000. Postharvest decay risk associated with hydrocooling tomatoes. Plant Dis. 84, 1314–1318.

Cooley, J.M., Carychao, D., Wiscomb, G.W., Sweitzer, R.A., Crawford-Miksza, L., Farrar, J.A., et al., 2007. *Escherichia coli* O157:H7 in feral swine near spinach fields and cattle, central California coast. Emerg. Infect. Dis. 13, 1908–1911.

Cooley, M.B., Miller, W.G., Mandrell, R.E., 2003. Colonization of *Arabidopsis thaliana* with *Salmonella enterica* and enterohemorrhagic *Escherichia coli* O157:H7 and competition by *Enterobacter asburiae*. Appl. Environ. Microbiol. 69, 4915–4926.

Cummings, K., Barret, E., Mohle-Boetani, J. C., Brooks, J. T., Farrar, J, Hunt, T., Fiore, A., Komatsu, S., Werner, S. B., and Slutsker, L. 2001. 200.

DACS, 2008. Tomato best practices manual—A guide to T-GAP and T-BMP. The Florida Department of Agriculture and Consumer Services, Tallahassee, FL.

Dong, Y., Iniguez, A.L., Ahmer, B.M., Triplett, E.W., 2003. Kinetics and strain specificity of rhizosphere and endophytic colonization by enteric bacteria on seedlings of Medicago sativa and Medicago truncatula. Appl. Environ. Microbiol. 69, 1783–1790.

Erickson, M.C., 2012. Internalization of fresh produce by foodborne pathogens. Ann. Rev. Fd. Sci. Tech. 3, 283–310.

Fattal, B., Wax, Y., Davies, M., Shuval, H.J., 1986. Health risks associated with wastewater irrigation: an epidemiological study. Amer. J. Pub. Health 76, 977–979.

FDA, 2003. The bad bug book. Foodborne Pathogenic Microorganisms and Natural Toxins Handbook. U.S. Department of Health and Human Services, Food and Drug Administration, Center for Food Safety and Applied Nutrition. Updated: 01-30-2003. www.cfsan.fda.gov/mow/intro.html.

Felkey, K.D., 2002. Optimization of chlorine treatments and the effects on survival of *Salmonella* spp. on tomato surfaces. M.S. Thesis, University of Florida, Gainesville, FL.

Gorny, J.R., 2006. Microbial contamination of fresh fruits and vegetables. In: Sapers, G.M., Gorny, J.R., Yousef, A.E. (Eds.), Microbiology of Fruits and Vegetables. CRC Press, Taylor and Francis, Boca Raton, FL, pp. 3–32.

Gorski, L., Parker, C.T., Liang, A., Cooley, M.B., Jay-Russell, M.T., Gordus, A.G., et al., 2011. Prevalence, distribution, and diversity of *Salmonella enterica* in a major produce region of California. Appl. Environ. Microbiol. 77, 2734–2748.

Goudeau, D.M., Parker, C.T., Zhou, Y., Sela, S., Kroupitski, Y., Brandl, M.T., 2013. The Salmonella transcriptome in lettuce and cilantro soft rot reveals a niche overlap with the animal host intestine. Appl. Environ. Microbiol. 79.

Gottwald, T.R., Graham, J.H., Schubert, T.S., 1997. Citrus canker in urban Miami: an analysis of spread and prognosis for the future. Citrus Industry, Aug. 5, 1997.

Greene, S.K., Daly, E.R., Talbot, E.A., Demma, L.J., Holzbauer, S., Patel, N.J., et al., 2008. Recurrent multistate outbreak of *Salmonella* Newport associated with tomatoes from contaminated fields, 2005. Epidemiol. Infect. 136, 157–165.

Guo, X., Chen, Brackett, J.E., Beuchat, L.R., 2001. Survival of salmonellae on and in tomato plants from the time of inoculation at flowering and early stages of fruit development through fruit ripening. Appl. Environ. Microbiol. 67, 4760–4764.

Hedberg, C.W., Angulo, F.J., White, et al., 1999. Outbreaks of salmonellosis associated with eating uncooked tomatoes: implications for public health. Epidemiol. Infect 122, 385–393.

Hirano, S.S., Upper, C.D., 2000. Bacteria in the leaf ecosystem with emphasis on *Pseudomonas syringae*—a pathogen, ice nucleus, and epiphyte. Microbiol. Mol. Biol. Rev. 64, 624–653.

Iniguez, A.L., Dong, Y.M., Carter, H.D., Ahmer, B.M.M., Stone, J.M., Triplett, E.W., 2005. Regulation of enteric endophytic bacterial colonization by plant defenses. Mol. Plant Microbe. Interact. 18, 169–178.

Islam, M., Morgan, J., Doyle, M.P., Phatak, S.C., Millner, P., Jiang, X., 2004. Persistence of *Salmonella enterica* serovar Typhimurium on lettuce and parsley and in soils on which they were grown in fields treated with contaminated manure composts or irrigation water. Foodborne Pathog. Dis. 1, 27–35.

Jablasone, J., Warriner, K., Griffiths, M., 2005. Interactions of Escherichia coli O157:H7, *Salmonella* Typhimurium and *Listeria monocytogenes* plants cultivated in a gnotobiotic system. Int. J. Food Microbiol. 99, 7–18.

Jay, M.T., Cooley, M., Carychao, Wiscomb, G.W., others, 2007. *Escherichia coli* O157:H7 in feral swine near spinach fields and cattle, Central California Coast. Emerg Infect Dis. 13, 1908–1911.

Johnson, J., 1947. Water-congestion in plants in relation to disease. Univ. Wis. Res. Bul, 160. Madison, WI.

King, A.D., Bolin, H.R., 1989. Physiological and microbiological storage stability of minimally processed fruits and vegetables. Food Technol. 43, 132–136.

Kisluk, G., Yaron, S., 2012. Presence and persistence *of Salmonella enterica* serotype Typhimurium in the phyllosphere and rhizosphere of spray-irrigated parsley. Appl. Environ. Microbiol. 78, 4030–4036.

Klerks, M.M., Franz, E., van Gent-Pelzer, M., Zijlstra, C., van Bruggen, A.H., 2007. Differential interaction of *Salmonella enterica* serovars with lettuce cultivars and plant-microbe factors influencing the colonization efficiency. ISME J. 1, 620–631.

Kroupitski, Y., Golberg, D., Belausov, E., Pinto, R., Swartzberg, D., Granot, D., et al., 2009a. Internalization of *Salmonella enterica* in leaves is induced by light and involves chemotaxis and penetration through open stomata. Appl. Environ. Microbiol. 75, 6076–6086.

Kroupitski, Y., Pinto, R., Belausov, E., Sela, S., 2011. Distribution of *Salmonella* Typhimurium in romaine lettuce leaves. Fd. Microbiol. 28, 990–997.

Kroupitski, Y., Pinto, R., Brandl, M.T., Belausov, E., Sela, S., 2009b. Interactions of *Salmonella enterica* with lettuce leaves. J. Appl. Microbiol. 106, 1876–1885.

Kroupitski, Y., Brandl, M.T., Pinto, R., Belausov, E., Tamir-Ariel, D., Burdman, S., et al., 2013. Identification of Salmonella enterica genes with a role in persistence on lettuce leaves during cold storage by recombinase-based in vivo expression technology. Phytopathology 103, 362–372.

Kyle, J.L., Parker, C.T., Goudeau, D., Brandl, M.T., 2010. Transcriptome analysis of *Escherichia coli* O157:H7 exposed to lettuce leaf lysates. Appl. Environ. Microbiol. 76, 1375–1387.

LeJeune, J.T., Besser, T.E., Hancock, D.D., 2001. Cattle water troughs as reservoirs of *Escherichia coli* O157. Appl. Environ. Microbiol. 67, 3053–3057.

Lindow, S.E., Brandl, M.T., 2003. Microbiology of the phyllosphere. Appl. Environ. Microbiol. 69, 1875–1883.

Mandrell, R., 2009. Enteric human pathogens associated with fresh produce: sources, transport, and ecology. In: Fan, X., Niemira, B.A., Doona, C.J., Feeherry, F.E., Gravani, R.B. (Eds.), Microbial Safety of Fresh Produce. Blackwell Publishing and the Institute of Food Technologies, Ames, Iowa.

Mandrell, R., Gorski, L., Brandl, M.T., 2006. Attachment of microorganisms to fresh produce. In: Sapers, G.M., Gorny, J.R., Yousef, A.E. (Eds.), Microbiology of fruits and vegetables. CRC Press, Taylor and Francis, Boca Raton, FL, pp. 33–73.

Marvasi, M., Cox, C., Xu, Y., Noel, J., Giovannoni, J., Teplitski, M., 2013. Differential regulation of *Salmonella* Typhimurium genes involved in O-antigen capsule production and their role in persistence within tomatoes. Mol. Plant. Microbe. Interact. E-pub ahead of print.

Maynard, D.N., Scott, J.W., Dunlap, A.M., 2000. Tomato variety evaluation Fall 1999. U.F. IFAS. GCREC Res. Rept. BRA 2000–2001.

Mermin, J., Hutwagner, L., Vugia, D., et al., 2004. Reptiles, amphibians, and human *Salmonella* infection: a population-based, case-control study. Clin. Infect. Dis. 38 (Suppl. 3), S253–S261.

Morris, C.E., Barnes, M.B., McLean, R.J.C., 2002. Biofilms on leaf surfaces: implications for the biology, ecology and management of populations of epiphytic bacteria. In: Lindow, S.E., Hecht-Poinar, E.I., Elliot, V.R. (Eds.), Phyllosphere Microbiology. APS Press, St. Paul, Mn, pp. 139–155.

Micallef, S.A., Rosenberg, R.E., Goldstein, G., Kleinfelter, A., Boyer, L., McLaughlin, M.S., et al., 2012. Occurrence and antibiotic resistance of multiple *Salmonella* serotypes recovered from water, sediment and soil on mid-Atlantic tomato farms. Environ. Res. 114, 31–39.

Noel, J.T., Arrach, N., Alagely, A., McClelland, M., Teplitski, M., 2010a. Specific responses of *Salmonella enterica* to tomato varieties and fruit ripeness identified by In Vivo Expression Technology. Plos One 5.

Noel, J.T., Joy, J., Smith, J.N., Fatica, M., Schneider, K.R., Ahmer, B.M., et al., 2010b. *Salmonella* SdiA recognizes N-acyl homoserine lactone signals from *Pectobacterium carotovorum* in vitro, but not in a bacterial soft rot. Mol. Plant Microbe. Interact. 23, 273–282.

Orozco, L., Rico-Romero, L., Escartín, E.F., 2008. Microbiological profile of greenhouses in a farm producing hydroponic tomatoes. J. Food Prot. 71, 60–65.

Quilliam, R.S., Williams, A.P., Jones, D.L., 2012. Lettuce cultivar mediates both phyllosphere and rhizosphere activity of *Escherichia coli* O157:H7. Plos One 7.

Rajabi, M., Jones, M., Hubbard, Rodrick, G., Wright, A.C., 2011. Distribution and genetic diversity of *Salmonella enterica* in the Upper Suwannee river. Int. J. Microbiol. 2011, 461321.

Saltveit, M.E., 2005. Fruit ripening and fruit quality. In: Heuvelink, E. (Ed.), Tomatoes. CABI Publishing, Cambridge, MA, pp. 145–170.

Samish, Z., Etinger-Tulczynska, R., 1963. Distribution of bacteria within the tissue of healthy tomatoes. Appl. Microbiol. 11, 7–10.

Sargeant, J.M., Hafer, D.J., Gillespie, J.R., Oberst, R.D., Flood, S.J., 1999. Prevalence of *Escherichia coli* O157:H7 in white-tailed deer sharing rangeland with cattle. J. Am. Vet. Med. Assoc. 215, 792–794.

Sargent, S.A., Brecht, J.K., Olczyk, T., 2005. Handling Florida Vegetables Series—Round and Roma Tomato Types. SS-VEC-928. Hort. SWci. Dept. FL. Coop. Ext. Ser., I.F.A.S., U. F. http://edis.ifas.ufl.edu.

Schikora, A., Virlogeux-Payant, I., Bueso, E., Garcia, A.V., Nilau, T., Charrier, A., et al., 2011. Conservation of Salmonella infection mechanisms in plants and animals. Plos One 6, e24112.

Semenov, A.M., Kuprianov, A.A., van Bruggen, A.H., 2010. Transfer of enteric pathogens to successive habitats as part of microbial cycles. Microb. Ecol. 60, 239–249.

Shi, X., Namvar, A., Kostrzynska, M., Hora, R., Warriner, K., 2007. Persistence and growth of different Salmonella serovars on pre- and postharvest tomatoes. J. Food Prot. 70, 2725–2731.

Schleker, S., Garcia-Garcia, J., Klein-Seetharaman, J., Oliva, B., 2012. Prediction and comparison of *Salmonella*-human and *Salmonella*-Arabidopsis interactomes. Chem. Biodivers. 9, 991–1018.

Smith, K.F., Yabsley, M.J., Sanchez, S., Casey, C.L., Behrens, M.D., Hernandez, S.M., 2012. *Salmonella* isolates from wild-caught Tokay Geckos (Gekko gecko) imported to the U.S. from Indonesia. Vector. Borne. Zoonotic. Dis. 12, 575–582.

Smith, L.A., White, P.C.L., Marion, G., Hutchings, M.R., 2009. Livestock grazing behavior and inter-versus intraspecific disease risk via the fecal-oral route. Behav. Ecol. 20 462–432.

Srikantiah, P., Bodager, D., Toth, B., et al., 2005. Web-based investigation of multistate Salmonellosis outbreak. CDC Emerging Infectious Diseases, vol. 11. www.cdc.gov/ncidod/eid/vol11no4/04-0997.htm.

Stephens, T.P., Loneragan, G.H., Thompson, T.W., Sridrara, A., Branham, L.A., Pitchiah, S., et al., 2007. Distribution of *Escherichia coli* O157 and *Salmonella* on hide surfaces, the oral cavity, and in feces of feedlot cattle. J. Food Prot. 70, 1346–1349.

Sundlin, G.W., 2002. Ultraviolet radiation on leaves; its influence and microbial communities and their adaptations. In: Lindow, S.E., Hecht-Poinar, E.I., Elliot, V.R. (Eds.), Phyllosphere Microbiology. APS Press, St Paul, Mn, pp. 27–41.

Switt, A.I.M., den Bakker, H.C., Cummings, C.A., et al., 2012. Identification and characterization of novel *Salmonella* mobile elements involved in the dissemination of genes linked to virulence and transmission. PLoS One 7, e41247.

Teplitski, M., Noel, J.T., Alagely, A., Danyluk, M.D., 2012. Functional genomics studies shed light on the nutrition and gene expression of non-typhoidal *Salmonella* and enterovirulent *E. coli* in produce. Food Res. Int. 45, 576–586.

Teplitski, M., Warriner, K., Bartz, J., Schneider, K.R., 2011. Untangling metabolic and communication networks: interactions of enterics with phytobacteria and their implications in produce safety. Trends Microbiol. 19, 121–127.

Todar, K., 2005. *Salmonella* and salmonellosis. In: Todar's online textbook of bacteriology. www.textbookofbacteriology.net/salmonella.html.

Umali, D.V., Lapuz, R.R., Suzuki, T., Shirota, K., Katoh, H., 2012. Transmission and shedding patterns of *Salmonella* in naturally infected captive wild roof rats (*Rattus rattus*) from a *Salmonella*-contaminated layer farm. Avian Dis. 56, 288–294.

USDA, ERS, 2013. North American Fresh-Tomato Market. U.S.D.A. Econ. Res. Serv. http://www.ers.usda.gov/topics/in-the-news/north-american-fresh-tomato-market.aspx.

VanSickle, J.J., 2008. Field study for analyzing the opportunities to change the box size for mature green tomatoes shipped from Florida to a 10 kg carton. In: Tomato Research Report for 2007–2008. U. of Florida, IFAS, Gainesville, pp. 40–60.

Vigneault, C., Bartz, J.A., Sargent, S.A., 2000. Postharvest decay risk associated with hydrocooling tomatoes. Plant Dis. 84, 1314–1318.

Wahlstrom, H., Tysen, E., Olsson Engvall, E., Brandstrom, B., Eriksson, E., Morner, T., et al., 2003. Survey of *Campylobacter* species, VTEC O157 and *Salmonella* species in Swedish wildlife. Vet. Rec. 153, 74–80.

Wei, C.I., Huang, J.M., Lin, W.F., Tamplin, M.L., Bartz, J.A., 1995. Growth and survival of Salmonella Montevideo on tomatoes and disinfection with chlorinated water. J. Food Prot. 8, 829–836.

Wells, J.M., Butterfield, J.E., 1997. *Salmonella* contamination associated with bacterial soft rot of fresh fruits and vegetables in the marketplace. Plant Dis. 81, 867–872.

Yamazaki, A., Li, J., Hutchins, W.C., Wang, L., Ma, J., Ibekwe, A.M., et al., 2011. Commensal effect of pectate lyases secreted from *Dickeya dadantii* on proliferation of *Escherichia coli* O157:H7 EDL933 on lettuce leaves. Appl. Environ. Microbiol. 77, 156–162.

Young, J.M., 1974. Effect of water on bacterial multiplication in plant tissue. N.Z. J. Agric. Res. 17, 115–119.

Zheng, J., Allard, S., Reynolds, S., Millner, P., Arce, G., Blodgett, R.J., et al., 2013. Colonization and internalization of *Salmonella enterica* in tomato plants. Appl. Environ. Microbiol. 79, 2494–2502.

Zottola, E.A., 1994. Microbial attachment and biofilm formation: a new problem for the food industry. Fd. Tech. 48, 107–114.

Tree Fruits and Nuts: Outbreaks, Contamination Sources, Prevention, and Remediation

13

Susanne E. Keller

FDA/CFSAN, Institute of Food Safety and Health, Bedford Park, IL, USA

CHAPTER OUTLINE

Introduction

According to the USDA, the U.S. had a per capita estimated consumption of over 300 g of fruit and tree nuts per day in 2012 (USDA, 2012). Consumption of produce, particularly fresh produce, continues to increase worldwide in part due to an increased recognition of health benefits and an increase in its year-round availability (WHO, 2013). Historically, tree fruits and nuts have not been associated with a high risk for causing foodborne disease. However, recent increases in foodborne illnesses associated with fresh produce in general have led to an increasing concern regarding the safety of all fresh fruits and vegetables that are not processed to eliminate any microbial hazard. As a result, the FDA promulgated rules and guidelines concerning not only the handling of fresh produce, but its growth and subsequent processing, including the processing and production of fresh juices in 2001. In 2013, the FDA published proposed changes to Good Manufacturing Practices in response to the Food Safety and Modernization Act passed in 2011 (FDA, 2013). The proposed changes would require new preventive controls in human food and produce safety that are meant to address risks inherent in their production. In this chapter, the risks

The Produce Contamination Problem. http://dx.doi.org/10.1016/B978-0-12-404611-5.00013-0
2014 Published by Elsevier Inc.

of foodborne illness, particularly as it affects the consumption of fresh or minimally processed tree fruits and nuts, will be discussed.

Organisms of concern

The overall incidence of foodborne illness for fruits and nuts remains low particularly with respect to meat products. Data from the Health Protection Agency related to vehicles of foodborne illness in England and Wales from 1992 through 2010 clearly show a majority of outbreaks are related to meats and poultry (HPA, 2011). However, fruits and nuts were attributed to 16% of foodborne illness outbreaks in the U.S. that were attributed to a single food vehicle in 2006 (CDC, 2009b). Data from the FDA's reportable food registry data indicate that raw agricultural commodities along with nuts/nut products and spices account for the largest portion of recalls for *Salmonella* in 2011 (FDA, 2012b). Consequently, although fewer foodborne illnesses may occur from tree fruits and nuts, clearly the potential for foodborne illness from these vehicles is not insignificant.

None of the usual human pathogens causing foodborne illness such as enterohemorrhagic *E. coli* (EHEC), *Salmonella* and *Shigella* species, *Cryptosporidium*, or *Listeria monocytogenes* are considered endogenous microflora of fruits and nuts, but all may occur as contaminants. In general, fruits and nuts provide an environment hostile to the growth and survival of these pathogens (Brandl, 2006; Winfield and Groisman, 2003). However, despite their low incidence, their presence on these foods can be particularly problematic since fruits and nuts are frequently consumed raw and are considered to be part of a healthy lifestyle. Consequently, any outbreaks related to these products could be viewed more negatively by the general public than an outbreak associated with a food that is typically cooked or processed.

In circumstances where pathogenic microorganisms are unlikely to grow or multiply such as may be found on most fruits and nuts, those pathogens with lowest infectious dose and the greatest propensity for survival are likely to be of greatest concern. Of those pathogens with low infectious dose and considerable ability to withstand harsh environments, pathogenic *E. coli* stands out. In particular EHEC is notable because of its association with hemorrhagic colitis, hemolytic uremic syndrome (HUS), and thrombotic thrombocytopenic purpura (TTP). HUS occurs primarily in children under 10 years of age and has a mortality of 3 to 5% (Buchanan and Doyle, 1997). The most common serotype for pathogenic *E. coli*, particularly in the United States, Canada, Great Britain, and parts of Europe, is *E. coli* O157:H7 (Buchanan and Doyle, 1997). Other types of pathogenic *E. coli* have also resulted in serious foodborne outbreaks, most notably an outbreak of the serotype O104:H7, an enteroaggregative *E. coli* (EAEC), that occurred in Europe due to contaminated sprouts (CDC, 2011c; FDA, 2012a). In the years from 1998 to 2000, the Centers for Disease Control (CDC) recorded 86 outbreaks attributed to *E. coli*. Of these, 68 were identified as outbreaks caused by *E. coli* O157:H7 (CDC, 2003). In 2006, there were 29 foodborne outbreaks attributed to Shiga toxin–producing *E. coli,* 27 attributed to the serotype O157 (CDC, 2009b). The majority of these outbreaks was either from meat products or had an unknown source; nonetheless outbreaks related to tree fruit or nuts have occurred.

A second group of pathogenic microorganisms with a low infectious dose is the *Shigella* species. As with *E. coli*, some strains produce enterotoxin and Shiga toxin (FDA, 2012a). Epidemics with fatalities as high as 5 to 15% have occurred in Africa and Central America (CDC, 2005b). Outbreaks related to tree fruits and nuts products caused by *Shigella* species are rare and appear to be more associated with poor hygiene (Castillo et al., 2006).

Salmonella species represents another group of foodborne pathogens that is also of concern with respect to tree fruits and nuts. *Salmonella* is divided into two species that can cause illness; *S. enterica* and *S. bongori* (FDA, 2012a). *S. enterica* is further divided into six subspecies and includes over 2000 serotypes that cause human disease. The infectious dose for some *Salmonella* serotypes may also be very low, with as little as one cell resulting in illness (FDA, 2012a). There are an estimated 1.4 million cases of salmonellosis annually, with an estimated 500 fatalities. Again, as with EHEC, these infections are more commonly associated with animal derived foods, such as meat, seafood, dairy, and egg products, rather than produce. Their occasional association with fruits and nuts is facilitated by their tolerance to some extreme conditions. *Salmonella* species is resistant to desiccation, which aids in its survival on the surface of fruits and nuts. Its survival on tree nuts and particularly almonds as well as in other low-moisture foods is now well documented (Beuchat and Heaton, 1975; Danyluk et al., 2007; Gruzdev et al., 2011; Keller et al., 2013; Kimber et al., 2012; Komitopoulou and Penaloza, 2009; Uesugi et al., 2006; Uesugi et al., 2007; Uesugi and Harris, 2006). *Salmonella* is also resistant to acids, a property it shares with both EHEC and *Shigella* species (Bagamboula et al., 2002; Lin et al., 1995).

In both *Salmonella* and *E. coli*, acid tolerance is inducible and increases when cells have been adapted either to acid conditions or are in a stationary phase (Benjamin and Datta, 1995; Buchanan and Edelson, 1996; Foster and Hall, 1990; Lin et al., 1995). In *E. coli*, tolerance to high acid levels involves three distinct inducible mechanisms and is enhanced in stationary cells (Benjamin and Datta, 1995; Buchanan and Edelson, 1996; Lin et al., 1995; Lin et al., 1996). For *S.* Typhimurium, two major acid tolerance systems have been identified, one associated with log phase and one associated with stationary phase (Bang et al., 2000). Not surprisingly, survival in acidic fruit juices for extended periods has been observed for both *E. coli* and *Salmonella* species (Goverd et al., 1979; Parish, 1997). Like both *Salmonella* species and *E. coli*, *Shigella* is also resistant to acids, however, to a somewhat lesser extent. It can survive at a pH as low as 2 to 2.5 and has some of the same acid tolerance mechanisms as does *E. coli* (Bagamboula et al., 2002; FDA, 2012a). Its less frequent appearance on tree fruits and nuts may be due to its lesser ability to survive harsh environments.

Listeria monocytogenes is yet another foodborne pathogen of concern with fruits and nuts, particularly fresh-cut products, which is also acid tolerant. *L. monocytogenes* is ubiquitous within the environment and frequently found on fruits and vegetables as well as nuts and nut products (Beuchat, 1995; Beuchat and Ryu, 1997; Cox et al., 1989; Eglezos, 2010; Fenlon et al., 1996; Gombas et al., 2003; Johnston et al., 2005a; Johnston et al., 2005b; Little and Mitchell, 2004; Sorrells et al., 1989). The minimum pH for growth of *L. monocytogenes* is dependant on the acidulant. For malic acid, the primary acid found in apple cider/juice, the lowest pH value for growth for some strains of *L. monocytogenes* is from 4.4 to 4.6 (Beuchat, 1995; Beuchat and Ryu, 1997; Cox et al., 1989; Fenlon et al., 1996; Gombas et al., 2003; Johnston et al., 2005b; Sorrells et al., 1989). *L. monocytogenes* will survive at lower pH similar to *E. coli* O157:H7 and *Salmonella* (Beuchat and Brackett, 1991; Sorrells et al., 1989). Although no foodborne outbreaks of listeriosis have been attributed specifically to tree fruits or nuts, *L. monocytogenes* has been isolated from unpasteurized apple juice (Sado et al., 1998). Its presence in fresh-cut produce in general has resulted in numerous product recalls with significant economic losses (FDA, 2012b).

In addition to the procaryotic pathogenic bacteria mentioned, fresh fruits and nuts are also susceptible to contamination by protozoa, particularly by *Cryptosporidium parvum*, a highly infectious protozoan parasite causing persistent diarrhea (Guerrant, 1997; FDA, 2012a). Although *Cryptopsoridium* cannot replicate in the environment, the thick-walled oocysts are resistant to acids and chlorine and persist in the environment. Infection does not always manifest itself in severe symptoms but can be dangerous for the immunocompromised population. Historically, the largest U.S. outbreak of cryptosporidiosis occurred in Milwaukee, Wisconsin in 1993 and affected an estimated 403,000 people (Guerrant, 1997).

The inability to grow and multiply in the environment clearly does not limit the ability of pathogens such as *Cryptosporidium parvum* to cause significant foodborne outbreaks such as those in apple cider in 1993, 1996, 2003, and 2004 (Blackburn et al., 2006; CDC, 1997; CDC, 2005a; Millard et al., 1994). However, *Cryptosporidium* is not the only microorganism to produce foodborne illnesses despite an inability to grow and multiply in the food environment. Viruses have also resulted in a significant number of outbreaks related to fresh produce.

Viruses transmitted by food or water fall into three groups: hepatovirus, enterovirus, and norovirus. Of these, the hepatovirus and norovirus appear to be of greatest concern with tree fruits and nuts. Viral outbreaks are frequently the result of poor sanitation or poor worker hygiene (Einstein et al., 1963; Herwaldt et al., 1994; Kassa, 2001; FDA, 2012a).

A final group of microorganisms with a propensity to result in foodborne hazards are molds, specifically molds associated with mycotoxin production (Murphy et al., 2006). In fresh apple juice, the mycotoxin patulin, which is considered toxic and is produced by the molds Penicillium, Aspergillus, and Byssochlamys, has been found to create a hazard (Harris et al., 2009; FDA, 2001b). Consequently, the FDA has set action limits for patulin in apple juice to 50 µg/kg. The same limits are set by the European Union and other countries (European Union, 2010; Kubo, 2012).

Mycotoxins, as a general rule, are less common in fresh fruit but can be a serious issue in dried fruit or tree nuts (Murphy et al., 2006; Nielsen et al., 2009). The mere presence of a mold that can produce a mycotoxin does not mean a mycotoxin will be present or will be produced (Bayman et al., 2002; Marín et al., 2008; Molyneux et al., 2007; Reis et al., 2012). Nonetheless, the presence of mycotoxins does create a serious health risk. Many mycotoxins such as aflatoxin B_1 are highly carcinogenic (U.S. Food and Drug Administration, 2012a). For this reason, tolerance levels are established in different countries for various mycotoxins in different foods (European Union, 2010; Kubo, 2012; FDA, 2000b).

Outbreaks associated with tree fruits

Despite the general perception that foodborne outbreaks related to tree fruits is a recent concern, incidents of foodborne illness associated with these products have been recorded as far back as the early- to mid-1900s (Duncan et al., 1946; Parish, 1997). In the period from 1995 through 2005, there were 21 juice-related outbreaks reported to the CDC (Vojdani et al., 2008). Outbreaks related to whole fresh tree fruits have also occurred, albeit somewhat less frequently (CDC, 2011a; CDC, 2012; Public Health Agency of Canada, 2012; Sivapalasingam et al., 2003). Of all recorded outbreaks, one juice outbreak is particular noteworthy. In October 1996, an outbreak of *E. coli* O157:H7 was traced to unpasteurized apple juice produced by Odwalla Inc. The outbreak involved three states and British Columbia, with more than 60 people sick and one death (CDC, 1996). Odwalla Inc. eventually pleaded guilty to violating Federal Food Safety laws and was fined $1.5 million for selling tainted apple juice (Belluck, 1998). As a consequence of this and other juice-related foodborne outbreaks, the FDA promulgated HACCP regulations for fresh juice in 2001 (FDA, 2001a). All juice in the United States is now required to undergo processing that will achieve a 5-log reduction in the most pertinent pathogen. The 5-log reduction standard was established based on recommendations by the National Advisory Committee on Microbiological Criteria for Foods (NACMCF). NACMCF considered worst-case scenarios, such as might occur if apples were contaminated directly with bovine feces, and included a 100-fold safety factor. Regulatory precedence was also considered when the 5-log pathogen reduction performance standard was established (FDA, 2001a).

The legally required 5-log reduction process must treat the whole juice, with the exception of citrus juice where the 5-log reduction can be achieved through treatments to the surface of the fruit before extraction of the juice. The exemption for citrus fruit was based on the relative low risk that microorganisms will become internalized within the fruit. With apples, internalization has been well documented in the literature (Buchanan et al., 1999; Burnett and Beuchat, 2001; Burnett et al., 2000; Eblen et al., 2004; Fatemi et al., 2006; Fleischman et al., 2001; Merker et al., 1999; Penteado et al., 2004; Soto et al., 2007). However, with citrus fruit, NACMCF determined that although internalization could occur, it was unlikely that such an event would occur, particularly with sound, tree-picked fruits. Consequently, to be eligible for this exemption, citrus fruit must be tree-picked, not wind-fall, or ground harvested, and must be clean and free of blemishes. In addition, if a citrus juice processor obtains a 5-log reduction based on surface treatments and not through treatment of the whole juice, an end-product testing program must be in place.

Since the implementation of the Juice HACCP rules, outbreaks in the United States related to juice have decreased but do continue to occur (Painter et al., 2013). Many of these outbreaks appear related to unpasteurized (not thermally processed) product, or product given an alternative disinfection treatment but not heat-pasteurized. One example of such an outbreak was attributed to ozonated apple cider in Ohio (Blackburn et al., 2006). In this case the processor attempted to destroy pathogens in juice through the use of ozone. In 2005, there was yet another incident concerning *S*. Typhimurium in unpasteurized orange juice (CDC, 2005a). In all these cases, problems occurred when novel disinfection treatments were used in lieu of a heat pasteurization treatment without appropriate validation of treatment efficacy.

Outbreaks related to tree fruits and juices are not limited geographically, although not all countries have a requirement for pasteurization as does the United States. In 2008, an outbreak of salmonellosis occurred in The Netherlands that was attributed to the consumption of fresh unpasteurized juice (Noël et al., 2010). This was reported to be the first such reported outbreak in fresh juice in Europe since 1922. However, the authors did not attribute the previous lack of such outbreaks to better food safety practices, rather they attributed less outbreaks to a lack of sensitivity of their surveillance systems (Noël et al., 2010). If a lack of sensitivity could result in under-reporting of foodborne illness related to fruit juice, then it is conceivable that foodborne illness related to whole fruit may suffer from even greater under-reporting.

Fresh juice, due to the nature of its production, is intuitively a more hazardous product than the original fruit from which it is produced. With juice, a single contaminated fruit can result in a widespread outbreak since the contamination will be spread through production and be consumed by multiple individuals. When consumed whole, a single contaminated fruit will likely result in illness of only a single consumer. Consequently, detection or characterization of outbreaks due to individual contaminated fruit may be difficult, resulting in greater under-reporting than that caused by fruit juices. Outbreaks, however, have been attributed to fresh fruit products other than juices made from fresh tree fruits. Fresh fruit salad has been implicated in outbreaks. Although it is not always possible to pinpoint the exact ingredient in a fruit salad responsible for the initial contamination, the association with fruit salads illustrates the vulnerability of fruits, particularly during preparation and processing. One outbreak of this type involved a variant strain of Norwalk virus (Herwaldt et al., 1994). Other minimally processed tree fruit products implicated in foodborne illnesses include frozen mamey. In 2010, an outbreak of typhoid fever, the largest in 10 years in the United States, was associated with frozen mamey fruit from Guatemala (Loharikar et al., 2012). The initial detection of this outbreak was attributed to the use of PulseNet, a molecular subtyping network for foodborne disease surveillance in the United States (Swaminathan et al., 2001).

Although outbreaks in whole fresh tree fruits may be more difficult to detect, several such outbreaks have occurred, particularly involving fresh topical fruit. Whole fresh mangoes were associated with an outbreak of salmonellosis in 1999 that sickened 78 people in 13 states, and resulted in two deaths (Sivapalasingam et al., 2003).

Whole fresh papaya were associated with an outbreak of *Salmonella* Agona that occurred in 2011 (CDC, 2011a). In that outbreak 106 individuals in 25 states were affected. An FDA import alert was issued for fresh papaya from Mexico. Although an intensive investigation followed this outbreak, the actual source of the contamination was not determined. In 2012, two outbreaks of salmonellosis, one caused by *S.* Braenderup and one caused by *S.* Worthington, were associated with fresh whole mangoes (CDC, 2012).

Outbreaks associated with tree nuts

As with fruits, tree nuts were not typically associated with the types of acute foodborne illnesses caused by bacteria. However, recent outbreaks in raw almonds and hazelnuts have shown that even low moisture levels will not protect this product from contamination with foodborne pathogens. The initial outbreak recorded involved S. Enteritidis and occurred from the late fall of 2000 to the spring of 2001 (Isaacs et al., 2005). Traceback investigations found *S.* Enteritidis PT30 on equipment surfaces used to process the suspect almonds and on the almond orchard floors. A second outbreak caused by *S.* Enteritidis occurred from September 2003 until April 2004 (CDC, 2004). In this outbreak there were 29 confirmed cases in 12 states with seven hospitalizations. The strain causing this outbreak was distinguished from the first outbreak strain by its pulsed-field electrophoresis (PFGE) pattern. Again, *Salmonella* was isolated from environmental samples collected at the packaging facilities and from huller-shellers that supplied the almonds.

These outbreaks resulted in considerable concern on the part of the industry and regulators, and in 2006, new proposed rules were published by the USDA outlining a mandatory program to reduce the potential for *Salmonella* in almonds (USDA, 2006). This regulation requires that handlers subject their almonds to a process that achieves a minimum 4-log reduction in *Salmonella* prior to shipment. Handlers are defined in 7 CFR 981.13 and exclude roadside sale of almonds, but the definition includes anyone who receives almonds from a grower for later sale. Exemptions were provided to handlers who ship untreated almonds to manufacturers within the United States, Canada, or Mexico who also agree to treat almonds to meet a 4-log pathogen reduction (under a directly verifiable program). These rules became effective March 31, 2007 and mandatory compliance began September 1, 2007.

Almonds are not the only nuts to have been associated with *Salmonella*. In 2009, a recall occurred due to the presence of several serotypes of *Salmonella* in pistachio nuts (CDC, 2009a). However, analysis by the CDC did not find a level of foodborne illness higher than typically expected related to the serotypes found, and consequently, could not attribute a specific outbreak to the pistachio nuts. On the other hand, a link to an outbreak of *S.* Enteritidis in 2011 was established for Turkish pine nuts (CDC, 2011b). In that outbreak, there were 43 illnesses reported in 5 different states.

Salmonella is not the only microorganism that has caused an acute foodborne illness linked to the consumption of tree nuts. In 2011, an outbreak of *E. coli* O157:H7 was traced to in-shell hazelnuts (Miller et al., 2012; Minnesota Department of Health, 2011).

A trace-back from sales receipts was successfully able to identify the suppliers and allow recall of contaminated lots (Miller et al., 2012).

Although outbreaks related to acute foodborne illnesses caused by *Salmonella* and *E. coli* O157:H7 on tree nuts are rare, mycotoxins, and particularly aflatoxins, on tree nuts are a major concern and their presence on nuts cannot be completely eliminated. Since some are teratogenic, mutagenic, or carcinogenic in susceptible animal species and may not produce readily identifiable acute symptoms, an outbreak as such would be difficult if not impossible to identify. Nonetheless, the ingestion of these mycotoxins would have a significant and deleterious effect. Consequently, aflatoxin levels are strictly controlled (FDA, 2011). Other legal levels are set by individual countries. Current standards are set at less then 4 ppb for some aflatoxins by the European Union (European Union, 2010; Kubo, 2012).

At higher levels mycotoxins can also have acute effects, although outbreaks as a result of such acute toxic effects are extremely rare. One example of such a recorded event occurred in New Zealand in a dog due to the consumption of moldy walnuts (Munday et al., 2008). Other outbreaks caused by mycotoxins are more commonly associated with grains (FDA, 2012a). Typically, levels of mycotoxins in dried tree fruits and tree nuts fruits are tightly controlled through routine analysis by various public agencies.

Routes of contamination

Foodborne pathogens are not typically considered to be part of the normal epiphytic populations of fresh tree fruits or nuts. The surfaces of fresh produce, such as tree fruits, have traditionally been viewed as hostile environments for human pathogens. However, favorable conditions may exist in microsites on plant surfaces where growth and survival may occur (Brandl, 2006). Abundant evidence exists for the survival and even growth of foodborne pathogens on the surface of fruits and nuts as well as all produce. Such growth and survival may occur as part of a biofilm firmly attached to the surface. In some fruits, such as apples, growth/survival may occur internally, particularly when the fruit is damaged (Dingman, 2000; Fatemi et al., 2006). Consequently, once introduced to the fruit surface, the pathogens are not easily removed.

The reservoir for *E. coli O157:H7* is generally considered to be domestic livestock (Keller and Miller, 2006). The same reservoir exists for *Cryptosporidium*. From this reservoir, these pathogens can spread to wildlife and ground water. *Salmonella* may occur in domestic livestock and in wild animal populations. Pathogens transferred to wild animals or that contaminate water that may be used in irrigation can result in contaminated produce (Cole et al., 1999; Hanning et al., 2009; Ingham et al., 2004; Jacobsen and Bech, 2012; Levantesi et al., 2012; Moore et al., 2003; Palacios et al., 2001; Sadovski et al., 1978; Steele and Odumeru, 2004). Pathogens may also become airborne in dust and be transferred by wind. Tree nuts such as almonds and hazelnuts are harvested using commercial harvesting equipment that sweeps the nuts from the ground along with dirt, leaves, and twigs, and that generates large amounts of airborne dirt and dust.

For many foodborne outbreaks related to tree fruits, the actual mechanism through which the contamination occurred remains unknown. In part, this can be attributed to the low incidence of occurrence of foodborne pathogens in general and on tree fruit and tree nuts in particular. A study examining the microbial quality of Golden Delicious apples in Spain did not find virulent *E. coli* strains or any *Salmonella* on any apples regardless of where they were selected in the production stream (Abadias et al., 2006). However, *E. coli* was identified on 3 out of 36 field samples. These results were similar to a study of U.S. orchards to identify potential sources of *E. coli* O157:H7 (Riordan et al., 2001). In this study 14 different orchards were surveyed during autumn of 1999 to determine the incidence and prevalence of *E. coli* O157:H7. *E. coli* was found in the soil and on 6% of fruit samples, but no *E. coli* O157:H7 was found. Pathogens were similarly difficult to find on tree nuts. In a survey of brazil nuts harvested in the Amazon, no *Salmonella* or *E. coli* were recovered from the harvested nuts (Arrus et al., 2005). Despite the more ubiquitous nature of molds, only one sample was found with *A. Flavus*, but no aflatoxins were detected (Arrus et al., 2005).

In the United States, Good Agricultural Practices (GAPs) have been published jointly by the USDA and FDA to help ensure the safe production of fresh produce (U.S. Food and Drug Administration, 1998). Adherence to GAPs should ensure an overall low incidence of foodborne pathogens on fresh tree fruit and nuts. However, since even a strict adherence to GAPs cannot reduce the incidence of foodborne pathogens completely, and the risk of an outbreak will be increased in a juice product, juice producers are also required to follow FDA's Juice HACCP regulations. Failure of producers to follow those regulations has resulted in outbreaks associated with juices.

Several studies have examined potential sources of microbial contamination during the manufacture of juices such as apple cider (Garcia et al., 2006; Keller et al., 2004; Keller et al., 2002). Risk factors cited included the use of ground harvested apples, improper cleaning, and contaminated wash water. A survey of apple cider produced in Michigan indicated higher than expected levels of microbial contamination leading to a conclusion that HACCP practices used were either inadequate or were improperly implemented (Bobe et al., 2007). Poor sanitation and hygiene practices can significantly increase risk and the level of foodborne pathogens present on fresh tree fruit and in juice. In a survey by Castillo et al. (Castillo et al., 2006), freshly squeezed orange juice and fresh oranges from public markets and street vendors in Guadalajara, Mexico were examined for the incidence of *Salmonella* and *Shigella*. *Salmonella* was isolated from 9% of orange juice samples collected and from 10% of orange surfaces tested. *Shigella* was isolated from 5% of juice and 8% of orange surfaces tested. Contamination of the juice was attributed to poor hygienic practices. Poor hygienic practices have been cited as the cause of foodborne illness in numerous outbreaks involving tree fruits. These include an outbreak of infectious hepatitis through contaminated orange juice, an outbreak of *Shigella* in orange juice, and an outbreak of Norwalk virus gastroenteritis from fruit salad on a cruise ship (Einstein et al., 1963; Herwaldt et al., 1994; Thurston et al., 1998).

Poor hygiene practices and sanitation are not the only risk factors contributing to outbreaks in fresh tree fruits. In the outbreak of *Salmonella* serotype Newport that occurred in 1999 associated with imported mangoes, traceback of the contaminated fruit pinpointed a farm in Brazil where hot water treatment was used to control Mediterranean Medfly (Sivapalasingam et al., 2003). The same mangoes were also exported to Europe where no outbreak occurred. The single difference in the fruit was the use of a USDA/APHIS-mandated hot water treatment (46.7°C for 75 to 90 minutes), implemented to replace the use of dibromide fumigation. The hot water was not routinely chlorinated but was followed by a chlorinated cold water rinse. Subsequent studies with a simulated process indicated that 80% of green mangoes internalized *Salmonella* when first dipped in hot water followed by a cool water rinse. Internalization facilitated by a temperature gradient has also been demonstrated in apples, oranges, and tomatoes (Buchanan et al., 1999; Merker et al., 1999; Zhuang et al., 1995). USDA/APHIS now recommends appropriate filtration and chlorination of hot water dips and a 20-minute period preceding any cold rinse. This single outbreak demonstrates the importance of a thorough investigation and evaluation of technologies used for food processing.

A thorough investigation and evaluation should not be limited to new technologies. Changes in normal procedures can also lead to unsafe practices. One example of this is in the use of browning inhibitors for fresh-cut apples. To conserve costs, processors may be tempted to extend the solution life beyond a single day or shift. Unfortunately, during extended use such solutions can become contaminated with suspended solids and microorganisms. Once contaminated, such solutions will now contain nutrients and provide an environment far more hospitable for survival and growth of pathogens. The survival of *Listeria innocua* has been demonstrated in calcium ascorbate solutions, particularly at higher pH (Karaibrahimoglu et al., 2004). Although no specific source of *L. monocytogenes* found in recalled apple slices was reported (FDA, 2001), contamination through reuse of antibrowning solutions does represent one possible route.

For nuts, as well as for fruits, harvesting or processing methods may play a significant role in their subsequent contamination. In both almond outbreaks of 2001 and 2004, the outbreak strains were isolated from orchard soil samples and equipment (CDC, 2004; Isaacs et al., 2005). Nonetheless, a source for the contamination was not identified. Persistence of *Salmonella* in the orchard environment was documented by Uesugi et al. (2007) who also found that the rate of isolation increased during months where harvesting occurred. In addition, the highest rate of isolation was found following a rain event while almonds were collected for harvest in windrows on the orchard floor. Previous work by Uesugi and Harris (Uesugi and Harris, 2006) has also suggested that rainfall during harvest when almonds are collected in windrows may play a role in their subsequent contamination. Higher rates of isolation during harvest were also explained by the presence of harvesting equipment and the generation of dust during harvest. The data suggests that *Salmonella* can survive for long periods of time on orchard floors, and almonds may become contaminated during harvesting operations.

Prevention

To reduce the risk of contamination, each step in the production of fresh fruits and nuts must be examined, beginning with field conditions. Ideally, even seemingly unimportant aspects such as drainage or type of soil may play a role in risk reduction. In 1998, the FDA, jointly with the USDA, issued the *Guide to Minimize Microbial Food Safety Hazards for Fresh Fruits and Vegetables* (FDA, 1998). This document outlined common best practices that may be employed to avoid contamination of fresh produce based on the available science. Fundamental issues addressed included the use of water, both for irrigation and subsequent processing. Since the plan is by necessity general in its outline, each grower/processor must tailor the plan more specifically to meet the requirements of each specific product. Since 1998, however, several large outbreaks related to produce have prompted additional legislation in the United States. In 2011, the Food Safety Modernization Act was passed by the U.S. Congress and in 2013 the FDA published proposed changes to GMPs in response to the Act. Many practices that were formally guidelines may become requirements (FDA, 2013).

For tree fruits meant for the fresh market, it is critical that GAP guidelines be followed. Domestic animals should not be allowed access to orchards, and an effort should be made to discourage foraging by wildlife (Cole et al., 1999). Irrigation water should be free of dangerous pathogens to reduce the risk of contamination on the field (Steele and Odumeru, 2004). Likewise any water used for the make-up of pesticides should be free of pathogens, and such pesticide solutions should be made fresh for each use. Contaminated pesticide was cited as the probable cause of contamination of Mandarin oranges (Poubol et al., 2006). In addition, pathogens have been demonstrated to grow in some pesticide solutions (Ng et al., 2005).

Fruits destined for the fresh market should be tree-picked, not ground harvested, and blemish-free. Several studies have linked ground-harvested fruit and damaged fruit with a greater risk of contamination (Dingman, 2000; Fatemi et al., 2006; Keller et al., 2004; Wells and Butterfield, 1997). *E. coli* O157:H7 has been shown to survive and grow in damaged apple tissue (Dingman, 2000; Fatemi et al., 2006). The extent of growth depended upon the apple variety (Dingman, 2000). Damaged and ground-harvested fruits should be directed to those products that will receive processing designed to destroy any pathogens present.

For tree nuts, an alternative to ground-harvesting may not be economically feasible; consequently, processing methods become increasingly important. Care should be taken to avoid further contamination during processing. Culling has been shown as one method to reduce contamination levels both in tree fruits and nuts (De Mello and Scussel, 2007; Jackson et al., 2003; Kadakal and Nas, 2002; Wells and Butterfield, 1997). Culling can aid in the reduction of any mycotoxins present. Although maintaining a "dry" low-moisture environment will not prevent contamination of tree nuts, it can prevent the growth of pathogens and can control the production of mycotoxins (Chen et al., 2009; Murphy et al., 2006). Unfortunately, for some nut products, such as almonds, prevention may not completely reduce risk. Consequently, a final pasteurization step is necessary (USDA, 2006).

Remediation

Once tree fruits and nuts have become contaminated with foodborne pathogens, they are extremely difficult to decontaminate while still retaining their "fresh" character. Numerous studies have examined the efficacy of various surface treatments for the removal of pathogens on produce of all types (Alvarado-Casillas et al., 2007; Annous et al., 2001; Beuchat et al., 1998; Bialka and Demirci, 2007; Das et al., 2006; Fatemi and Knabel, 2006; Jin and Niemira, 2011; Kenney and Beuchat, 2002; Kim et al., 2006; Kondo et al., 2006; Kumar et al., 2006; McWatters et al., 2002; Pao and Davis, 2001; Pao et al., 2000; Park et al., 2011; Parnell et al., 2005; Pierre and Ryser, 2006; Pirovani et al., 2000; Raiden et al., 2003; Sy et al., 2005; Venkitanarayanan et al., 2002). Typical reductions range from 1 to 4 log depending on the produce type, method of inoculation, level of inoculation, and method of pathogen recovery.

None of the surface methods listed would remove any internally occurring pathogens. Pathogens may also exist in a biofilm or other protected state rendering surface decontamination methods ineffective (Burnett and Beuchat, 2001). Only treatment methods with greater penetration can be expected to destroy these hard to reach pathogens. Currently, the only method able to achieve such penetration other than thermal processing to "cook" produce is irradiation. Several studies have investigated the effects of radiation, either alone or in combination with other treatments, on the decontamination of fresh produce (Fan et al., 2006; Nthenge et al., 2007; Saroj et al., 2006; Schmidt et al., 2006; Young Lee et al., 2006). Although irradiation may be effective on some types of produce, it is not currently approved as a food additive for use on fresh produce in the United States (FDA, 2000a). Irradiation at up to 1 kiloGray (kGy) may be used for growth and maturation inhibition of fresh foods and for deinfestation of arthropod pests; however, these levels would likely be insufficient for reduction or elimination of some microbial pathogens. Irradiation may be used on produce in the European Union but must be labeled.

Surface decontamination, although unable to destroy pathogens borne internally, can still be viewed as efficacious for tree fruits and nuts for which internalization is unlikely to occur. Such fruits would include citrus fruits, for which the FDA has provided an exemption in the HACCP regulation, allowing a 5-log reduction in the pertinent pathogen to be applied to the surface of the fruit as opposed to the whole fruit. As with any processing step aimed at pathogen reduction, efficacy of the process is predicated on assumed initial microbial loads. Therefore, processes applied to the surface must be applied to clean fruit without visible blemishes or damage. Blemishes or damage to the protective peel could allow pathogens to become internalized, rendering processing treatments ineffective. Consequently, culling or quality sorting can be viewed as a critical step prior to the application of surface treatments (Keller, 2006).

In the production of fruit juices, a variety of processing methods are available to achieve an appropriate 5-log reduction in pertinent pathogens. The most commonly used method remains thermal pasteurization; however, UV-light treatment and ozone treatment have also been explored (Duffy et al., 2000; Harrington and Hills, 1968; Koutchma et al., 2004; Williams et al., 2004; Wright et al., 2000). Regardless of the method used, the juice processor must validate that the process used will achieve the prescribed 5-log reduction with the juice processed in the specific equipment used. Failure to do so has resulted in recent outbreaks. In addition, it should be noted that such treatments will result in a processed, not fresh product, and must be labeled as such.

Conclusions

Although foodborne illnesses associated with tree fruits and nuts are not common, their occurrence is particularly troublesome since they are associated with a healthful diet. Once tree fruits or nuts become contaminated with foodborne pathogens, their removal is extremely difficult while still maintaining their "fresh" state. Consequently, a fundamental key in the prevention of foodborne illness in fresh produce is to remove the vectors that may transmit the pathogen from the animal source to the produce. Unfortunately, when the vectors may be wild animals, wind, or irrigation water, risk can be reduced, but cannot be eliminated. Adherence to the FDA/USDA *Guide to Minimize Microbial Food Safety Hazards for Fresh Fruits and Vegetables* is critical to ensuring the safety of these foods. In addition, although adherence to GAPs is important, the role of postharvest contamination cannot be overlooked. A principal cause of foodborne outbreaks related to fresh produce of all types remains poor worker and facility hygiene. Attention to the entire production chain, from the farm, throughout processing and transport, to the table of the consumer is required to ensure that tree fruits and nuts do not become contaminated with foodborne pathogens. Such attention should result in the continued safety of tree fruits and nuts.

References

Abadias, M., Canamas, T.P., Asensio, A., Anguera, I.V., 2006. Microbial quality of commercial 'Golden Delicious' apples throughout production and shelf-life in Lleida (Catalonia, Spain). Int. J. Food Microbiol. 108, 404–409.

Alvarado-Casillas, S., Ibarra-Sanchez, S., Rodriguez-Garcia, O., Martinez-Gonzales, N., Castillo, A., 2007. Comparison of rinsing and sanitizing procedures for reducing bacterial pathogens on fresh cantaloupes and bell peppers. J. Food Prot. 70, 655–660.

Annous, B.A., Sapers, G.M., Mattrazzo, A.M., Riordan, D.C.R., 2001. Efficacy of washing with a commercial flatbed brush washer, using conventional and experimental washing agents, in reducing populations of *Escherichia coli* on artificially inoculated apples. J. Food Prot. 64, 159–163.

Arrus, K., Blank, G., Clear, R., Holley, R.A., Abramson, D., 2005. Microbiological and aflatoxin evaluation of Brazil nut pods and the effects of unit processing operations. J. Food Prot. 68, 1060–1065.

Bagamboula, C.F., Uyttendaele, M., Debevere, J., 2002. Acid tolerance of *Shigella sonnei* and Shigella flexneri. J. Appl. Microbiol. 93, 479–486.

Bang, I.S., Kim, B.H., Foster, J.W., Park, Y.K., 2000. OmpR regulates the stationary-phase acid tolerance response of *Salmonella enterica* serovar Typhimurium. J. Bacteriol. 182, 2245–2252.

Bayman, P., Baker, J.L., Mahoney, N.E., 2002. *Aspergillus* on tree nuts: incidence and associations. Mycopathologia 155, 161–169.

Belluck, P., 1998. Juice-poisoning case brings guilty plea and a huge fine. The New York Times, The New York Times, New York.

Benjamin, M.M., Datta, A.R., 1995. Acid tolerance of enterohemorrhagic *Escherichia coli*. Appl. Environ. Microbiol. 61, 1669–1672.

Beuchat, L.R., 1995. Pathogenic microorganisms associated with fresh produce. J. Food Prot. 59, 204–216.

Beuchat, L.R., Brackett, R.E., 1991. Behavior of *Listeria monocytogenes* inoculated into raw tomatoes and processed tomato products. Appl. Environ. Microbiol. 57, 1367–1371.

Beuchat, L.R., Heaton, E.K., 1975. *Salmonella* survival on pecans as influenced by processing and storage conditions. Appl. Microbiol. 29, 795–801.

Beuchat, L.R., Nail, B.V., Clavero, M.R.S., 1998. Efficacy of chlorinated water in killing pathogenic bacteria on raw apples, tomatoes, and lettuce. J. Food Prot. 61, 1305–1311.

Beuchat, L.R., Ryu, J.-H., 1997. Produce handling and processing practices. Emerg. Infect. Dis. 3.

Bialka, K.L., Demirci, A., 2007. Efficacy of aqueous ozone for the decontamination of *Escherichia coli* O157:H7 and *Salmonella* on raspberries and strawberries. J. Food Prot. 70, 1088–1092.

Blackburn, B.G., Mazurek, J.M., Hlavsa, M., Park, J., Tillapaw, M., Parrish, M., et al., 2006. Cryptosporidiosis associated with ozonated apple cider. Emerg. Infect. Dis. 12, 684–686.

Bobe, G., Thede, D.J., Eyck, T.A.T., Bourquin, L.D., 2007. Microbial levels in Michigan apple cider and their association with manufacturing practices. J. Food Prot. 70, 1187–1193.

Brandl, M.T., 2006. Fitness of human enteric pathogens on plants and implications for food safety. Annu. Rev. Phytopathol. 44, 367–392.

Buchanan, R.L., Doyle, M.E., 1997. Foodborne disease significance of *Escherichia coli* O157:H7 and other enterohemorrhagic *E. coli*. Food Technol. 51, 69–76.

Buchanan, R.L., Edelson, S.G., 1996. Culturing enterohemorrhagic *Escherichia coli* in the presence and absense of glucose as a simple means of evaluating the acid tolerance of stationary-phase cells. Appl. Environ. Microbiol. 62, 4009–4013.

Buchanan, R.L., Edelson, S.G., Miller, R.L., Sapers, G.M., 1999. Contamination of intact apples after immersion in an aqueous environment containing *Escherichia coli* O157:H7. J. Food Prot. 62, 444–450.

Burnett, S.L., Beuchat, L.R., 2001. Human pathogens associated with raw produce and unpasteurized juices, and difficulties in decontamination. J. Ind. Microbiol. Biotechnol. 27, 104–110.

Burnett, S.L., Chen, J., Beuchat, L.R., 2000. Attachment of *Escherichia coli* O157:H7 to the surfaces and internal structures of apples as detected by confocal scanning laser microscopy. Appl. Environ. Microbiol. 66, 4679–4687.

Castillo, A., Villarruel-Lopez, A., Navarro-Hidalgo, V., Martinez-Gonzalez, N.E., Torres-Vitela, M.R., 2006. *Salmonella* and *Shigella* in freshly squeezed orange juice, fresh oranges, and wiping cloths collected from public markets and street booths in Guadalajara, Mexico: Incidence and comparison of analytical routes. J. Food Prot. 69, 2595–2599.

Centers for Disease Control and Prevention, 1996. Outbreaks of *Escherichia coli* O157:H7 infections associated with drinking unpasteurized commercial apple juice-October, 1996. Morb. Mortal. Wkly. Rep. 45, 975.

Centers for Disease Control and Prevention, 1997. Update: Outbreaks of Cyclosporiasis–United States and Canada, 1997. MMWR. Morb. Mortal. Wkly. Rep. 46, 521–523.

Centers for Disease Control and Prevention, 2003. U.S. Foodborne disease outbreakes, annual listing 1990–2000. http://www.cdc.gov/foodborneoutbreaks/report_pub.htm (accessed 06.03.03.).

Centers for Disease Control and Prevention, 2004. Outbreak of Salmonella Serotype Enteritidis Infections Associated with Raw Almonds –- United States and Canada, 2003–2004. Morb. Mortal. Wkly. Rep. 53, 484–487.

Centers for Disease Control and Prevention, 2005a. Preventing health risks associated with drinking unpasteurized or untreated juice. http://www.cdc.gov/foodborne/juice_spotlight. htm (accessed 26.02.13.).

Centers for Disease Control and Prevention, 2005b. Shigellosis. http://www.cdc.gov/ncidod/ dbmd/diseaseinfo/shigellosis_t.htm (accessed 26.02.13.).

Centers for Disease Control and Prevention, 2009a. *Salmonella* in pistachio Nuts, 2009. U.S. Department of Health and Human Services. www.cdc.gov/salmonella/pistachios/update. html (accessed 14.01.13.).

Centers for Disease Control and Prevention, 2009b. Surveillance for foodborne disease outbreaks - United States, 2006. Morb. Mortal. Wkly. Rep. 58, 609–615.

Centers for Disease Control and Prevention, 2011a. Investigation update: Multistate outbreak of human *Salmonella* Agona infections linked to whole, fresh imported papayas. U.S. Department of Health and Human Services. http://www.cdc.gov/salmonella/agona-papayas/082911/index.html (accessed 14.01.13.).

Centers for Disease Control and Prevention, 2011b. Investigation update: Multistate outbreak of human *Salmonella* Enteritidis infections linked to Turkish pine nuts. U.S. Department of Health and Human Services. http://www.cdc.gov/salmonella/pinenuts-enteriditis/1117 11/index.html (accessed 14.01.13.).

Centers for Disease Control and Prevention, 2011c. Investigation update: Outbreak of Shiga toxin-producing E. coli O104 (STEC O104:H4)infections associated with travel to Germany. http://www.cdc.gov/ecoli/2011/ecolio104/ (accessed 26.02.13.).

Centers for Disease Control and Prevention, 2012. Multistate outbreak of Salmonella Braenderup infections associated with mangoes (final update). U.S. Department of Health and Human Services. http://www.cdc.gov/salmonella/braenderup-08-12/index.html accessed.

Chen, Y., Scott, V.N., Freier, T.A., Kuehm, J., Moorman, M., Meyer, J., et al., 2009. Control of *Salmonella* in low-moisture foods III: Process validation and environmental monitoring. Food Prot. Trends 29, 493–508.

Cole, D.J., Hill, V.R., Humenik, F.J., Sobsey, M.D., 1999. Health, safety and environmental concerns of farm animal waste. Occup. Med. 14, 423–428.

Cox, L.J., Kleiss, T., Cordier, J.L., Cordellana, C., Konel, P., Pedrazzini, C., et al., 1989. *Listeria* spp. in food processing, non-food and domestic environments. Food Microbiol. 6, 49–61.

Danyluk, M.D., Jones, T.M., Abd, S.J., Schlitt-Dittrich, F., Jacobs, M., Harris, L.J., 2007. Prevalence and amounts of *Salmonella* found on raw California almonds. J. Food Prot. 70, 820–827.

Das, E., Gurakan, G.C., Bayindirli, A., 2006. Effect of controlled atmosphere storage, modified atmosphere packaging and gaseous ozone treatment on the survival of *Salmonella* Enteritidis on cherry tomatoes. Food Microbiol. 23, 430–438.

De Mello, F.R., Scussel, V.M., 2007. Characteristics of in-shell Brazil nuts and their relationship to aflatoxin contamination: criteria for sorting. J. Agric. Food Chem. 55, 9305–9310.

Dingman, D.W., 2000. Growth of *Escherichia coli* O157:H7 in bruised apple (Malus domestica) tissue as influences by cultivar, date of harvest, and source. Appl. Environ. Microbiol. 66, 1077–1083.

Duffy, S., Churey, J., Worobo, R., Schaffner, D.W., 2000. Analysis and modeling of the variability associated with UV inactivation of *Escherichia coli* in apple cider. J. Food Prot. 63, 1587–1590.

Duncan, T.G., Doull, J.A., Miller, E.R., Bancroft, H., 1946. Outbreak of typhoid fever with orange juice as the vehicle, illustrating the value of immunization. Am. J. Public Health 36, 34–36.

Eblen, B.S., Walderhaug, M.O., Edelson-Mammel, S., Chirtel, S.J., De Jesus, A., Merker, R.I., et al., 2004. Potential for internalization, growth and survival of *Salmonella* spp. and *Escherichia coli* O157:H7 in oranges. J. Food Prot. 67, 1578–1584.

Eglezos, S., 2010. The bacteriological quality of retail-level peanut, almond, cashew, hazelnut, brazil, and mixed nut kernels produced in two Australian nut-processing facilities over a period of 3 years. Foodborne Pathog. Dis. 00 00.

Einstein, A.B., Jacobsohn, W., Goldman, A., 1963. An epidemic of infectious hepatitis in a general hospital: probable transmission by contaminated orange juice. J. Am. Med. Assoc. 185, 171–174.

European Union, 2010. Summaries of EU legislation: Maximum levels for certain contaminants. http://europa.eu/legislation_summaries/food_safety/contamination_enviro nmental_factors/l21290_en.htm (accessed 01.03.13.).

Fan, X., Annous, B.A., Sokorai, K.J.B., Burke, A., Mattheis, J.P., 2006. Combination of hot-water surface pasteurization of whole fruit and low-dose gamma irradiation of fresh-cut cantaloupe. J. Food Prot. 69, 912–919.

Fatemi, P., Knabel, S.J., 2006. Evaluation of sanitizer penetration and its effect on destruction of *Escherichia coli* O157:H7 in golden delicious apples. J. Food Prot. 69, 548–555.

Fatemi, P., LaBorde, L.F., Patton, J., Sapers, G.M., Annous, B.A., Knabel, S.J., 2006. Influence of punctures, cuts, and surface morphologies of golden delicious apples on penetration and growth of *Escherichia coli* O157:H7. J. Food Prot. 69, 267–275.

Fenlon, D.R., Wilson, J., Donachie, W., 1996. The incidence and level of *Listeria monocytogenes* contamination of food scources at primary production and initial processing. J. Appl. Bacteriol. 81, 641–650.

Fleischman, G.J., Bator, C., Merker, R., Keller, S.E., 2001. Hot water immersion to eliminate *Escherichia coli* O157:H7 on the surface of whole apples: thermal effects and efficacy. J. Food Prot. 64, 451–455.

Foster, J.W., Hall, H.K., 1990. Adaptive acidification tolerance response of *Salmonella typhimurium*. J. Bacteriol. 172, 771–778.

Garcia, L., Henderson, J., Fabri, M., Moustapha, O., 2006. Potential sources of microbial contamination in unpasteurized apple cider. J. Food Prot. 69, 137–144.

Gombas, D.E., Chen, Y., Clavero, R.S., Scott, V.N., 2003. Survey of *Listeria monocytogenes* in ready-to-eat foods. J. Food Prot. 66, 559–569.

Goverd, K.A., Beech, F.W., Hobbs, R.P., Shannon, R., 1979. The occurrence and survival of colifoms and salmonellas in apple juice and cider. J. Appl. Bacteriol. 46, 521–530.

Gruzdev, N., Pinto, R., Sela, S., 2011. Effect of Desiccation on tolerance of *Salmonella enterica* to multiple stresses. Appl. Environ. Microbiol. 77, 1667–1673.

Guerrant, R.L., 1997. Cryptosporidiosis: An emerging highly infectious threat. Emerg. Infect. Dis. 3, 51–57.

Hanning, I.B., Nutt, J.D., Ricke, S.C., 2009. Salmonellosis outbreaks in the United States due to fresh produce: sources and potential intervention measures. Foodborne Pathog. Dis. 6, 635–648.

Harrington, W.O., Hills, C.H., 1968. Reduction of the microbial population of apple cider by ultraviolet irradiation. Food Technol. 22, 117–120.

Harris, K.L., Bobe, G., Bourquin, L.D., 2009. Patulin surveillance in apple cider and juice marketed in Michigan. J. Food Prot. 72, 1255–1261.

Health Protection Agency, 2011. Foodborne outbreaks reported to the Health Protection Agency, England and Wales, 1992–2010 (by food vehicle). http://www.hpa.org.uk/Topics/Infectiou sDiseases/InfectionsAZ/FoodborneOutbreakSurveillanceAndRiskAssessment/Foodborne Outbreaks/eFOSSFoodborneoutbreakspathogenbyfood19922010gi/ (accessed 26.02.13.).

Herwaldt, B.L., Lew, J.F., Moe, C.L., Lewis, D.C., Humphrey, C.D., Monroe, S.S., et al., 1994. Characterization of a variant strain of norwalk virus from a foodborne outbreak of gastroenteritis on a cruise ship in Hawaii. J. Clin. Microbiol. 32, 861–866.

Ingham, S.C., Losinski, J.A., Andrews, M.P., Breuer, J.E., Breuer, J.R., Wood, T.M., et al., 2004. *Escherichia coli* contamination of vegetables grown in soils fertilized with noncomposted bovine manure: Garden-scale studies. Appl. Environ. Microbiol. 70, 6420–6427.

Isaacs, S., Aramini, J., Ciebin, B., Farrar, J.A., Ahmed, R., Middleton, D., et al., 2005. An international outbreak of salmonellosis associated with raw almonds contaminated with a rare phage type of *Salmonella enteritidis*. J. Food Prot. 68, 191–198.

Jackson, L.S., Beacham-Bowden, T., Keller, S.E., Adhikari, C., Taylor, K.T., Chirtel, S.J., et al., 2003. Apple quality, storage, and washing treatments affect patulin levels in apple cider. J. Food Prot. 66, 618–624.

Jacobsen, C.S., Bech, T.B., 2012. Soil survival of *Salmonella* and transfer to freshwater and fresh produce. Food Res. Int. 45, 557–566.

Jin, T., Niemira, B.A., 2011. Application of polylactic acid coating with antimicrobials in reduction of *Escherichia coli* O157:H7 and *Salmonella* Stanley on apples. J. Food Sci. 76, M184–M188.

Johnston, L.M., Jaykus, L., Moll, D., Anciso, J., Mora, B., Moe, C.L., 2005a. A field study of the microbiological quality of fresh produce of domestic and Mexican origin. Int. J. Food Microbiol. 112, 83–96.

Johnston, L.M., Jaykus, L., Moll, D., Martinez, A.C., Anciso, J., Mora, B., et al., 2005b. A field study of the microbiological quality of fresh produce. J. Food Prot. 68, 1840–1847.

Kadakal, C., Nas, S., 2002. Effect of apple decay proportion on the patulin, fumaric acid, HMF and other apple juice properties. J. Food Saf. 22, 17–25.

Karaibrahimoglu, Y., Fan, X., Sapers, G.M., Sokorai, K., 2004. Effect of pH on the survival of *Listeria innocua* in calcium ascorbate solutions and on quality of fresh-cut apples. J. Food Prot. 67, 751–757.

Kassa, H., 2001. An outbreak of Norwalk-like viral gastroenteritis in a frequently penalized food service operation: a case for mandatory training of food handlers in safety and hygiene. J. Environ. Health 64, 9–12 33.

Keller, S.E., 2006. Chapter 16: Effect of quality sorting and culling on the microbiological quality of fresh produce. In: Sapers, G.M., Gorny, J.R., Yousef, A.E. (Eds.), Microbiology of Fruits and Vegetables. CRC Press, Boca Raton, Fl, pp. 365–373.

Keller, S.E., Chirtel, S.J., Merker, R.I., Taylor, K.T., Tan, H.L., Miller, A.J., 2004. Influence of fruit variety, harvest technique, quality sorting, and storage on the native microflora of unpasteurized apple cider. J. Food Prot. 67, 2240–2247.

Keller, S.E., Merker, R.I., Taylor, K.T., Tan, H.L., Melvin, C.D., Chirtel, S.J., et al., 2002. Efficacy of sanitation and cleaning methods in a small apple cider mill. J. Food Prot. 65, 911–917.

Keller, S.E., Miller, A.J., 2006. Chapter 9: Microbiological safety of fresh citrus and apple juices. In: Sapers, G.M., Gorny, J.R., Yousef, A.E. (Eds.), Microbiology of Fruits and Vegetables. CRC press, Boca Raton, FL, pp. 211–230.

Keller, S.E., Van Doren, J.M., Grasso, E.M., Halik, L.A., 2013. Growth and survival of *Salmonella* in ground black pepper (*Piper nigrum*). Food Microbiol. 34, 182–188.

Kenney, S.J., Beuchat, L.R., 2002. Comparison of aqueous commercial cleaners for effectiveness in removing *Escherichia coli* O157:H7 and *Salmonella muenchen* from the surface of apples. Int. J. Food Microbiol. 74, 47–55.

Kim, H., Ryu, J.-H., Beuchat, L.R., 2006. Survival of *Enterobacter sakazakii* on fresh produce as affected by temperature, and effectiveness of sanitizers for its elimination. Int. J. Food Microbiol. 111, 134–143.

Kimber, M.A., Kaur, H., Wang, L., Danyluk, M.D., Harris, L.J., 2012. Survival of *Salmonella, Escherichia coli* O157:H7, and *Listeria monocytogenes* on inoculated almonds and pistachios stored at -19, 4, and 24°C. J. Food Prot. 75, 1394–1403.

Komitopoulou, E., Penaloza, W., 2009. Fate of *Salmonella* in dry confectionery raw materials. J. Appl. Microbiol. 106, 1892–1900.

Kondo, N., Murata, M., Isshiki, K., 2006. Efficiency of sodium hypochlorite, fumaric acid, and mild heat in killing native microflora and *Escherichia coli* O157:H7, *Salmonella* Typhimurium DT104, and *Staphylococcus aureus* attached to fresh-cut lettuce. J. Food Prot. 69, 323–329.

Koutchma, T., Keller, S.E., Parisi, B., Chirtel, S.J., 2004. Ultraviolet disinfection of juice products in laminar and turbulent flow reactors. Innovative Food Sci. Emerg. Technol. 5, 179–189.

Kubo, M., 2012. Mycotoxins Legislation Worldwide European Mycotoxins Awareness Network. http://services.leatherheadfood.com/eman/FactSheet.aspx?ID=79 (accessed 01.03.13.).

Kumar, M., Hora, R., Kostrzynska, M., Waites, W.M., Warriner, K., 2006. Inactivation of Escherichia coli O157:H7 and *Salmonella* on mung beans, alfalfa, and other seed types destined for sprout production by using an oxychloro-based sanitizer. J. Food Prot. 69, 1571–1578.

Levantesi, C., Bonadonna, L., Briancesco, R., Grohmann, E., Toze, S., Tandoi, V., 2012. *Salmonella* in surface and drinking water: Occurrence and water-mediated transmission. Food Res. Int. 45, 587–602.

Lin, J., Lee, I.S., Frey, J., Slonczewski, J.L., Foster, J.W., 1995. Comparative analysis of extreme acid survival in *Salmonella typhimurium, Shigella flexneri*, and *Escherichia coli*. J. Bacteriol. 177, 4097–4104.

Lin, J., Smith, M.P., Chapin, K.C., Baik, H.S., Bennett, G.N., Foster, J.W., 1996. Mechanisms of acid resistance in Enterohemorrhagic *Escherichia coli*. Appl. Environ. Microbiol. 62, 3094–3100.

Little, C.L., Mitchell, R.T., 2004. Microbiological quality of pre-cut fruit, sprouted seeds, and unpasteurised fruit and vegetable juices from retail and production premises in the UK, and the application of HACCP. Commun. Dis. Public Health 7, 184–190.

Loharikar, A., Newton, A., Rowley, P., Wheeler, C., Bruno, T., Barillas, H., et al., 2012. Typhoid fever outbreak associated with frozen mamey pulp imported from Guatemala to the western United States, 2010. Clin. Infect. Dis. 55, 61–66.

Marín, S., Hodzić, I., Ramos, A.J., Sanchis, V., 2008. Predicting the growth/no-growth boundary and ochratoxin A production by *Aspergillus carbonarius* in pistachio nuts. Food Microbiol. 25, 683–689.

McWatters, K.H., Doyle, M.P., Walker, S.L., Rimal, A.P., Venkitanarayanan, K., 2002. Consumer acceptance of raw apples treated with an antibacterial solution designed for home use. J. Food Prot. 65, 106–110.

Merker, R., Edelson-Mammel, S., Davis, V., Buchanan, R.L., 1999. Preliminary experiments on the effect of temperature differences on dye uptake by oranges and grapefruit. U.S. Food and Drug Aministration.

Millard, P.S., Gensheimer, K.F., Addiss, D.G., Sosin, D.M., Beckett, G.A., Houck-Jankoski, A., et al., 1994. An outbreak of Cryptosporidiosis from fresh-pressed apple cider. J. Am. Med. Assoc. 272, 1592–1596.

Miller, B.D., Rigdon, C.E., Ball, J., Rounds, J.M., Klos, R.F., Brennan, B.M., et al., 2012. Use of traceback methods to confirm the source of a multistate *Escherichia coli* O157:H7 outbreak due to in-shell hazelnuts. J. Food Prot. 75, 320–327.

Minnesota Department of Health, 2011. MDA laboratory testing confirms *E. coli* O157:H7 in recalled hazelnuts. http://www.health.state.mn.us/news/pressrel/2011/ecoli030911.html (accessed 30.10.12.).

Molyneux, R.J., Mahoney, N., Kim, J.H., Campbell, B.C., 2007. Mycotoxins in edible tree nuts. Int. J. Food Microbiol. 119, 72–78.

Moore, B.C., Martinez, E., Gay, J.M., Rice, D.H., 2003. Survival of *Salmonella enterica* in freshwater and sediments and transmission by the aquatic midge *Chironomus tentans* (Chironomidae: Diptera). Appl. Environ. Microbiol. 69, 4556–4560.

Munday, J.S., Thompson, D., Finch, S.C., Babu, J.V., Wilkins, A.L., di Menna, M.E., et al., 2008. Presumptive tremorgenic mycotoxicosis in a dog in New Zealand, after eating mouldy walnuts. N. Z. Vet. J. 56, 145–148.

Murphy, P.A., Hendrich, S., Landgren, C., Bryant, C.M., 2006. Food mycotoxins: An update. J. Food Sci. 71, R51–R65.

Ng, P.J., Fleet, G.H., Heard, G.M., 2005. Pesticides as a source of microbial contamination of salad vegetables. Int. J. Food Microbiol. 101, 237–250.

Nielsen, K.F., Mogensen, J.M., Johansen, M., Larsen, T.O., Frisvad, J.C., 2009. Review of secondary metabolites and mycotoxins from the *Aspergillus niger* group. Anal. Bioanal. Chem. 395, 1225–1242.

Noël, H., Hofhuis, A., De Jonge, R., Heuvelink, A.E., De Jong, A., Heck, M.E., et al., 2010. Consumption of fresh fruit juice: how a healthy food practice caused a national outbreak of *Salmonella* Panama gastroenteritis. Foodborne Pathog. Dis. 7, 375–381.

Nthenge, A.K., Weese, J.S., Carter, M., Wei, C., Huang, T., 2007. Efficacy of gamma radiation and aqueous chlorine on *Escherichia coli* O157:H7 in hydroponically grown lettuce plants. J. Food Prot. 70, 748–752.

Painter, J.A., Hoekstra, R.M., Ayers, T., Tauxe, R.V., Braden, C.R., Angulo, F.J., et al., 2013. Attribution of foodborne illnesses, hospitalizations, and deaths to food commodities by using outbreak data, United States, 1998–2008. Emerg. Infect. Dis. http://dx.doi.org/10.3 201/eid1903.111866 (accessed 27.02.13.).

Palacios, M.P., Lupiola, P., Tejedor, M.T., Del-Nero, E., Pardo, A., Pita, L., 2001. Climatic effects on *Salmonella* survival in plant and soil irrigated with artificially inoculated wastewater: preliminary results. Water Sci. Technol. 43, 103–108.

Pao, S., Davis, C.L., 2001. Maximizing microbiological quality of fresh orange juice by processing sanitation and fruit surface treatments. Dairy, Food Environ. Sanit. 21, 287–291.

Pao, S., Davis, C.L., Kelsey, D.F., 2000. Efficacy of alkaline washing for the decontamination of orange fruit surfaces inoculated with *Escherichia coli*. J. Food Prot. 63, 961–964.

Parish, M.E., 1997. Public health and nonpasteurized fruit juices. Crit. Rev. Microbiol. 23, 109–119.

Park, S.H., Choi, M.R., Park, J.W., Park, K.H., Chung, M.S., Ryu, S., et al., 2011. Use of organic acids to inactivate *Escherichia coli* O157:H7, *Salmonella* Typhimurium, and *Listeria monocytogenes* on organic fresh apples and lettuce. J. Food Sci. 76, M293–M298.

Parnell, T.L., Harris, L.J., Suslow, T.V., 2005. Reducing *Salmonella* on cantaloupes and honeydew melons using wash practices applicable to postharvest handling, foodservice, and consumer preparation. Int. J. Food Microbiol. 99, 59–70.

Penteado, A.L., Eblen, B.S., Miller, A.J., 2004. Evidence of *Salmonella* internalization into fresh mangos during simulated postharvest insect disinfestation procedures. J. Food Prot. 67, 181–184.

Pierre, P.M., Ryser, E.T., 2006. Inactivation of *Escherichia coli* O157:H7, *Salmonella* Typhimurium DT104, and *Listeria monocytogenes* on inoculated alfalfa seeds with a fatty acid–based sanitizer. J. Food Prot. 69 582–290.

Pirovani, M.E., Guemes, D.R., Di Pentima, J.H., Tessi, M.A., 2000. Survival of *Salmonella hadar* after washing disinfection of minimally processed spinach. Lett. Appl. Microbiol. 31, 143–148.

Poubol, J., Tsukada, Y., Sera, K., Izumi, H., 2006. On-the-farm contamination of Satsuma mandarin fruit at different orchards in Japan. Acta. Hortic. 712, 551–560.

Public Health Agency of Canada, 2012. Public health notice: Outbreak of *Salmonella* illness related to mangoes. Health Canada. http://www.phac-aspc.gc.ca/fs-sa/phn-asp/osm-esm-eng.php (accessed 14.01.13.).

Raiden, R.M., Sumner, S.S., Eifert, J.D., Pierson, M.D., 2003. Efficacy of detergents in removing *Salmonella* and *Shigella* spp. from the surface of fresh produce. J. Food Prot. 66, 2210–2215.

Reis, T.A., Oliveira, T.D., Baquião, A.C., Gonçalves, S.S., Zorzete, P., Corrêa, B., 2012. Mycobiota and mycotoxins in Brazil nut samples from different states of the Brazilian Amazon region. Int. J. Food Microbiol. 159, 61–68.

Riordan, D.C.R., Sapers, G.M., Hankinson, T.R., Magee, M., Mattrazzo, A.M., Annous, B.A., 2001. A study of U.S. orchards to identify potential sources of *Escherichia coli* O157:H7. J. Food Prot. 64, 1320–1327.

Sado, P.N., Jinneman, K.C., Husby, G.J., Sorg, S.M., Omiecinski, C.J., 1998. Identification of *Listeria monocytogenes* from unpasteurized apple juice using rapid test kits. J. Food Prot. 61, 1199–1202.

Sadovski, A.Y., Fattal, B., Goldberg, D., Katzenelson, E., Shuval, H.I., 1978. High levels of microbial contamination of vegetables irrigated with wastewater by the drip method. Appl. Environ. Microbiol. 36, 824–830.

Saroj, S.D., Shashidhar, R., Pandey, M., Dhokane, V., Hajare, S., Sharma, A., et al., 2006. Effectiveness of radiation processing in elimination of *Salmonella* Typhimurium and *Listeria monocytogenes* from sprouts. J. Food Prot. 69, 1858–1864.

Schmidt, H.M., Palekar, M.P., Maxim, J.E., Castillo, A., 2006. Improving the microbiological quality and safety of fresh-cut tomatoes by low-dose electron beam irradiation. J. Food Prot. 60, 575–581.

Sivapalasingam, S., Barrett, E., Kimura, A., Van Duyne, S., De Witt, W., Ying, M., et al., 2003. A multistate outbreak of *Salmonella enterica* Serotype Newport infection linked to mango consumption: impact of water-dip disinfestation technology. Clin. Infect. Dis. 37, 1585–1590.

Sorrells, K.M., Enigl, D.C., Hatfield, J.R., 1989. Effect of pH, acidulant, time, and temperature on the growth and survival of *Listeria monocytogenes*. J. Food Prot. 52, 571–573.

Soto, M., Chavez, G., Baez, M., Martinez, C., Chaidez, C., 2007. Internalization of *Salmonella typhimurium* into mango pulp and prevention of fruit pulp contamination by chlorine and copper ions. Int. J. Environ. Health Res. 17, 453–459.

Steele, M., Odumeru, J., 2004. Irrigation water as source of foodborne pathogens on fruit and vegetables. J. Food Prot. 67, 2839–2849.

Swaminathan, B., Barrett, T.J., Hunter, S.B., Tauxe, R.V., 2001. PulseNet: the molecular subtyping network for foodborne bacterial disease surveillance, United States. Emerg. Infect. Dis. 7, 382–389.

Sy, K.V., McWatters, K.H., Beuchat, L.R., 2005. Efficacy of gaseous chlorine dioxide as a sanitizer for killing Salmonella, yeasts, and molds on blueberries, strawberries, and raspberries. J. Food Prot. 68, 1165–1175.

Thurston, H., Stuart, J., McDonnell, B., Nicholas, S., Cheasty, T., 1998. Fresh orange juice implicated in an outbreak of *Shigella flexneri* among visitors to a South African game reserve. J. Infect. 36, 350.

U.S. Department of Agriculture, 2006. 7 CFR Part 981: Almonds grown in California; Outgoing quality control requirements and request for approval of new information collection. Fed. Regist. 71, 70683–70692.

U.S. Department of Agriculture, 2012. Commodity consumption by population characteristics. http://www.ers.usda.gov/data-products/commodity-consumption-by-population-characteristics.aspx (accessed 26.02.13.).

U.S. Food and Drug Administration, 1998. Guide to minimize microbial food safety hazards for fresh fruits and vegetables. Washington, D.C. http://www.fda.gov/downloads/Food/GuidanceComplianceRegulatoryInformation/GuidanceDocuments/ProduceandPlanProducts/UCM169112.pdf (accessed 27.08.12.).

U.S. Food and Drug Administration, 2000a. 21 CFR Part 179. Irradiation in the production, processing and handling of food. Fed. Regist. 65, 71056–71058.

U.S. Food and Drug Administration, 2000b. Action levels for poisonous or deleterious substances in human food and animal feed. U.S. Food Drug Adm.

U.S. Food and Drug Administration, 2001a. 21 CFR Part 120. Hazard analysis and critical control point (HACCP); Procedures for the safe and sanitary processing and importing of juice. Final rule. Fed. Regist. 66, 6137–6202.

U.S. Food and Drug Administration, 2001b. Patulin in apple juice, apple juice concentrates and apple juice products. http://www.fda.gov/Food/FoodSafety/FoodContaminantsAdulteration/NaturalToxins/ucm212520.htm (accessed 01.03.13.).

U.S. Food and Drug Administration, 2011. Food defect levels handbook. http://www.fda.gov/food/guidancecomplianceregulatoryinformation/guidancedocuments/sanitation/ucm056174.htm (accessed 24.08.12.).

U.S. Food and Drug Administration, 2012a. The bad bug book, second edition. Foodborne pathogenic microorganisms and natural toxins handbook. http://www.fda.gov/Food/FoodSafety/FoodborneIllness/FoodborneIllnessFoodbornePathogensNaturalToxins/BadBugBook/default.htm (accessed 23.10.12.).

U.S. Food and Drug Administration, 2012b. Reportable food registry annual report: FDA foods and veterinary medicine program: the reportable food registry: Targeting inspection resources and identifying patterns of adulteration: Second annual report: September 8, 2010 – September 7, 2011. http://www.fda.gov/Food/FoodSafety/FoodSafetyPrograms/RFR/ucm200958.htm (accessed 25.05.12.).

U.S. Food and Drug Administration, 2013. Current good manufacturing practice and hazard analysis and risk-based preventive controls for human food. Fed. Regist. 78, 3646–3824.

Uesugi, A.R., Danyluk, M.D., Harris, L.J., 2006. Survival of *Salmonella* Enteritidis phage type 30 on inoculated almonds stored at -20, 4, 23, and 35 °C. J. Food Prot. 69, 1851–1857.

Uesugi, A.R., Danyluk, M.D., Mandrell, R.E., Harris, L.J., 2007. Isolation of *Salmonella* Enteritidis Phage Type 30 from a single almond orchard over a 5-year period. J. Food Prot. 70, 1784–1789.

Uesugi, A.R., Harris, L.J., 2006. Growth of *Salmonella* Enteritidis phage type 30 in almond hull and shell slurries and survival in drying almond hulls. J. Food Prot. 69, 712–718.

Venkitanarayanan, K.S., Lin, C., Bailey, H., Doyle, M.E., 2002. Inactivation of *Escherichia coli* O157:H7, *Salmonella* Enteritidis and *Listeria monocytogenes* on apples, oranges, and tomatoes by lactic acid and hydrogen peroxide. J. Food Prot. 65, 100–105.

Vojdani, J.D., Beuchat, L.R., Tauxe, R.V., 2008. Juice-associated outbreaks of human illness in the United States, 1995 through 2005. J. Food Prot. 71, 356–364.

Wells, J.M., Butterfield, J.E., 1997. *Salmonella* contamination associated with bacterial soft rot of fresh fruits and vegetables in the marketplace. Plant Dis. 81, 867–872.

Williams, R.C., Sumner, S.S., Golden, D.A., 2004. Survival of *Escherichia coli* O157:H7 and *Salmonella* in apple cider and orange juice as affected by ozone and treatment temperature. J. Food Prot. 67, 2381–2386.

Winfield, M.D., Groisman, E.A., 2003. Role of nonhost environments in the lifestyles of *Salmonella* and *Escherichia coli*. Appl. Environ. Microbiol. 69, 3687–3694.

World Health Organization, 2013. Food safety: General information related to microbiological risks in food. http://www.who.int/foodsafety/micro/general/en/ (accessed 26.02.13.).

Wright, J.R., Sumner, S.S., Hackney, C.R., Pierson, M.D., Zoecklein, B.W., 2000. Efficacy of ultraviolet light for reducing *Escherichia coli* O157:H7 in unpasteurized apple cider. J. Food Prot. 63, 563–567.

Young Lee, N., Jo, C., Hwa Shin, D., Geun Kim, W., Woo Byun, M., 2006. Effect of g-irradiation on pathogens inoculated into ready-to-use vegetables. Food Microbiol. 23, 649–656.

Zhuang, R.Y., Beuchat, L.R., Angulo, F.J., 1995. Fate of *Salmonella montevideo* on and in raw tomatoes as affected by temperature and treatment with chlorine. Appl. Environ. Microbiol. 61, 2127–2131.

Berry Contamination: Outbreaks and Contamination Issues

14

Kalmia E. Kniel, Adrienne E.H. Shearer

Department of Animal and Food Sciences, University of Delaware, Newark, DE

CHAPTER OUTLINE

Introduction

Fruits commonly referred to as berries are popular for many reasons including taste, nutrition, and convenience. Berries include strawberries, brambles (raspberries, blackberries, and associated hybrids), blueberries, cranberries, currants, grapes, gooseberries, and elderberries. However, although these are all technically considered berries, only blueberries and grapes are true berries, as the fruits are multi-seeded and derived from a single ovary (Bowling, 2000). This diverse group of fruits has been a source of sustenance throughout history beginning with the earliest hunting and gathering people (Bowling, 2000) and remains an important crop today. Recently the nutritional importance of berries has been suggested, including their high levels of antioxidants and anticancer compounds (Liu, 2007).

Berry production and consumption have increased steadily over the past decade. In the United States, berry production has increased dramatically; whereas only 5.3 million pounds of berries (fresh weight equivalent) were produced in 1970, more than 14.4 million pounds were produced in 2006 (ERS, 2008). Raspberries have

The Produce Contamination Problem. http://dx.doi.org/10.1016/B978-0-12-404611-5.00014-2

seen the greatest increase; production values have increased seven-fold from 1991 to 2006. During that same time period strawberry production doubled. Berry consumption is also growing in the United States (ERS, 2008). Over the past six years the consumption of fresh blueberries has more than doubled, raspberry consumption has increased four-fold, and strawberry consumption has steadily increased over that time. These numbers don't account for the imported berries that are shipped into the United States during much of the year. In 2005 more than 6.3 million tons of berries were produced worldwide (Strik, 2007). These numbers continue to grow with production on large and small levels. The local foods movement has increased the number of smaller producers as well as increased the enthusiasm for U-pick farms.

Berry crops are produced and harvested through three marketing channels, including customer-harvested through U-pick farms, fresh sales via local stores or distant domestic and international markets, and processed as frozen fruit, puree, dried fruit, or juice (here processed berries may be sold directly to retail). From a food-safety standpoint, the berries picked fresh and sold internationally or at local markets are of the greatest concern; however, as discussed later, further processing steps including freezing do not kill pathogenic microorganisms, and cases of foodborne illness have been associated with frozen berries.

In terms of impact on food safety, the most significant berries to date have been blackberries, raspberries, blueberries, and strawberries. As with all other produce commodities, the facts that berries are often consumed raw and unwashed may influence the removal of potential contaminants and pathogenic microorganisms. It is necessary to prevent the initial contamination by utilizing Good Agricultural Practices, including using treated or composted manures, using high-quality irrigation water at the production level and potable water for making ice after harvesting, using clean and sanitized equipment and transportation vehicles, providing proper sanitation systems for workers, assuring worker health, and maintaining a proper cold chain through delivery to the final customer.

Nearly two dozen documented outbreaks of foodborne illness associated with berries are discussed within this chapter. Additionally, it is likely that numerous other outbreaks have gone undetected or unconfirmed. Several smaller outbreaks that have been traced to viruses and protozoa that have occurred over the past 5 years are listed in Table 14.1. There may be several reasons why berries have been involved in a myriad of outbreaks; contamination by farm and packing-plant workers, use of unsafe agricultural practices, and global sourcing to provide yearlong availability. In general, berry crops are best watered using trickle irrigation, as this allows the grower to apply water at the critical period of fruit development and avoids wetting the fruit, which could foster the development of disease and rot (Bowling, 2000). However, this does not preclude the possibility that non-potable or even potentially contaminated water might be used for spray irrigation or for pesticide or fertilizer applications. This point is mentioned later in the discussion on contamination of raspberry plants with the protozoan parasite *Cyclospora cayetanensis*. For the most part berries are highly perishable and require minimal handling. In order to do this, sorting and packing into the shipping containers are often performed in the field by the pickers (Ryall and Pentzer, 1982).

Table 14.1 Efficacy of Intervention Strategies to Reduce or Eliminate Microbial Contamination of Berries

	Target Microbe	Treatment	Conditions	Effectiveness	Reference
Raspberry	*Eimeria acervulina* (*Cyclospora cayetanensis* surrogate)	Wash	Flowing, cold tap water, 5 min	Incomplete removal, duodenal lesions detected in natural host	Lee and Lee, 2001
Raspberry	*Eimeria acervulina* (*Cyclospora cayetanensis* surrogate)	Freezing	−18°C	No duodenal lesions detected in natural host	Lee and Lee, 2001
Raspberry	*Eimeria acervulina* (*Cyclospora cayetanensis* surrogate)	Heat	Water bath, minimum internal temperature of berry of 80°C maintained for 1h	No duodenal lesions detected in natural host	Lee and Lee, 2001
Raspberry	*Eimeria acervulina* (*Cyclospora cayetanensis* surrogate), 10^4 and 10^6 initial inocula levels	HP	550 MPa, 2 min, 40°C	No symptoms of infectivity in natural host	Kniel et al., 2007
Raspberry	*Eimeria acervulina* (*Cyclospora cayetanensis* surrogate), 10^4 and 10^6 initial inocula levels	UV Light	80, 160, or 261 mW/cm^2	80 mW/cm^2 for 10^6 inoculum: reduced severity of intestinal lesions in natural host (chicken); 160 mW/cm^2 for 10^4 inoculum: asymptomatic but shed oocysts	Kniel et al., 2007
Raspberry	*E. coli* O157:H7 (5 strains), *Salmonella enterica* (5 serotypes), 10^5 cfu/g initial inoculum	Pulsed UV Light	72 J/cm^2	Approximately 3.5 and 4.5 log reductions in *Salmonella* and *E. coli* O157:H7, respectively	Bialka et al., 2008

The impact of major outbreaks

Over the past two decades, several outbreaks associated with contaminated berries stand out from the rest. Interestingly, these are not associated with bacteria. The protozoan parasite *Cyclospora* caused illness throughout North America associated with raspberries (Herwaldt, 2000), and hepatitis A virus caused numerous illnesses associated with strawberries (Niu et al., 1992). At the time, these outbreaks initiated discussion on the role of "emerging" foodborne organisms in produce contamination, issues of detection, and the need for improved diagnostic methods (Tauxe et al., 1997). These are still important issues today as we seek better diagnostic and prevention methods. We have a greater understanding of the process our food undergoes along the farm-to-fork continuum, but we are still missing some basic information concerning the growth, survival, and transmission of these pathogenic organisms along this route. Historically, these berry-associated outbreaks showed the necessity for reacting to strong epidemiologic data without laboratory confirmation in order to have better control over the course of an epidemic (Tauxe et al., 1997). Although epidemiological data is always important in an outbreak investigation, detection of viruses and protozoa is traditionally difficult and forces investigators to rely on epidemiological data for traceback. Additionally, two factors, relating to these organisms in particular, increase our need for greater understanding, and these are the extent of the *Cyclospora* outbreaks (1996–2001) and the long incubation period of hepatitis A virus (10–50 days, average 30 days).

The economic effects of these outbreaks are long-lasting and often linger more with the perception that specific foods (i.e., imported Guatemalan raspberries and imported Mexican strawberries, contaminated either in Mexico or in the United States) are suspect (Calvin, 2004). The outbreaks discussed as follows provide multiple examples of the importance of good epidemiology. The epidemiological investigations were able to implicate the berry-containing food item and then led to proper recalls and environmental investigations. This is not always the case. In the spring of 1996 with the first reports of illness attributed to *Cyclospora*, the Texas Department of Health erroneously issued a report identifying the source of the problem as strawberries from California (Herwaldt and Beach, 1999). This was disastrous for the California strawberry industry that was in peak production at this time. However, when the CDC issued a statement that the source was Guatemalan raspberries (Herwaldt and Beach, 1999), the Guatemalan spring export season had just concluded, and the growers suffered few effects. Although this led to a huge economic loss in California, the California strawberry growers were able to develop an enhanced food safety system after this problem. This system was tested in 1997 when there was a problem associated with hepatitis A contamination of Mexican strawberries, and consumers questioned California-grown strawberries for the second year in a row (Calvin, 2004). The California produce industry was able to survive these issues; however, after more than two years of repeated outbreaks involving Guatemalan raspberries in outbreaks in North America, the industry there never recovered.

History of viral contamination of berries

Both raspberries and strawberries (raw and frozen) have been associated with out-breaks of hepatitis A virus and norovirus. Hepatitis A, a virus spread by human feces, is thought to have contaminated the berries by contact with infected farm workers during harvest or contaminated irrigation water. Frozen and fresh raspberries have also been associated with illness due to norovirus, also spread through contact with human feces and infected food handlers. Processing berries, including freezing and mild cooking, may be an important issue in the case of virally contaminated berries. These processing steps do not necessarily clear berries from viral contamination. The stems of strawberries destined for freezing are removed in the field, either using a metal device or a thumbnail. The berries are then transported at ambient temperature to a processing facility where they are washed with water, sliced if applicable, and often mixed with up to 30% sucrose before freezing. The extra human handling dur-ing harvesting and comingling in the processing facility is believed to place these berries at greater risk for viral contamination (CFSAN, 2001).

An increased awareness that berries play a role in the transmission of viruses led to epidemiological surveys that in turn increased awareness of contaminated berries associated with illness (Butot et al., 2007). For example, approximately 15 outbreaks of viral illness involving berries were recently identified in Finland between 1998 and 2001 (Ponka et al., 1999). Similarly, frozen berries imported from Poland were found to be responsible for more than 1100 illnesses in Europe in six different outbreaks (Cotterelle et al., 2005; Falkenhorst et al., 2005; Korsager et al., 2005). While the majority of outbreaks associated with fresh berries are still quite small, a large outbreak occurred in the fall of 2012 caused by norovirus contamination of frozen strawberries that originated from China and were believed to be prepared by one catering operator in ten different locations in Germany. The strawberries were supplied to a distributor who provides foods and beverages to schools. Certainly outbreaks, like this one, tend to be quite large when linked to institutional food pro-duction. The resulting outbreak led to a peak of over 11,000 cases of norovirus with more than 30 people hospitalized that occurred between September 25 to 27 in five federal states within Germany (Whitworth, 2012). During the outbreak investigation, epidemiological studies showed that dishes made from deep-frozen strawberries were likely the cause of the outbreak. Once these foods were removed the outbreak was considered over.

Interestingly, few human norovirus outbreaks have been documented as associ-ated with strawberries; while the FAO reports that strawberry production exceeds raspberry production about ten times (FAO, 2009; Verhaelen et al., 2012). The lack of virus testing and difficulties with detection likely impact this in the United States and across Europe and Asia. While one may consider strawberries to be at greater risk since they are grown closer to the ground and may be more susceptible to contamination with irrigation water and soil and are likely to come in contact with food handlers during harvest, attribution of norovirus illness to strawberries is quite challenging (Verhaelen et al., 2012).

Hepatitis A outbreaks with raspberries and strawberries

One of the earlier recorded berry outbreaks associated with viral contamination was an outbreak of hepatitis A in Scotland linked to consumption of raspberry mousse prepared from frozen raspberries (Reid and Robinson, 1987). Raspberries previously had been noted in epidemiological investigations as potential carriers of virus (Noah, 1981). The raspberry mousse was prepared specifically for a banquet held for 10 people at a large hotel in Aberdeen. The mousse was prepared from two 3-lb tubes of frozen raspberries, gelatin, sugar, and pasteurized cream. Some of the leftover mousse was sent home with the staff or was served on the "sweet trolley" in the dining room the next day. Twenty-four individuals were diagnosed with jaundice, deranged liver functions, fever, malaise, nausea, and flu-like symptoms approximately 24 to 28 days after consuming the mousse. The raspberries were blast-frozen at a distribution center. The raspberries had been obtained from several farms, including small holdings and large private gardens. Three of these growers were implicated indirectly in a previous outbreak (Noah, 1981). Contamination of raspberries apparently occurred at the time of picking or packing, probably by a food handler who was unknowingly shedding hepatitis A virus. A local physician reported that one of the pickers had a hepatitis A infection at the time of picking (Reid and Robinson, 1987). This restates the importance and impact of good personal hygiene and sanitation practices, along with the need for good education in food safety of food handlers at each stage from the farm to the consumer.

A multistate outbreak of hepatitis A was traced to frozen strawberries processed in a single plant in California in 1990 (Niu et al., 1992). Nine-hundred students, teachers, and staff in Georgia and Montana developed hepatitis A infection from eating strawberry shortcake and other desserts. Epidemiological data indicated that contamination did not occur from an infected worker within the processing plant but most likely from an infected picker, perhaps when the stems were being removed by hand rather than with a metal tool. Strawberries still are often destemmed prior to being brought into the processing facility.

In the months of February and March of 1997, in Michigan and Maine, there was a similar outbreak, at first linked to frozen raspberries and strawberries (Hutin et al., 1998). More conclusive epidemiological evidence from case-control studies determined that the illness was associated only with frozen strawberries, and the cases involved school children and employees. In Michigan, as in the outbreak in 1990, the frozen strawberries were consumed in strawberry shortcake desserts served in the school cafeterias. A total of 287 cases of hepatitis A were reported from 23 schools in Michigan and 13 schools in Maine. Traceback analysis implicated strawberries grown in Mexico, and processed and distributed through a California processing facility. There was no indication of specific lots that were contaminated as these records were not maintained by the schools at this time. A thorough investigation of the California processing facility did not identify any problems, showed good sanitation and manufacturing practices, limited hand contact from the employees with the berries, and no record of employees with illnesses at the time the strawberries were processed. The FDA also conducted an investigation in the strawberry fields in Mexico. These fields were drip-irrigated rather than spray-irrigated, which eliminated the likelihood that berries were contaminated by contaminated water.

This investigation revealed several potential problems, including limited slit latrines for the workers and limited access to hand-washing facilities that were on trucks circulating through the fields. Although no records were maintained on worker illnesses, the workers did not wear gloves and removed the stems from the strawberries with their fingernails in the fields. Direct hand contact with the berries combined with poor hygienic practices was a possible source of contamination. Other strawberries from the same distributor were placed on hold, and of the 13 other states that received the frozen strawberries, only two cases of hepatitis A were reported in Tennessee, nine cases in Arizona, five cases in Wisconsin, and four cases in Louisiana. All these cases with the exception of those from Louisiana were associated with state school-lunch programs. The Louisiana cases were traced back to consumption of a commercially prepared smoothie drink. For these clusters of cases no epidemiological studies were conducted. The viruses isolated from the majority of cases of hepatitis A described in this multistate outbreak showed high genetic similarity. Due to the relatively low number of cases compared to the large quantity of frozen strawberries that were consumed, it is likely that contamination was not uniform and perhaps at low levels. The findings of this investigation played a role in many of the food safety initiatives designed by Congress and the Clinton Administration within the United States (Lindsay, 1997).

Hepatitis A continues to be a problem in fresh and frozen berries, where like norovirus attribution is a significant issue impacted by the long incubation period, as mentioned above. In May 2013, there was multistate outbreak associated with a frozen berry blend. The U.S. Food and Drug Administration investigated 19 confirmed of 30 cases across Colorado, New Mexico, Nevada, Arizona, and California. The implicated food product was a frozen berry and pomegranate seed mix. While the product was removed from shelves, at the time of writing no known cause has been determined. As stated previously, detection is quite difficult. In this case the HAV was genotype 1B, a strain rarely seen in the Americas but that circulates in North Africa and in the Middle East, where the fruits may have been grown (CDC, 2013).

Norovirus-associated outbreaks with raspberries

Over the past seven years, several outbreaks were associated with the consumption of raspberries contaminated with norovirus. Perhaps as virus detection methods improve, more outbreaks associated with these viruses will be detected. The eight outbreaks discussed here occurred in Europe. In November 2001, an outbreak of norovirus in 30 individuals involved baked raspberry cakes (Le Guyader et al., 2004). At first there was an apparent association with both pear and raspberry cakes, but epidemiological evidence indicated raspberry cakes with a stronger association with illness. The pink cakes were made with a cream topping containing whole frozen raspberries. Multiple norovirus strains were detected in the raspberries after implementing a complex series of extractions coupled with polymerase chain reaction and genetic sequencing methods.

In France in March 2005, 75 students and teachers reported symptoms of nausea, vomiting, and diarrhea lasting for one to two days (Cotterelle et al., 2005). Epidemiology showed that the illness was strongly associated with consumption of raspberries blended with *fromage blanc*, a fresh cheese similar to cottage cheese, that was served with lunch in the school cafeteria. Stool samples were positive for norovirus, Musgrove strain, but the virus was not successfully isolated from raspberry samples. As in the previous cases, the raspberries in this outbreak were deep frozen and were blended with the cheese while frozen. The blended desserts were topped with individual frozen berries placed by hand; however, the workers were not ill prior to or at the time of the outbreak.

In May 2005, nearly 200 patients and employees at two hospitals in Denmark fell ill with symptoms of norovirus (Korsager et al., 2005). Again, epidemiology linked illness to consumption of a *fromage blanc* cheese dessert made with raspberries. Again, fecal samples were found to be positive for norovirus. When these illnesses occurred, the Regional Food Inspectorate called for withdrawal of the frozen raspberries. Unfortunately the recall did not happen quickly enough, and just shortly after the recall in early June, nearly 300 cases of norovirus were associated with the same dessert served to approximately 1100 people in a "meals on wheels" system. As earlier, many of the fecal samples were positive for norovirus. The three outbreaks described earlier were not believed to be linked to each other since the raspberries came from a different producer in the outbreak in France compared to those in Denmark. The exact cause of the outbreak was not determined, but it is clear that contamination was spotty due to the relative low number of illnesses compared to the numbers that consumed the frozen raspberries.

During the summer months of 2006, 43 individuals became ill with norovirus associated with the consumption of contaminated raspberries in four outbreaks (Hjertqvist et al., 2006). A homemade cake containing raspberries and cream was the cause of one outbreak. Another was associated with cheesecake and raspberries. Norovirus was detected only in the fecal samples of these patients. The raspberries were of the same brand and imported from China. In a third outbreak at a school, drinks made from the same brand of imported raspberries caused 30 illnesses. In the fourth outbreak a homemade raspberry parfait, made from the same brand of imported raspberries, was served to nine participants of a meeting who all became ill with norovirus.

It is not clear at this time why there appears to have been a sudden increase in outbreaks in Europe associated with fresh or frozen raspberries. This may be a real increase due to contaminated irrigation water, farm workers, or food handlers. Alternatively, this may be an artificial increase due to an increase in reporting or detection in connection with the Foodborne Viruses in Europe network (FBVE).

Viruses are able to survive a variety of environmental pressures (Pirtle and Beran, 1991; Koopmans and Duizer, 2004), including those related to the processing of foods as evident from the outbreaks of hepatitis A and norovirus just described. As discussed previously, several berry-associated outbreaks linked to viruses involved frozen berries. Simple removal or inactivation by washing and freezing varies depending on berry and virus type. Rinsing berries or other soft fruits has been shown to remove bacteria; however, poliovirus was not removed from strawberries after rinsing with warm water (Lukasik et al., 2003), whereas a cool water rinse removed about 2 logs of the norovirus surrogate, feline calicivirus (Gulati et al., 2001). Butot et al. (2008) showed limited effectiveness of washing in removing enteric viruses altogether from berries with either cold or warm water. It is important to note that berries are not washed in the field nor in the packaging plant as this could induce tissue decay, making this an important step for the consumer. As stated earlier, frozen berries have been responsible for numerous cases of illness. Enteric viruses, with the exception of feline calicivirus, were reduced by less than 1 log during freezing on strawberries and raspberries (Butot et al., 2008). Feline calicivirus was reduced by 2 logs in this study; however, it has been noted previously that as a respiratory virus, feline calicivirus is not likely to be the ideal surrogate for norovirus (Cannon et al., 2006).

The role of *Cyclospora cayetanensis* in berry-associated outbreaks

Raw raspberries and blackberries imported from Guatemala have been associated with several large outbreaks of gastrointestinal illness attributed to *Cyclospora cayetanensis*, a food and waterborne protozoan parasite that infects the upper small intestine of humans and can cause severe diarrhea, stomach cramps, and nausea, which may be accompanied by fever (Dawson, 2005; Shields et al., 2003). Cyclosporiasis is treatable with trimethoprim-sulamethoxazole (Hoge et al., 1995). *Cyclospora* oocysts first were observed in stool samples in Papua, New Guinea 30 years ago (Ashford, 1979), but interestingly, it is still referred to as an emerging pathogen due to the many unknowns regarding its transmission (Chacin–Bonilla, 2008). Cyclosporiasis is not thought to be associated primarily with immunocompromised individuals like other human protozoan pathogens. It was identified as a new coccidian species in 1993 by Ortega et al., when they successfully induced oocyst sporulation and excystation of the sporozoites *in vitro* (Ortega et al., 1993).

Cyclospora cayetanensis oocysts are quite large at 7.5 to 10 μm in diameter. These oocysts have a strong outer membrane composed of complex carbohydrates and lipids that make the oocysts acid fast. The oocyst membrane protects two oblong sporocysts that surround the infective life stages, with four sporozoites in each sporocyst. The oocyst and sporocyst membranes are strong structures that provide great stability to environmental pressures and ensure that the sporozoites remain viable along their journey to the small intestine. Like many protozoa, *Cyclospora* oocysts are shed unsporulated and sporulate outside the host within 7 to 10 days under favorable environmental conditions (Ortega et al., 1994). In comparison *Cryptosporidium* oocysts are shed already sporulated and infectious, whereas *Toxoplasma gondii* oocysts sporulate within 48 to 72 hours of being in the environment. The infection process begins when the oocysts are ingested by the host. Coccidian oocyst outer membranes respond to the acidic pH of the stomach. When the sporocysts reach the intestinal tract of their hosts, the sporocyst wall breaks down and the sporozoites are released to invade host epithelial cells and undergo multiple cycles of asexual multiplication followed by sexual development for the formation of the unsporulated oocysts that are shed in the host feces.

In total, the 10 events that have involved *C. cayetanensis* and contaminated raspberries accounted for 2864 illnesses. Subsequently, eight traceback investigations were conducted, including five farm investigations, four of these in Guatemala and one in Chile (Timbo et al., 2007). The first reported outbreaks of cyclosporiasis associated with raspberries were in New York and Florida in 1995. These outbreaks did not involve traceback investigations of any kind, and approximately 71 individuals were involved. In New York, drinking water from portable coolers at a country club and raspberries that were served during the outbreak period were both suspected (Carter, 1996). In Florida, raspberries were suspected, but were a component of a fruit cup and desserts served at various social events (Koumans et al., 1998). During 1996, over 1660 individuals in the United States became ill from raspberries contaminated with *Cyclospora*. Together these events involved 20 states and the District of Columbia (Herwaldt et al., 1997, 1999; Careres et al., 1998). There were traceback and farm investigations associated with these three large outbreaks. Raspberries were traced back to Guatemala for many of these events, and berries that were implicated were harvested from between three and 30 farms. In the large multistate outbreak in 1996, a majority of the raspberries were traced to one exporter; however, nothing concrete came of the farm investigations, as exporters included raspberries from different farms in a single shipment (Timbo et al., 2007; Herwaldt et al., 1999). Another large multistate outbreak occurred the following year, again associated with Guatemalan raspberries (Careres et al., 1998). Several farms were identified during the investigation by the FDA (Timbo et al., 2007). In 1998, an outbreak of cyclosporiasis occurred in Massachusetts, but no farm investigation was pursued as it was not possible to determine whether the raspberries originated in Chile or Guatemala (Catherine et al., 1998).

In 2000, raspberries associated with a cake were involved in an outbreak, and the berries were traced to three possible sources (a Guatemalan farm, a Chilean farm, and an unknown U.S. farm) (Ho et al., 2000). The farm in Guatemala was later implicated in an outbreak in Pennsylvania associated with a cake served with cream and raspberries where over 50 people became ill (Ho et al., 2000), and also found to be associated with an outbreak in the state of Georgia the same year when raspberries were served with other fruit over ice cream at a bridesmaids' luncheon (Marrow et al., 2002; Timbo et al., 2007). Raspberries from Chile also were suspected in this latter outbreak in 2000 (Murrow et al., 2002). In 2002, raspberries from Chile again were suspected in an outbreak that involved 22 individuals (CFSAN, 2003).

Transmission of *Cyclospora* oocysts and the role of foods

The first reported outbreak in the United States of cyclosporiasis involved contaminated water in Chicago, Illinois, in 1990 (Timbo et al., 2007). Cyclosporiasis has been associated with fresh fruits, vegetables, and herbs, likely contaminated by water, soil, or handlers. In particular, raspberries, basil, parsley, snow peas, and leafy greens have been implicated as probable transmission vehicles in 19 outbreaks of cyclosporiasis in the United States (Timbo et al., 2007; Dawson, 2005; CDC 2004; Shields et al., 2003; Lopez et al., 2001). As with hepatitis A virus, *C. cayetanensis* oocysts are shed by humans, and contamination can occur at both preharvest points (soil, feces, irrigation water, dust, insects, or animals) and postharvest points (human handling, equipment, or transport containers) (Beuchat, 2002).

The role of water has been questioned in the transmission of oocysts to berries and other foods, including basil. The water used to mix pesticides was previously identified as a possible source of contamination in the outbreaks of cyclosporiasis associated with contaminated raspberries (Herwaldt, 2000; Herwaldt and Ackers, 1997). Water was found to be a main vehicle of transmission in a study in Egypt assessing irrigation canals, groundwater, and finished water (El-Karamany et al., 2005). This study also named contact with soil and poultry litter as potential risks for transmission. Water and soil are of concern in many parts of the world where cyclosporiasis is endemic, and individuals' shedding of oocysts may be asymptomatic (Chacin–Bonilla, 2008; El-Karamany et al., 2005; Katz and Taylor, 2001; Sturbaum et al., 1998; Hoge et al., 1993, 1995; Eberhard et al., 1999; Bern et al., 1999; Chacin-Bonilla et al., 2003). Interestingly, several recent studies focused on risks associated with soil. In Peru, contact with the soil was observed to be an important risk factor among children (Mansfield and Gajadhar, 2004). Contact with soil among healthcare and farm workers in Guatemala was a risk factor for cyclosporiasis infection (Bern et al., 1999). In similar studies in communities in Haiti and Venezuela, statistical models showed that contact with soil was an important mode of transmission for *Cyclospora* oocysts (Chacin–Bonilla, 2008; Lopez et al., 2001). In these same studies, poverty was identified as a risk factor for the prevalence of infection in endemic areas (Chacin–Bonilla, 2008).

In several documented outbreaks, raspberries were the likely vehicle of contamination, but they were associated with other foods, including wedding cake (Herwaldt and Ackers, 1997) and lemon tart (Herwaldt, 2000). Although studies attempting to determine the infectious dose have not been successful (Alfano–Sobsey et al., 2004), epidemiological investigations have suggested that the infective dose is low or that in the berry-associated outbreaks, the number of oocysts per berry was high (Herwaldt, 2000). Compared to other coccidians, *Cyclospora* oocysts require a large amount of time to sporulate in the environment. However, the rate of sporulation does not appear to be fixed, as sporulation increased with increasing temperatures from 4 to 22°C (Smith et al., 1997). *Cyclospora cayetanensis* is difficult to study in the laboratory, as humans are its only known host, making access to oocysts and methods to evaluate viability difficult and limited. Experimental work using the related poultry coccidian parasite *Eimeria acervulina* as a surrogate for *Cyclospora* has shown possible interactions between the oocyst and the raspberry (Kniel et al., 2007). Phylogenetic analysis supports the conclusion that *Eimeria* and *Cyclospora* could belong to the same family (Relman et al., 1996), and one author even suggested that *Cyclospora* should be considered a member of the genus *Eimeria*, based on small subunit ribosomal RNA gene alignment (Pieniazek and Herwaldt, 1997).

Regarding oocyst breakdown, slight differences were observed in oocysts recovered from raspberries; however, this difference was not observed during excystation, but rather in the number of sporocysts present before bead-beating, which is used with *Eimeria* in the laboratory to mechanically disrupt the oocyst membrane. This phenomenon was obvious by microscopy as there was no clumping of sporocysts. There was a visual increase in the number of sporocysts recovered from raspberries, compared to oocysts in suspension or those recovered from basil; where more than 2.2 times the sporocysts were observed from oocysts that had contact with intact raspberries compared to those that did not. The increased release of sporocysts observed after interactions with raspberries may be a combination of factors including pH (pH 3.4 ± 0.2), flavonoids, and other plant phenols. The greater release of sporocysts in the presence of raspberries as compared to other matrices may influence the apparent infectivity of oocysts. An increase in sporulation coupled with an increase in the breakdown of the oocyst membranes on acidic berries like raspberries (Kniel et al., 2007) could lead to a higher infection rate. It is important to note that sporulation can be inactivated by exposing oocysts to extreme temperatures that would be used at home in food preparation or by the food industry (Sathyanarayanan and Ortega, 2006).

The berry surface topography certainly plays a large role, as seen in the interaction of *Toxoplasma gondii* oocysts with the hair-like projections on the raspberry surface, as compared to the relatively smooth surface of a blueberry (Figure 14.1) (Kniel et al., 2002). *Toxoplasma* is a protozoan parasite related to both *Eimeria* and *Cyclospora* and similar in size and shape to both. Due to the facts that cyclosporiasis is often associated with imported produce, and unknown reservoirs or routes of contamination exist, alternative treatment methods for fresh berries should be examined. These are discussed further in this chapter. There are many questions that still need to be addressed, including the potential seasonality of prevalence of infections and outbreaks (Herwaldt, 2000).

Bacterial contamination of berries

Whereas other types of produce have been more frequently implicated as vehicles for foodborne outbreaks of bacterial origin, eight outbreaks of bacterial etiology have been attributed to berries in the United States since 1973 (Sivapalasingam et al., 2004; CDC, 2013). These include strawberries contaminated with *Staphylococcus aureus* (1985), *Salmonella* Group B (2003), and strawberries/blueberries contaminated with enterohemorrhagic *E. coli* O26 (2006), blueberries contaminated with *Salmonella enterica* serovar Muenchen (2009) and *S.* Newport (2010), and strawberries contaminated

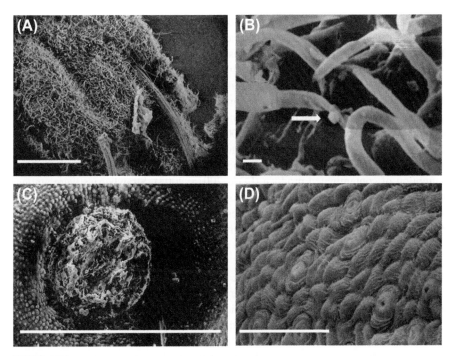

FIGURE 14.1

Scanning electron micrographs of raspberry and blueberry surfaces inoculated with 2.0×10^4 *Toxoplasma* oocysts. Hair-like structures on the raspberry likely aid in retention of oocyst contamination (A, bar = 1 mm). One oocyst is visibly attached to one of these structures in B (bar = 10 μm). The blueberry surface lacks these structures (C, bar = 1 mm and D, bar = 0.1 mm).

(Adapted from Kniel et al., 2002. Copyright permission from Allen Press and Copyright Clearance Services.)

with *E. coli* O157:H7 (2011) (CDC, 2013). The most recent outbreaks are noted in Table 14.1. Red grapes (2000) and green grapes (2001) were identified as the vehicles for outbreaks of enterohemorrhagic illnesses and *Salmonella* Senftenberg, respectively. Collectively, these outbreaks involved 121 known cases of illness. The outbreaks were sourced to berries from a variety of locations including a grocery store, home, roadside stands, and restaurant/daycare/school/delicatessen. A number of other outbreaks have been attributed to fruit salad, but did not single out berries as the original source (CDC, 2013). Two additional outbreaks were attributed to berries, but the etiology was not determined (Sivapalasingam, 2004). The limited number of traced outbreaks does not mean that additional bacterial illnesses did not result from consumption of contaminated berry products. The food vehicle in outbreaks cannot always be determined, and it is assumed that foodborne illnesses are generally underreported.

This outbreak history does not reveal a pattern in type or source of bacterial contamination for berry products. Contamination of strawberries in the *E. coli* O157:H7 2011 outbreak in Oregon, was traced back to the growing field; a match to the outbreak strain was recovered from deer droppings (Terry, 2011). The sporadic outbreaks of bacterial illness that were not necessarily traced to farm-level contamination raises the possibility that the berries were cross-contaminated postharvest by other foods or surfaces. This also highlights challenges for outbreak investigations with perishable products that are not easily traced to source and receive substantial handling prior to receipt by the end customer.

Surveys of produce have provided some information on the incidence of bacterial contamination of berries. In 1999 and 2000, the FDA initiated surveys of imported and domestic produce, respectively, to determine the incidence of contamination and the research and education needs to reduce foodborne illnesses resulting from consumption of contaminated produce (FDA, 2001a, 2003). Strawberries were among the produce samples analyzed for *Salmonella* and *E. coli* O157:H7. For the survey of imported produce, 143 strawberry samples from five countries (Argentina, Belgium, Canada, Mexico, and New Zealand) were analyzed, and only one sample (0.7%) was found positive for *Salmonella*. None tested positive for *E. coli* O157:H7 (FDA, 2001a). For domestic strawberries, 136 samples from two states (California and Florida) were tested, and all were negative for *Salmonella* and *E. coli* O157:H7 (FDA, 2003). Another published survey (Mukherjee et al., 2006) of bacterial contamination of berries involved sampling of strawberries, blueberries, and raspberries from farms in the Upper Midwest region of the United States. Samples were collected preharvest over a 2-year period (2003–2004) from farms that used conventional, organic, and semiorganic (organic practices used, but not certified organic) growing practices. Berries were tested for coliforms, *E. coli, Salmonella*, and *E. coli* O157:H7. Coliform counts on berries were approximately 1 to 2 $\log_{10}$ MPN/g, and 2 of 194 (1%) berries tested positive for *E. coli*. None of the berries tested positive for *Salmonella* or *E. coli* O157:H7 (Mukherjee et al., 2006). Results from this very limited number of publicly available studies would suggest that the incidence of contamination of berries with pathogenic bacteria is low.

Berry extracts are reported to have antibacterial properties (Ryan et al., 2001; Puupponen-Pimiaet al., 2005) owing to their acidity and phenolic compounds (Puupponen-Pimia et al., 2005), and research on the antimicrobial properties of berry extracts continues (Caillet et al., 2012; Lacombe et al., 2013; Lacombe et al., 2012; Soare et al., 2012). However, several studies with artificially contaminated berries suggest these antibacterial properties should not be relied upon for safety. Research studies have been conducted to determine, in the event of contamination, the fate of microbes after exposure to various intrinsic stresses as well as intervention strategies to remove or inactivate pathogens.

Contamination reduction strategies

The common processing strategy of thermal pasteurization as a means to inactivate viruses, parasites, and bacteria on berry products has received limited attention. Typical conditions for hot-filled, shelf-stable, single-strength white grape juice (pH 3.9) were calculated to achieve at least a 5-log reduction of *E. coli* O157:H7, *S. enterica*, and *L. monocytogenes* (Mazzotta, 2001). However, mild heat treatment (75°C for 15s) of raspberry puree yielded less than a 3-log reduction in murine norovirus 1 and less than a 4-log reduction of *E. coli* and *B. fragilis* HSP40 infecting phage B40-8 (Baert et al., 2008). Heat processing renders berries as different products from their fresh counterparts, which may be more acceptable for berries used for juices, cereals, and purees than whole berries. Alternatively, pasteurization to achieve a 5-log reduction of bacterial pathogens in strawberry juice has been demonstrated with pulsed electric fields in conjunction with other hurdles (Mosqueda-Melgar et al., 2008; Gurtler et al., 2011). *S.* Enteritidis and *E. coli* O157:H7 were inactivated by high-intensity PEF of 35 kV/cm, 4 μs pulse length in bipolar mode at or below 37°C when used in conjunction with cinnamon bark oil concentrate (0.05%) or citric acid (0.5%) (Mosqueda-Melgar et al., 2008). *E. coli* O157:H7 in strawberry juice was also susceptible to PEF treatment of 18.6 kV/cm for 150 μs when combined with citric acid (2.7%), sodium benzoate (750 ppm), and potassium sorbate (350 ppm) (Gurtler et al., 2011).

Nonthermal processes have been explored as potential means of rendering berries microbiologically safe while maintaining their fresh-like characteristics. Technologies evaluated for application to berries include washing with and without disinfectants, frozen storage, high hydrostatic pressure, exposure to ultraviolet (UV) light, and irradiation. Details of these studies are summarized in Table 14.1.

Washing produce in plain water is generally recommended to consumers as a means of removing visible soil from produce and reducing microbial populations. However, washing is not a means of rendering fresh produce completely free of risk from potential pathogen contamination. A study with strawberries, raspberries, and blueberries, artificially contaminated with various viruses, demonstrated that washing these berries with either cold or warm water yielded viral population reductions of less than 1.5 $\log_{10}$ units (Butot et al., 2008). Likewise, washing inoculated strawberries with water at 22 or 43°C, either with or without scrubbing, yielded less than 1-log reductions in populations of *E. coli* O157:H7, *S.* Montevideo, poliovirus 1, and three different bacteriophages (Lukasik et al., 2003). In another study with protozoa, raspberries contaminated with *Eimeria acervulina* as a surrogate for *C. cayetanensis* were washed with cold tap water and fed to chickens. Washing was not a fail-safe method for removing protozoa as some chickens were symptomatic of infection (Lee and Lee, 2001).

The addition of sanitizers or disinfectants to water washes is one of the most commonly studied strategies to remove or inactivate pathogens on berries. Those agents evaluated have included chlorine, chlorine dioxide, ozone, peroxyacetic acid, quaternary ammonium compounds, trisodium phosphate, and hydrogen peroxide, among others. Chlorine is a common sanitizer used in food-processing facilities with broad efficacy against many foodborne pathogens and spoilage organisms. It is approved for use in the United States as a wash-water additive for the produce industry (Seymour and Appleton, 2001). Chlorine washes have been applied at various concentrations (20 to 300 ppm) to strawberries, blueberries, and raspberries inoculated with several different types of viruses and bacteria (Butot et al., 2008; Casteel et al., 2008; Gulati et al., 2001; Lukasik et al., 2003; Rodgers et al., 2004; Udompijitkul et al., 2007; Wei et al., 2007). Chlorine washes of berries have generally yielded 1- to 2-log unit reductions in bacteria and viruses, although Pangloli and Hung (2013) reported a greater than 4-$\log_{10}$ reduction of *E. coli* O157:H7 by chlorine (100 mg/L) on surface-contaminated blueberries. The degree of inactivation observed was incomplete for the inoculation levels studied and would not provide a 5-log reduction that serves as the standard for processing technologies utilized to render juices made from fruits safe from pertinent pathogens (typically *Salmonella* or *E. coli* O157:H7) (FDA, 2001).

It has been reported that upper levels of free available chlorine that are at least partially effective for disinfection (200 ppm) can also cause bleaching and off-flavors in produce (Hurst and Schuler, 1992); however, these effects may be product specific and need to be evaluated for each type of produce (Wei et al., 2007). At levels normally used in food processing, chlorine is typically not as effective on protozoan parasites (King and Monis, 2007), which have been implicated in outbreaks associated with consumption of contaminated berries. Chlorine efficacy is affected by the presence of organic material and pH. Additionally, some by-products resulting from chlorine reactions with organic material are considered potentially mutagenic or carcinogenic (Sapers, 2001; Wu and Kim, 2007), and alternatives have been sought.

Chlorine dioxide has been evaluated as an alternative to chlorine as it appears to be less affected by pH changes and the presence of organic material (Seymour and Appleton, 2001). Chlorine dioxide can be applied in both aqueous and gaseous form. Aqueous systems are easier to administer whereas gaseous systems offer good penetration without residual surface moisture to support subsequent growth of spoilage organisms (Wu and Kim, 2007). The reported efficacy of chlorine dioxide for berry decontamination varies widely from approximately 1-log unit reduction (Butot et al., 2008) to 5-log unit reductions (Yuk et al., 2006) depending on the method of delivery and produce type. Generally, delivery in gaseous systems has been reported to have greater efficacy.

Ozone can also be delivered in gaseous or aqueous phase. Ozone has had GRAS status since 2001 and is approved as an antimicrobial treatment (Wei et al., 2007). Ozone safely decomposes to oxygen and water (Wei et al., 2007), although safety precautions during use are needed. Ozone applied to raspberries inoculated with *S. enterica* or *E. coli* O157:H7 reduced the pathogen populations by approximately 3.5 to 5.5 $\log_{10}$ cfu/g (Bialka and Demirci, 2007a, 2007b) while ozone (1.5 mg/L) reduced *E. coli* O157:H7 on blueberries by 2.3 to 3.5 $\log_{10}$ (cfu/g) (Pangloli and Hung, 2013). Less dramatic reductions were obtained on strawberries (Bialka and Demirci, 2007a, 2007b). Shorter treatment times with ozone on strawberries only yielded approximately 1-log reductions in the natural microbial flora (Wei et al., 2007). Further analysis of the inactivation kinetics of *E. coli* O157:H7 and *S. enterica* on raspberries and strawberries by ozone indicated that the Weibull model was more suitable for estimation of microbial inactivation than first-order kinetics (Bialka et al., 2008). Ozone applied to grapes has also been shown to increase shelf-life by reducing fungal decay (Sarig et al., 1996).

Several other disinfectants have been tested against pathogens on berries (Table 14.1), but with fewer studies than have been conducted for chlorine, chlorine dioxide, or ozone. Results are variable, with 1- to 4.4-$\log_{10}$ cfu/g reductions reported (Gulati et al., 2001; Lukasik et al., 2003; Rodgers et al., 2004; Udompijitkul et al., 2007; Table 14.1). Quality attributes were not the focus of these studies, but it was noted that hydrogen peroxide caused slight discoloration of strawberries (Lukasik et al., 2003).

The efficacy of any disinfectant wash or external treatment would be impacted by microbial inaccessibility. Pathogens that are irreversibly attached to surfaces or protected in aggregations or biofilms may be more resistant to inactivation. Additionally, pathogens may be less accessible if they reside in surface crevices, which are abundant on raspberries and strawberries, or are internalized through surface damage (Bassett and McClure, 2008; Sapers, 2001). A microbial risk assessment of various produce types suggested that the irregular topography of certain berries, and consequent difficulty in washing, makes them high-risk products despite their low pH values, which are believed to be below the growth boundaries of pathogenic bacteria (Bassett and McClure, 2008). The enhanced survival of microbes to disinfection as well as freezing, when located in scars or puncture wounds, has been demonstrated (Flessa et al., 2005; Knudsen et al., 2001; Yuk et al., 2006). Internalization of pathogens by other means has been reported for other types of produce in some laboratory-simulated growth conditions (Guo et al., 2002; Solomon et al., 2002) or processing conditions, such as due to temperature differentials between the fruits and wash water (Bassett and McClure, 2008; Sapers, 2001). It remains to be determined whether or not these latter causes of internalization occur with fresh berries. Despite the known and theoretical limitations of washes for inactivation of pathogens on berry surfaces, disinfectant washes can be effective in the processing environment and thereby serve an important role in minimizing cross-contamination (Sapers, 2001).

The persistence of pathogenic microorganisms on berries or in berry juices during storage at different temperatures has been studied. The survival of *Salmonella* spp., *E. coli* O157:H7, and *L. monocytogenes* on surfaces of unwashed and intact or cut strawberries was evaluated at ambient, refrigerated, and frozen storage. Bacteria were allowed to dry on the strawberry surfaces, and this drying step resulted in a slight population reduction. Populations of *L. monocytogenes* declined by 1 to 3 logs during storage at 24°C for 48 h (dependent on initial inocula of either 10^6 to 10^8 per sample; Flessa et al., 2005), but no changes were observed in *E. coli* or *Salmonella* (Knudsen et al., 2001). During refrigerated storage for seven days, all pathogen cocktail populations declined (1–3 logs) on intact strawberries, but no population reduction was observed on cut surfaces. Frozen storage for one month resulted in population reductions of approximately 2 logs or less, and this reduction was incomplete, even including the reductions observed during drying. These studies indicate that these bacterial pathogens are capable of survival through the normal shelf-life of strawberries at 24 and 4°C and at least one month in frozen storage. Survival on cut surfaces was greater than on intact strawberry surfaces. In another study, survival of stationary phase and acid-adapted *Salmonella, E. coli*, and *L. monocytogenes* was evaluated in concentrated cranberry juice at 0°C (Enache and Chen, 2007). The population reduction of the three bacterial pathogen cocktails was dependent on growth phase, °Brix, and pathogen type. Stationary-phase cells were more resistant than acid-adapted cells for all three pathogens. A °Brix of 18 to 46 (pH 2.2–2.5) resulted in population reductions of 5 logs within 24 hours, but at 14 °Brix (pH 2.5), a 5-log reduction took longer (96 hours) for *E. coli* O157:H7 (Enache and Chen, 2007). Frozen storage (-18°C for 24 h) of blackberry juice concentrate (pH 1.8) yielded a 7-$\log_{10}$ cfu/ml reduction of *Salmonella* spp. and *Shigella* sp. (Wong et al., 2010). Freeze/thaw treatment of strawberry juice was effective for *S*. Enteritidis inactivation, whereas temperature stress in conjunction with essential oils was needed for similar inactivation of *E. coli* O157:H7 (Duan and Zhao, 2009). A lack of persistence of *S*. Typhimurium and *Campylobacter jejuni* was demonstrated in blueberry juice stored at an elevated temperature of 37°C for 24 h (Biswas et al., 2012).

As mentioned previously, the effects of freezing and frozen storage on persistence of viruses are similarly troublesome. Unwashed strawberries, raspberries, and blueberries inoculated with 10^4 to 10^6 TCID$_{50}$ or PCRU (RT-PCR units) with norovirus (NV), hepatitis A virus (HAV), rotavirus (RV), or feline calicivirus (FCV) generally showed less than a 1-log unit reduction after frozen storage at -20°C for two days (Butot et al., 2008). Somewhat greater reductions were observed with FCV on strawberries and raspberries and with RV on blueberries. Even prolonged frozen storage of three months' duration had minimal effect on HAV and RV survival in all berries types.

The survival of a protozoan parasite on frozen raspberries was evaluated in one study (Lee and Lee, 2001). Raspberries were inoculated with the poultry parasite *Eimeria acervulina* as a surrogate for the human parasite *C. cayetenensis*. Berries were frozen at −18°C and then fed to chickens to determine infectivity. Lesions were not found in the duodenal loop of the intestinal tract of chickens fed previously frozen berries containing *E. acervulina*, indicating that freezing may damage oocysts to some degree.

High hydrostatic pressure processing (HPP) has been studied at length for the inactivation of microbial pathogens on various food products, and often, with minimal to no detriment to fresh characteristics. HP has also been applied to various berry products, and the conditions required for inactivation of microbes depend on berry type and microbial target, as described later. Strawberry puree inoculated with hepatitis A virus in excess of 10^6 PFU was pressure-treated at 375 MPa for 5 minutes, and a 4.3-log PFU reduction was obtained (Kingsley et al., 2005). *L. innocua* (approximately 10^6 cfu/sample inoculum), as a surrogate for *L. monocytogenes*, could be inactivated with 450 MPa on strawberries and blueberries for 1.5 minutes and on grapes with 3 minutes' treatment (Chauvin et al., 2005). *Eimeria acervulina* (10^6 oocysts/sample inoculum) as a surrogate for *C. cayetanensis* was inactivated within 2 minutes of treatment at 550 MPa and 40°C (Kniel et al., 2007). The results of these studies suggest that pressure processing is promising for enhancing the microbial safety of berries. Although of greater significance for quality than safety, HP is also effective against *Saccharomyces cerevisiae* on grapes, strawberries, and blueberries (Chauvin et al., 2005) at pressures less than those required for the previously mentioned human pathogens and their surrogates. Organoleptic properties of pressure-treated berries were not reported in the aforementioned studies, although other studies have demonstrated that pressure levels affect the pigment stability of red raspberries with greatest color retention at low (200 MPa) and high (800 MPa) pressures (Suthanthangjai et al., 2005), and pressure treatment of berry purees better preserves volatiles as compared to heat processing (Dalmadi et al., 2007).

Because ultraviolet light (UV-C) in sufficient doses can cause irreparable damage to genetic material, it has broad efficacy against microorganisms. UV light also offers advantages of being relatively low in cost and without irritating or toxic by-products (Fino and Kniel, 2008). Its potential application for food safety, however, depends in part on whether the UV light adequately reaches microorganisms that could be protected by shadows or turbidity. Thus, for whole fruits, UV treatment bears similarity with external washes in that its application would be limited to accessible surface microorganisms. UV light treatment has been studied for the inactivation of bacteria, viruses, a protozoan, and fungi on berries. The degree of inactivation of *S. enterica* and *E. coli* O157:H7 populations depends on berry type as greater reductions were observed on raspberries (3.5–4.5 logs) than on strawberries (< 2.5 logs) with 72 and 64.4 J/cm^2 dosages, respectively (Bialka et al., 2008). These researchers further noted that population reductions were better characterized by the Weibull model than log-linear estimations (Bialka et al., 2008).

Similar reductions in viruses on strawberries were achieved with UV light (240 mW s/cm^2) with 1.9, 2.3, and 2.6 log TCID$_{50}$/ml reductions in aichi virus, feline calicivirus (as a surrogate for norovirus), and hepatitis A virus, respectively (Fino and Kniel, 2008). Research from the same laboratory on raspberries inoculated with *E. acervulina* (as a surrogate for *C. cayetanensis*) demonstrated that UV treatment (261 mW/cm^2) can inactivate the protozoa, but inoculum level was an important factor, with populations of 10^4 to 10^6 oocysts partially inactivated (Kniel et al., 2007). In each of these studies, the researchers noted that accessibility was a likely factor in efficacy. UV light has been evaluated for inactivation of fungal microorganisms of significance to the quality and shelf-life of berries. UV light has yielded partial inactivation of fungal populations in strawberry nectar (Keyser et al., 2008), and variable results have been reported for UV effects on extension of berry shelf-life (0–2 days) (Fino and Kniel, 2008; Marquenie et al., 2003).

Irradiation is approved for use in the United States for several food commodities for the control of insects and microorganisms and to delay maturation of fresh commodities (FDA, 2008). The effects of irradiation on the quality of berry products have been studied; however, fewer studies have been conducted to fully characterize irradiation needs for enhanced microbial safety of berries. Bidawid and colleagues (Bidawid et al., 2000) determined that gamma irradiation at a dose of 10 kilograys (kGy) reduced titers of hepatitis A virus inoculated onto strawberries by approximately 3 log$_{10}$ (PFU/ml). An irradiation dosage of 1 kGy is permitted for fresh foods for inhibition of growth and maturation (FDA, 2008); at this dose, there was virtually no reduction in HAV titers on strawberries

(Bidawid et al., 2000). Irradiation has been reported to extend the shelf-life of berries, but effects on quality depend on the berry type and dosage. Electron-beam irradiation of strawberries at doses of 1 and 2 kGy provided shelf-life extension of two and four days, respectively, although a decrease in firmness and increase in off-flavors were reported during storage of strawberries (Yu et al., 1995). Similarly, raspberries treated with 1.5 kGy gamma radiation reduced microbial load, but with effects on firmness, phenolic content, and antioxidant capacity during storage (Cabo Verde et al., 2013). Studies on the qualities of various types of blueberries receiving doses above 1 kGy reported decreases in ascorbic acid content (Moreno et al., 2008) and decreases in berry firmness (Moreno et al., 2007; Miller et al., 1994), but the berries were not necessarily deemed unacceptable. Higher doses up to 3.2 kGy are reported to affect flavor and color of blueberries (Morena et al., 2007). Gamma irradiation (2 kGy) of grapes reduced fungal spoilage with no apparent effects on berry firmness or soluble solids, but with reduction in titratable acids and ascorbic acid (Thomas et al., 1995). Low levels (ng/g) of furan were detected in gamma-irradiated (5 kGy) grapes (Fan and Sokorai, 2008).

Use of bacteriophage as a means to control bacterial contamination of foods has been suggested as a possible intervention strategy at the farm level to protect crops from plant diseases and to minimize contamination of animal products. Approvals for various bacteriophage products have been granted by the EPA or the FDA for growing or processing tomatoes and poultry products since 2006 (Hagens and Offerhaus, 2008). To our knowledge, no studies have been published to date on application of bacteriophage technology for berry production.

A number of other approaches have been tested for the improvement of microbiological quality of berries, including hot water for blueberries (Fan et al., 2008), modified atmosphere packaging in conjunction with natural antimicrobials for grapes (Guillén et al., 2007), and carbonate or bicarbonate salts for grapes (Gabler and Smilanick, 2001). These studies did not address microbiological safety for human pathogens.

In summary

Prevention of produce contamination is a preferred practice, compared to decontamination especially with products such as berries, for which postharvest intervention strategies are limited, provide incomplete protection, or compromise quality attributes. Since the increase in foodborne illness outbreaks associated with produce, additional measures have been taken to determine and implement good practices aimed at reducing the incidence of contamination on the farm and in handling. In 1998, the *Guide to Minimize Microbial Food Safety Hazards for Fresh Fruits and Vegetables* was released. This guide was developed by the Food and Drug Administration (FDA) and U.S. Department of Agriculture (USDA) in conjunction with the produce industry. Subsequently, the Produce Safety Action Plan of 2004 was implemented. These documents serve as guides rather than regulatory requirements and have broad application to all types of produce. Specific guidance documents have been written to expand and customize insight for certain types of produce including leafy greens, tomatoes, and fresh-cut produce. Such specific guidelines have not been prepared for berries at this writing. Given the diversity of growing methods for berries from ground level to bush to vine, specific guidelines that suit all berry products would be particularly challenging to devise.

Good Agricultural, Manufacturing, and Management Practices address major potential sources of contamination throughout the production and processing of produce including water, crop treatment, handling, facilities, and transportation. The reader is referred to other chapters of this book for a detailed discussion of recommendations to minimize microbial hazards in produce production, some of which may apply to berries.

References

Alfano-Sobsey, E.A., Eberhard, M.L., Seed, J.R., Weber, D.J., Won, K.Y., Nace, E.K., et al., 2004. Human challenge pilot study with Cyclospora cayetanensis. Emerg. Infect. Dis. 10, 726–728.

Ashford, R.W., 1979. Occurrence of an undescribed coccidian in man in Papue New Guinea. Ann. Trop. Med. Parasitol. 73, 479–500.

Baert, L., Uyttendaele, M., Van Coillie, E., Debevere, J., 2008. The reduction of murine norovirus 1, *B. fragilis* HSP40 infecting phage B40-8 and *E. coli* after a mild thermal pasteurization process of raspberry puree. Food Microbiol. 25, 871–874.

Bassett, J., McClure, P., 2008. A risk assessment approach for fresh fruits. J. Appl. Microbiol. 104, 925–943.

Bern, C., Hernandez, B., Lopez, M.B., Arroword, M., 1999. Epidemiological studies of *cyclospora cayteanensis* in Guatemala. Emerg. Infect. Dis. 5, 766–774.

Beuchat, L.R., 2002. Ecological factors influencing survival and growth of human pathogns on raw fruits and vegetables. Microbes. Infect. 4, 413–423.

Bialka, K.L., Demirci, A., 2007. Utilization of gaseous ozone for the decontamination of *Escherichia coli* O157:H7 and *Salmonella* on raspberries and strawberries. J. Food Prot. 70, 1093–1098.

Bialka, K.L., Demirci, A., Puri, V.M., 2008. Modeling the inactivation of *Escherichia coli* O157:H7 and *Salmonella enterica* on raspberries and strawberries resulting from exposure to ozone or pulsed UV-light. J. Food Eng. 85, 444–449.

Bidawid, S., Farber, J.M., Sattar, S.A., 2000. Inactivation of hepatitis A virus (HAV) in fruits and vegetables by gamma irradiation. Int. J. Food Microbiol. 57, 91–97.

Biswas, D., Wideman, N.E., O'Bryan, C.A., Muthaiyan, A., Lingbeck, J.M., Crandall, P.G., et al., 2012. Pasteurized blueberry (*Vaccinium corymbosum*) juice inhibits growth of bacterial pathogens in milk but allows survival of probiotic bacteria. J. Food Saf. 32, 204–209.

Butot, S., Putallaz, T., Sanchez, G., 2008. Effects of sanitation, freezing and frozen storage on enteric viruses in berries. Int. J. Food Microbiol. 126, 30–35.

Cabo Verde, S., Trigo, M.J., Sousa, M.B., Ferreira, A., Ramos, A.C., Nunes, I., et al., 2013. Effects of gamma radiation on raspberries: Safety and quality issues. J. Toxicol. Environ. Health, Part A 76, 291–303.

Caillet, S., Cote, J., Sylvain, J.F., Lacroix, M., 2012. Antimicrobial effects of fractions from cranberry products on the growth of seven pathogenic bacteria. Food Control 23, 419–428.

Calvin, L., Response to US Foodborne Illness outbreaks associated with imported produce, 2004. Economic Research Service. Isues in Diet, Safety, and Health/Agricultruer Information Bulletin Number. 789–5.

Cannon, J.L., Papafragkou, E., Park, G.W., Osborne, J., Jaykus, L.A., Vinje, J., 2006. Surrogates for the study of norovirus stability and inactivation in the environment: a comparison of murine norovirus and feline calicivirus. J. Food Prot. 69, 2761–2765.

Carter, R., Guido, F., Jacquette, G., Rapoport, M., 1996. Outbreak of cyclosporiasis associated with drinking water. In: Program of the 30th Interscience Conference on Antimicrobial Agents and Chemotherapy, New Orleans, September 15–18, 1996. American Society for Microbiology, Washington, D.C., p. 259.

Casteel, M.J., Schmidt, C.E., Sobsey, M.D., 2008. Chlorine disinfection of produce to inactivate hepatitis A virus and coliphage MS2. Int. J. Food Microbiol. 125, 267–273.

Center for Food Safety and Applied Nutition, FDA, 2001. Analysis and Evaluation of Preventive Control Measures for the Control and Reduction/Elimination of Microbial Hazards on Fresh and Fresh-Cut Produce. Chapter IV Outbreaks associated with fresh and fresh-cut produce.

Center for Food Safety and Applied Nutrition, FDA, 2003. Food Emergency Response (FER) outbreak folders, raspberry-associated outbreak, Vermont.

Centers for Disease Control and Prevention, 2004. Outbreak of cyclosporiasis associated with snow peas–Pennsylvania, 2004. Morb. Mortal. Wkly. Rep. 53, 876–878.

Centers for Disease Control and Prevention, CDC 2013 Available at www.cdc. gov/foodborneoutbreaks/ (accessed 22.05.13.).

Centers for Disease Control and Prevention, CDC, 2013. Multistate outbreak of HAV potential associated with frozen berry blen food product. Available at http://www.cdc.gov/hepatitis /Outbreaks/2013 (accessed 31.05.13.).

Chacin-Bonilla, L., 2008. Transmission of *Cyclospora cayetanensis* infection: a review focusing on soil-borne cyclosporiasis. Trans. Trop. Med. Hyg. 102, 215–216.

Chauvin, M.A., Lee, S.Y., Chang, S., Gray, P.M., Kang, D.H., Swanson, B.G., 2005. Ultra high pressure inactivation of *Saccharomyces cerevisiae* and *Listeria innocua* on apples and blueberries. J. Food Processing Preservation 29, 424–435.

Cotterelle, B., Drougard, C., Rolland, J., Becamel, M., Boudon, M., Pinede, S., et al., 2005. Outbreak of norovirus infection associated with the consumption of frozen raspberries, France, March 2005. Euro. Surveill. 10, E050428.1 http://www.eurosurveillance.org/ew /2005/050428.asp#1.

Dawson, D., 2005. Foodborne protozoan parasites. Int. J. Food Microbiol. 103, 207–227.

Eberhard, M.L., Nace, E.K., Freeman, A.R., Streit, T.G., Lammie, P.J., 1999. *Cyclospora cayetananesis* infection sin Haiti: a common occurrence in the absence of watery diarrhea. Am. J. Trop. Med. Hyg. 60, 584–586.

Falkenhorst, G., Krusell, L., Lisby, M., Madsen, S.B., Bottiger, B., Molbak, K., 2005. Imported frozen raspberries cause a series of norovirus outbreaks in Denmark, 2005. Euro. Surveill. 10, E050922.2. http://www.eurosurveillance.org/ew/2005/050922.asp#2.

Fan, L., Forney, C.F., Song, J., Doucette, C., Jordan, M.A., McRae, K.B., et al., 2008. Effect of hot water treatments on quality of highbush blueberries. J. Food Sci. 73, M292–M297.

Fan, X., Sokorai, K.L.B., 2008. Effect of ionizing radiation on furan formation in fresh-cut fruits and vegetables. J. Food Sci. 73, C79–C83.

FDA (U.S. Food and Drug Administration), 2008. Foods permitted to be irradiated under FDA's regulations (21 CFR 179.26). Available at http://www.cfsan.fda.gov/~dms/irrafood.html (accessed 11.09.08.).

Fino, V.R., Kniel, K., 2008. UV light inactivation of hepatitis A virus, aichi virus, and feline calicivirus on strawberries, green onions, and lettuce. J. Food Prot. 71, 908–913.

Flessa, S., Lusk, D.M., Harris, L.J., 2005. Survival of *Listeria monocytogenes* on fresh and frozen strawberries. Int. J. Food Microbiol. 101, 255–262.

Food and Drug Administration (FDA) 2001. FDA Survey of Imported Fresh Produce. Available at www.cfsan.fda.gov/~;dms/prodsur6.html (accessed 21.07.08.).

Food and Drug Administration (FDA) 2001 The Juice HACCP Regulation, Questions and Answers. Available at www.cfsan.fda.gov/~;comm/juiceqa.html#F (accessed 03.09.08.).

Food and Drug Administration (FDA) 2003 FDA Survey of Domestic Fresh Produce. Available at www.cfsan.fda.gov/~;dms/prodsu10.html (accessed 21.07.08.).

Guillén, F., Zapata, P.J., Martínez-Romero, D., Castillo, S., Serrano, M., Valero, D., 2007. Improvement of the overall quality of table grapes stored under modified atmosphere packaging in combination with natural antimicrobial compounds. J. Food Sci. 72, S185–S190.

Gulati, B.R., Allwood, P.B., Hedberg, C.W., Goyal, S.M., 2001. Efficacy of commonly used disinfectants for the inactivation of Calicivirus on strawberry, lettuce, and a food-contact surface. J. Food Prot. 64, 1430–1434.

Guo, X., van Iersel, M.W., Chen, J., Brackett, R.E., Beuchat, L.R., 2002. Evidence of association of Salmonellae with tomato plants grown hydroponically in inoculated nutrient solution. Appl. Environ. Microbiol. 68, 3639–3643.

Gurtler, J.B., Bailey, R.B., Geveke, D.J., Zhang, H.Q., 2011. Pulsed electric field inactivation of *E. coli* O157:H7 and non-pathogenic surrogate *E. coli* in strawberry juice as influenced by sodium benzoate, potassium sorbate, and citric acid. Food Control 22, 1689–1694.

Hagens, S., Offerhaus, M.L., 2008. Bacteriophages – new weapons for food safety. Food Technol. 62 (4), 46–54.

Herwaldt, B.L., Ackers, M.L., the Cyclospora Working Group, 1997. An outbreaks in 1996 of cyclosporiasis associated with imported raspberries. N. Engl. J. Med. 329, 1504–1505.

Herwaldt, B.L., Beach, M.J., the Cyclospora Working Group, 1999. The Return of Cyclospora in 1997: Another Outbreak of Cyclosporiasis in North America Associated with Imported Raspberries. Ann. Intern. Med. 130, 210–220.

Herwaldt, B.L., 2000. *Cyclospora cayetanensis*. A review, focusing on the outbreaks of cyclosporiasis in the 1990s. Clin. Infect. Dis. 5, 1040–1057.

Hjertqvist, M., Johansson, A., Svensson, N., Abom, P.E., Magnusson, C., Olsson, M., et al., 2006. Four outbreaks of norovirus gastroenteritis after consuming raspberries, Sweden, June-August 2006. Euro. Surveill. 11 pii=3038.

Ho, A.Y., Lopez, A.S., Eberhart, M.G., Levenson, R., Finkel, B.S., da Silva, A.J., et al., 2000. Outbreaks of cyclopsporiasis associated with imported raspberries, Philadelphia, Pennsylvania, 2000. Emerg. Infect. Dis. 8, 783–788.

Hoge, C.W., Shlim, D.R., Rajan, R., Triplett, J., Shear, M., Rabold, J.G., 1993. Epidemiology of diarrheal illness associated with coccidian like organisms among travelers and foreign residents in Nepal. Lancet 341, 1175–1179.

Hoge, C.W., Echeverria, P., Rajan, R., Jacobs, J., Malt, S., Chapman, E., 1995. Prevalance of *Cyclospora* spears and the enteric pathogens mong children less than 5 years of age in Nepal. J. Clin. Microbiol. 33, 3958–3960.

Hurst, W.C., Schuler, G.A., 1992. Fresh produce processing – an industry perspective. J. Food Prot. 55, 824–827.

Hutin, Y.J.F., Pool, V., Cramer, E.H., Nainan, O.V., Weth, J., Williams, I., et al., 1998. A multistate foodborne outbreak of hepatitis A. N. Engl. J. Med. 340, 595–602.

Katz, D.E., Taylor, D.N., 2001. Parasitic infection of the gastro-intestinal tract. Gastroenterol. Clin. 30, 635–653.

Keyser, M., Müller, I.A., Cilliers, F.P., Nel, W., Gouws, P.A., 2008. Ultraviolet radiation as a non-thermal treatment for the inactivation of microorganisms in fruit juice. Innovat. Food Sci. Emerg. Technol. 9, 348–354.

King, B.J., Monis, P.T., 2007. Critical processes affecting *Cryptosporidium* oocyst survival in the environment. Parasitology 134 (3), 309–323.

Kingsley, D.H., Guan, D., Hoover, D.G., 2005. Pressure inactivation of hepatitis A virus in strawberry puree and sliced green onions. J. Food Prot. 68 (8), 1748–1751.

Kniel, K.E., Lindsay, D., Sumner, S., Hackney, C.R., Pierson, M., Dubey, J.P., 2002. Examination of attachment and survival of *Toxoplasma gondii* oocysts on raspberries and blueberries. J. Parasitol. 88, 790–793.

Kniel, K.E., Shearer, A.E.H., Cascarino, J.L., Wilkins, G.C., Jenkins, M.C., 2007. High hydrostatic pressure and UV light treatment of produce contaminated with *Eimeria acervulina* as a *Cyclospora cayetanensis* surrogate. J. Food Prot. 70, 2837–2842.

Knudsen, D.M., Yamamoto, S.A., Harris, L.J., 2001. Survival of *Salmonella* spp. and *Escherichia coli* O157:H7 on fresh and frozen strawberries. J. Food Prot. 4, 1483–1488.

Koopmans, M., Duizer, E., 2004. Foodborne viruses: an emerging problem. Int. J. Food Microbiol. 90, 23–41.

Korsager, B., Hede, S., Boggild, H., Bottiger, B., Molbak, K., 2005. Two outbreaks of norovirus infections associated with the consumption of imported frozen raspberries, Denmark, May-June 2005. Euro. Surveill. 10, E050623.1. http://www.eurosurveillance. org/ew/2005/050623.asp#1.

Koumans, E.H., Katz, D.J., Malecki, J.M., Kumar, S., Wahlquist, S.P., Arrowood, M.J., et al., 1998. An outbreak of cyclosporiasis in Florida in 1995: a harbinger of multistate outbreaks in 1996 and 1997. Am. J. Trop. Hyg. 59, 235–242.

Lacombe, A., McGivney, C., Tadepalli, S., Sun, X.H., Wu, V.C.J., 2013. The effect of American cranberry (*Vaccinium macrocarpon*) constituents on the growth inhibition, membrane integrity, and injury of *Escherichia coli* O157:H7 and *Listeria monocytogenes* in comparison to *Lactobacillus rhamnosus*. Food Microbiol. 34, 352–359.

Lacombe, A., Wu, V.C.H., White, J., Tadepalli, S., Andrew, E.E., 2012. The antimicrobial properties of the lowbush blueberry (*Vaccinium angustifolium*) fractional components against foodborne pathogens and the conservation of probiotic *Lactobacillus rhamnosus*. Food Microbiol. 30, 124–131.

Le Guyader, F.S., Mittelholzer, C., Haugarreau, L., Hedlund, K., Alsterlund, R., Pommepuy, M., et al., 2004. Decetion of norovirus in raspberries associated with a gastroenteritis outbreak. Int. J. Food Microbiol. 97, 179–186.

Lee, M.B., Lee, E.H., 2001. Coccidial contamination of raspberries: mock contamination with *Eimeria acervulina* as a model for decontamination treatment studies. J. Food Prot. 64, 1854–1857.

Liu, R.H., 2007. The potential health benefits of phytochemicals in berries for protecting against cancer and coronary heart disease. In: Zhao, Y. (Ed.), Berry fruit value added products for health promotion. CRC Press, Boca Raton, FL, pp. 187–204.

Lopez, A.S., Dodson, D.R., Arrowood, M.J., Orlandi, P.A., da Silva, A.J., Bier, J.W., et al., 2001. Outbreak of cyclosporiasis associated with basil in Missouri in 1999. Clin. Infect. Dis. 32, 1010–1017.

Lukasik, J., Bradley, M.L., Scott, T.M., Dea, M., Koo, A., Hsu, W.Y., et al., 2003. Reduction of Poliovirus 1, bacteriophages, *Salmonella* Montevideo, and *Escherichia coli* O157:H7 on strawberries by physical and disinfectant washes. J. Food Prot. 66, 188–193.

Marrow, L.B., 2002. Outbreak of cyclosporiasis in Fulton County, Georgia. Ge. Epidemiol. Rep. 18, 1–2.

Mazzotta, A.S., 2001. Thermal inactivation of stationary-phase and acid-adapted *Escherichia coli* O157:H7, *Salmonella*, and *Listeria monocytogenes* in fruit juices. J. Food Prot. 64, 315–320.

Miller, W.R., McDonald, R.E., McCollum, T.G., Smittle, B.J., 1994. Quality of climax blueberries after low-dosage electron-beam irradiation. J. Food Qual. 17, 71–79.

Moreno, M.A., Castell-Perez, M.E., Gomes, C., Da Silva, P.F., Moreira, R.G., 2007. Quality of electron beam irradiation of blueberries (*Vaccinium corymbosum* L.) at medium dose levels (1.0-3.2 kGy). LWT-Food Sci. Technol. 40, 1123–1132.

Moreno, M.A., Castell-Perez, M.E., Gomes, C., Da Silva, P.F., Moreira, R.G., 2008. Treatment of cultivated highbush blueberries (*Vaccinium corymbosum* L.) with electron beam irradiation: Dosimetry and product quality. J. Food Process. Eng. 31, 155–172.

Mosqueda-Melgar, J., Raybaudi-Massilia, R.M., Martín-Belloso, O., 2008. Non-thermal pasteurization of fruit juices by combining high-intensity pulsed electric fields with natural antimicrobials. Innov. Food Sci. Emerg. Technol. 9, 328–340.

Mukherjee, A., Speh, D., Jones, A.T., Buesing, K.M., Diez-Gonzalez, F., 2006. Longitudinal microbiological survey of fresh produce grown by farmers in the Upper Midwest. J. Food Prot. 69, 1928–1936.

Murrow, L.B., Blake, P., Kreckman, L., 2002. Outbreak of cyclosporiasis in Fulton County, Georgia. Ga. Epidemiol. Rep. 18, 1–2.

Niu, M.T., Polish, L.B., Robertson, B.H., Khanna, B.K., Woodruff, B.A., Shapiro, C.N., et al., 1992. Multistate outbreak of hepatitis A associated with frozen strawberries. J. Infect. Dis. 166, 518–524.

Noah, N.D., 1981. Foodborne outbreaks of hepatitis A. Med. Lab. Sci. 38, 428.

Ortega, Y.R., Sterling, C.R., Gilman, R.H., Cama, V.A., Diaz, F., 1993. *Cyclospora* species. A new protozoan pathogen of humans. N. Engl. J. Med. 328, 1308–1312.

Ortega, Y.R., Gilman, R.H., Sterling, C.R., 1994. A new coccidian parasite (Apicomplexa: Eimeriidae) from humans. J. Parasitol. 80, 625–629.

Pangloli, P., Hung, Y.-C., 2013. Reducing microbiological safety risk on blueberries through innovative washing technologies. Food Control 32, 621–625.

Pieniazek, N.J., Herwaldt, B.L., 1997. Reevaluationg the molecular taxonomy: Is human-associated *Cyclospora* a mammalian *Eimeria* species? Emerg. Infect. Dis. 3, 381–383.

Pirtle, E.C., Beran, G.W., 1991. Virus survival in the environment. Rev. Sci. Technol. 10, 733–748.

Ponka, A., Maunula, L., von Bonsdorff, C.H., Lyytikainen, O., 1999. An outbreak of calicivirus associated with consumption of frozen raspberries. Epidemiol. Infect. 123, 469–474.

Puupponen-Pimiä, R., Nohynek, L., Alakomi, H.L., Oksman-Caldentey, K.M., 2005. Bioactive berry compounds – novel tools against human pathogens. Appl. Microbiol. Biotechnol. 67, 8–18.

Reid, T.M.S., Robinson, H.G., 1987. Frozen raspberries and hepatitis A. Epidemiol. Infect. 98, 109–112.

Relman, D.A., Schmidt, T.M., Gajadhar, A., Sogin, M., Cross, J., Yoder, K., 1996. Molecular phylogeneit analysis of *Cyclospora*, the human intestinal pathogen, suggests that it is closely related to *Eimeria* species. J. Infect. Dis. 173, 440–445.

Rodgers, S.L., Cash, J.N., Siddiq, M., Ryser, E.T., 2004. A comparison of different chemical sanitizers for inactivating *Escherichia coli* O157:H7 and *Listeria monocytogenes* in solution and on apples, lettuce, strawberries, and cantaloupe. J. Food Prot. 67, 721–731.

Ryan, T., Wilkinson, J.M., Cavanagh, J.M.A., 2001. Antibacterial activity of raspberry cordial in vitro. Res. Vet. Sci. 71, 155–159.

Sapers, G.M., 2001. Efficacy of washing and sanitizing methods for disinfection of fresh fruit and vegetable products. Food Tech. Biotech. 39, 305–311.

Sarig, P., Zahavi, T., Zutkhi, Y., Yannai, S., Lisker, N., Ben-Arie, R., 1996. Ozone for control of post-harvest decay of table grapes caused by *Rhizopus stolonifer*. Phys. Mol. Plant Path. 48, 403–415.

Seymour, I.J., Appleton, H., 2001. Foodborne viruses and fresh produce. J. Appl. Microbiol. 91, 759–773.

Shields, J.M., Olson, B.H., 2003. *Cyclospora cayetanensis*: a review of an emerging parasitic coccidian. Int. J. Parasitol. 33, 371–391.

Smith, H.V., Paton, C.A., Mtambo, M.M.A., Girdwood, R.W.A., 1997. Sporulation of *Cyclospora* sp. Oocysts. Appl. Environ. Microbiol. 63, 1631–1632.

Soare, L.C., Ferdes, M., Stefanov, S., Denkova, Z., Nicolova, R., Denev, P., et al., 2012. Antioxidant and antimicrobial properties of some plant extracts. Rev. de Chim. 63, 432–434.

Solomon, E.B., Yaron, S., Matthews, K.R., 2002. Transmission of *Escherichia coli* O157:H7 from contaminated manure and irrigation water to lettuce plant tissue and its subsequent internalization. Appl. Environ. Microbiol. 68, 397–400.

Sturbaum, G.D., Ortega, Y.R., Gilman, R.H., Sterling, C.R., Cabrera, L., Klein, D.A., 1998. Detection of *Cyclospora cayetamensis*in wastewater. Appl. Environ. Microbiol. 64, 2284–2286.

Suthanthangjai, W., Kajda, P., Zabetakis, I., 2005. The effect of high hydrostatic pressure on the anthocyanins of raspberry (*Rubus idaeus*). Food Chem. 90, 193–197.

Terry, L., 2011. Oregon confirms deer droppings caused *E. coli* outbreak tied to strawberries. Available at http://www.oregonlive.com/washingtoncounty/index.ssf/2011/08/tests_reveal_ e_coli_in_deer_dr.html (accessed 30.05.13.).

Timbo, B., Ross, M., Street, D., Guzewich, J., 2007. FDA's use of epidemiological data, trace-back investigations and farm investigations as regulatory tools during outbreaks of *Cyclospora cayetanensis* infections associated with produce in the U.S. 1995–2005. IAFP Ann. Mtg. *Abstracts Book* P3–4.

Udompijitkul, P., Daeschel, M.A., Zhao, Y., 2007. Antimicrobial effect of electrolyzed oxidizing water against *Escherichia coli* O157:H7 and *Listeria monocytogenes* on fresh strawberries (*Fragaria* x *ananassa*). J. Food Sci. 72, M397–M406.

Yuk, H.-G., Bartz, J.A., Schneider, K.R., 2006. The effectiveness of sanitizer treatments in inactivation of *Salmonella* spp. from bell pepper, cucumber, and strawberry. J. Food Sci. 71, M95–M99.

Wei, K., Zhou, H., Zhou, T., Gong, J., 2007. Comparison of aqueous ozone and chlorine as sanitizers in the food processing industry: impact on fresh agricultural produce quality. Ozone: Sci. Eng. 29, 113–120.

Whitworth, J., 2012. Outbreak of norovirus in Germany 'over'. Food Prod Daily.com. 10-Oct-2012.

Wong, E., Vaillaint, F., Pérez, A., 2010. Osmosonication of blackberry juice: Impact on selected pathogens, spoilage microorganisms, and main quality parameters. J. Food Sci. 75, M468–M474.

Wu, V.C.H., Kim, B., 2007. Effect of a simples chlorine dioxide method for controlling five foodborne pathogens, yeasts and molds on blueberries. Food Microbiol. 24, 794–800.

Contamination Avoidance Pre and Postharvest

Produce Contamination Issues in Mexico and Central America

15

J. Fernando Ayala-Zavala, Miguel A. Martínez-Téllez, Leticia Felix-Valenzuela, Verónica Mata-Haro

Centro de Investigación en Alimentación y Desarrollo, AC, Hermosillo, Sonora, Mexico

CHAPTER OUTLINE

Introduction

Mexico and Central America (Guatemala, Nicaragua, Honduras, El Salvador, Belize, Panama, and Costa Rica) are important countries exporting considerable volumes of fresh fruits and vegetables to different parts of the globe, especially the United States (Figures 15.1–15.3). From 2007 to 2013, the dollar value of exported vegetables, fresh fruits (other than bananas), and bananas and plantains from all these countries increased 44.4%, 52.4%, and 83.3% to the United States, respectively (Department of Commerce, U.S. Census Bureau, Foreign Trade Statistics, 2013). Mexico was the main exporter by far on vegetables and fruits other than bananas, compared to Central America. Among the most important fresh vegetables traded by these regions

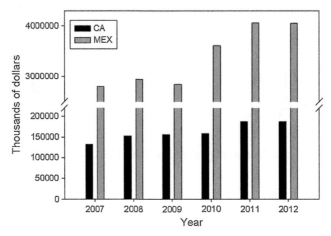

FIGURE 15.1

Fresh Vegetables (January–December, values in thousands of dollars) Imported by United States from Central American Countries (CA) and Mexico (MEX) from 2007 to 2012.

Source: Department of Commerce, U.S. Census Bureau, Foreign Trade Statistics, 2013.

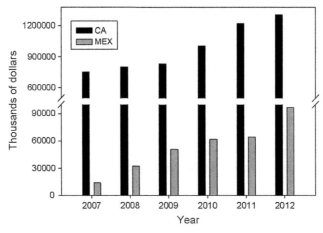

FIGURE 15.2

Bananas and Plantains (January–December, values in thousands of dollars) Imported by United States from Central American Countries (CA) and Mexico (MEX) from 2007 to 2012.

Source: Department of Commerce, U.S. Census Bureau, Foreign Trade Statistics, 2013.

were tomatoes, beans, cassava, chili pepper, peas, squash, dasheens, onion, yams, bell pepper, cucumbers, and broccoli; and fresh fruit like avocado, grapes, berries, lime, melons, mangoes, papaya, and pineapples.

The export industries from these countries grow products to meet foreign-market consumer demand, retail preferences, and governmental restrictions (limits on

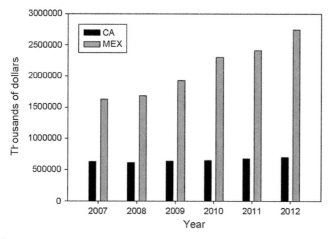

FIGURE 15.3

Fruits other than bananas (January–December, values in thousands of dollars) Imported by United States from Central American Countries (CA) and Mexico (MEX) from 2007 to 2012.

Source: Department of Commerce, U.S. Census Bureau, Foreign Trade Statistics, 2013.

chemical and pesticide residues, programs to deal with quarantine pests, etc.). The technology is quite similar to that used in the United States, as U.S. firms are active in Mexican export industries. Producers for Mexico's domestic market tend to be more labor-intensive than in the United States and employ more traditional methods of cultivation and harvesting. The overall sanitary quality of crops during production is dependent primarily on the growing environment. However, harvesting introduces human and mechanical contact that has an impact on the microbiological safety of fresh produce. The degree of farm workers' personal hygiene can have an important influence on the transmission of pathogenic bacteria to produce being harvested. Farm workers often come from diverse cultural backgrounds, not all of which stress proper personal hygiene as an important behavioral value. These characteristics impact directly the risk of contamination, and special attention is being paid in assuring safety of exported fresh produce.

Several large foodborne outbreaks have been linked to fresh produce imported from these regions, including crops such as cantaloupe, tomatoes, peppers, and papaya (CSPI, 2013). Although contributing factors have not been determined in all cases, quite a few notable causes have been proposed. In particular, cross-contamination with fecal matter of both domestic and wild animals has been suggested. In addition, contact with contaminated water also has been identified as a source of contamination. Moreover, the use of untreated manure or sewage as fertilizer, lack of field sanitary toilet facilities, poorly or unsanitized transportation vehicles, and contamination by handlers are also suggested as potential contributing factors. These cases and the importance of the economic revenues in trading fresh fruits and vegetables are triggering the implementation and validation of Good Agricultural Practices, Good

Manufacturing Practices, and Hazard Analysis and Critical Control Points in Mexico and Central America regions. In this context, the present chapter describes the main sources of contamination and the main approaches to assure safety of fresh produce in Mexico and Central America.

Sources of contamination

Irrigation water

Water is used in diverse agricultural activities. Water used to apply fertilizers, as well as pesticides, has to meet both chemical and microbiological requirements before its use (Siller–Cepeda et al., 2002). When water is in contact with fruits and vegetables, the risk of contamination will depend on the microbiological quality of the water source. Irrigation of crops with poor quality water is one way that fruits and vegetables can become contaminated with pathogenic microorganisms. To avoid contamination, wells and all other sources of water must be submitted periodically for chemical and microbiological examination, have the results recorded, and correct existing problems. Bathing and grazing of animals near water resources should be prohibited, to prevent fecal contamination of water and reduce risks to human health from consumption of contaminated fresh produce. In the particular case of irrigation, contamination is associated with the irrigation type and the kind of crop. Flooding irrigation represents the greatest possibility of contamination if it is used on creeping crops such as lettuce or strawberries by permitting contact with the soil. With the sprinkler irrigation technique, the spray provides a rapid means to contaminate the product. In both cases, water quality is important. With the drip irrigation technique, the risk of contamination is smaller.

Most of the large areas dedicated to growing and exporting vegetables in the winter in northwest Mexico make use of water stored behind dams for irrigation; however, some areas in Sonora, Baja California, Coahuila, and Nuevo Leon States depend mostly on water extracted from wells. Water from dams is conducted to the fields through irrigation canals. Conditions vary widely across the country; some are protected by concrete to avoid water leaks and growth of weeds, others are just abandoned. Areas in Central and South Mexico, as well as some regions in Central American countries use different water sources for irrigation. Microbiological quality has been always a concern in some of these areas, especially areas in Central Mexico and in Central America that use waste water coming from sewage treatment for irrigation of crops intended for the domestic market. Use of waste water can contaminate produce with pathogenic microorganisms by direct contact. Sadovski et al. (1978) noticed that the use of waste water was of greater concern if applied immediately before harvest rather than during the early stages of the production cycle.

Among the types of irrigation used in Mexico and Central American countries, drip-irrigation is the most common type used for agriculture dedicated to exportation, particularly in Northwest Mexico, reducing risk of contamination. However, this represents only around 15% of the land being irrigated. Overhead-irrigation with sprinklers is utilized mostly to irrigate fruit crops, especially citrus and apple crops. Drip-irrigation and sprinklers are mainly used to make more efficient use of water and apply fertilizer properly dosed. However, most tropical crops in these regions will depend on seasonal rainy periods for water availability. Flooding-irrigation is the type of irrigation more widely used in countries where water is conducted from sources impounded by dams to fields. This type of irrigation represents higher possibilities of contaminating produce, especially when it is applied to creeping crops and where water is in contact with contaminated soil or runoff from large cities or cattle-raising facilities. On crops that are staked and raised above the soil, such as those vegetables mostly grown for exportation, the risk of contamination is low.

In Mexico, few studies have been conducted on the quality of irrigation water in general. One study revealed the presence of pathogens such as *Cryptosporidium* oocyst (48%) and *Giardia* cyst (50%) in surface water coming from water impounded by dams. Water samples were collected from the distribution canal system that irrigates the Culiacan valley, located at the northwestern state of Sinaloa, Mexico. In addition wash-water tanks filled with water coming from rivers or impounded by dams and used in selected packinghouses tested positive for *Cryptosporidium* oocysts and *Giardia* cysts with concentrations ranging from 1 to 133 oocysts and 100 to 533 cysts per 100 liters, respectively (Chaidez et al., 2005). The occurrence of these parasites in irrigation waters (rivers and canals) used to grow crops usually eaten raw has been also investigated in some Latin American countries. Thurston-Enriquez et al. (2002) reported a geometric mean of 6,426, 227, and 693 *Giardia* cysts per 100 liters for Costa Rica, Mexico, and Panamá, respectively, and a geometric mean of 150, 612, and 190 *Cryptosporidium* oocysts per 100 liters for Costa Rica, Mexico, and Panamá, respectively. This suggests that there may be a risk of contamination of fresh produce as protozoan oocysts/cysts might come in contact with and attach to produce surfaces, posing a risk of infection to consumers.

A few *Cryptosporidium* oocysts, *Giardia* cysts, *E. coli,* and coliphages have also been identified in irrigation water from El Valle del Yaqui in the northwest of Mexico. This agricultural valley comprises approximately 220,000 ha, which are irrigated by a canal network of 2,774 km. There are no direct sewage discharges into this canal systems, so enteric organisms could contaminate the water through runoff from rain events, septic tanks, human bathing or recreation, presence of domestic animals, and return flows from irrigation (Gortares-Moroyoqui et al., 2011).

A quantitative microbial risk assessment (QMRA) was conducted to evaluate the public-health impact of protozoan-laden water used for irrigating produce in Northwest Mexico. Specifically, a QMRA was conducted to address the human health impact associated with consumption of tomatoes, bell peppers, cucumbers, and lettuce irrigated with water contaminated with *Cryptosporidium* and *Giardia*. Yearly infection risks were estimated based on the assumption of a 120-day exposure in a given year. Annual risks range from 9×10^{-6} for *Cryptosporidium* at the lowest concentration associated with bell peppers to almost 2×10^{-1} for exposure to *Giardia* on lettuce at the highest detected concentration. With the relatively high number of illnesses resulting from produce-related outbreaks, addressing pre- and postharvest points of contamination for fruits and vegetables consumed raw should be a food industry priority. This research shows how QMRA can be used to interpret microbial contamination data for public health significance and subsequently provide the foundation for guideline development (Mota, 2004).

Agricultural water was assessed to determine the presence of pathogens such as *Salmonella* and *Escherichia coli* from January to May of 2005 in four regions of the Culiacan Valley located in Northwest Mexico, an area known for its vast agricultural production, and as one of the major exporters of fresh produce, worldwide. Samples from what is known as the interphase water-sediment of water used to irrigate agricultural crops were positive for 20 strains of *Salmonella*. Serotyping revealed the presence of 13 strains of *Salmonella* Typhimurium two of Infantis, and one each of Anatum, Agona, Oranienburg, Minnesota, and Give. Ninety-eight percent of the analyzed water samples were contaminated with *E. coli*, averaging 1.6×10^4 cfu/100 ml (Lopez-Cuevas et al., 2005).

Irrigation of crops with waste water is a common practice in some regions of Mexico. Mezquital Valley, located in the Central part of Mexico, receives approximately 50 m^3/s of untreated waste water from Mexico City, which is used to irrigate about 45,214 ha (Lesser-Carrillo et al., 2011); crops grown at the Mezquital Valley include lettuce, tomato, carrot, coriander, and spinach, and they are mainly marketed to Pachuca-City, Hidalgo, and Mexico City (Castro-Rosas et al., 2012). Recent studies have evaluated the microbiological quality of vegetables grown in this area; 130 ready to eat salads (RTE) and 280 fresh carrot juice acquired at various restaurants at Pachuca city were analyzed. Results showed that 99% of the analyzed RTE salads were contaminated with fecal coliforms; 85% harbored *E. coli* and 7% diarrhoeagenic *E. coli* pathotypes (DEPs) (Castro-Rosas et al., 2012). Moreover, 96.8% of analyzed juices were contaminated with fecal coliforms, 53.6% harbored *E. coli*, 8.9% DEPs, and 8.6% were positives for *Salmonella* (Torres-Vitela et al., 2013). The fact that almost all salads and juices samples were contaminated with fecal matter regardless of the status of hygiene of the restaurants, it may be because most vegetables consumed in Pachuca-City are grown in the Mezquital Valley. Thus, contamination most likely originated from irrigation water rather than from poor sanitation practices at the restaurants (Torres-Vitela et al., 2013).

A study performed in the Xochimilco agricultural area in Mexico, located at the central part of the country and irrigated with waste water, showed a variety of microorganisms such as Enterobacteriaceae, *Escherichia coli, Enterobacter cloacae, Klebsiella pneumoniae, K. oxytoca, Citrobacter freundii*, and *Salmonella* spp., as well as non-fermenting microorganisms such as *Pseudomonas* spp. and *Acinetobacter* spp. Some species are not native to the natural environment and may represent exogenous microorganisms, further indicating a human or animal fecal source. The observed patterns of irregular urban area settlements and presence of animals such as cows or sheep grazing in some areas provide suggestive evidence of the source of non-native microorganisms. Mexican guidelines for use of residual water for irrigation (SEMARNAT, 1996) specify 1000 or less cfu/100 ml as the fecal coliform limit for acceptable irrigation water for crops likely to be eaten uncooked and for sport fields and parks; this limit is exceeded in areas of Xochimilco (Mazari-Hiriart et al., 2008). Perhaps the health authorities and government may introduce measures to ban the use of untreated waste water, or that do not meet the specifications of the Mexican standard guidelines to irrigate vegetable crops, especially those that are eaten raw. However, the above is not an easy task, since irrigation with untreated waste water is a custom rooted by some farmers because it can be used as fertilizer which represents an important economic activity for them, and also because waste water is more accessible than other sources. Considering the above, the disinfection of the fresh produce in packinghouses should be an important concern to prevent outbreaks.

Runoff
Potential movement of fecal matter during rainy season

An important environmental factor affecting microbial movement is rainfall. Rain events may carry fecal contamination from wild and domestic animals in rivers, canals, and wells that serve as sources of irrigation water. Also, rain events can result in pathogen spread by runoff from places where manure or biosolids have been applied as fertilizer or by leaching through the soil profile. It is known that bacterial and viral groundwater contamination increase during heavy rainfall. The presence of coliforms was monitored for 9.4-meter and 153.3-meter wells. Coliforms were detected in both shallow and deep wells, with bacterial contamination coinciding with the heaviest rainfalls (Gerba and Bitton, 1984). In Quebec, Canada, human and pig enteroviruses were isolated from 70% of the samples collected from a river. The contamination source was attributed to a massive pig-raising activity in the area (Payment, 1989). In contrast, some authors have reported a decrease of pathogen concentrations during the rainy season. Cazarez-Diarte et al. (2004) observed a reduction from 41,493 cfu/100 mL during the dry season to 8,525 cfu/100 mL during the rainy season. Tyrrel and Quinton (2003) stated that coliform transport is mediated by water density and turbidity; hence, water-body volumes when pluvial precipitations exist are diluted and so are the coliforms present. In Costa Rica, Calvo et al. (2004) evaluated the presence of *Cyclospora* spp., *Cryptosporidium* spp., microsporidia, and fecal coliforms on lettuce, celery, cilantro, strawberries, and blackberries. Fifty samples of each product were analyzed, 25 in the dry season and 25 during the rainy season. At least one of the parasites in research was found in all of the products tested. The lettuce had the highest prevalence of *Cryptosporidium* spp., with a statistically significant difference between the rainy and dry season. One-hundred percent of the vegetable samples had fecal coliform, and the highest prevalence was obtained during the rainy season with lettuce having significant differences between the rainy and dry season. The contamination of these products with these microorganisms were attributed to various factors such as poor sanitary quality of water used for irrigation, agricultural practices, transport, and storage of the products; however, due to high levels of fecal coliforms, especially in the rainy season (average range of $7.9 \times 10^5 - 3.4 \times 10^7$ cfu/g), fecal runoff from rain events should not be discarded.

Flooding

Another factor that could affect the bacteriological quality of croplands is a history of flooding. This situation can become a problem when floodwaters cover areas on which farm animals have been grazed or confined, upstream from vegetable production areas. Floodwaters can become polluted with animal waste and carry the contaminants downstream, where they may also flood over croplands. Major flooding has also caused rivers to cover or damage sewage treatment plants. Either the floodwaters or effluents from the plants then become contaminated with human, municipal, and industrial wastes. Again, such events can subsequently contaminate downstream croplands. Microorganisms deposited on flooded croplands may remain for months or years after the flood (Beuchat and Ryu, 1997). Therefore, it is recommendable that fresh produce that were covered by flood waters, especially leafy vegetables, should not be consumed. The chlorinated water usually used to disinfect them perhaps fails completely to eliminate pathogens that can remain on the surface of vegetables. Information regarding how flooding affects the bacteriological quality of crops is scarce in Mexico and Central America.

Inadequate disinfection processes at packinghouses

The disinfection processes at packinghouses in Mexico and Central America are among the main analyzed procedures to reduce or eliminate contamination. The more common disinfectant agent in these regions is chlorine; however, other options are being adopted including peroxyacetic acid and ozone.

Chlorination is widely practiced as a disinfection process for microbial control in water used to wash fruits and vegetables at most of the packinghouses dedicated for exportation, but this practice is not used in packing of vegetables for domestic market. When properly applied, chlorine-based products are efficient. However, several drawbacks have been identified, including the protection exerted by the organic and inorganic matter to chlorine disinfection (Gomez-Lopez, 2012). Chlorination prevents the buildup of decay-producing organisms in the water of the dump tank, washer, hydrocooler, among other surfaces in the packinghouse.

Most Mexican packinghouses that export fruits are large, sophisticated, high-volume operations. For instance, in the tomato line production, upon transfer to the packing line, tomatoes are washed, pre-sized, waxed, sorted and graded, sized, packed into shipping containers, and unitized for shipment while in the packinghouse. Water dump tanks are used routinely for receiving tomatoes at the packinghouse. Pallet bins are emptied into the dump tank while tomatoes are water flumed from gondolas into the dump tank. In each case, tomatoes in the dump tank are flumed to an elevator where they are spray washed and conveyed to the packing lines. Serious losses due to decay occur periodically in shipments during transit or at destination. Poor dump-tank and wash-water management practices can be major contributors to decay or contamination problems.

Some products are received and maintained dry, whereas others are received by submersion in a chlorinated water tank at concentrations fluctuating from 100 to 300 ppm of total chlorine or from 50 to 75 ppm of free chlorine. When immersing the product, parameters such as turbidity, water temperature, disinfectant concentration, and pH are constantly monitored in order to maintain the optimum conditions for the disinfectant. pH is usually maintained at values between 6.5 to 7.0. Water temperature is one of the critical parameters because if a temperature differential of 10°F (~6°C) or more between water and pulp exists, the produce will tend to absorb water and the possible bacterial contamination.

In the case of the products maintained dry, the washing and disinfection process occurs further on, while the product is transported on selection rollers. These selection rollers are specially designed to cause the product to spin, ensuring a homogeneous brushing, washing, and disinfection process. Most of these systems use gaseous chlorine directly dispersed in the water through a sparger, wetting and disinfecting produce as it is spinning. Contact time of produce in chlorinated water is short, usually between 15 and 30 seconds (Gomez-Lopez, 2012).

A study was undertaken by Chaidez et al. (2003) to determine the efficacy of three commonly used disinfectants in packinghouses of Sinaloa, Mexico: sodium hypochlorite (NaOCl), trichlor-s-triazinetrione (TST), and trichlormelamine (TCM). Even though TST is approved only for swimming pool area disinfection, sometimes it is used for application to produce contact surfaces in Mexico. Each microbial challenge consisted of water containing approximately 8 $\log_{10}$ bacterial cfu mL^{-1}, and 8 $\log_{10}$ viral PFU mL^{-1} treated with 100 and 300 mg L^{-1} of total chlorine with modified turbidity. Water samples were taken after two minutes of contact. They found that chemical disinfectants inactivate *E. coli* and *S.* Typhimurium in water by greater than 6 $\log_{10}$ at the initial test point. It is known that under conditions of high water-quality, waterborne vegetative bacteria are highly susceptible to relatively low doses of chlorine. Factors such as the amount of organic matter surrounding the target organisms are likely to influence the adhesion characteristics of cells and the lethal effect of sanitizers.

Results of this study showed, however, that the amount of organic material present in the wash water influenced the efficacy of disinfectants. Results also show that similar and significant reductions in populations of *E. coli* and *S.* Typhimurium occur in water used to wash fruits and vegetables at the packinghouses using 100 or 300 mg L^{-1} of NaOCl or TST in both average and worst-case water conditions. TST (300 mg L^{-1}) and NaOCl (300 mg L^{-1}) after an exposure of two minutes were found to effectively reduce the number of bacterial pathogens and viral indicators by 8 $\log_{10}$ and 7 $\log_{10}$, respectively (P = 0.05). The highest inactivation rate was observed when the turbidity was low and the disinfectant was applied at 300 mg L^{-1}. TCM did not show effective results when compared with TST and NaOCl (P ≤ 0.05). Significant reduction of MS2 phage only was achieved using 300 mg L^{-1} of NaOCl or TST in average-case water, whereas in the worst-case water challenge, neither NaOCl nor TST or TCM were effective. These findings suggest that turbidity created by the organic and inorganic material present in the water tanks that were carried in by the fresh produce may affect the efficacy of the chlorine-based products.

Martinez-Tellez et al. (2009) studied the effect of different disinfectants including chlorine, hydrogen peroxide, and lactic acid in bunches of green asparagus and green onions. The produce was immersed for 1 min in a bacterial suspension containing approximately 6.0 $\log_{10}$ cfu mL^{-1} of *Salmonella* Typhimurium. The samples were drained and air-dried at ambient temperature (about 25°C) for 30 minutes. The sanitizers utilized were 6% sodium hypochlorite, 20% hydrogen peroxide, and 85% lactic acid. Solutions were prepared and applied with sterile deionizer water at 10°C. Concentrations included were chlorine solutions (pH 6.5, at 200 and 250 mg L^{-1}), hydrogen peroxide (1.5 and 2.0%), and lactic acid (pH 6.5, at 1.5 and 2.0%). The sanitation procedure was developed simulating a typical packinghouse washing process, which included a spray application of sanitizer solutions. After draining for 30 minutes, groups of four bunches of green onions and green asparagus were selected for each treatment application; the bunches were placed on a sterile wire screen and were sprayed during 40, 60, and 90 s either with sterile distilled water, or with the sanitizers mentioned above at 10°C.

Chlorine sanitation showed better efficacy at higher concentration (inhibition of *Salmonella* at 1.36–1.74 $\log_{10}$ cfu g^{-1}), however, no significant differences between 200 and 250 mg L^{-1} Cl$_2$ were observed. Chlorine showed a more effective *Salmonella* inhibition on inoculated fresh green onions compared to asparagus spears; nevertheless, no significant effect of exposure time was observed. Hydrogen peroxide sanitation showed better efficacy at higher concentration (inhibition of *Salmonella* at 1–1.43 $\log_{10}$ cfu g^{-1}); on the other hand, no significant differences between 1.5 and 2% H$_2$O$_2$ were observed. No significant differences among exposure times and between fresh produce were observed, respectively. Comparing the efficacy of hydrogen peroxide with chlorine sanitation, it was observed that chlorine showed a more effective *Salmonella* inhibition on inoculated fresh green onions and asparagus spears. Lactic acid sanitation showed better efficacy at the highest concentration, 2% (inhibition of *Salmonella* at 2.9 $\log_{10}$ cfu g^{-1}); however, significant differences between sanitized fresh produce was observed. Lactic acid showed a more effective *Salmonella* inhibition on inoculated fresh asparagus spears compared to green onions; nevertheless, no significant effect of exposure time was observed. Comparing the efficacy of lactic acid with chlorine sanitation, it was observed that lactic acid showed a more effective *Salmonella* inhibition on inoculated fresh asparagus spears.

Conditions for agricultural workers

The innocuity in the chain of fresh vegetables production can be affected by the health of workers handling these products in field and packing. In order to avoid affecting their quality, many of the open field grown vegetables are not disinfected before they are packed for commercial distribution. The health of agricultural workers is closely related to the working conditions where most of the activities require a lot of physical effort and are performed outdoors in cold, warm, or hot weather, with inadequate sanitary facilities. Hence, a lack of protection against rough weather conditions along with insufficient tools and equipment may often lead to work disabilities (CDC, 2002).

Most agricultural workers are immigrant people who have left home to earn a livelihood for themselves and their families in the large agricultural fields, which lack adequate living places during growing seasons, causing instability and stress for the workers and their families and consequently, increasing health risks among workers. The employees of socially responsible companies are protected by a series of basic indicators of potential social risks (GLOBALG.A.P., 2011) and, among other things, workers and their families are provided with appropriate housing with drinking water, adequate clean sanitary facilities, and health care; elementary conditions to ensure the health of workers in agricultural environments.

Prevalence and incidence of bacterial-, viral-, and protozoan-related gastroenteritis

The degree of farm workers' personal hygiene can have an important influence on the transmission of pathogenic bacteria to produce being harvested. Farm workers often come from diverse cultural backgrounds. Infectious diarrhea is an important cause of serious morbidity in developed nations, in hospitalized patients, and in travelers to tropical or subtropical regions of the world. Most important, it is a major public health problem in developing countries, where it is an important cause of morbidity and mortality in children. Bacterial pathogens are responsible for more than 50% of diarrheal diseases in developing countries. In Mexico, gastrointestinal infectious diseases have been the first cause of mortality especially in children from 1 to 4 years, most of them living in marginated zones with malnutrition (Gomez-Dantes et al., 2011; SUIVE/DGE/SSA, 2011).

Rotavirus is considered the greatest, worldwide cause of viral gastroenteritis in children. Its prevalence is similar in developed and developing countries (Payne et al., 2012). For instance, the incidence of rotavirus in Paraguayan children from 2004 to 2005 was seasonal, with the highest incidence during the coolest and driest months of the year; rotavirus incidence was 23.8% and 14.9% for children and adults, respectively (Amarilla et al., 2007). In the U.S., before vaccination was introduced, four of five children had rotavirus infection before age five. A highly effective vaccine was introduced in 2006, and the burden of the disease has diminished ever since (Payne et al., 2012).

Norovirus (NoV), which belongs to the Caliciviridae family, is now recognized as the leading cause of epidemic and endemic nonbacterial gastroenteritis. In industrialized countries, NoV may be responsible for 68 to 93% of nonbacterial gastroenteritis outbreaks (Fankhauser et al., 2002). Studies in developing countries have shown that NoV is a major nonbacterial pathogen that causes acute diarrhea in children (Talal et al., 2000).

Long et al. (2007) provided unique information about the epidemiology of NoV infection and the effect that vitamin A supplementation has on this infection among children living in peri-urban communities of Mexico City. First, they found a high prevalence of NoV infections during the summer months. NoV was isolated from 114

(30.5%) of 374 stool samples collected during the summer months. NoV GI and NoV GII were found in 62 (54.4%) and 52 (45.6%) of the 114 positive samples, respectively. Twenty-five (21.9%) of the 114 NoV positive samples had coinfections: 7 (6.1%) were coinfected with EPEC, 10 (8.8%) with ETEC, and 7 (6.1%) with *G. lamblia*.

Intestinal parasites remain extremely common worldwide. In developing countries, intestinal protozoans are important causes of childhood diarrhea. Cryptosporidiosis is a common cause of chronic diarrhea in patients with AIDS. With the advent of current active antiretroviral therapy, the prevalence of cryptosporidiosis in AIDS has decreased. By contrast, *Cryptosporidium, Cyclospora*, and *Giardia* outbreaks continue to be associated with contamination of food or water (Okhuysen and White, 1999). *Cryptosporidium* spp. and *Giardia* spp. are intestinal protozoan parasites that are recognized as prevalent and widespread pathogens of humans and many species of mammals. They constitute a common cause of gastroenteritis that manifests as a watery diarrhea in humans, and are the third most common protozoan infections in humans worldwide (Paziewska et al., 2007).

Hygiene facilities

Workers involved in farming can have an important impact on the microbial safety of produce they handle. Most of the foodborne illnesses are transmitted by humans. It is therefore very important to have agricultural workers adhere to proper sanitary procedures. Workers who either fail to practice or refuse to apply important hygienic practices, such as handwashing, constitute a risk for contaminating the produce they touch with human pathogens. An outbreak of cholera associated with sliced melon was traced to agricultural workers as the source of contamination (Ackers et al., 1997). Several critical points must be considered to reduce the contribution of agricultural workers to pathogen transmission. Among them, adequate sanitary facilities must be provided to workers. Portable toilets and handwashing facilities are the minimum requirements to be implemented. This practice is applied on most operations; however, there are still a lot of operations that do not comply at 100% with these practices. It is also important that such facilities be placed in a relatively convenient location, in close proximity to work areas, but outside the packinghouse, in order to reduce cross-contamination by pathogenic microorganisms. These facilities must exist for both sexes, sufficient to the number of employees who are working at the field. It is also important to provide instructions explaining the handwashing process to all personnel. Potable water, bactericide soap, and paper towels must always be available. When sanitary facilities are next to or near packinghouse areas, a sanitary rug must be placed on the bathroom's exit door to disinfect the soles of shoes (Siller-Cepeda et al., 2007). Another important issue is that workers must be trained in the importance of proper personal hygiene, specifically handwashing after using restrooms. Ideally, such training would motivate the workers to willingly conform to the required sanitary practices. Training should be in the first language of the worker to ensure the real transmission of knowledge (Siller-Cepeda et al., 2007).

A study to demonstrate the importance of good handwashing as well as the transfer of pathogens between hands and produce was done by Jimenez et al. (2007); they assessed the effectiveness of hand-hygiene techniques and quantified the amount of *Salmonella* Typhimurium transferred from volunteers' hands (bare or gloved) to green bell peppers and vice versa. Their results showed that the efficiency of transmission of *Salmonella* from green bell peppers to hands was high, whereas transfer rate from hands to the fresh produce was low. A combination of handwashing and handrubbing with alcohol gel significantly reduces the presence of *Salmonella* on hand surfaces, and it should be considered as part of routine packinghouse activity. However, the primary method to avoid the presence of *Salmonella* during packinghouse operations would be the strict adherence to the GAPs, which means that the best strategy, undoubtedly, is prevention.

Good Agricultural Practices

The concept of Good Agricultural Practices (GAP) has evolved in recent years in the context of a rapidly changing and globalized food economy, and as a result of the concerns and commitments of a wide range of stakeholders regarding food production and security, food safety and quality, and the environmental sustainability of agriculture. These stakeholders have representatives in the supply dimension (farmers, farmers' organizations, workers), the demand dimension (retailers, processors, consumers), and those institutions and services (education, research, extension, input supply) that support and connect demand and supply and who seek to meet specific objectives of food security, food quality, production efficiency, livelihoods, and environmental conservation in both the medium and long term.

Broadly defined, a GAP approach aims at applying available knowledge to addressing environmental, economic, and social sustainability dimensions for on-farm production and postproduction processes, resulting in safe and quality food and nonfood agricultural products. Based on generic sustainability principles, it aims at supporting locally developed optimal practices for a given production system based on a desired outcome, taking into account market demands and farmers' constraints and incentives to apply practices.

However, the term GAP has different meanings and is used in a variety of contexts. For example, it is a recognized terminology used in international regulatory frameworks as well as in reference to private, voluntary, and non-regulatory applications that are being developed and applied by governments, civil society organizations, and the private sector. Recently, the Food Safety Modernization Act in the United States indicates that the processors of all types of food will now be required to evaluate the hazards in their operations, implement and monitor effective measures to prevent contamination, and develop a plan to take any necessary corrective actions. Also, the U.S. Food and Drug Administration (FDA) will have much more effective enforcement tools for ensuring those plans are adequate and properly implemented, including mandatory recall authority when needed to swiftly remove contaminated food from the market (FDA, 2013).

For many developing countries, the export of fruits and vegetables accounts for significant income from hard-currency earnings. However, rejection of fresh produce has been related to overall quality, presence of non-authorized pesticides, pesticide residues, and contaminants exceeding permissible limits. Other causes include inadequate labeling and packaging, not having the required nutritional information, and bacterial contamination. An international effort to standardize protocols implementing the GAP and Good Manufacturing Processess (GMP) has been the Global Food Safety Initiative (GFSI), a business-driven initiative for the continuous improvement of food safety management systems to ensure confidence in the delivery of safe food to consumers worldwide. GFSI provides a platform for collaboration between some of the world's leading food safety experts from retailer, manufacturer, and food-service companies, and service providers associated with the food supply chain, international organizations, academia, and government (GFSI, 2013). Efforts are underway by governments to develop and apply GAPs, GMPs, and Hazard Analysis and Critical Control Points (HACCP) throughout the food chain. The challenges to the system are the lack of or weak coordination between the public and private sectors, training programs targeting appropriate stakeholders, the needs to harmonize national standards with international standards, and especially, the lack of political concern and incentives for adoption of programs at the farmers' level. The quality and safety programs and initiatives implemented are targeting mainly production supplying export markets, with little or no emphasis on production supplying domestic markets (national consumers' protection aims). Also, food control systems in some countries do not have a clear distinction regarding responsibilities and roles of the government and institutions involved in quality and safety issues at the production level. There is a clear need to define institutional roles in terms of quality and safety for primary production. There is a need for enforcing pesticide regulations. Preventing the misuse of pesticides and emphasizing the use of approved pesticides, applied to effectively control pests and diseases in conformance with the approved Minimum Risk Levels (MRLs) and the International Code of Conduct for Distribution and Use of Pesticides, are important actions that need to be enforced.

In U.S. and EU markets, private initiatives have been implemented (EurepGAP, Safe and Quality Food-SQF Code, BRC Global Standard Packaging, ProSafe Certified Program, GAP Certification); however, in Latin America, initiatives are taken over by the private and public sectors (ChileGAP, PIPAA Program-Guatemala, SENASICA–Mexico, SENASA–Argentina, SENA–Colombia, PRMPEX–Peru, OIRSA–El Salvador).

In Mexico, the Agriculture Department (SAGARPA), a federal institution, supports growers in developing GAP Manuals for main crops. Within its organic structure, the Servicio Nacional de Sanidad, Inocuidad y Calidad Agroalimentaria (SENASICA) offers certification of agricultural companies under the GMP through the volunteer program, "Reductions System Risk Contamination." Its aim is to avoid having food-safety barriers become an obstacle for national produce in international markets. This program consists of a system that minimizes any risk in the production and packing of fruits and vegetables. As an example, federal authorities (SENASICA) and state authorities (Government of Baja California) in conjunction with the green onion export industry and growers developed the Green Onion Protocol (GAPs and GMPs) based on FDA guidelines.

The "Mexico Calidad Suprema" program is an official mark of identification that guarantees good sanitation, food safety, and a high quality for Mexican products. This label seeks to identify products that comply with the following regulations: Mexican Official Norms (NOMs), Mexican Norms (NMX), and International Rules in a confident and transparent system for the benefit of producers, packers, and distributors. One of the control points in this protocol is the certification of the "Reductions System Risk Contamination" of the Mexican Federal Government dictated by SENASICA.

Outbreak-related cases in Mexico and Central America

According to data from the CDC and published in Center for Science in the Public Interest reports, over a 10-year period from 2001 to 2010, produce-related outbreaks were in decline; however, there was a sharp increase in produce-related foodborne illnesses in 2008 due to a large multi-state *Salmonella* outbreak from peppers and tomatoes that sickened over 1,500 people. During the same period, produce was linked to the largest number of foodborne outbreaks and illnesses associated with them, constituting 23% of all illnesses reported to the CDC (CSPI, 2013; DeWaal et al., 2002). The most common identified pathogens contaminating produce were norovirus and *Salmonella* spp. Some of the outbreaks in which the pathogen could be identified were traced to produce imported from Mexico and Central American countries, and they are described in the following.

The cantaloupe melon is ranked as the sixth highest cause of fresh produce-related foodborne disease, resulting in nearly 2462 cases during the period of 1990 to 2010 (CSPI, 2013). *Salmonella* infections due to cantaloupe consumption have been reported since at least 1990. In that year, cantaloupes presumably originating in Mexico or Guatemala were found to be contaminated with *S.* Chester and caused 245 cases of infections. Between 1990 and 1991, FDA personnel isolated several *Salmonella* serotypes from 1% of cantaloupe and watermelon samples collected at the border (CDC, 1991). In 1991, in California, 25 people were infected with *S.* Saphra due to consumption of cantaloupe imported from Altamirano, Guerrero, Mexico (Mohle-Boetani et al., 1999). Between May and June 1998, again the Mexican cantaloupe was involved in an outbreak in Ontario, Canada where 22 illnesses were reported due to the presence of *S.* Oranienburg. During this outbreak, cantaloupes were imported from Mexico and Central America (Sewell and Farber, 2001). The recurring presence of *Salmonella*-contaminated cantaloupes triggered the establishment of a surveillance program at field and packinghouse levels to source track the points of contamination and to implement corrective actions to remedy the problem. After a series of visits and training, four Mexican firms implemented GAP and GMP programs and were allowed to export again (CDC, 2002; FDA, 2001a). In 2001, 29 fruits were examined, and none presented *Salmonella* spp., *Shigella* spp., or *Escherichia coli* O157:H7 (CDC, 2002).

Between 2000 and 2002, three *Salmonella* outbreaks occurred in the United States, and they were associated with the consumption of Mexican cantaloupe melons. In the first outbreak, samples were shown to contain *S.* Poona; in total, 155 cases, 28 hospitalizations, and two deaths were reported. *Salmonella* was confirmed by serotyping and PFGE (CDC, 2002). The second outbreak occurred between April and June 2000 where 47 salmonellosis cases were reported. The cases were, again, associated with cantaloupe consumption. The third outbreak was reported between April and May 2001, and 50 salmonellosis cases (*S.* Poona) were confirmed; 10 patients developed septicemia, and two deaths occurred. Another outbreak occurred between March and May, 2002, and 58 cases were confirmed with 10 hospitalizations (CDC, 2002). Although these three cases were associated with consumption of Mexican cantaloupe, studies conducted by the FDA with a sample of 115 cantaloupes cultivated in United States showed the presence of *Salmonella* and *Shigella* in 2.6% and 0.9% of the melons, respectively (FDA, 2001b). Castillo et al. (2004) collected and analyzed 1735 samples, including cantaloupe melon, irrigation and surface waters from six farms and packinghouses in south Texas, and three in Colima Mexico. A total of 1.8% of the samples were positive for *Salmonella* spp. However, the levels of contamination were similar in Mexican and U.S. farms. The Mexican cantaloupe melon industry has not yet recovered to even 10% of the volumes exported before these events occurred. Another important outbreak occurred in 2008, which affected 51 people in the U.S. and 9 in Canada, and the source of the infection was traced to cantaloupe grown in Honduras.

Despite the fact that many outbreaks have implicated cantaloupes, the largest produce-related *Salmonella* outbreak in U.S. history was caused by Mexican hot peppers (serrano and jalapeño type). This outbreak affected 1500 people and caused 308 hospitalizations in nearly all 50 U.S. states. This incident dealt an economic blow to the produce industry in both countries, and challenged the consumers' confidence in the safety of the food supply. *Salmonella* Saintpaul, the causative agent of the outbreak, was isolated from serrano and jalapeño pepper samples from two packinghouses in Tamaulipas, Mexico; after several unproductive weeks of inspecting and sampling in Mexican tomato packinghouse operations (Mody et al., 2011). To date, the mechanism by which these vegetables were contaminated has not yet been determined (CDC, 2008).

In 2011, papayas grown in Mexico were implicated in an outbreak caused by *S.* Agona. A total of 106 people became ill in 25 states between January and August 2011. From May to August 2011, the FDA together with Mexican authorities analyzed papayas from different regions, and found 15.6% positive samples. The FDA banned the import of papayas from Mexican origin unless the importer shows evidence of a *Salmonella*-free shipment (CDC, 2011). More recently, Canadian health investigators announced an outbreak of *Salmonella* serotype Braenderup associated with consumption of mangoes in August 2012. The outbreak affected 22 patients. In late August 2012, the CDC announced this same strain was identified in the U.S., affecting 127 patients, and 33 of them were hospitalized. The outbreak traced the source of infection to a single producer; in October 2011, the FDA announced the recall of mangoes from this particular brand. In addition to whole mangoes, products containing mangoes such as pineapple/mango pico de gallo were also recalled. The FDA isolated *Salmonella* from mangoes from a supplier with multiple plantations and a single packinghouse located in Sinaloa, Mexico.

Another organism of great concern is *Cyclospora cayetanensis* because nearly all the outbreaks in the U.S. have been linked to berries imported from Central American countries (Hall et al., 2011). In the spring of 1996, an outbreak occurred in the United States and Canada where a total of 1465 cases of cyclosporiasis were identified in 20 states and two Canadian provinces. Florida initiated an investigation, led by the Florida Department of Health, because the largest number of clusters occurred in this state. The investigation determined the size of the outbreak, identified the vehicle of transmission, and discovered more regarding the morbidity associated with cyclosporiasis. The researchers conducted a case-control study, looking at the clusters of cases associated with a common food item, and attempted to trace that food item back to its country of origin. It was found that the consumption of raspberries was strongly associated with cyclosporiasis, and that Guatemalan raspberries were the source of the cyclosporiasis outbreak. This conclusion was supported by information from 19 other states as well (Calvin et al., 2003).

The occurrence of a second and similar outbreak, described by Herwaldt and Beach (1999), prompted another look at this foodborne illness and what must be done to prevent it. They confirmed Guatemalan raspberries as the vehicle for *Cyclospora cayetanensis*. Many of their findings are similar to those reported in the 1996 outbreak investigation. One notable finding is that case exposures generally consisted of only a few raspberries, but the median attack rate among people who ate raspberries was 91.7%. This suggests a very low infectious dose for *C. cayetanensis* and relatively uniform contamination of the implicated raspberry lots. It is unlikely that such contamination of raspberries would result from contact by an infected worker; rather, it seems more likely that an environmental reservoir was responsible. Contaminated water used for irrigation or pesticide spraying continues to be an important consideration. Contamination of raspberries through exposure to bird or insect droppings on packing material stored on open contaminated space also remains a possibility.

It is clear that the control measures instituted after the 1996 outbreak were inadequate, since importation of fresh Guatemalan raspberries into Canada in the spring of 1998 caused another outbreak (CDC, 1998). Following the 1997 outbreak, Guatemala suspended exports of fresh raspberries to the United States. Despite this measure, another cyclosporiasis outbreak affecting 50 people was reported from May through June 2004; the cases were linked to consumption of raw Guatemalan snow peas at five special events. Pasta salad was the only food item statistically associated with illness. The pasta salad included multiple types of raw produce, but only snow peas were used in all batches of event pasta salads, and the same lot of snow peas was used to make the salads (Crist et al., 2004).

In 2005, two different outbreaks were linked to the consumption of fresh basil. The first outbreak was reported in Florida where health authorities began receiving reports of gastroenteritis caused by *Cyclospora* between March and June 2005. There were a total of 592 cases, 365 confirmed by laboratory testing. A total of 493 of the ill were residents of Florida, 89 from other states, and 10 were residents of Canada. All of the cases outside Florida reported visits to Florida 2 to 3 weeks prior to the onset of symptoms. Food questionnaires and other epidemiological studies indicated fresh basil was the most likely source of the outbreak. The FDA initiated a traceback and found that the basil had originated in Peru (Hammond and Bodager, 2006). The other outbreak involving basil was originated in a Quebec restaurant in June 2005; this outbreak affected 142 workers who became ill an average of 7 days after eating at the restaurant. A retrospective cohort study was performed to see if any food item was associated with the disease. The results showed that 94% (133/141) of those exposed to the appetizer and 0% of those unexposed (0/4) were sick. The investigation concluded that this outbreak was caused by consumption of uncooked appetizer garnished with fresh basil. The basil was imported from a Mexican farm (Milord et al., 2012).

Other organisms have also been implicated in produce-related foodborne outbreaks. In the fall of 2003, large outbreaks of hepatitis A in the United States (Tennessee, North Carolina, and Georgia) were associated with consumption of raw or undercooked green onions from Mexico, though the source of the green onions associated with the outbreak was never determined by the FDA. Between October and early November, before the FDA's first announcement regarding contaminated green onions, another very large outbreak of hepatitis A occurred in Pennsylvania among diners at one restaurant. Over 500 people contracted hepatitis A, and three died. Later, the FDA announced that this outbreak was also associated with green onions from Mexico and named the four farms that grew the product associated with the outbreak. Before the 2003 outbreaks of hepatitis A in the United States, many growers in Mexico already used third-party certification for GAPs and GMPs. Despite survey results suggesting that most growers have an interest in food safety, a lack of concern by only a few growers can affect the entire industry (Calvin et al., 2004).

Conclusions

Mexico and Central America are important producers of fresh fruits and vegetables and their production is distributed in several countries, and this fact highlights the relevance of well-implemented procedures (GAP, GMP, HACCP) to assure safety of their products. In addition, it is imperative that these procedures be applied not only to production directed for international trading, but also to produce for domestic consumption.

Produce safety in Mexico and Central America requires continued monitoring and recommendations to achieve desired outcomes including implementation strategies and pilot-scale activities. Private and government agencies need to identify mechanisms to ensure products comply with international standards, partnering with foreign agencies in the development and implementation of a Good Agriculture Practice approach. Having Good Agricultural and Management practices in place ensures that appropriate steps are being used to enhance microbial safety of commodities. It is important to understand where products are coming from and where they are going. It requires development, implementation, and verification of specifications. Collaborations are needed throughout the entire food chain from farm-to-fork. Other important considerations are recognition of the power of risk perception, and the understanding that there is no zero risk, but that risk assessments and evaluations are needed. Additionally, the need for awareness planning before a crisis occurs, the need for auditing of suppliers, and the proper use of documents such as letters of guarantee and traceability are important matters that must also be considered.

References

Ackers, M., Pagaduan, R., Hart, G., et al., 1997. Cholera and sliced fruit: Probably secondary transmission from an asymptomatic carrier in the United States. Int. J. Infect. Dis. 1, 212–214.

Amarilla, A., Espindola, E.E., Galeano, M.E., et al., 2007. Rotavirus infection in the Paraguayan population from 2004 to 2005: High incidence of rotavirus strains with short electropherotype in children and adults. Med. Sci. Monit. 13, 333–337.

Beuchat, L.R., Ryu, J.H., 1997. Produce handling a processing practice. Emerg. Infect. Dis. 3, 459–465.

Calvin, L., Flores, L., Foster, W., 2003. Case study: Guatemalan raspberries and *Cyclospora*. Food. Saf. Food. Secur. Food. Trade. http://www.ageconsearch.umn.edu/bitstream/123456789/19445/1/fo031007.pdf.

Calvin, L., Avendano, B., Schwentesius, R., 2004. The economics of food safety: The case of green onions and hepatitis A outbreaks. Electronic. Outlook. Rep. Econ. Res. Serv. www.ers.usda.gov/publications/vgs/nov04/VGS30501/VGS30501.

Calvo, M., Carazo, M., Arias, M.L., et al., 2004. Prevalence of *Cyclospora* sp., *Cryptosporidium* sp, microsporidia and fecal coliform determination in fresh fruit and vegetables consumed in Costa Rica. Arch. Latinoam. Nutr. 54, 428–432.

Castillo, A., Mercado, I., Lucia, L.M., et al., 2004. *Salmonella* contamination during production of cantaloupe: A binational study. J. Food Prot. 67, 713–720.

Castro-Rosas, J., Cerna-Cortés, J.F., Méndez-Reyes, E., et al., 2012. Presence of faecal coliforms, *Escherichia coli* and diarrheagenic *E. coli* pathotypes in ready-to-eat salads, from an area where crops are irrigated with untreated sewage water. Int. J. Food Microbiol. 156, 176–180.

Cazarez-Diarte, J.G., 2004. Presencia y supervivencia de coliformes fecales, Salmonella spp y Listeria spp. en agua de uso agrícola del Valle de Culiacán. Master's Degree Thesis. Centro de Investigación en Alimentación y Desarrollo, A. C. Unidad Culiacán.

CDC, 1991. Multistate outbreaks of *Salmonella* Poona infections—United States and Canada, 1991. J. Am. Med. Assoc. 266, 1189–1190.

CDC, 1998. Outbreak of cyclosporiasis—Ontario, Canada, May 1998. Morb. Mortal. Wkly. Rep. 47, 806–809.

CDC, 2002. Multistate outbreaks of *Salmonella* serotype Poona infections associated with eating cantaloupe from Mexico—United States and Canada, 2000–2002. Morb. Mortal. Wkly. Rep. 51, 1044–1047.

CDC, 2008. Outbreak of *Salmonella* serotype Saintpaul infections associated with multiple raw produce items—United States, 2008. Morb. Mortal. Wkly. Rep. 57, 929–934.

CDC, 2011. Investigation Update: Multistate Outbreak of Human *Salmonella* Agona Infections Linked to Whole, Fresh Imported Papayas. http://www.cdc.gov/salmonella/agona-papayas/082911/ last (accessed 22.04.13.).

Chaidez, C., Moreno, M., Rubio, W., et al., 2003. Comparison of the disinfection efficacy of chlorine-based products for inactivation of viral indicators and pathogenic bacteria in produce wash water. Int. J. Environ. Health Res. 13, 295–302.

Chaidez, C., Soto, M., Gortares, P., et al., 2005. Occurrence of *Cryptosporidium* and *Giardia* in irrigation water and its impact on the fresh produce industry. Int. J. Environ. Health Res. 15, 339–345.

Crist, A., Morningstar, C., Chambers, R., et al., 2004. Outbreak of cyclosporiasis associated with snow peas—Pennsylvania. Morb. Mortal. Wkly. Rep. 53, 876–878.

CSPI, Center for the Science in the Public Interest. (2013). http://www.cspinet.org/foodsafety/outbreak_report.html

Department of Commerce, U.S. Census Bureau Foreign Trade Statistics. (2013). http://www.fas.usda.gov/gats/ExpressQuery1.aspx.

DeWaal, C.S., Barlow, K., Alderton, L., et al., 2002. Outbreak alert. Center for Science in the Public Interest, Washington, D.C. 57.

Fankhauser, R.L., Monroe, S.S., Noel, J.S., et al., 2002. Epidemiologic and molecular trends of "Norwalk-like viruses" associated with outbreaks of gastroenteritis in the United States. J. Infect. Dis. 186, 1–7.

FDA, 2001a. FDA survey of imported fresh produce FY 1999 field assignment. www.cfsan.fda.gov/~dms/prodsur6.html.

FDA, 2001b. Survey of domestic fresh produce: Interim results. www.cfsan.fda.gov/~dms/prodsur9.html.

FDA, 2013. Food Safety Modernization Act: Putting the Focus on Prevention. http://www.foodsafety.gov/news/fsma.html (accessed 21.04.13.).

Gerba, C.P., Bitton, G., 1984. Microbial pollutants: Their survival and transport pattern to groundwater. In: Bitton, G., Gerba, C.P. (Eds.), Groundwater pollution microbiology. Wiley, New York, pp. 39–54.

GFSI. (2013). http://www.mygfsi.com/ (accessed 20.04.13.).

GLOBALG.A.P. 2011. Risk Assessment on Social Practice. General Regulations, V 1.1. http://www1.globalgap.org/cms/upload/The_Standard/GRASP/GR/V1_1/110208-GRASP-GR-ENG-V1_1_Jan11.pdf. (accessed 18.04.13.).

Gomez-Dantes, O., Sesma, S., Becerril, V.M., et al., 2011. Sistema de salud de Mexico. Salud. Publica. Mex. 53 (suppl. 2), S220–S232.

Gomez-Lopez, V.M., 2012. Decontamination of fresh and minimally processed produce, first ed. Wiley-Blackwell, Ames, Iowa, USA.

Gortares-Moroyoqui, P., Castro-Espinoza, L., Naranjo, J.E., et al., 2011. Microbiological water quality in a large irrigation system: El Valle del Yaqui, Sonora Mexico. J. Environ. Sci. Health Part A. 46, 1708–1712.

Hall, R.L., Jones, J.L., Herwaldt, B.L., 2011. Surveillance for Laboratory-confirmed Sporadic Cases of Cyclosporiasis: United States, 1997–2008. US Department of Health and Human Services, Centers for Disease Control and Prevention.

Hammond, R., Bodager, D., 2006. Surveillance and investigation of a large statewide *Cyclospora* foodborne disease outbreak involving an imported stealth ingredient. In: Institute of Medicine. Addressing foodborne threats to health: policies, practices, and global coordination. National Academies Press, Washington, D.C., pp. 115–124. 133–40.

Herwaldt, B.L., Beach, M.J., 1999. The return of *Cyclospora* in 1997: Another outbreak of cyclosporiasis in North America associated with imported raspberries. The *Cyclospora* Working Group. Ann Intern. Med. 130, 210–220.

Jimenez, E.M., Siller, J.H., Valdez, J.B., et al., 2007. Bidirectional *Salmonella enterica* serovar Typhimurium transfer between bare/glove hands and green bell pepper and its interruption. Int. J. Environ. Health Res. 17, 381–388.

Lesser-Carrillo, L.E., Lesser-Illades, J.M., Arellano-Islas, S., et al., 2011. Balance hídrico y calidad del agua subterránea en el acuífero del Valle del Mezquital, Mexico central. Rev. Mex. Cienc. Geol. 28, 323–336.

Long, K.Z., Garcia, C., Santos, J.I., et al., 2007. Vitamin A supplementation has divergent effects on norovirus infections and clinical symptoms among Mexican children. J. Infect. Dis. 196, 978–985.

Lopez-Cuevas, O., 2005. Resistencia antimicrobiana de *Escherichia coli* y serotipos de *Salmonella* aisladas de agua y suelo de uso agrícola. Master's Degree Thesis Centro de Investigacion en Alimentacion y Desarrollo, A.C. Unidad Culiacan: Culiacan, Sinaloa, Mexico.

Martinez-Tellez, M.A., Rodriguez-Leyva, F.I., Espinoza-Medina, I.E., et al., 2009. Sanitation of Fresh Green Asparagus and Green Onions Inoculated with *Salmonella*. Czech. J. Food Sci. 27, 454–462.

Mazari-Hiriart, M., Ponce-de-Leon, S., Lopez-Vidal, Y., et al., 2008. Microbiological implications of periurban agriculture and water reuse in Mexico city. PLoS ONE 3, 1–8.

Milord, F., Lampron-Goulet, E., St-Amour, M., et al., 2012. *Cyclospora cayetanensis*: a description of clinical aspects of an outbreak in Quebec, Canada. Epidemiol. Infect. 140, 626–632.

Mody, R.K., Greene, S.A., Gaul, L., et al., 2011. National Outbreak of *Salmonella* Serotype Saintpaul Infections: Importance of Texas Restaurant Investigations in Implicating Jalapeño Peppers. PLoS ONE 6 (2), e16579.

Mohle-Boetani, J., Reporter, R., Werner, S.B., et al., 1999. An outbreak of *Salmonella* serogroup Saphra due to cantaloupe from Mexico. J. Infect. Dis. 180, 1361–1364.

Mota, A., 2004. Risk assessment from *Giardia* and *Cryptosporidium* in irrigation water from Culiacan River University of Texas at El Paso: Master Thesis.

Okhuysen, P.C., White, A.C., 1999. Parasitic infections of the intestines. Curr. Opin. Infect. Dis. 12, 467–472.

Payment, P., 1989. Presence of human and animal viruses in surface and ground water. Water Sci. Technol. 21, 283–285.

Payne, D.C., Wikswo, M., Parashar, U.D., 2012. Chapter 13 Rotavirus. Centers for Disease Control and Prevention. fifth ed. Manual for the surveillance of vaccine-preventable diseases.

Paziewska, A., Bednarska, M., Nieweglowski, H., et al., 2007. Distribution of *Cryptosporidium* and *Giardia* spp. in selected species of protected and game mammals from north-eastern Poland. Ann. Agric. Environ. Med. 14, 265–270.

Sadovski, A., Fattal, Y.B., Goldberg, D., 1978. Microbial contamination of vegetables irrigated with sewage effluent by the drip method. J. Food Prot. 41, 336–340.

SEMARNAT (Secretaria de Medio Ambiente, Recursos Naturales y Pesca), 1996. Norma Oficial Mexicana NOM-001-ECOL-1996. Límites máximos permisibles de contaminantes en las descargas de aguas residuales en aguas y bienes nacionales.

Sewell, A.M., Farber, J.M., 2001. Foodborne outbreaks in Canada linked to produce. J. Food Prot. 64, 1863–1877.

Siller-Cepeda, J.H., Baez-Sanudo, M.A., Sanudo-Barajas, A., et al., 2002. Manual de Buenas Prácticas Agrícolas. Guía para el Agricultor Centro de Investigación en Alimentación y Desarrollo: A.C. SAGARPA, 1ra edicion.

Siller-Cepeda, J.H., Baez-Sanudo, M., Chaidez-Quiroz, C., et al., 2007. Produccion poscosecha en la industria fruticola. first ed. Buenas Practicas en la Produccion de alimentos Trillas, pp. 171–221.

SUIVE/DGE/SSA. Sistema Único de Información para la Vigilancia Epidemiológica. Dirección General de Epidemiología. Secretaría de Salud, 2011. Información Epidemiológica de Morbilidad, Anuario versión ejecutiva.

Talal, A.H., Moe, C.L., Lima, A.A., et al., 2000. Seroprevalence and seroincidence of Norwalk-like virus infection among Brazilian infants and children. J. Med. Virol. 61, 117–124.

Thurston-Enriquez, J.A., Watt, P., Dowd, S.E., et al., 2002. Detection of protozoan parasites and microsporidia in irrigation waters used for crop production. J. Food Prot. 65, 378–382.

Torres-Vitela, M.D.R., Gomez Aldapa, C.A., Cerna-Cortes, J.F., et al., 2013. Presence of indicator bacteria, diarrhoeagenic *Escherichia coli* pathotypes and *Salmonella* in fresh carrot juice from Mexican restaurants. Lett. Appl. Microbiol. 56, 180–185. http://dx.doi.org/10.1111/lam.12030.

Tyrrel, S.F., Quinton, J.N., 2003. Overland flow transport of pathogens from agricultural land receiving faecal wastes. Appl. Microbiol. 94, 87–93.

Regulatory Issues in Europe Regarding Fresh Fruit and Vegetable Safety

16

Gro S. Johannessen, Kofitsyo S. Cudjoe

Section for Bacteriology – Food and GMO, Norwegian Veterinary Institute, Oslo, Norway

CHAPTER OUTLINE

The Produce Contamination Problem. http://dx.doi.org/10.1016/B978-0-12-404611-5.00016-6

Introduction

In the last decade there appears to have been an increase in the numbers of outbreaks of foodborne disease associated with fruit and vegetables (Doyle and Erickson, 2008). Several factors may have influenced the notion of these apparent increases in numbers. In outbreak situations there has been an increased awareness that fresh produce could be the cause of disease. An increased consumption of fruits and vegetables due the expanded import and export trade has led to an increase in choices available to consumers. Due to improved technology, fresh produce can now be transported over long distances within a short period of time, thus leading to import of products from areas faraway from the consumption centers. Furthermore, the trendy habit of demanding fresh and "natural" products may also be an important factor.

In 2002 the Scientific Committee on Food in the EU published a microbiological risk profile of fruits and vegetables eaten raw (Food, 2002). The report concludes that "the most efficient way to improve safety of fruits and vegetables is to rely on a proactive system of reducing risk factors during production and handling. Apart from washing, other methods of decontamination seem to have a limited influence on safety." The report recommends among other things that a more robust traceability system would improve epidemiological investigation of suspected foodborne illness; that there is a need for production measures for fruits and vegetables based on Good Handling Practices (GHP), Good Agricultural Practice (GAP) and Hazard Analysis of Critical Control Points (HACCP); and that water and organic fertilizers should be of such quality that they do not contaminate the products with harmful microorganisms.

Early in 2008, the FAO/WHO published a report on microbiological hazards in fresh fruit and vegetables (FAO/WHO, 2008). This report gave top-priority ranking to leafy green vegetables (including fresh herbs) as commodities of global concern. This ranking was due to the large volume of production and export, the association of this product type with numerous outbreaks with different agents, and the considerable complexity of growing and processing. A level 2 priority was given to berries, green onions, melons, sprouted seeds, and tomatoes, while level 3 priority was given to a large group, comprising carrots, cucumbers, almonds, baby corn, sesame seeds, onion and garlic, mango, paw paw, and celery.

The European Food Safety Authority (EFSA) published in 2013 a scientific opinion of the risk posed by pathogens in food of non-animal origin (Part 1: Outbreak Data Analysis and Risk Ranking of Food/Pathogen Combinations) (EFSA, 2013) (see section on EFSA for more information). In this opinion official data on foodborne outbreaks collected between 2007 and 2011 were analyzed, and data from other sources were excluded. The food/pathogen combination ranked on top was *Salmonella* spp. and leafy greens eaten raw. Ranked second, with equal rank, were *Salmonella* spp. and bulb and stem vegetables, *Salmonella* spp. and tomatoes, *Salmonella* spp. and melons, and pathogenic *E. coli* and fresh pods, legumes, and grains. It must be taken into account that there are shortcomings with such an analysis, including variable reporting routines, influence of rare events leading to bias, etc.

Although reporting of outbreaks to EFSA is harmonized, it is important to note that the outbreak investigation systems are not harmonized at a national level within the EU-member states. In the model applied for this exercise, seven criteria were used: strength of associations between food and pathogen, incidence of illness, burden of disease, dose-response relationship, consumption, prevalence of contamination, and pathogen growth during potential shelf life.

Europe has a large production of fruit and vegetables, but the demand for fresh fruit and vegetables during all seasons necessitates import from other areas. This is perhaps particularly important in the northern areas where the growing season is rather short, and greenhouse production is not sufficient to meet all consumer needs.

This chapter is mainly limited to regulatory issues in Europe regarding fresh fruits and vegetables, but we will also discuss some dried products due to problems with mycotoxins. Grain crops will not be discussed. We will also only focus on microbiological issues of human concern in fresh produce. Discussion on plant diseases and health is beyond the scope of this chapter.

The European Union
Basic facts

The European Union (EU) began its activities as the European Coal and Steel Union (ECSC) in 1951. The six nations of Belgium, France, Germany, Italy, Luxembourg, and the Netherlands signed the Paris Treaty, which entered into force in 1952. In 1953 the first Common Market for coal and iron was set into place. The treaty of Rome, establishing among other things the European Economic Community, was signed in 1957 and entered into force on January 1, 1958. On July 1, 1968, the six member states removed custom duties on goods imported from each other, allowing free cross-border trade for the first time. On January 1, 1973, the six member states were formally joined by Denmark, Ireland, and the United Kingdom. Expansion of the EU member states has continued to reach its current level of 27 members (EU-27) (Table 16.1). Croatia became the 28th EU member country on July 1, 2013. Five candidate countries, namely Iceland, Montenegro, Serbia, the Former Yugoslav Republic of Macedonia, and Turkey, are awaiting approval of their membership application. In addition Albania, Bosnia, and Herzegovina and Kosovo are potential candidates.

The European Free Trade Association and the European Economic Area

The European Free Trade Association (EFTA) was founded in 1960 by Austria, Denmark, Norway, Portugal, Sweden, Switzerland, and the UK. EFTA is an intergovernmental organization set up for the promotion of free trade and economic integration to the benefits of its member states. In 1992 the community, and the member states and the then seven members of EFTA negotiated and signed the agreement that

Table 16.1 Member States in the European Union (April, 2013)	
Member State (Year of Joining)	
Austria (1995)	Latvia (2004)
Belgium (ECSC,1951)	Lithuania (2004)
Bulgaria (2007)	Luxembourg (ECSC, 1951)
Cyprus (2004)	Malta (2004)
Czech Republic (2004)	The Netherlands (ECSC, 1951)
Denmark (1973)	Poland (2004)
Estonia (2004)	Portugal (1986)
Finland (1995)	Romania (2007)
France (ECSC, 1951)	Slovakia (2004)
Germany (ECSC, 1951)	Slovenia (2004)
Greece (1981)	Spain (1986)
Hungary (2004)	Sweden (1995)
Ireland (1973)	United Kingdom (1973)
Italy (ECSC, 1951)	

created the European Economic Area (EEA). With the exception of Switzerland, the current EFTA countries of Iceland, Norway, Liechtenstein, and Switzerland are also a part of EEA. Today EFTA maintains management of the EFTA Convention (intra-EFTA trade), the EEA Agreement (EFTA-EU relations), and the EFTA Free Trade Agreements (third-country relations).

Although Switzerland decided not to take part in the EEA, the agreement was maintained because the remaining countries wished to take part in EU's internal market, while not assuming the full responsibilities of a membership. The EEA countries have the right to be consulted during formulation of community legislation, but they do not have the right to a voice in the decision-making. All the new community legislation that is covered in the EEA is integrated into the Agreement through an EEA Joint Committee decision and subsequently becomes a part of the national legislation of the EEA states. The function of the EEA Joint Committee is to adopt decisions extending community regulation and directives to the EEA States. In this process of adopting community legislation, the EEA states have to speak as one voice. As a result of the EEA Agreement, EU legislation, such as the Food Law, will also enter into force in the EEA states, and the same regulation with respect to import from third countries (countries outside EU and the EEA) applies to the EEA states.

European fruit and vegetable production

In 2011 the production of fruit and vegetables in the EU-27 was 36 and 58 million tons, respectively (Anon., 2012b). Fruits, vegetables, and horticultural products accounted for approximately 20% of the gross agricultural output at basic prices

Table 16.2 Non-community Countries with Largest Import (Metric Tons) of Fruit and Vegetables into EU-27 in 2000 and 2007 (from Martinez-Palou and Rhoner-Thielen, 2008)

Country	Import (% of total) 2000	Import (% of total) 2007
Thailand	20.3	7.4
Canada	6.3	-*
Turkey	6.0	7.3
Ecuador	5.9	6.2
Brazil	5.7	6.2
South Africa	5.2	5.7
Costa Rica	5.0	8.2
USA	5.0	-
Morocco	3.9	4.5
China	-	6.8
Colombia	-	5.3
Others	36.7	42.4

*Not on top 9-list in this year.

in EU-27 in 2008 (Martinez-Palou, 2011). The most important vegetables in terms of production in 2011 were tomatoes, carrots, and onions, while apples, oranges, and peaches were the major fruits (Anon., 2012a). Typical fruit and vegetable farms in the member states were rather small, less than 10 hectares on average (Martinez-Palou and Rhoner-Thielen, 2008) (Table 16.2). The production of fruit and vegetables tends to be concentrated in only a few member states. In 2010, Italy and Spain had the largest fruit and vegetable production among the EU-27. For example, almost 60% of EU-27's apple production took place in Italy, Poland, and France, while about two-thirds of the tomatoes were produced in Italy and Spain (Anon., 2012b).

In 2010, the import of fruit and vegetables accounted for 26.5% of the total imports of food and beverage (Martinez–Palou, 2011). According to the European Fresh Produce Association (Freshfel), bananas and tomatoes are the largest fruit and vegetable category imported into the EU (Freshfel, 2012).

Fresh produce contamination problems in Europe

As in all other parts of the developed world, Europe has also encountered problems and outbreaks with respect to microbiological contamination of fruits and vegetables. There have been large multinational outbreaks and small national outbreaks involving numerous different commodities, thus reflecting the variety and complexity of production and distribution systems.

In the scientific opinion published by EFSA (EFSA, 2013) the most frequent combinations reported between 2007 and 2011 were norovirus and raspberries with 27 outbreaks, followed by norovirus and leafy greens eaten raw as salads, *Salmonella* spp. and sprouted seeds, and *Salmonella* spp. and leafy greens eaten raw as salads. This report also highlights the variation of reporting practices between countries, as for example, the Nordic countries, Denmark, Finland, Norway, and Sweden reported 51% of the foodborne outbreaks with strong evidence. This implies that the evidence implicating a specific food vehicle is strong, based on an assessment of all available evidence, and that a detailed data set is reported.

Foodborne bacteria of foodborne disease

In 2011 Germany encountered one of the largest outbreaks reported in Europe with more than 3800 people ill, including 54 deaths (Frank et al., 2011). The infection was caused by *E. coli* O104:H4, and the consumption of sprouts was incriminated as the most likely source of infection. Simultaneously, a smaller outbreak caused by a related *E. coli* O104:H4 strain took place in southwest France (Gault, 2011). Several other countries also experienced cases of infection and deaths caused by this agent, many of which had connections to Germany (for example, visiting Germany prior to the onset of disease). This outbreak showed the potential of fresh produce-related outbreaks in terms of severity and spread.

One of the first recognized multinational outbreaks associated with fresh vegetables in Europe was in 1994 when an increasing number of domestic cases of *Shigella sonnei* was observed in May to June in Norway (Kapperud et al., 1995). A similar increase was also independently observed in other European countries. Epidemiological investigations in Norway, Sweden, and the UK incriminated iceberg lettuce imported from Spain as the source. Since then several outbreaks have occurred involving produce imported from third countries (i.e., countries outside the EU) and fruits and vegetables cultivated and sold within the EU/EEA.

In November 2007, Eurosurveillance published information on an outbreak of *Salmonella* Weltewreden infections in Norway, Denmark, and Sweden associated with alfalfa sprouts (Emberland et al., 2007). It was concluded from this outbreak investigation that alfalfa sprouts grown from contaminated seeds was the source of the outbreak. The seeds used in Denmark and Norway were part of the same batch and were traced, according to invoices, to retailers in Germany and the Netherlands. The seeds used in Finland came from the same Dutch supplier, but were not part of the same batch. Further investigations showed that the seeds originated from Pakistan (RASFF 2007.0760, RASFF 2007.0760-add01, -add02 and -add03). This outbreak showed that the import and trade routes can be rather complicated and hard to follow. However, European legislation on tracing requirements has proven to be effective and important in tracing the source of products incriminated in outbreak cases. After the sprout-related outbreak in Germany and other countries, there has been increased focus on traceability.

For an overview of outbreaks reported to the Zoonoses Database between 2007 and 2011 where foods of non-animal origin are implicated, see the EFSA opinion (2013).

Parasites

Fruits and vegetables that are eaten raw and without peeling have been demonstrated to harbor a range of protozoan parasites such as *Giardia*, *Cryptosporidium*, *Cyclospora*, and the helminth parasite *Ascaris*. Some of these organisms, particularly the protozoan parasites, have caused infections characterized by prolonged diarrhea (Dawson, 2005; Döller et al., 2002; Hoang et al., 2005). These parasites are essentially derived from waste-water reuse and are therefore of primary public health concern. In a recently published report from WHO/FAO (Anon., 2012d; Robertson, et al., 2013) parasites associated with fresh produce and derived products were among the top-20 foodborne parasites ranked, with *Taenia solium* ranked as number one. Among the others listed that could be related to fresh produce were *Echinococcus granulosis* and *multilocularis*, *Toxoplasma gondii*, *Entamoeba histolytica* and *Fasciola* spp. Only a few outbreaks from parasites associated with fresh produce have been reported in Europe. These include a *Cryptosporidium parvum* outbreak in Finland in 2008 linked to a salad mixture (Pönkä et al., 2009), a *Cryptosporidium hominis* outbreak linked to a salad bar in Denmark in 2005 (Ethelberg et al., 2009), and an outbreak of cyclosporiasis in Germany in 2000 associated with the consumption of salad (Döller et al., 2002).

Foodborne human pathogenic virus

Viruses cannot replicate in or on foods but might sometimes be present on fresh produce as a result of human fecal contamination. This contamination can originate in the growing and harvesting area from contact with polluted water and inadequately or untreated sewage sludge used for irrigation and fertilization. Alternatively, fruits or vegetables handled by an infected person might become contaminated with a virus and transmit infection. The most frequently reported foodborne viral infections are viral gastroenteritis (norovirus) and hepatitis A. In the EFSA opinion (2013) norovirus and raspberries and norovirus and leafy greens eaten raw as salads were the most frequently reported combinations of foodborne disease and also the most frequently reported foodborne viral infection with 27 and 24 outbreaks, respectively. In 2010 several outbreaks of gastroenteritis from norovirus associated with consumption of Lollo Bionda lettuce were reported in Denmark and Norway (EFSA, 2013; Ethelberg et al., 2010).

Several epidemiological studies have associated viral hepatitis A infections with the consumption of fecally contaminated raw vegetables or drinking water (Hjertqvist et al., 2006; Long et al., 2002; Nygard et al., 2001; Cotterelle et al., 2005; Hernandez et al., 1997). In 2010 and 2011 two outbreaks of foodborne hepatitis A were reported in the Netherlands and England, respectively (Carvalho et al., 2012;

Petrignani et al., 2010). These outbreaks were associated with semi-dried tomatoes in oil and semi-dried tomatoes.

Currently there is increased control of frozen strawberries imported from China with a sampling frequency of 5% (Annex 1, EU regulation 2009/669, see below for explanation).

Molds and mycotoxins

The major problems of mold contamination of fruits and vegetables are economic with a significant loss of useful food materials (Moss, 2008). There are a few examples implicating a role for mycotoxins in the safety of fresh fruits and fruit juices. Undoubtedly, the most important is patulin, mainly produced by *Penicillium expansum*. Patulin is especially important in apple juice and apple products. Some members of the black-spored *Aspergillus niger* group, particularly *A. carbonarius*, may cause bunch rot of grapes, and ochratoxin has been detected in both fresh fruits and raisins as well as in wine and grape juice (Belli, et al., 2004). The mold *Alternaria alternata* grows on a wide range of fruits and vegetables and is a major pathogen of fresh tomatoes, in which it can produce tenuazonic acid. According to Moss (2008), unlike patulin and ochratoxin A, there are no regulatory limits set for tenuzonic acid or other *Alternaria* metabolites, reflecting the lack of any evidence implicating them in human illness.

According to the Healthy Nut Initiative (Healthy Nut Initiative, 1998), nuts should be classified as fruits, and in particular, as so-called shell fruits. This classification thus includes fruits with edible kernels contained in inedible shells. Nut consumption in Europe is on the rise. However, edible nuts, dried figs, and spices can be associated with aflatoxins and other mycotoxins. This contamination can be uneven and spasmodic.

In 2011 there were in total 685 notifications of mycotoxins to RASFF (Rapid Alert System for Food and Feed; see below) (Anon., 2013b), of which approximately 75% were from products of concern in this chapter. This was a moderate decrease in notifications, mostly due to a decrease in reported notifications of aflatoxins. There were 78 and 10 notifications for aflatoxins and ochratoxin A, respectively, from the category fruit and vegetables, 51 and 17 for aflatoxins and ochratoxin A from herbs and spices, and 320 notifications for aflatoxins in nuts, nut products, and seeds. Most of the notifications on aflatoxins are related to product/country of origin combinations. This has led to increased frequency of control at import (see section on Import from countries outside EU and EEA).

In the United Kingdom, the Contaminants in Food (Amendment) Regulations (*United Kingdom 1999 (Great Britain):* which were made under sections of the Food Safety Act of 1990, set limits for aflatoxins in groundnuts, nuts, dried fruit, and cereals. A higher limit is provided for groundnuts, nuts, and dried fruit intended for further processing before human consumption. The higher limit for these commodities recognizes that processing and sorting can reduce the levels of aflatoxin contamination in consignments below that of the lower limit. The limits are low and were set on the basis that they represent the lowest level technologically achievable, consistent with meeting food safety objectives. The regulations are targeted at those products which surveillance has indicated may be most highly or most frequently contaminated with aflatoxin. These make the greatest contributions to consumer exposure, and controls targeted at these products are the most effective way of reducing exposure.

What do the Europeans do – European regulations

In January 2000, the Commission of European Communities launched their white paper on food safety (Commission, The European, 2000) in which a radical new approach to food safety was proposed. This process was driven by the need to ensure that food safety was and is a priority of the EU. The key points in the white paper were the establishment of an independent European Food Authority and the set-up of a new legal framework covering the whole food chain, including animal feed production. Focus was also directed at food safety controls, especially controls of imports at the borders of the community, consumer information, and an international dimension with respect to an effective presentation of the actions to trading partners.

EU central regulations

On January 28, 2002, the Food Law (Regulation (EC) No. 178/2002) was passed in the EU, and became immediately applicable in the member states (Regulation (EC) No. 178/2002 of the European Parliament and of the Council of 28 January 2002, laying down the general principles and requirements of food law, establishing the European Food Safety Authority, and laying down procedures in matters of food safety). The Food Law establishes common principles and responsibilities and the means to provide a strong science base, efficient organizational arrangements, and procedures to underpin decision-making in matters of food and feed safety. It is stated in Article 1, paragraph 2 that "this regulation lays down the general principles governing food and feed safety in particular, at community and national level. This regulation applies to all stages of production, processing and distribution of food and feed, but shall not apply to primary production for private domestic use, domestic preparation, handling or storage of food for private domestic consumption."

Under this legislation, food operators shall not place on the market unsafe food. The key obligations of food business operators are:

- Operators are responsible for the safety of the food and feed which they produce, transport, store, or sell.
- Operators shall be able to rapidly identify any supplier or consignee.
- Operators shall immediately inform the competent authorities if they have a reason to believe that their food or feed is not safe.
- Operators shall immediately withdraw food or feed from the market if they have a reason to believe that it is not safe.

- Operators shall identify and regularly review the critical points in their processes and ensure that controls are applied at these points.
- Operators shall cooperate with the competent authorities in actions taken to reduce risks.

European Food Safety Authority

The European Food Safety Authority (EFSA) was set up in January 2002 as part of a program to improve food safety, ensure a high level of consumer protection, and restore and maintain confidence in the food supply within the EU (www.efsa.europa.eu). In Europe, risk assessments are done independently of the risk management, and EFSA's role is to assess and communicate all risks associated with the food chain. EFSA produces scientific opinions and advice in close cooperation and open consultation with national authorities and other stakeholders. These are important in providing a sound foundation for European policy and legislation making and in supporting the European Commission, European Parliament, and EU member states in taking effective and timely risk management decisions. EFSA consists of a scientific committee and scientific panels that are composed of highly qualified experts in risk assessment. EFSA collects data on a mandatory basis for several zoonotic agents; including *Salmonella* spp., thermotolerant *Campylobacter* spp., *Listeria monocytogenes,* and verotoxigenic *E. coli.* This data collection also includes data on foodborne outbreaks, and from 2007, harmonized specifications on the outbreaks reporting at the EU level have been applied. The Food Law states that EFSA should cooperate closely with the competent bodies (i.e., national food safety authorities) in the member states if it is to operate effectively. During the sprout-related outbreak in Germany in 2011, experts from the EFSA, the Commission, and ECDE (European Centre for Disease Prevention and Control) assisted the German authorities in the outbreak investigation.

Rapid alert system for food and feed

An important tool for the rapid exchange of information with respect to different contaminants of food and feed is the Rapid Alert System for Food and Feed (RASFF) (Anon., 2012e). The purpose of the RASFF is to provide the control authorities with an effective tool for exchange of information on measures taken to ensure food safety. The RASFF has been in operation since 1979. The legal basis for this rapid information exchange system is Article 50 in the Food Law. RASFF systematically informs countries outside the EU (third countries of origin) of notifications concerning products manufactured in, distributed to, or dispatched from these countries through the Commission delegates. Although a country may be mentioned as the origin of a product that does not necessarily imply that the hazard originated in the country concerned. However, if serious problems are detected several times, a letter is sent to the competent authority of the country. The relevant country is then expected to take appropriate measures to rectify the situation.

The member states may also intensify their import checks. In addition, the Food and Veterinary Office (FVO) uses the information provided by RASFF when prioritizing their inspection program.

The notifications are listed under four headings: Alert notification, which is sent when a food or feed presenting a serious risk is on the market and when immediate action is required; information notification, which is sent when a risk has been identified for food or feed where it is on the market, but the other members of the network do not need to take immediate actions; border rejections, where food and feed consignments have been tested and rejected at the external borders of the EU and the EEA when a health risk was found; and news, which is "any information related to the safety of food and feed products which has not been communicated as an alert or an information notification, but which is judged to be interesting for the control authorities." The RASFF contact points in the member states receive alert notifications and additional information regarding the alert notifications via e-mail. If there are other special situations, notifications of these are also sent via e-mail. Lists of all alerts, additional and information notifications, border rejections, and news are distributed daily to the RASFF contact points. The RASFF contact points go through the messages, and if there is anything that is particularly interesting, further information can be collected from the CIRCA-database of the EU Commission. Weekly overviews of the RASFF notifications are posted on the RASFF-website (http://ec.europa.eu/food/food/rapidalert/index_en.htm).

Hygiene and control rules

In April 2004 new hygiene rules were adopted, and these became applicable in the member states on January, 1 2006. These hygiene rules comprise three regulations and one directive; most important for the fruit and vegetable chain is Regulation (EC) No 852/2004 of the European Parliament and of the Council of 29 April 2004 on the hygiene of food stuffs. The other two regulations and the directive, Regulation (EC) No 853/2004, Regulation (EC) No 854/2004, and Directive 2004/41/EC, are more concerned with food of animal origin and are thus not that important in the context discussed here. The hygiene rules particularly focus on the following points:

- That the food business operator has primary responsibility for food safety
- That food safety is ensured throughout the food chain starting with primary production
- That general procedures based on HACCP principles must be implemented
- That basic common hygiene requirements must be applied
- That certain food establishments must be registered or approved
- That guides to good practice for hygiene or for the application of HACCP principles should be developed as valuable instruments to aid food business operators at all levels of the food chain to comply with the new rules
- That flexibility is provided for food produced in remote areas and for traditional products and methods

The control rule (Regulation (EC) No. 882/2004) were also adopted at the same time as the hygiene rules. This regulation states that member states shall ensure that official controls are carried out regularly by the competent authority, on a risk basis and with appropriate frequency. The compliance can be verified in several ways, such as by audits, inspections, monitoring, surveillance, sampling, and testing. The controls shall be carried out at any stage of production, processing, and distribution and may be carried out without prior warning except when prior notification is necessary. The competent authorities are also responsible for regular official control of food of non-animal origin that is imported from third countries. It must be noted that the control rule is currently under revision (as of summer 2013).

The appointment of community reference laboratories for food and feed (European Reference Laboratory EURL) (Anon., 2013a) and the subsequent nomination of National Reference Laboratories (NRL) are important elements of the control rule. The EURL are responsible for providing the NRL with details of analytical methods, including reference methods; coordinating application of the methods by the NRL by arranging for example ring trials and follow-ups; coordinating and assisting in implementation of new analytical methods and informing about advances in the relevant fields; conducting initial and further training courses for the NRL and experts from developing countries; providing scientific and technical assistance to the Commission; and collaborating with laboratories responsible for analyzing food and feed in third countries. EURL have been appointed in the areas of biological risks (including *Campylobacter,* analysis and testing of Zoonoses (*Salmonella*), *Listeria monocytogenes, Escherichia coli*, including Verotoxigenic *E. coli* (VTEC), parasites (in particular *Trichinella, Echinocoocus,* and *Anisakis*), GMO, feed additives, food contact materials, pesticides, contaminants (including mycotoxins), and residues.

The EURL are important within the context of assisting laboratories in the European countries. During the sprout-related outbreak in 2011, the appropriate EURL provided NRL with methods and reference material enabling European countries to use the methods for the detection of *E. coli* O104:H4 if needed. The EURL for *Listeria monocytogenes* has developed the guidelines for shelf-life studies for *Listeria monocytogenes* in ready-to-eat foods (EURL, 2008). This shows that the EURLs play an important role in crisis and outbreak investigations and in mitigating food safety problems.

Another important regulation is Regulation (EC) No 2073/2005 of 15 November on microbiological criteria for foodstuffs. This regulation is directed at food operators and gives food safety criteria and process hygiene criteria. A food safety criterion means a "criterion defining the acceptability of a product or a batch of foodstuff applicable to products placed on the market." A food safety criterion for fresh produce is absence of *Salmonella* in five 25-g samples of sprouted seeds, pre-cut, ready-to-eat fruit and vegetables, and unpasteurized fruit and vegetable juices. For the sprouted seeds, samples should be collected either of the seeds prior to sprouting or in the production process when the possibility is greatest to detect *Salmonella*.

A second, perhaps more general, food safety criterion applies to *Listeria monocytogenes* in ready-to-eat foods "able to support the growth of *L. monocytogenes*, other than those intended for infants and for special medicinal purposes." This also includes fresh produce products. This criterion requires either that each of five samples have less than 100 cfu/g in products placed on the market during their shelf-life and applies when the manufacturer is able to demonstrate to the competent authority's satisfaction, that the product will not exceed the limit of 100 cfu/g during the shelf-life. In this case the manufacturer may fix intermediate limits during the process that should be low enough to guarantee that the limit of 100 cfu/g is not exceeded at the end of the shelf life. If the food business operator is not able to demonstrate, to the competent authority's satisfaction, that the product will not exceed the limit of 100 cfu/g throughout the shelf-life, a second *L. monocytogenes* criterion may be applied. This criterion requires absence of *L. monocytogenes* in five samples of 25 g, taken before the food has left the immediate control of the food business operator, who has produced it.

After the sprout-related outbreak in 2011, a specific food safety criterion for Shiga toxin-producing *E. coli* and sprouts was added. This criterion requires absence of Shiga toxin-producing *E. coli* (STEC) O157, O26, O111, O103, O145, and O104:H4 in five samples of 25 g taken from products placed on the market during their shelf-life.

A process hygiene criterion means "a criterion indicating the acceptable functioning of the production process (Regulation (EC) No 2073/2005). Such a criterion is not applicable to products placed on the market. It sets an indicative contamination value above which corrective actions are required in order to maintain the hygiene of the process in compliance with food law." The process hygiene criteria for fresh produce are shown in Table 16.3.

This means that if all sample units have values < m the results are satisfactory, if maximum c/n sample units have numbers between m and M and the rest of the sample units are < M, the results are acceptable, or they are unsatisfactory if one of the values is > M or more than c/n values are between m and M.

Table 16.3 Process Hygiene Criteria

Product	Microorganism	Sampling Plan*		Criteria		When in Process
		n	c	m	M	
Pre-cut ready-to-eat fruit and vegetables	*E. coli*	5	2	100 cfu/g	1000 cfu/g	During production
Unpasteurized fruit and vegetable juice	*E. coli*	5	2	100 cfu/g	1000 cfu/g	During production

*n = number of units making up the sample, c = number of sample units with results that can be between m and M.

In addition, the EU has set limits for maximum levels for certain contaminants in foodstuffs (Regulation (EC) No 1881/2006 of 19 December 2006 setting maximum levels for certain contaminants in food stuffs). This regulation concerns among other contaminants, mycotoxins, in particular, aflatoxins, ochratoxin A, and patulin in product types such as spices, dried fruits, and apples. In 2010, this regulation was changed (Regulation (EU) No 165/2010 of 26 February 2010 amending Regulation (EC) No 1881/2006) wherein the maximum levels for aflatoxins in almonds, hazelnuts, pistachios, and Brazil nuts were aligned with the Codex Alimentarius maximum levels.

Follow-up on the sprout-associated outbreak in 2011

After the very large sprout-associated outbreak in 2011 and also previous outbreaks related to sprouts, the European Commission asked the Panel on Biological Hazards to issue a scientific Opinion on the public health risk of Shiga toxin-producing *E. coli* (STEC) and other pathogenic bacteria that may contaminate seeds and sprouted seeds (EFSA, 2011). The Panel was mandated: "to recommend possible specific mitigation options, and to assess their effectiveness and efficiency to reduce the risk throughout the food chain and to recommend, if considered relevant, microbiological criteria for seeds and sprouted seeds, water and other material that may contaminate the seeds and sprouts throughout the production chain." A thorough opinion was published in November 2011, and a second edition was published in March 2012 with several conclusions and recommendations particularly directed at mitigation options and microbiological criteria. Consequently, in March 2013, the Commission published four regulations directed towards the safe production of sprouts. These regulations cover traceability requirements (Regulation (EU) No 208/2013), additional microbiological criteria for sprouts (Regulation (EU) No 209/2013), approval of establishments producing sprouts (Regulation (EU) No 210/2013), and certification requirements for imports into the Union (Regulation (EU) No 211/2013). In addition to the specific STEC criteria for sprouts, Regulation (EU) No 209/2013 also provides sampling rules for sprouts, including preliminary testing of seeds, sampling and testing of the sprouts and the spent irrigation water, and sampling frequency.

Good handling practices and quality assurance in fruit and vegetable production

In the EU, primary production, i.e., "production, rearing or growing of plant products such as grains, fruits, vegetables and herbs as well as their transport within and storage and handling of products (without substantially changing their nature) at the farm and their further transport to an establishment," is covered by Regulation (EC) No 852/2004 on the hygiene of food stuffs (2012c). According to the regulation, "HACCP is not yet generally feasible, but guides to good practice should encourage the use of appropriate hygiene practices at farm level." However, if operations, that may take place on a farm, that are likely to alter a product and/or introduce new hazards to the products are carried out, further food safety requirements must be satisfied (implementation of HACCP). This includes, for example, peeling, slicing, or bagging of salads with the application of packaging gases. Operations like packaging without further treatment, washing and removing leaves from vegetables, and sorting of fruit, etc., are considered normal routine operations at primary production level and must not lead to further food safety requirements.

The regulation does not apply to small quantities of primary products. This means that farmers are allowed to sell their products directly to the consumer, to local retail shops for direct sale to the consumer, and to local restaurants. It is up to the member states to define "small quantities" depending on the local situation. The member states must also lay down national rules in order to ensure that the safety of such foods is guaranteed.

According to the 2003 Common Agricultural Policy (CAP) reform, all farmers receiving direct payment from the EU must respect "cross compliance," i.e., farmers must comply with all legislation affecting their business (Anon., 2003). This was made compulsory in 2005 (Anon., 2008). This means that those who receive direct payment are obliged to keep land in good agricultural and environmental condition. The cross-compliance concept links direct payments to the farmers to their respect of, among other things, environmental requirements at both EU and national levels. Since protection of soil and water from pollution and contamination is imperative, this will also be a positive additional factor for the production of safe fruit and vegetables. The reason for this is that by complying with the agricultural and environmental legislation, both at EU and national levels, the pollution and contamination of soil and water will, at least theoretically, be reduced, thus resulting in, for example, cleaner water that may be used for irrigation.

By respecting cross-compliance (agricultural legislation) and implementing and maintaining HACCP procedures (food legislation), complemented by national legislation in the respective fields, a sound basis for the production of safe fresh fruit and vegetables is laid.

Import from countries outside EU and EEA (third countries)

Article 11 of the Food Law (178/2002) states that food imported in to the community "shall comply with the relevant requirements of food law or conditions recognised by the community to be at least equivalent thereto or, where a specific agreement exists between the community and the exporting country, with requirements contained therein." There is also a demand for traceability; however, the regulations do not have an extra-territorial effect outside the EU. That means that the requirements extend from the importer to the retailer. However, after the sprout-associated outbreak in 2011, specific traceability requirements for sprouts and seeds intended for the production of sprouts entered into force on July 1, 2013 (Regulation (EU) No 208/2013 of 11 March 2013).

There are food business's contractual arrangements that exist. The responsibility is placed on the food business operator, and in the case of imports, this is the importer. It is important to be aware that food business operators in third countries need to respect the relevant requirements with regard to the hygiene of food as stated in Article 3-6 of Regulation (EC) No 852/2004 (Anon., 2006b). This means that there is a general obligation to monitor food safety of products and processes under the operator's responsibility, that there needs to be general hygiene provisions for primary production, that there are detailed requirements after primary production, and that for certain products there are microbiological requirements. It is the responsibility of the importer to ensure compliance with these requirements.

For example, the EU has established a maximum tolerance for aflatoxin in almonds shipped to its member countries. Handlers who choose to ship almonds to the EU must comply with EU specifications. However, in the United States, there are no mandatory requirements pertaining to aflatoxin. The absence of official, specific outgoing quality requirements for shipments to the EU has forced the hands of the almond industry to develop their own voluntary aflatoxin testing protocol for handlers to follow when shipping almonds to the EU.

For import of foods of animal origin, registration of food businesses is necessary. This is different for plant foodstuffs. Here it is usually sufficient that exporting companies in third countries are known and accepted as suppliers by importers in the EU.

Further, according to the EU rules on food hygiene, food business operators in third countries intending to export foodstuffs into the EU must put in place, implement, and maintain procedures based on the HACCP principles after primary production (Anon., 2006b). The Commission is responsible for requesting third countries to provide accurate and up-to-date information on the general organization and management of sanitary control systems. The contact point for the Commission in third countries is the competent authority. It is also noteworthy that it is incumbent upon the importer to ensure compliance with the relevant requirements or food law or with conditions recognized as equivalent.

As a part of the Control Rule (Regulation (EC) No 669/2009 of 24 July 2009), regarding the increased level of official controls on imports of certain feed and food of non-animal origin and amending Decision (EC) 504/2006, was implemented in 2009. This regulation allows known or emerging risks in feed and food of non-animal origin to be countered more effectively. It requires the member states to increase controls on certain imports of food and feed of non-animal origin from specified countries. The enhanced control includes systematic checks of documents accompanying consignments, as well as physical checks, including laboratory analysis at a frequency related to the risk that is identified.

The list of imports (Annex 1 to the regulation) is updated quarterly. In the update that applies from April 1, 2013, the control frequency for coriander leaves and basil from Thailand for the possible presence of pesticide residues was reduced from 20% to 10%. In the amendments that applied from 1 January 2013, watermelons from Brazil were included (control frequency of 10% for the possible presence of *Salmonella*). It should be noted that several of the products with increased import control from specific countries should be tested for mycotoxins, i.e., Ochratoxin A and Aflatoxins. These products comprise dried grapes, nuts, and derived products and dried spices (Regulation (EC) No 669/2009, Annex 1).

GlobalGAP

The Good Agricultural Practices (GAP) concept is used in GlobalGAP (formerly known as EurepGAP), which is a private sector body that sets voluntary standards for the certification of agricultural products around the world (www.globalgap.org). GlobalGAP serves as a practical manual of GAP anywhere in the world. Members of GlobalGAP are both retailers and producers, and the standard is a "pre-farm-gate" standard that covers the production of a certified product from farm inputs to when the product leaves the farm. GlobalGAP is a single integrated standard with modular applications for different product groups, i.e., fruit and vegetables have their own specific module (www.globalgap.org/cms/front_content.phpidcat=3). In the control point and compliance criteria for fruit and vegetables, there are specific points with respect to microbiological quality of irrigation water and hygiene risk analysis at several points during the process, which also includes worker hygiene.

In addition to GlobalGap there are other private standards and certification schemes that are used; for example, BRC (http://www.brcglobalstandards.com/), Food Safety System Certification 2200 (www.fssc22000.com/en/index.php), SQF (www.sqfi.com/), and others.

Differences from U.S. regulations

In January 2011, the FDA Food Safety Modernization Act (FSMA) became law. FSMA aims to ensure that the U.S. food supply is safe by shifting the focus to prevention of contamination instead of responding to it. In this scheme, the FDA has proposed a rule on Standards for Produce Safety (FDA, 2013). The proposed rule builds on previous industry guidelines, which many producers have already implemented. All fruits and vegetables, except those rarely consumed raw, produced for personal consumption, or commercially processed to reduce microorganisms of public health concern, are covered. The proposed rule is science and risk analysis based and focuses on risk areas, in particular, agricultural water, biological soil amendments, health and hygiene, domesticated and wild animals, and equipment, tools, and buildings. The proposed rule sets standards and thus differs from the more general European legislation where the food business operator is responsible for the safety of his/her products and where Good Hygiene Practices and HACCP are implemented at farms and in further processing of the products. For example, there are no common requirements and microbiological criteria for testing of irrigation water. There are food safety criteria and process hygiene criteria for sprouts, pre-cut, ready-to-eat fruit and vegetables, and unpasteurized fruit and vegetable juices. For the production of sprouts, four new regulations with more specific requirements have been enforced in 2013.

Funding of food safety research in Europe

The Frame Programmes (FP) have been the main financial tools through which the EU supports research and development activities in almost all scientific disciplines, and it is estimated that 5 to 10% of research in the EU is financed through the FPs. The current FP (FP7) was fully operational on January 1, 2007 and will expire in 2013. FP7 is organized into four basic components of European Research; namely, Ideas, People, Capacities, and Cooperation. Cooperation is defined as collaboration between industry and academia in key technology areas. International cooperation between EU and third countries is an integral part of this action and is encouraged. One of the themes identified in FP7 is "Food, agriculture and fisheries and biotechnology," and the primary aim of funding in this theme is to build a Knowledge-Based Bio-Economy (KBBE). EU has earmarked more than 1.9 billion € for funding of this theme over the duration of FP7, and work programs for the theme are published annually. The next EU Framework Programme for Research and Innovation is Horizon 2020, which will run from 2014 to 2020 (http://ec.europa.eu/research/horizon2020/index_en.cfm?pg=h2020). Horizon 2020 aims to

- strengthen the EU's position in science,
- strengthen industrial leadership in innovation, and
- address major concerns shared by the Europeans, amongst these, climate change, sustainable transport and mobility, and ensuring food safety and security.

Another feature of European research is the European Research Area (ERA) (http://ec.europa.eu/research/era/index_en.htm). As described on the website, "ERA is a unified research area open to the world, based on the Internal market, in which researchers, scientific knowledge and technology circulate freely." ERA focuses on five key priorities: more effective national research systems, optimal transnational co-operation and competition, an open labour market for researchers, gender equality and gender mainstreaming in research, and optimal circulation and transfer of scientific knowledge.

A challenge for research is that many programs and projects are carried out in an isolated way. Within the European Research Area, Joint Programming (http://ec.europa.eu/research/era/joint-programming_en.html) aims to remedy this. The main aim of Joint Programming is "to pool national research efforts in order to make better use of Europe's precious public R&D resources and to tackle common European challenges more effectively in a few key areas." One such Joint Programming Initiative is for agriculture, food security, and climate change (FACCE-JPI), and their strategic research agenda was published in October 2012. The Strategic Agenda identifies five core research themes, defines short-, medium-, and long-term research priorities, and sets out joint actions. It aims at infrastructure and platforms, training, and capacity building and knowledge transfer.

COST (European Cooperation in Science and Technology) was founded in 1971 and is the widest framework for transnational coordination of nationally funded research activities (COST, 2012). The key features of COST are building capacity by connecting high-quality scientific communities throughout Europe and worldwide, providing networking opportunities for early career investigators, and increasing the impact of research on policymakers, regulatory bodies, and national decision makers in the private sector as well. COST has one single instrument, the COST Action. This is a science and technology network, with a duration of four years, a minimum of five participating COST member countries, and is organized through networking tools. The average COST Action support is around approximately 130,000 € per year, based on an average of 19 participating countries.

Food safety research is also financed through national research councils/agencies and other sources, but it is difficult to estimate the extent of this funding.

Sources for further information

More information on the EU, the Common Agricultural Policy (CAP), and the European Food Law can be found on the websites listed below. There is also useful information on the RASFF pages, and statistics from the EU can be found on the web pages of Eurostat. Information on European research and development can be found on the Cordis website.

> http://ec.europa.eu/food/food/index_en.htm
> http://ec.europa.eu/agriculture/foodqual/index_en.htm
> http://www.efsa.europa.eu/EFSA/efsa_locale-1178620753812_home.htm
> http://ec.europa.eu/eurostat
> http://ec.europa.eu/agriculture/index_en.htm
> http://ec.europa.eu/external_relations/eea/
> http://cordis.europa.eu/home_en.html
> http://www.efta.int
> http://www.globalgap.org

References

Anon., 2003. Cross compliance. (Reprinted from: In File.)

Anon., 2006b. Guidance document. Key questions related to import requirements and the new rules on food hygiene and official controls. Brussels: European Commission Health & Consumer Protection Directorate General, Reprinted from: In File.

Anon., 2008. Agriculture and the environment: Cross compliance. (Reprinted from: In File.)

Anon., 2012a. Agriculture, fishery and forestry statistics. Main results - 2010-11 Eurostat Pocketbooks, (pp. 228). doi: http://epp.eurostat.ec.europa.eu/cache/ITY_OFFPUB/KS-CD-12-001/EN/KS-CD-12-001-EN.PDF.

Anon., 2012b. Europe in figures - Eurostat yearbook. doi: http://epp.eurostat.ec.europa.eu/cache/ITY_OFFPUB/KS-CD-12-001/EN/KS-CD-12-001-EN.PDF.

Anon., 2012c. Guidance document on the implementation of certain provisions of Regulation (EC) No 852/2004 on the hygiene of foodstuffs (pp. 1–20). Brussels: European Commission Health and Consumers Directorate-General.

Anon., 2012d. Multicriteria-based ranking for risk management of foodborne parasites. Report of a joint FAO/WHO expert meeting, 3–7 september, 2012, (Preliminary report ed., pp. 57). Rome, Italy: FAO/WHO.

Anon., 2012e, 20.07.2012. Rapid Alert System for Food and Feed. Retrieved 15.05.2013, from http://ec.europa.eu/food/food/rapidalert/index_en.htm.

Anon., 2013a. EU Reference laboratories - Biological Risks. Retrieved 15.05.2013, 2013, from http://ec.europa.eu/food/food/biosafety/laboratories/bio_risks_en.htm.

Anon., 2013b. The Rapid Alert System for Food and Feed 2011 Annual Report (pp. 52). Office for Official Publications of the European Communities, Luxembourg: European Communities; 2012.

Belli, N., Ramos, A.J., Sanchis, V., Marin, S., 2004. Incubation time and water activity effects on ochratoxin A production by Aspergillus section Nigri strains isolated from grapes. Lett. Appl. Microbiol. 38 (1), 72–77.

Carvalho, C., Thomas, H.L., Balogun, K., Tedder, R., Pebody, R., Ramsay, M., et al., 2012. A possible outbreak of hepatitis A associated with semi-dried tomatoes, England, July–November 2011. Euro. Surveill. 17 (6), 14–17.

Commission, The European, 2000. White paper on Food Safety.

COST. (2012). About COST How to join a COST Action In E. C. i. S. a. Technology (Ed.), (2012-2013 ed.). Brussels: COST Office.

Cotterelle, B., Drougard, C., Rolland, J., Becamel, M., Boudon, M., Pinede, S., et al., 2005. Outbreak of norovirus infection associated with the consumption of frozen raspberries, France, March 2005. Euro. Surveill. 10 (17).

Dawson, D., 2005. Foodborne protozoan parasites. Int. J. Food. Microbiol. 103 (2), 207–227.

Doyle, M.P., Erickson, M.C., 2008. Summer meeting 2007 – the problems with fresh produce: an overview. J. Appl. Microbiol. 105 (2), 317–330.

Döller, P.C., Dietrich, K., Filipp, N., Brockmann, S., Dreweck, C., Vonthein, R., et al., 2002. Cyclosporiasis outbreak in Germany associated with the consumption of salad. Emerg. Infect. Dis. 8 (9), 992–994.

EFSA, 2011. Scientific Opinion on the risk posed by Shiga toxin-producing *Escherichia coli* (STEC) and other pathogenic bacteria in seeds and sprouted seeds. (E. P. o. B. H. (BIOHAZ), Trans.). EFSA J. (vol. 9, pp. 2424 (2101pp)): EFSA.

EFSA, 2013. Scientific Opinion in the risk posed by pathogens in food of non-animal origin. Part 1 (outbreak data analysis and risk ranking of food/pathogen combinations). (E. P. o. B. H. B. Panel, Trans.) EFSA J. 11, 138.

Emberland, K.E., Ethelberg, S., Kuusi, M., Vold, L., Jensvoll, L., Lindstedt, B.A., et al., 2007. Outbreak of Salmonella Weltevreden infections in Norway, Denmark and Finland associated with alfalfa sprouts, July–October 2007. Euro. Surveill. 12 (11), E071129.

Ethelberg, S., Lisby, M., Bottiger, B., Schultz, A.C., Villif, A., Jensen, T., et al., 2010. Outbreaks of gastroenteritis linked to lettuce, Denmark, January 2010. Euro. Surveill. 15 (6).

Ethelberg, S., Lisby, M., Vestergaard, L.S., Enemark, H.L., Olsen, K.E.P., Stensvold, C.R., et al., 2009. A foodborne outbreak of Cryptosporidium hominis infection. Epidemiol. Infect. 137 (3), 348–356. http://dx.doi.org/10.1017/S0950268808001817.

EURL, 2008. Technical Guidance Document on shelf-life studies for Listeria monocytogenes in ready-to-eat foods (pp. 31). EU Community Reference Laboratory for Listeria monocytogenes, Paris.

FAO/WHO, 2008. Microbiological hazards in fresh fruits and vegetables. In: FAO/WHO (Ed.), Microbiological Risk Assessment Series, pp. 1–28. Rome/Geneva.

FDA, 2013. Overview of the FSMA proposed rules on Produce Safety Standards and Preventive Controls for Human Food. Retrieved 16.05.2013, from http://www.fda.gov/Food/Guidance Regulation/FSMA/ucm334120.htm.

Food, The Scientific Committee on, 2002. Risk profile on the microbiological contamination of fruits and vegetables eaten raw. Brussels.

Freshfel, 2012. Activity report 2011–2012 (pp. 36). Brussels: The European Fresh Producs Association.

Gault, G., Weill, F.W., Mariani-Kurkdjian, P., Jourdan-da Silva, N., King, L., Aldabe, B., et al., 2011. Outbreak of haemolytic uraemic syndrome and bloody diarrhoea due to *Escherichia coll* O104:H4, south-west France, June 2011. Euro. Surveill. 16 (26) pii=19905.

Hernandez, F., Monge, R., Jimenez, C., Taylor, L., 1997. Rotavirus and hepatitis A virus in market lettuce (Latuca sativa) in Costa Rica. Int. J. Food. Microbiol. 37 (2-3), 221–223.

Hjertqvist, M., Johansson, A., Svensson, N., Åbom, P.E., Olsson, M., Hedlund, K.O., et al., 2006. Four outbreaks of norovirus gastroenteritis after consuming raspberries, Sweden, June–August 2006. Euro. Surveill. 11 (36).

Hoang, L.M., Fyfe, M., Ong, C., Harb, J., Champagne, S., Dixon, B., et al., 2005. Outbreak of cyclosporiasis in British Columbia associated with imported Thai basil. Epidemiol. Infect. 133 (1), 23–27.

Initiative, Healthy Nut, 1998. Nuts and Aflatoxins. Data, facts, background. Hamburg, Germany.

Kapperud, G., Roervik, L.M., Hasseltvedt, V., Hoeiby, E.A., Iversen, B.G., Staveland, K., et al., 1995. Outbreak of Shigella sonnei Infection traced to imported Iceberg lettuce. J. Clin. Microbiol. 33 (3), 609–614.

Long, S.M., Adak, G.K., O'Brien, S.J., Gillespie, I.A., 2002. General outbreaks of infectious intestinal disease linked with salad vegetables and fruit, England and Wales, 1992–2000. Commun. Dis. Public Health 5 (2), 101–105.

Martinez-Palou, A., Rhoner-Thielen, E., 2008. Fruit and vegetables: fresh and healthy on European tables. In: Eurostat (Ed.), Eurostat - Statistics in focus, pp. 1–8.

Martinez Palou, A., Rohner-Thielen, E., 2011. From farm to fork - a statistical journay alond the EU's food chain., Eurostat. Statistics in focus.: Vol. 27/2011 (pp. 12): European Union.

Moss, M.O., 2008. Fungi, quality and safety issues in fresh fruits and vegetables. J. Appl. Microbiol. 104 (5), 1239–1243.

Nygard, K., Andersson, Y., Lindkvist, P., Ancker, C., Asteberg, I., Dannetun, E., et al., 2001. Imported rocket salad partly responsible for increased incidence of hepatitis A cases in Sweden, 2000–2001. Euro. Surveill. 6 (10), 151–153.

Petrignani, M., Harms, M., Verhoef, L., van Hunen, R., Swaan, C., van Steenbergen, J., et al., 2010. Update: A food-borne outbreak of hepatitis A in the Netherlands related to semi-dried tomatoes in oil, January–February 2010. Euro. Surveill. 15 (20).

Pönkä, A., Kotilainen, H., Rimhanen-Finne, R., Hokkanen, P., Hänninen, M.-L., Kaarna, A., et al., 2009. A foodborne outbreak due to *Cryptosporidium parvum* in Helsinki, November 2008. Euro. Surveill. 14 (28).

Regulation (EC) No 178/2002 of the European Parliament and of the Council of 28 January 2002 laying down the general principles and requirements of food law, establishing the European Food Safety Authority and laying down procedures in matters of food safety., EC No 178/2002, European Union, Health & Consumer Protection, Directorate General (2002).

Regulation (EC) No 852/2004 of the European Parliament and of the Council of 29 April 2004 on the hygiene of food stuffs., 2004, European Union, Health & Consumer Protection, Directorate General (2004).

Regulation (EC) No 882/2004 of the European Parliament and of the Council of 29 April 2004 on official controls performed to ensure the verification of compliance with feed and food law, animal health and animal welfare rules., EC No 882/2004, European Union, Health & Consumer Protection, Directorate General (2004).

Regulation (EC) No 2073/2005 of 15 November 2005 on microbiological criteria for foodstuffs., EC No 2073/2005, European Union, Health & Consumer Protection, Directorate General (2005).

Regulation (EC) No 1881/2006 of 19 December 2006 setting maximum levels for certain contaminants in foodstuffs., 1881/2006 C.F.R. (2006a).

Regulation (EC) No 669/2009 of 24 July 2009 implementing Regulation (EC) No 882/2004 of the European Parliament and the Council as regards the increased level of official controls on imports of certain feed and food of non-animal origin and amending Decision 2006/504/EC, 669/2009 C.F.R. (2009).

Regulation (EU) No 165/2010 of 26 February 2010 amending Regulation (EC) No 1881/2006 setting maximum levels for certain contaminants in foodstuffs as regards aflatoxin., 165/2010 C.F.R. (2010).

Regulation (EU) No 208/2013 of 11 March 2013 on traceability requirements for sprouts and seeds intended for the production of sprouts 208/2013 C.F.R. (2013).

Robertson, L.J., van der Giessen, J.W.B., Batz, M.B., Kojima, M., Cahill, S., 2013. Have foodborne parasites finally become a global concern? Trends Parasitol. 29 (3), 101–103. http://dx.doi.org/10.1016/J.Pt.2012.12.004.

United Kingdom (Great Britain): Contaminants in Food (Amendment) Regulations 1999., S.I. No. 1603 of 1999, United Kingdom (1999).

Technology for Reduction of Human Pathogens in Fresh Produce

Disinfection of Contaminated Produce with Conventional Washing and Sanitizing Technology

17

Gerald M. Sapers

Eastern Regional Research Center, Agricultural Research Service, U.S. Department of Agriculture (Retired), Wyndmoor, PA

CHAPTER OUTLINE

The Produce Contamination Problem. http://dx.doi.org/10.1016/B978-0-12-404611-5.00017-8

Introduction

Prepackaged fresh and fresh-cut fruits and salad vegetables represent a major segment of the fresh produce industry. In preceding chapters, we have seen how such products can become contaminated with human pathogens, resulting in outbreaks of foodborne illness. With the exception of vegetable products normally cooked by the consumer, these items are not subjected to a final inactivation step prior to consumption. However, they are washed with water or sanitizing agents, primarily to remove soil and pesticide residues, but also to remove or inactivate human pathogens and spoilage-causing microorganisms. In this chapter we will examine the efficacy of produce disinfection treatments that are based on washing and sanitizing technology.

In addition to populations of epiphytes, freshly harvested produce may contain localized, heavy loads of microbial contaminants, including plant and human pathogens, often associated with soil, decay, and mechanical injury. Washing such produce can transfer microbial contaminants to the wash water and thence to other, uncontaminated raw material as well as the conveying, packing, and processing equipment. Addition of a sanitizing agent to the wash water can greatly reduce the population of planktonic bacterial cells and thus lower the risk of cross-contamination. Such reductions can improve product safety and shelf-life, thereby enabling grower/packers to ship their products nationwide or to overseas markets.

However, cleaning and sanitizing agents are much less effective in removing or inactivating human pathogens and other microorganisms that have attached to produce surfaces. This is a consequence of strong microbial attachment as well as attachment to inaccessible sites such as pores, punctures, surface irregularities, and cut surfaces, which limits contact between sanitizer solutions and the targeted microorganisms. In a comparison of the inactivation of *E. coli* O157:H7 on fresh-cut apples and cantaloupe rinds by acidic electrolyzed water, peroxyacetic acid and chlorine, Wang et al. (2006) observed dual phase kinetics for each of these treatments which they attributed to fruit surface topography, a determinant of bacterial distribution. Wang et al. (2009) quantified the relationship between surface roughness and retention and removal of *E. coli* O157:H7 on selected fruits that were spot inoculated and then washed with water or sanitizing agents. Additionally, bacterial incorporation within biofilms will confer greater resistance to microcidal agents (see Chapter 2 for more information on microbial attachment). Thus, the level of pathogen reduction obtained by washing and sanitizing may be inadequate to assure food safety.

In this chapter we will review the characteristics of conventional and alternative washing and sanitizing agents suitable for produce packing and fresh-cut processing, regulatory restrictions regarding their use, the characteristics of commercial equipment used for washing and disinfecting produce, and disinfection treatments suitable for food service and consumer use. We will examine the efficacy of disinfection treatments in reducing pathogen levels on commodities that have a history of association with outbreaks of foodborne illness such as leafy vegetables, tomatoes, cantaloupes, apples, and sprouts. Previously, these topics have been reviewed by Beuchat (1998), Parish et al. (2003), Sapers (2003, 2005), and Gil et al. (2009). Also, recommendations regarding washing and sanitizing appear in the FDA's *Guide to Minimize Microbial Food Safety Hazards of Fresh-cut Fruits and Vegetables* (FDA, 2007a).

Washing and sanitizing agents
Detergent products

A number of surfactants, including sodium n-alkylbenzene sulfonate, sodium dodecylbenzene sulfonate, sodium mono- and dimethyl naphthalene sulfonates, sodium 2-ethylhexyl sulfate, and others are permitted by the FDA for washing fruits and vegetables (21CFR173.315). Various detergent formulations for washing fresh produce are commercially available, including products prepared at a neutral pH, acidified with citric or phosphoric acid, or made alkaline with sodium or potassium hydroxide (see www.Decco.US.com, www.microcide.com, www.stepan.com, and www.afcocare.com for details).

Detergents reduce microbial populations on produce surfaces by detachment rather than inactivation. Studies of the efficacy of various commercial detergent formulations in reducing populations of human pathogens on inoculated fruits and vegetables and comparisons with other treatments have been reported for apples (Sapers et al., 1999; Wright et al., 2000; Kenney and Beuchat, 2002), strawberries (Raiden et al., 2003), cantaloupe (Sapers et al., 2001), tomatoes (Raiden et al., 2003; Sapers and Jones, 2006), and lettuce (Raiden et al., 2003). The results of these studies indicate that detergent washes sometimes achieve population reductions as great as 2 to 3 logs, equaling or surpassing sodium hypochlorite, but in other cases showed no greater efficacy than water (Raiden et al., 2003; Samadi et al., 2009). In a study to determine the efficacy of adding detergents to sanitizer solutions for disinfection of inoculated Romaine lettuce, Keskinen and Annous (2011) reported that most of the wash treatments achieved reductions in the population of *E. coli* O157:H7 of less than 1 log; however, washing with an experimental short chain fatty acid formulation resulted in a 5-log reduction. This result requires confirmation and elucidation.

Chlorine

Because of its microbiocidal activity and low cost, chlorine (as sodium or calcium hypochlorite or Cl_2 gas) is the agent most widely used to sanitize fresh produce

(Suslow, 2000). Typically, sodium hypochlorite concentrations of 50 to 200 ppm are used. Concentrations may be expressed as total available chlorine (the calculated amount of chlorine present in the sanitizer solution, which includes both free and combined forms of chlorine) or as free chlorine, which depends on the actual chlorine, hypochlorous acid, and hypochlorite ion concentrations. At the pH used in packinghouse water systems, the elemental chlorine concentration is near 0, and free available chlorine is the sum of hypochlorous acid and hypochlorite ion. The chlorine concentration (total available or free) can be monitored using test kits based on colorimetric measurements (www.chemetrics.com, www.emdmillipore.com, www.hach.com), or by measurement of the oxidation-reduction potential (ORP; www.pulseinstruments.com).

Depending on the pH, hypochlorite solutions contain varying proportions of hypochlorite ion and hypochlorous acid, the latter having the most bactericidal activity. To enhance the antimicrobial activity of hypochlorite solutions, the solution pH may be reduced from the alkaline range (about pH 9) to the slightly acidic range of 6 to 7 by addition of citric acid, a mineral acid, or a buffer (available commercially from www.Decco.US.com). A chlorine stabilizer, marketed as SmartWash (also designated as T-128), was shown to decrease the rate of free chlorine depletion in the presence of soil and lettuce extract. Additionally, the survival of bacterial pathogens in wash solutions with high organic loads and potential for cross-contamination were significantly reduced by T-128. T-128 in chlorinated wash solutions also enhanced the inactivation of *Salmonella enterica* serovars and *Pseudomonas fluorescens* biofilms on stainless steel (Shen et al., 2012a). However, T-128 did not enhance the efficacy of chlorinated wash solutions in reducing microbial populations on contaminated iceberg lettuce (Nou et al., 2011; Christie, 2010a). Additional benefits may be realized by adding a surfactant to the hypochlorite solution to improve contact with the microbial surface (Segall, 1968; Spotts, 1982). Kondo et al. (2006) reported that a disinfection treatment using 200 ppm sodium hypochlorite, applied to inoculated iceberg lettuce leaves with mild heating (50°C), was unable to achieve reductions in *E. coli* O157:H7 and *S.* Typhimurium DT104 populations greater than 1.2–1.7 logs.

Chlorine is highly reactive with organic species originating in soil and debris or leached from damaged produce into the process water, resulting in rapid chlorine depletion and greater survival of the targeted microflora when the organic load is high (Suslow, 2000; Shen et al., 2013). Hence, the chlorine level in wash water should be monitored continuously by measuring the ORP and replenished to maintain the desired concentration using automated commercial systems (see www.pulse instruments.net or www.globalspec.com).

Although chlorine has a broad spectrum of antimicrobial activity and is highly efficacious in inactivating planktonic microorganisms in wash water (depending on the organic load), it is far less effective against bacteria attached to produce surfaces. Population reductions reported in the literature for indigenous microflora and for human pathogens on inoculated samples rarely exceed 2 logs (99%) (Brackett, 1987; Zhuang et al., 1995; Beuchat et al., 1998; Garcia et al., 2003). Such reductions may have a large impact on the incidence of spoilage and will significantly reduce the risk of foodborne illness by reducing the load of attached pathogens on produce and cross-contamination. However, because of the low infectious dose of some pathogens, one cannot assure safety.

To provide an acceptable level of safety for fresh juice, the FDA has mandated a 5-log reduction (99.999%) in the population of pathogens in the juice product. It is clear that this cannot be accomplished solely by use of chlorine (or any other sanitizer); in the case of apple juice, this may be accomplished by heat pasteurization or UV-treatment of the finished juice. With citrus juices, the 5-log reduction can be apportioned between surface treatment of the fruit and treatment of the juice (FDA, 2001).

Chlorine solutions are considered to be highly corrosive, especially at low pH, and will shorten the life of tanks and other stainless steel equipment used in produce packing/processing operations. Also, because of reports in the literature indicating potential mutagenicity and carcinogenicity from exposure to reaction products of chlorine with food constituents, there is some concern in the food industry regarding future regulatory restrictions on the use of this sanitizer (Chang et al., 1988; Hidaka et al., 1992). Consequently, a number of alternatives to chlorine have been developed or are under study for use by the food industry.

Alternatives to chlorine
Electrolyzed water

Electrolyzed water (also known as electrolyzed oxidizing or EO water) has received much attention as a replacement for chlorine in sanitizing produce. In principle, it represents an alternative means of generating hypochlorous acid (Izumi, 1999). Hypochlorous acid is formed at the anode during electrolysis of water that contains some sodium chloride. Depending on the sodium chloride concentration, the available chlorine level can reach or exceed 100 ppm. If the electrolyzed water generator has a membrane separating the electrodes, highly acidic (pH < 3.0) water will be produced at the anode, and alkaline water (pH ≥ 11.0) will be produced at the cathode. Electrolyzed water is considered to be an effective sanitizing agent at low pH with an oxidation-reduction potential greater than 1000 mV. At high pH with a redox potential less than 800 mV, it can be used as a cleaning agent (Deza et al., 2003; Yang et al., 2003; Ozer and Demirci, 2006).

Electrolyzed water is highly effective in reducing the population of planktonic cells (Venkitanarayanan et al., 1999a), but like chlorine, its efficacy in reducing bacterial populations attached to produce surfaces is generally limited to 1 to 3 logs (Izumi, 1999; Park et al., 2001; Pangloli et al., 2009; Ding et al., 2011). Other studies have yielded population reductions between 1 and 7 logs, depending on the commodity, method of inoculation and recovery, inoculation site, time interval between inoculation and treatment, application method, and strength of the electrolyzed water (Koseki et al., 2003, 2004; Yang et al., 2003; Deza et al., 2003; Bari et al., 2003; Paola et al., 2005). Rodriguez–Garcia et al. (2011) reported 4- to 5-log reductions in populations of *S. enterica*, *E. coli* O157:H7, and *L. monocytogenes* on inoculated Hass avocados by spraying with alkaline electrolyzed oxidized water and then acid electrolyzed oxidizing water. Hao et al. (2011) reported reductions in excess of 4 logs on cilantro inoculated with *E. coli* O78 using slightly acidic electrolyzed water. It is questionable whether some of the larger reductions in attached microbial populations reported above can be realized in a packing or processing plant situation.

Aqueous chlorine dioxide and acidified sodium chlorite

Solutions of chloride dioxide (ClO_2), at residual concentrations not to exceed 3 ppm, are permitted by the FDA for sanitizing fresh fruits and vegetables (21CFR173.300). Such treatment shall be followed by a potable water rinse. ClO_2 must be generated on-site by such means as reaction of sodium chlorite with either chlorine gas or a mixture of sodium hypochlorite and hydrochloric acid. Information on ClO_2 generators can be obtained from the many vendors of this equipment: see Aquapulse Systems (www.aquapulsesystems.com), Vulcan Chemical (800–873–4898); CH2O Inc. (Fresh-Pak 2; www.ch2o.com); Rio Linda Chemical Co., Inc. (916–443–4939); Bio-Cide International, Inc. (Oxine; www.bio-cide.com); http:/// DuPont (www2.dupont.com/Chlorine_Dioxide/en_US/index.html); http:/// CDG Environmental (www.cdgenvironmental.com), and others. ClO_2 gas also can be generated from sachets containing a dry mixture of sodium chlorite or sodium chlorate and an activator (www.icatrinova.com). ClO_2 can be applied as an aqueous solution or in the gas phase. In contrast to hypochlorite, ClO_2 is claimed to be more effective at neutral pH, less reactive with organic substances, less corrosive, and less able to form chlorinated byproducts (Anon, 2001). However, ClO_2 gas is unstable, and at partial pressures greater than 120 mm Hg (15.8% by volume at atmospheric pressure), it becomes explosive (see CDG Environmental *Guidelines and Recommendations for the Transport, Handling, and Application of CDG Solution 3000*™ at www.cdgenvironmental.com).

Pathogen reductions obtained with ClO_2 solutions vary widely from study to study, depending on the target organism, the commodity, and to a lesser extent, on the ClO_2 concentration (1–5 ppm), but most reported reductions are in the range of 1 to 3 logs (Zhang and Farber, 1996; Wisniewsky et al., 2000; Han et al., 2001; Huang et al., 2006). However, Rogers et al. (2004) reported much larger reductions in the populations of *E. coli* O157:H7 and *L. monocytogenes* on inoculated apples, lettuce, strawberries, and cantaloupe. Treatments with gaseous ClO_2 were reported to be highly effective in reducing populations of foodborne pathogens on fresh and fresh-cut produce. Pathogen reductions as high as 3 to 5 logs were obtained with inoculated fresh-cut cabbage and carrots, apples, tomatoes and peaches. However, pathogen reductions with onions and fresh-cut lettuce were less than 2 logs (Sy et al., 2005). Using ClO_2 gas generated from dry reactant sachets, Lee et al. (2006) obtained 4.5-log reductions in *Alicyclobacillus acidoterrestris* spores, inoculated on apple surfaces; this organism is responsible for spoilage of apple juice. Popa et al. (2007) reported reductions greater than 3 logs in populations of *Salmonella*, *E. coli*

O157:H7, and *L. monocytogenes* on inoculated blueberries by treatment with ClO_2 gas generated from dry reactant sachets. A system for in-package generation of ClO_2 from chlorite has been incorporated into an absorbent pad used in packaging produce (www.biovation.com).

Acidified sodium chlorite (ASC) is produced by mixing a solution of sodium chlorite with any GRAS acid and is considered to be a source of chlorous acid ($HClO_2$), the primary active antimicrobial agent, although some ClO_2 is produced gradually as ASC decomposes (Warf, 2001). ASC is permitted by the FDA for use as an antimicrobial agent at concentrations of 500 to 1200 ppm (pH 2.3–2.9) in the water applied to processed fruits and vegetables by spraying or dipping, provided that the treatment is followed by a potable water rinse and a 24-hour waiting period prior to consumption. With leafy vegetables, only application by dipping is permitted (21CFR173.325). Most efficacy studies indicate that treatment of fresh produce with up to 1200 ppm ASC can reduce the natural microflora and *Salmonella, Staphylococcus aureus, E. coli* O157:H7, and *L. monocytogenes* by about 1 to 3 logs on inoculated produce (lettuce, cucumbers, bell peppers, tomatoes, Chinese cabbage, cantaloupes, strawberries, apples, alfalfa sprouts) (Park and Beuchat, 1999; Conner, 2001; Fett, 2002; Caldwell et al., 2003; Yuk et al., 2005, 2006; Inatsu et al., 2007). In side-by-side comparisons, ASC was just as efficacious if not superior to hypochlorite. However, Gonzalez et al. (2004) obtained reductions in the population of *E. coli* O157:H7 as great as 5.25 logs on inoculated shredded carrot by treatment with 1000 ppm ASC.

Aqueous and gaseous ozone

Ozone is a highly effective, broad spectrum antimicrobial agent, effective at low concentrations and short contact times (Wickramanayake, 1991; Restaino et al., 1995). Ozone is highly unstable and decomposes to nontoxic products. However, it is corrosive to equipment and can cause physiological injury to produce and degrade product color and flavor. Ozone is toxic and an irritant to workers at concentrations in air greater than 0.1 ppm (29CFR1910.1000). It must be adequately vented to avoid worker exposure (Anon., 2001). Ozone is approved for food use by the FDA (21CFR173.368). Food applications of ozone have been reviewed by Graham (1997), Kim et al. (1999), Xu (1999), Khadre et al. (2001), Smilanick (2003), Suslow (2004), Sharma (2005), and Karaca and Velioglu (2007). Ozone must be generated on-site by passing air or oxygen through a corona discharge or UV light (Xu, 1999). Information about commercial ozone generators is available online from Praxair, Inc. (www.praxair.com), Ozone Safe Foods, Inc. (www.ozonesafefood.com Ozonelab (www.ozonelab.com/products), Ozonia North America, Inc. (www.ozonia.com), Lynntech, Inc. (www.lynntech.com), and others.

Ozonation can reduce bacterial populations in flume and wash water; typical use rates for disinfection of postharvest water are 2 to 3 ppm (Suslow, 2004). The efficacy of ozone treatment of fresh produce is generally similar to that of chlorine and other chlorine alternatives (Kim et al., 1999; Garcia et al., 2003). However, ozone treatment was ineffective in reducing populations of *E. coli* O157:H7 in the stem and calyx regions of inoculated apples (Achen and Yousef, 2001), reducing postharvest fungal decay of pears (Spotts and Cervantes, 1992), and decontaminating alfalfa seeds inoculated with *E. coli* O157:H7 (Sharma et al., 2002) and *L. monocytogenes* (Wade et al., 2003), probably because of the difficulty in contacting and inactivating bacteria attached in inaccessible sites (see Chapter 2). In contrast, Rodgers et al. (2004) reported much higher population reductions with several commodities. Conditions for obtaining a 5-log reduction of *E. coli* O157:H7 in apple cider by treatment with ozone gas were described by Steenstrup and Floros (2004). However, an outbreak of cryptosporidiosis associated with ozonated apple cider suggests that this application is not feasible, perhaps because of the inherent inadequacy of ozone in inactivating *Cryptosporidium* or improper application of ozone. The FDA advised juice processors not to use ozone unless they can demonstrate a 5-log reduction through ozonation (FDA, 2004; Blackburn et al., 2006).

Bialka and Demerci (2007a) obtained large reductions in the populations of *E. coli* O157:H7 and *Salmonella* on inoculated raspberries and strawberries, commodities that are difficult to treat because of their fragility, by treatment with aqueous ozone (1.7–8.9 mg/liter), but treatment times were as long as 64 minutes. This study did not report treatment effects on sample shelf-life. One might expect some fungal spoilage during storage of washed small fruits unless the fungal population was greatly reduced by exposure to ozone. Application of gaseous ozone to inoculated blueberries reduced levels of *S. enterica* and *E. coli* O157:H7 by 3.0 and 2.2 logs, respectively (Bialka and Demirci, 2007b).

Gaseous ozone treatments have been applied to spinach leaves inoculated with *E. coli* O157:H7 as part of a vacuum cooling process (SanVac) and during simulated transportation (SanTrans). Population reductions for the optimized SanVac and SanTrans treatment were 1.8 and 1.0 logs, respectively. However, sequential application of the two optimized treatments resulted in 4.1- to 5-log reductions, depending on treatment time (Vurma et al., 2009).

Peroxyacetic acid

Peroxyacetic acid (PAA), a highly effective antimicrobial agent (Block, 1991), is actually an equilibrium mixture of PAA, hydrogen peroxide, and acetic acid. This product is approved by the FDA (21CFR173.315) for addition to wash water at concentrations not to exceed 80 ppm, and, under EPA regulations, is exempt from the requirements of a tolerance for residues resulting from treatment of fruits and vegetables with PAA solutions at concentrations up to 100 ppm (40CFR180.1196). PAA is a strong oxidizing agent, and handling at high concentrations may be hazardous. PAA is available at various strengths from Ecolab, Inc. (www.ecolab.com), FMC Corp. (www.fmcchemicals.com), and Solvay Chemicals North America (www.solvaychemicals.us/EN/homepage.aspx). PAA formulations are recommended for treating process water and are also claimed to substantially reduce microbial populations on fruit and vegetable surfaces (www.ecolab.com/initiatives/foodsafety). Lower concentrations of PAA are effective in killing pathogenic bacteria in aqueous suspension than would be required with chlorine (Block, 1991). However, population reductions for the indigenous microflora and human pathogens on inoculated produce are generally no greater than 1 or 2 logs (Sapers et al., 1999; Wisniewsky et al., 2000; Wright et al., 2000; Lukasik et al., 2003; Caldwell et al., 2003; Nascimento et al., 2003; Beuchat et al., 2004; Oh et al., 2005; Yuk et al., 2005, 2006; Shiron et al., 2009; Vandekinderen et al., 2009) with few exceptions (Park and Beuchat, 1999; Rodgers et al, 2004; Allwood et al., 2004). Formulations of PAA containing octanoic acid were more effective in killing yeasts and molds in fresh-cut vegetable process waters but had little effect on populations attached to fresh-cut vegetables (Hilgren and Salverda, 2000).

Efficacy of combination of treatments

Certain combinations or sequences of treatments may show synergism or an additive effect in reducing populations of microbial contaminants on produce. Combinations of lactic acid with chlorine (Zhang and Farber, 1996; Escudero et al., 1999; Materon, 2003) or hydrogen peroxide (Venkitanarayanan et al., 1999c, 2002; Lin et al., 2002; Rupasinghe et al., 2006), acetic acid with hydrogen peroxide (Liao et al., 2003), and ozone with chlorine (Garcia et al., 2003) show promise. However, a combination of chlorine with lactic acid bacteria and modified atmospheres achieved minimal reductions in populations of *E. coli* O157:H7 and *Clostridium sporogenes* on inoculated spinach (Brown et al., 2011). Treatment of inoculated romaine lettuce and spinach with combinations of lactic acid and peroxyacetic acid yielded larger reductions in the population of *E. coli* K-12, *Salmonella* and *L. inocua* than were obtained with these agents applied individually. This technology has been validated using pathogenic strains and is being marketed as Fresh Rinse™ (Ho et al., 2011; Christie, 2010b). Combinations of levulinic acid and sodium dodecyl sulfate have achieved reductions in excess of 6.7 logs for *E. coli* O157:H7 and *Salmonella* on inoculated romaine lettuce (Zhao et al., 2009). The combination of levulinic acid and sodium dodecyl sulfate is being licensed for use in a reformulated FIT Fruit and Vegetable Wash (see FreshCUT, July 2009; www.freshcut.com). However, combinations of levulinic acid and sodium dodecyl sulfate were ineffective against oocytes of *Cryptosporidium parvum* and microsporidian spores of *Encephalitozoon intestinalis* (Ortega et al., 2011). Additionally, treatment of fresh-cut iceberg lettuce with a combination of levulinic acid and sodium acid sulfate, with 0.05% sodium dodecyl sulfate, was detrimental to product quality (Guan et al., 2010).

The combination of aqueous ozone and UV reduced the microbial flora of vegetable wash waters, generated during fresh-cut processing, by as much as 6.6 logs after 60 min (Selma et al., 2008c). Enhanced inactivation of *E. coli* O157:H7 on inoculated blueberries by sequential treatment with gaseous ozone at 4000 mg/L for 1 min, followed by UV light at 20 mW/cm^2 for 2 min, was reported by Kim and Hung (2012). Huang et al. (2006) reported enhancement of ClO_2 treatment by sonication. Sequential washing of whole cantaloupes with an acidic detergent, followed by a 2000 ppm chlorine wash, reduced the total aerobic plate count of the fresh-cut melon pieces initially and delayed outgrowth of survivors during storage at 4°C, suggestive of injury to survivors (Sapers et al., 2001). On the other hand, addition of surfactants did not enhance efficacy of chlorine or chlorine dioxide solutions (Zhang and Farber, 1996; Escudero et al., 1999; Keskinen and Annous, 2011) or hydrogen peroxide (Sapers and Jones, 2006). Further research in this area may yield treatment combinations that show greater efficacy.

Other approved sanitizing agents for produce
Hydrogen peroxide

Hydrogen peroxide (HP), a highly effective antimicrobial agent (Block, 1991), may be a potential alternative to chlorine for sanitizing fresh produce (Sapers, 2003), although HP's regulatory status in the United States requires clarification. Use of HP as a wash for raw agricultural commodities is covered under regulations of the U.S. Environmental Protection Agency (EPA), and such applications are exempt from the requirements of a tolerance if the concentration used is 1% or less per application (40CFR180.1197). However, use of HP in fresh-cut processing operations would be regulated by the FDA, and although HP is considered GRAS for certain specified applications, its use as a produce wash is not addressed by current FDA regulations (21CFR184.1366).

Numerous studies have demonstrated the efficacy of dilute hydrogen peroxide in sanitizing fresh produce, including mushrooms (Sapers et al., 2001), apples (Sapers et al., 1999, 2000, 2002), melons (Sapers et al., 2001; Ukuku et al., 2004), and eggplant and sweet red pepper (Fallik et al., 1994). Hydrogen peroxide treatments were ineffective in decontaminating sprouts (Fett, 2002) or the seeds used to produce sprouts (Weissinger and Beuchat, 2000). Contact with stainless steel and aluminum alloy equipment can destabilize hydrogen peroxide solutions, and such equipment must be passivated by treatment with nitric or citric acid solution prior to exposure to H_2O_2 to render it less reactive (Sapers, 2003). Information on hydrogen peroxide applications can be obtained from FMC Corp. (www.fmcchemicals.com), Solvay Chemicals North America (www.solvaychemicals.us/EN/homepage.aspx), U.S. Peroxide (www.h2o2.com), Evonik Degussa Corp. (http://h2o2.evonik.com/product/h2o2/en), and BioSafe Systems (www.biosafesystems.com).

Organic acids

Organic acids such as citric, lactic, and acetic acids are effective antibacterial agents (Doores, 2005) and have been classified by the FDA as GRAS (21CFR184.1005, 1033, 1061). Information about applications of lactic acid and lactates can be obtained from Purac America, Inc. (www.purac.com). Numerous studies have demonstrated the efficacy of organic acids, used in combination with other sanitizing agents, in reducing pathogen levels on fresh produce (see page 397). However, use of organic acids alone in wash water has been less effective, resulting in pathogen reductions of

1 log or less for *L. monocytogenes* in shredded lettuce treated with acetic and lactic acids (Zhang and Farber, 1996), 1 to 2 log for *E. coli* and *L. monocytogenes* on iceberg lettuce treated with 0.5% lactic or citric acid (Akbas and Olmez, 2007), 1 log or less for *E. coli* O157:H7 and *Salmonella* in apples treated with vinegar (Lukasik et al., 2003; Liao et al., 2003), 1.6 to 2 log *E. coli* O157:H7 in cantaloupe treated with 2% lactic acid at 55°C (Alvarado-Casillas et al., 2007), less than 3 logs for *S. enterica* and *E. coli* O157:H7 on inoculated strawberries by rinsing with solutions of lactic, acetic, and citric acid, applied individually or in combinations (Gurtler et al., 2012), and less than 1 log *S.* Typhimurium on tomatoes treated with 2% lactic acid (Ibarra-Sánchez et al., 2004). However, greater reductions were reported for organic acid treatment of *Yersinia enterocolitica* on lettuce (Escudero et al., 1999), *E. coli* O157:H7 on cantaloupes (Materon, 2003) and apples (Wright et al., 2000), *E. coli* (CDC1932) on iceberg lettuce (Vijayakumar and Wolf–Hall, 2002), mesophilic aerobes on lettuce (Nascimento et al., 2003), and *E. coli* O157:H7, *S.* Typhimurium, and *L. monocytogenes* on organic lettuce and apples (Park et al., 2011). Differences in results between studies on the same commodity using comparable treatments probably reflect differences in methodology.

Alkaline products

Sodium metasilicate (AvGard®XP) has been marketed by Danisco A/S (www.danisco.com) as an antimicrobial rinse to reduce human pathogen populations on processed beef and poultry, and this product has been approved by the FDA for produce washing (21CFR184.1769a). The antimicrobial activity of alkaline products such as sodium metasilicate and trisodium phosphate (TSP, AvGard®) is probably due to their high pH (11–12), which disrupts the cytoplasmic membrane (Mendonca et al., 1994; Sampathkumar et al., 2003). Population reductions of 1 to 3 logs have been reported with alkaline produce washes (Zhuang and Beuchat, 1996; Pao et al., 2000; Lukasik et al., 2003), although treatment of shredded lettuce with TSP was ineffective in killing *L. monocytogenes* (Zhang and Farber, 1996). TSP was highly effective in inactivating *E. coli* O157:H7 in biofilms but less effective against biofilms of *S. typhimurium* and *L. monocytogenes* (Somers et al., 1994). TSP also was effective in reducing levels of human norovirus surrogates on inoculated lettuce and Jalapeno peppers (Su and D'Souza, 2011). However, TSP has fallen out of favor because of phosphate disposal issues (http://meatupdate.csiro.au/new/Trisodium%20Phosphate.pdf).

Iodine

An iodine-based system (Isan®) for treatment of fruits and vegetables has been claimed to provide a high kill rate, require no pH adjustment, and be less corrosive than other sanitizers (www.ioteq.com; Klein and Morris, 2004). However, data demonstrating efficacy of this treatment against human pathogens have not been published. This system is approved for use in Australia and New Zealand.

Sanitizing agents for organic crops

Packers and processors of organic crops must conform to special USDA regulations regarding use of nonagricultural substances for washing and sanitizing processed organic products, if these products are to be labeled and marketed as organic. Approved antimicrobial agents, identified in the USDA National Organic Program List of Allowed and Prohibited Substances (7CFR205.605), include chlorine materials (calcium hypochlorite, sodium hypochlorite, ClO_2) and PAA for disinfecting and sanitizing food contact surfaces, and HP, ozone, and PAA for use in wash or rinse water in accordance with FDA limitations. Additional restrictions placed on chlorine materials state that "residual chlorine levels in the water shall not exceed the maximum residual disinfectant limit under the Safe Drinking Water Act." According to Suslow (2000), this is interpreted to be "10 ppm residual chlorine measured downstream of the wash step."

Anti-viral treatments

Relatively few studies have addressed the problem of pathogenic viral contamination of leafy vegetables and other produce items. Bae et al. (2011) compared the efficacy of various water-washing techniques and a commercial detergent product in removing human norovirus from inoculated iceberg lettuce; population reductions were less than 1 log. Similar results were reported when shredded iceberg lettuce, inoculated with murine norovirus 1, was washed with sodium hypochlorite and PAA solutions (Baert et al., 2009); the value of these treatments was in preventing cross-contamination rather than in reducing the pathogen load on lettuce. In a related study, Li et al. (2011) reported reductions no greater than 1 log in populations of norovirus surrogates on shredded iceberg lettuce from treatment with vaporized 2.52% HP and UV light. In contrast, Casteel et al. (2008) reported 1- to 2-log reductions in hepatitis A virus inoculated on leaves of head lettuce by treatment with sodium hypochlorite (pH7). Two-log reductions also were reported by Hirneisen et al. (2011) for norovirus surrogates inoculated on lettuce and green onion by treatment with ozonated wash water. Predmore and Li (2011) reported the enhanced removal of murine norovirus 1 from lettuce, cabbage, and raspberries by treatment with the combination of 200 ppm chlorine and 50 ppm sodium dodecyl sulfate, polysorbates or other surfactants. The combination treatments produced a 3-log reduction, compared to the 1-log reduction obtained by chlorine alone. The combination of levulinic acid and sodium dodecyl sulfate achieved 3- to 4-log reductions in human norovirus surrogates inoculated onto stainless steel (Cannon et al., 2012).

Surprisingly, a wash with acidic electrolyzed water increased the binding of human norovirus to inoculated romaine lettuce and raspberries. A prewash with acidic, alkaline, or neutral electrolyzed water prior to inoculation also increased the binding of the virus to these commodities. Thus, this technology offered no advantage over a simple water wash that reduced the virus population by less than 1 log (Tian et al., 2011). At this time, we must conclude that existing disinfection technologies are incapable of reducing pathogenic virus levels in contaminated produce sufficiently to assure safety, although some novel treatment combinations show promise.

Novel sanitizing agents

A number of experimental sanitizing wash treatments employ agents that have not been approved by the FDA for this purpose but may show promise for future consideration. Bacteriophages that target foodborne human pathogens have been evaluated for disinfection of cantaloupes and lettuce (Sharma et al., 2009), tomato, spinach, broccoli (Abuloadze et al., 2008), and produce-harvesting equipment (Patel et al., 2010). Various essential oils (Fisher et al., 2009; Obaidat and Frank, 2009; Erkman, 2010; Lu and Wu, 2010; Pérez–Conesa et al., 2011; Yossa et al., 2012), plant extractives (Moore et al., 2011; Jaroni and Ravishankar, 2012), and other compounds with antimicrobial properties (Molinos et al., 2005; Osman et al., 2006; Gopal et al., 2010) have been examined as potential sanitizing agents for produce. While some of these treatments have achieved pathogen reduction levels superior to those obtained with conventional sanitizing agents such as sodium hypochlorite, they seem incapable of producing a 5-log reduction. Additionally, some agents seem likely to impart an off-flavor to the treated produce, depending on the use level required. However, Lu and Wu (2010) reported reductions in the population of *S. enterica* serovars on inoculated grape tomatoes approaching 5 logs following treatment with thymol solutions, without affecting taste, aroma, or visual quality. Chen et al. (2012) obtained population reductions exceeding 5 logs for *Salmonella* on inoculated cantaloupe by application of chitosan coatings containing allyl isothiocyanate; the addition of nisin to this coating synergistically increased the antibacterial effect. Using vaporized ethyl pyruvate, Durak et al. (2012) obtained reductions in *E. coli* O157:H7 on inoculated green onions and baby spinach approaching or exceeding 5 logs; however, this treatment adversely affected the sensory attributes of green onions.

Physical antimicrobial treatments including ionizing radiation, pulsed UV, cold plasma, and microwave heating are discussed in other chapters of this book and will not be included in this chapter unless the treatments are carried out in conjunction with a sanitizing wash treatment.

Expectations for sanitizing agents

Numerous studies have demonstrated that use of chlorine and other sanitizing agents permitted by the FDA and EPA cannot achieve better than 1- to 3-log reductions in microbial populations attached to fresh produce. Some incremental improvements in efficacy may be possible. It is clear that washing and sanitizing treatments represent a hurdle, accomplishing some good by reducing the microbial load and reducing cross-contamination, but not enough to assure safety. When infectious doses are small (e.g., as few as 10 cells for *E. coli* O157:H7) (FDA, 2003), a 1-, 2-, or even 3-log reduction may not be enough to prevent significant numbers of people from getting sick.

Washing and sanitizing equipment

Types of washers

Many types of washers are available to the produce industry, designed according to the characteristics (shape, size, and fragility) and special requirements of specific commodities, for removal of soil, debris, and pesticide residues from harvested produce. Such equipment generally is not designed specifically to remove microorganisms attached to fruit and vegetable surfaces. Design criteria are reviewed by Saravacos and Kostaropoulos (2002). Types of commercial washers for fresh produce include flumes, dump tanks, flatbed and U-bed brush washers, reel washers, pressure washers, hydro-air agitation wash tanks, and immersion pipeline washers. Suppliers of such equipment are listed in buyers' guides published online at sites such as the United Fresh Produce Association (www.unitedfresh.org/programs/special) and FreshCUT magazine (www.freshcut.com). Major suppliers include FTNON USA, Inc. (www.ftnon.com/en/processes/washing/), Heinzen Manufacturing International (www.heinzen.com), Jarvis Products Corp. (www.jarvisproducts.com), Kronen Corp. (www.kronencorp.com), Lyco Manufacturing, Inc. (www.lycomfg.com), Sormac B.V. (www.sormac.nl), Turatti North America (www.turatti-us.com), Vanmark Equipment LLC (www.vanmarkequipment.com), and others.

Application of sanitizing agents

Sanitizing treatments can be applied by addition of disinfecting agents to wash water or as a post-washing spray or dip. However, re-use of lettuce wash water and increasing the product-to-water ratio will result in increased chemical and biological demand and greater declines in free and total chlorine levels (Luo, 2007). Addition of chlorine stabilizer T-128 (SmartWash) has been shown to enhance chlorine efficacy (Nou et al., 2011). Luo et al. (2011) reported that a minimum free chlorine concentration of 10 mg/liter was needed to prevent cross-contamination of uninoculated fresh-cut lettuce by transfer from lettuce inoculated with E. coli O157:H7.

Tomas–Callejas et al. (2012a) evaluated operating conditions for disinfection of tomatoes by chlorine dioxide. They concluded that ClO_2 could be effective in flume and spray-wash systems but not in dump tanks. Operating conditions for the electrolytic disinfection of process wash water from the fresh-cut industry are described in a recent paper by Gómez-López et al. (2013).

Biosafe Systems has developed a fog tunnel for the application of an activated peroxygen solution as an aerosol (http://thefogtunnel.com/about-the-fogtunnel/). Recent studies in which aerosolized organic acids were applied to lettuce and spinach that had been inoculated with human pathogens have yielded promising results (Choi et al., 2011; Ganesh et al., 2012). Huang et al. (2012) applied aerosolized antimicrobials including allyl isothiocyanate, hydrogen peroxide, acetic acid, and lactic acid as a post-washing treatment to baby lettuce, inoculated with E. coli O157:H7, and reported reductions approaching, and in some cases exceeding, 5 logs.

Efficacy of commercial washers

Studies conducted by Annous et al. (2001) with dip-inoculated apples demonstrated that the population of attached *E. coli* (strain K12) could be reduced by about 1 log (90%) by passage of the apples through a dump tank containing water with minimal agitation. However, further cleaning of the apples in a flat-bed brush washer (rotating brushes in a horizontal plane under spray) had little additional effect on the remaining *E. coli* population, irrespective of whether the washing agent used was water, a detergent, or a biocide (Table 17.1). Similar results were obtained in experiments with a U-bed brush washer (rotating brushes in U-shaped configuration causing tumbling action under spray) (Sapers, 2002). Survival of *E. coli* was attributed to attachment at inaccessible surfaces in the stem and blossom ends of the apples, infiltration within the latter region, and incorporation into resistant biofilms. Greater efficacy was obtained when the apples were washed by full immersion in a sanitizing solution with vigorous agitation (Sapers et al., 2002).

Garcia et al. (2006) identified the washing step in commercial apple cider production as a potential source of contamination, possibly because of excessive microbial build-up in dump tanks and improper cleaning and sanitizing of washing equipment. In a study of commercial washing practices in the Rio Grande River Valley of Texas, Gagliardi et al. (2003) reported little or no reduction and some significant increases in populations of coliforms and enterococci in cantaloupes cleaned in a "spray-propulsioned" immersion wash tank, hydrocooled, and then spray rinsed on a conveyor line. Much of the contamination was traced to the wash tank or hydrocooler, perhaps resulting from soil accumulation and chlorine depletion. Cantaloupes may be especially difficult to disinfect, even if fully immersed in the sanitizing solution, because of microbial attachment within inaccessible sites in the netting and stem

Table 17.1 Decontamination of Apples Inoculated with *E. coli* (Strain K12) with Sanitizing Washes Applied in a Flat-Bed Brush Washer[a]

Wash Treatment	Temp. (°C)	*E. coli* ($\log_{10}$ cfu/g)[b]		
		Before Dump Tank	After Dump Tank	After Brush Washer
Water	20	5.49 ± 0.09	4.92 ± 0.37	4.81 ± 0.26
	50	5.49 ± 0.09	5.03 ± 0.15	4.59 ± 0.08
200 ppm Cl_2	20	5.87 ± 0.07	5.45 ± 0.05	5.64 ± 0.23
8% Na_3PO_4	20	5.49 ± 0.09	5.02 ± 0.43	4.98 ± 0.02
	50	5.49 ± 0.09	5.02 ± 0.08	4.75 ± 0.45
1% acidic detergent	50	5.87 ± 0.07	5.49 ± 0.03	5.42 ± 0.50
5% H_2O_2	20	5.87 ± 0.07	5.54 ± 0.31	5.49 ± 0.10

[a]*From Annous, B.A. et al. (2001). Reprinted with permission from the* Journal of Food Protection. *Copyright held by the International Association for Food Protection, Des Moines Iowa.*
[b]*Mean of 4 determinations ± standard deviation.*

scar (Richards and Beuchat, 2004). Hassenberg et al. (2007) reported only a small decrease in the population of microorganisms in lettuce washed with ozonated water in a commercial lettuce-washing facility.

Produce washes for food service and home use

Many washing and sanitizing agents that can achieve pathogen reduction levels as great as 3 logs under commercial treatment conditions are not suitable for food service or home use because the users lack the technical skills, knowledge, and equipment to apply treatments safely and effectively. However, because of greater awareness by food-service managers and many consumers of the increasing risk of produce-associated foodborne illness, there has been an explosion of interest in produce washes that can be used safely by food-service workers or in the kitchen. A sampling of websites describing such products is provided in Table 17.2. Most of these fruit and vegetable washes contain mixtures of surfactants, and in some cases, are combined with chelating agents, buffers, and antioxidants. Product descriptions on their websites claim that the washes are capable of removing dirt, pesticide residues,

Table 17.2 Fruit and Vegetable Wash Products Available on the Internet[a]

Product Name	Composition	Website
Clean Greens	Surfactants, chelating agents, buffers, antioxidants	www.cleangreensinc.com
Earth Friendly Fruit & Vegetable Wash	Surfactant, citric acid	www.kalyx.com
Fit Fruit & Vegetable Wash	Citric acid, oleic acid, glycerol, ethyl alcohol, baking soda, potassium hydrate, distilled grapefruit oil	www.tryfit.com
Fruit & Vegetable Wash	Surfactant blend	www.vegiwash.com
Mom's Veggie Wash	Surfactant blend	www.veggiewash.com
PRO-SAN Fruit and Vegetable Wash	Unspecified	www.microcide.com
Sprout Spray Fruit & Veggie Wash	Unspecified	www.handypantry.com
Veggie Wash	Unspecified ingredients from citrus, corn, and coconut	www.citrusmagic.com
Veggi Wash Fruit Too	Plantaren, sucrose esters, cocoyl glutamate, trisodium citrate, glycerine	www.goodnessdirect.co.uk

[a]Listing of products in Table 17.2 does not constitute an endorsement by the author, and the products listed therein are not recommended over other products of a similar nature not identified by the author.

waxes, animal waste, and bacteria from fruit and vegetable surfaces. However, only three of these products, PRO-SAN (www.microcide.com), FIT (www.tryfit.com), and Victory (www.ecolab.com) are claimed to be bactericidal and appear to be suitable for the institutional market. Victory is a peroxyacetic acid-based antimicrobial produce wash designed specifically for the food-service industry.

There are few scientific studies validating the use of produce washes marketed for home use. Lukasik et al. (2003) obtained population reductions of 1 to 2 logs on strawberries inoculated with *E. coli* O157:H7, *S.* Montevideo, and several viruses, by treatment with Fit® and Healthy Harvest (a nonionic surfactant product). Much larger reductions were reported for tomatoes inoculated with *Salmonella* serotypes and washed with Fit, using a standardized method of testing (Harris et al., 2001). A reformulated Fit would employ the highly effective combination of levulinic acid and sodium dodecyl sulfate as an antibacterial agent (Zhao et al., 2009). Smith et al. (2003) reported reductions of only 1 log in the microbial load on lettuce by treatment with Victory. A study by Drury (2011) showed that Pro-San™, a biodegradable foodgrade sanitizer, was superior to bleach solution in inactivating *S. enterica* on Romaine lettuce leaves. Other studies have evaluated diluted vinegar and lemon juice (Vijayakumar and Wolf-Hall, 2002; Parnell and Harris, 2003; Nascimento et al., 2003; Kilonzo-Nthenge et al., 2006) as a produce wash for consumer use. Population reductions generally were no greater than 1 to 3 logs (Kilonzo-Nthenge et al., 2006). Neither the FDA (www.fda.gov/ForConsumers/ConsumerUpdates/ucm256215.htm) nor the USDA (www.fsis.usda.gov/Fact_Sheets/Does_Washing_Food_Promote_Fo od_Safety/index.asp) recommends that consumers wash fruits and vegetables with soap, detergents, or commercial produce washes. They *do* recommend, however, washing under cold running tap water to remove any lingering dirt and scrubbing with a clean produce brush if the produce has a firm surface. However, reductions in the bacterial load on cantaloupe obtained by washing with water and scrubbing were poor, only 70%, but not much worse than the 90% reduction obtained by dipping in 150 ppm sodium hypochlorite (Barak et al., 2003).

Small-scale systems for applying electrolyzed water, ozone, and chlorine dioxide are now being marketed (Table 17.3). Some of these may have application for food-service use, but treatment control and safety issues must be addressed before such equipment can be recommended. Venkitanarayanan et al. (1999b) reported that an electrolyzed water treatment was effective in inactivating foodborne pathogens on plastic cutting boards.

Efficacy of washing and sanitizing methods for problem commodities

Leafy vegetables

Leafy vegetables and herbs, including lettuce (romaine, iceberg, mesclun), spinach, parsley, and cilantro have been implicated in numerous outbreaks of food poisoning caused by *E. coli* O157:H7, *Salmonella*, Norwalk-like virus, hepatitis A, and

Table 17.3 Equipment for Small Scale Application of Commercial Sanitizing Agents for Fresh Produce

Sanitizing Agent	Product Name	Website
Electrolyzed water	ElectroCide System	www.electrolyzercorp.com
	Sterilox Food Safety Generator	www.puricore.com
Ozone	Applied Ozone Systems	www.appliedozone.com
	The Aqua Clean AQ-20	www.ozonesafefood.com
	ClearWater Tech Dissolved Ozone Systems	www.cwtozone.com
	Pacific Ozone	www.pacificozone.com
Chlorine Dioxide	Engelhard Aseptrol (sachets and tablets)	www.engelhard.com www.idspackaging.com/packaging/us
	Quiplabs MB 10 Tablet	www.quiplabs.com

other human pathogens in recent years (DeWaal and Barlow, 2002). Fresh-cut salad vegetables are subjected to triple-wash treatments and, since the inception of this industry, have been claimed to be safe. Yet bagged or fresh-cut spinach and romaine and iceberg lettuce have been associated with major outbreaks. In view of these outbreaks, we must question whether washing and sanitizing treatments are capable of disinfecting contaminated leafy vegetables.

Numerous studies have indicated the limited ability of chlorine to reduce populations of human pathogens on inoculated lettuce, typically reporting 1- to 2-log reductions, with regrowth during post-treatment storage (Beuchat and Brackett, 1990; Zhang and Farber, 1996; Beuchat et al., 1998; Delaquis et al., 2002; Lang et al., 2004; Beuchat et al., 2004; Hellström et al., 2006; Keskinen et al., 2009). Lang et al. (2004) reported somewhat greater log reductions with parsley. Francis and O'Beirne (1997) found greater regrowth of *L. innocua* at 8°C in shredded lettuce that had been dipped in 100 ppm chlorine than in water-dipped controls. They suggested that this might be due to a reduction in the population of indigenous microflora, thereby giving *L. innocua* a competitive advantage. Beuchat et al. (2004) reported large reductions in the concentration of free chlorine as the ratio of lettuce to solution and treatment time increased, the largest decreases occurring with shredded iceberg lettuce and iceberg pieces. These losses were attributed to release of chlorine-consuming tissue juices from the cut lettuce. Pirovani et al. (2001) modeled depletion of chlorine during washing of fresh-cut spinach at different chlorine concentrations, water-to-produce ratios, and treatment times, all of which affected extent of depletion. Reductions in the total microbial populations were only 2 to 3 logs. Zhang et al. (2009) demonstrated the impact of organic load on extent of cross-contamination by *E. coli* O157:H7 during washing of inoculated and non-inoculated lettuce leaves with peroxyacetic acid and other antimicrobial agents. Improvements in chlorine efficiency during fresh-cut processing can be obtained by washing whole lettuce leaves in sanitizer solution prior to cutting, thereby reducing the release of organic matter which would deplete chlorine and reduce microbial inactivation. This strategy is claimed to increase population reductions of attached human pathogens and background microflora by 1 log over that obtained by a post-cutting sanitizer wash; also, a reduction in cross-contamination is obtained (Nou and Luo, 2010).

Treatments combining acidified sodium hypochlorite with germicidal UV and mild heat (50°C) were reported to be highly effective in inactivating *E. coli* O157:H7 (> 5 logs) on surface-contaminated green onions but were less effective with dip-inoculated green onions and spot- and dip-inoculated baby spinach (<3 logs); these results were attributed to the limited efficacy of the combined treatments against infiltrated, attached, or protected pathogen cells (Durak et al., 2012).

Electrolyzed water has been evaluated as a sanitizing agent for lettuce with mixed results. This may be explained in part by differences in methodology used by investigators, especially the method of inoculation. Population reductions of *E. coli* O157:H7 and *Salmonella* spp. on inoculated head lettuce, treated with acidic electrolyzed water (200 ppm free available chlorine), were only about 1 log for dip inoculation, compared to reductions of about 2.5 logs for samples spot inoculated on the inner surface, and 4.5 logs for samples spot inoculated on the outer surface (Koseki et al., 2003). E.-J. Park et al. (2008) reported similarly large reductions in populations of *E. coli* O157:H7, *S.* Typhimurium, and *L. monocytogenes* on spot-inoculated spinach and lettuce leaves by treatment with acidic electrolyzed water for 3 min. However, the fact that the time interval between inoculation and treatment was brief (~1 h) may have contributed to the large reductions by precluding opportunities for bacterial internalization and biofilm formation that might otherwise have occurred if the samples had been stored prior to treatment. These results may provide some insight into the survival of human pathogens on contaminated lettuce leaves cleaned and sanitized with commercial washing equipment. Similar reductions were reported previously by C.-M. Park et al. (2001), with iceberg lettuce spot inoculated with *E. coli* O157:H7 and *L. monocytogenes*. Abadias et al. (2008) reported 1- to 2-log reductions with neutral electrolyzed water. However, Guentzel et al. (2008), also using neutral electrolyzed water, obtained 4 to 5 log reductions for spinach inoculated with. *E. coli*, *S. typhimurium,* and other bacterial species. Using "low concentration" electrolyzed water, Rahman et al. (2010) reported 1.64- to 2.80-log reductions in populations of *E. coli* O157:H7 and *L. monocytogenes* on inoculated spinach leaves. Other studies with inoculated romaine and iceberg lettuce reported 2-log reductions (Yang et al., 2003; Koseki et al., 2004).

Other FDA-approved alternatives to chlorine, including ozone, ClO_2, PAA, organic acids, and detergents have been evaluated for use as sanitizing agents for lettuce. Treatment of shredded or fresh-cut iceberg lettuce with ozone resulted in 1- to 2-log reductions in counts of the natural microflora (Kim et al., 1999; Garcia et al., 2003); ozone-chlorine combinations were more effective than the individual treatments (Garcia et al., 2003). Treatment of inoculated shredded lettuce with 1 to 5 ppm ClO_2 resulted in minimal reductions in *L. monocytogenes* population ($\leq$ 1 log) (Zhang and Farber, 1996), but treatment of inoculated romaine lettuce leaves with 5 to 40 ppm ClO_2 in combination with ultrasonification reduced *Salmonella* and *E. coli* O157:H7 populations by 2 to 3 logs (Huang et al., 2006). Treatment of fresh-cut iceberg lettuce with 3 ppm ClO_2 preserved product quality without formation of trihalomethanes (López-Gálvez et al., 2010). Application of 80 ppm PAA (Tsunami 100) to lettuce reduced the population of mesophilic aerobes and total coliforms by 1.85 and 1.44 logs, respectively (Nascimento et al., 2003), and reduced the population of *L. monocytogenes* on inoculated iceberg and romaine lettuce by 0.7 to 1.8 logs (Beuchat et al., 2004). Application of 40 ppm Tsunami 200 as an aerosol to spot-inoculated iceberg lettuce leaves reduced populations of *E. coli* O157:H7, *S.* Typhimurium, and *L. monocytogenes* by 2.2, 3.3, and 2.7 logs after 30-minute exposure (Oh et al., 2005). Mahmoud and Linton (2008) obtained 5-log reductions in populations of *E. coli* O157:H7 and *S. enterica* by treatment of inoculated lettuce leaves with ClO_2 gas; however, this treatment adversely affected leaf appearance.

Allende et al. (2009) reported reductions of more than 3 logs in populations of *E. coli* O157:H7 on inoculated cilantro by treatment with 0.1% acidified sodium chlorite. Similar results were obtained by Stopforth et al. (2008) using acidified sodium chlorite to disinfect mixed greens inoculated with *E. coli* O157:H7, *Salmonella*, and *L. monocytogenes*. However, treatment of inoculated spinach leaves with 15 mg/L acidified sodium chlorite reduced the population of *E. coli* O157:H7 by only 1 to 2 logs (Nei et al., 2009).

Various combinations of organic acids with 1% hydrogen peroxide and mild heat (40°C) were superior to 200 ppm chlorinated water in inactivating *E. coli* O157:H7 on baby spinach, but population reductions were no greater than 2.7 logs (Huang and Chen, 2011). Combining organic acid treatment with ultrasound gave population reductions in inoculated lettuce leaves approaching 3 logs for *E. coli* O157:H7, *S.* Typhimurium, and *L. monocytogenes*, representing a 1-log gain over the treatments applied individually (Sagong et al., 2011).

It is apparent that the efficacy of commercially available, FDA-approved sanitizing agents against human pathogens on contaminated lettuce and other leafy vegetables is limited to population reductions of 1 to 3 logs at best. Presumably, this is a consequence of the strong attachment of contaminants to inaccessible sites on the leaf surface and cut edges, internalization of bacterial cells within the leaf (Solomon et al., 2002), and incorporation of cells within resistant biofilms (Carmichael et al., 1999) so that contact between the human pathogens and sanitizing agent is insufficient for inactivation (see Chapter 2). Additionally, depletion of chlorine and other sanitizing agents by reaction with tissue juices at cut edges of leaves during washing (Pirovani et al., 2001; Beuchat et al., 2004) may contribute to their limited efficacy. However, sanitizing agents such as chlorine and Tsunami may be beneficial in preventing cross-contamination of produce during processing (López-Gálvez et al., 2009, 2010). While addition of ClO_2 to wash water used in fresh-cut processing of red chard prevented cross-contamination by *E. coli* O157:H7 on inoculated leaves, it did not prevent cross-contamination from *Salmonella* (Tomas-Callejas et al., 2012b).

Leafy greens are also vulnerable to contamination by parasites. Studies by Ortega et al. (2008) have shown that gaseous ClO_2 is capable of reducing loads of

Cryptosporidium parvum on inoculated lettuce and basil leaves by 2.6 and 3.3 logs, respectively; however, *Cyclospora cayetanensis* oocysts were not affected by this treatment.

Tomatoes

Tomatoes have a history of association with outbreaks of *Salmonella* food poisoning (Cummings et al., 2001; CDC, 2002, 2005, 2006). Research conducted in the 1990s demonstrated that *Salmonella* could be inactivated on the unbroken skin of tomatoes by washing with chlorinated water (100 ppm free Cl_2) but could survive if attached in the stem scar, in core tissue, or within growth cracks (Wei et al., 1995; Zhuang et al., 1995). Furthermore, internalization of bacteria by infiltration through the stem scar into the core could be driven by a temperature differential between the tomato fruit (warm) and the wash water (cold) or by a hydrostatic pressure differential depending on the depth of immersion of tomatoes in a wash tank (Bartz and Showalter, 1981; Bartz, 1982). Guo et al. (2002) demonstrated infiltration of *Salmonella* through the stem scar when tomatoes were placed stem-scar-down in contact with inoculated moist soil. However, this condition would not be likely to occur under field conditions unless accidentally detached or dropped fruits were subsequently harvested from the ground. Xia et al. (2012) showed that tomato variety and interactions between varieties and post-stem removal times, and between temperature differentials and post-stem removal times, had significant effects on populations of internalized *S. enterica*. Various sanitizing agents and methods of application have been employed in attempts to improve the disinfection of tomatoes inoculated with *Salmonella*. However, population reductions generally were in the range of 1 to 3 logs for a spray application of 2000 ppm chlorine (Beuchat et al., 1998; Chaidez et al., 2007), a dip treatment with 15% trisodium phosphate (Zhuang and Beuchat, 1996), a dip or spray treatment with lactic acid solution at 55°C (Ibarra–Sanchez et al., 2004), a dip treatment with 5% HP at 60°C (Sapers and Jones, 2006), treatment with 200 ppm sodium hypochlorite, 1200 ppm acidified sodium chlorite, 87 ppm PAA, ClO_2 gas, or a combination of these treatments (Yuk et al., 2005), exposure to 5 ppm ozone and UV-C (Bermúdez-Aguirre and Barbosa-Cánovas, 2013), in-package ozonation (Fan et al., 2012), and treatment of air-dried tomatoes with 5 ppm aqueous ClO_2 (Pao et al., 2007). However, treatment with ClO_2 gas for 2 h was highly effective in reducing populations of *S.* Typhimurium inoculated into wounds on tomato surfaces (Mahovic et al., 2009). Treatments in which the tomatoes were spot inoculated on smooth skin away from the stem-scar area (Harris et al., 2001; Venkitanarayanan et al., 2002; Bari et al., 2003, Park et al., 2009) tended to yield greater population reductions (4–7.5 logs) than when the tomatoes were spot inoculated at the stem scar or dip inoculated (Raiden et al., 2003; Ibarra-Sanchez et al., 2004; Yuk et al., 2005; Pao et al., 2007). Jin et al. (2012) reported that antimicrobial coatings comprising chitosan and three organic acids reduced the population of *Salmonella* serovars inoculated on tomato stem scars by as much as 6 logs; addition of allyl isothiocyanate to this coating did not significantly increase the treatment efficacy.

Some of the larger population reductions reported in studies where the interval between inoculation and treatment was brief (< 60 min) may be indicative of treatment efficacy when contamination occurs on the packing or processing line; for example, in dump tanks, hydrocoolers, or flumes. However, such treatments might be substantially less effective when the interval between contamination and treatment is longer (days), as would be the case with preharvest contamination where pathogens might be protected by attachment to protected sites (growth cracks, punctures, or other surface irregularities) or by biofilm formation. In experiments with dip-inoculated tomatoes, Sapers and Jones (2006) obtained 1.8- and 2.6-log reductions in the *Salmonella* population when the inoculated tomatoes were held at 20°C for one hour prior to treatment with 150 ppm chlorine at 20°C or 5% hydrogen peroxide at 60°C, respectively, but reductions were less than 1.5 log when the tomatoes were held for 24 hours at 20°C prior to treatment (Table 17.4). Similar results were obtained when tomatoes were spot-inoculated with *S. enterica* and air dried for 24 h prior to treatment with aqueous ClO_2 (Pao et al., 2007).

We can conclude from these studies that the efficacy of washing and sanitizing treatments in decontaminating tomatoes is greatly limited when the site of contamination is in punctures, cracks, or the stem-scar area, and if contamination occurred preharvest. More emphasis must be placed on avoidance of contamination. The FDA has initiated a collaborative effort to "identify practices or conditions that potentially lead to product contamination" (FDA, 2007b).

Table 17.4 Efficacy of Wash Treatments in Reducing Population of *Salmonella* on Dip-Inoculated Tomatoes[a,b]

Treatment	Storage at 20 °C (h)	Population Reduction ($\log_{10}$ cfu/g)[c]
Rinsed control at 20°C for 2 min	1	1.11 ± 0.18 C
	24	0.40 ± 0.32 D
150 ppm Cl_2 at 20°C for 2 min	1	1.78 ± 0.49 B
	24	1.34 ± 0.39 BC
5% H_2O_2 at 60°C for 2 min	1	2.59 ± 0.74 A
	24	1.45 ± 0.33 BC
H_2O at 60°C for 2 min	1	1.75 ± 0.11 B
	24	0.99 ± 1.00 CD

[a]*From Sapers and Jones (2006). Reprinted with permission from the* Journal of Food Science. *Copyright held by the Institute of Food Technologists, Chicago, IL.*
[b]*Inoculum prepared from cocktail containing S. Montevideo (G4639) and S. Baildon (61–99); mean inoculum population was 10.13 ± 0.04 for the treatment comparisons.*
[c]*Mean population reductions ± standard deviations based on corresponding control means for 2 or 3 independent experiments, each with duplicate trials; control means were 5.61 ± 0.27 and 5.42 ± 0.26 $\log_{10}$ cfu/g for the 0-time and 24 h treatment time comparisons, respectively. Means not followed by the same letters are significantly different (P < 0.05).*

Cantaloupe

Because of the association of cantaloupes with large *Salmonella* outbreaks (CDC, 1991, 2002, 2008, 2011a, 2012a; DeWaal and Barlow, 2002), much attention has been given to the efficacy of washing and sanitizing treatments in reducing pathogen populations on cantaloupe surfaces. Cantaloupes were also the source of a large outbreak of Listeriosis (CDC, 2011b). Cantaloupe is inherently difficult to disinfect because of the roughness of its surface, which provides many protected sites for attachment (Wang et al., 2009). It is believed that contamination of fresh-cut melon results from transfer of human pathogens on the surface to the interior flesh during cutting (Ukuku and Sapers, 2001; Ukuku and Fett, 2002; Vadlamudi et al., 2012). As with lettuce and tomatoes, efforts to decontaminate whole cantaloupes by application of sanitizers have achieved limited success (Akins et al., 2008; Fan et al., 2009). Typically, 1–3-log reductions have been reported for cantaloupes inoculated with *Salmonella* spp. and treated with 150 to 200 ppm chlorine (Park and Beuchat, 1999; Barak et al., 2003), 80 ppm PAA (Park and Beuchat, 1999), 1 to 5% HP (Park and Beuchat, 1999; Ukuku and Sapers, 2001; Ukuku et al., 2004), 10,000 ppm gaseous ozone (Selma et al., 2008a and 2008b), and 2% lactic acid or 30 ppm aqueous ozone (Vadlamudi et al., 2012). Other studies have reported similar reductions in the populations of the native microflora (Sapers et al., 2001; Ukuku et al., 2001) and in populations of a nonpathogenic *E. coli* (Ukuku et al., 2001) and *L. monocytogenes* (Ukuku and Fett, 2002) on inoculated cantaloupes washed with 1000 ppm chlorine and 5% HP. However, larger reductions in *Salmonella* were reported with 850 ppm acidified sodium chlorite (Park and Beuchat, 1999) and in *E. coli* O157:H7 and *L. monocytogenes* with 80 ppm PAA, 100 and 200 ppm chlorinated trisodium phosphate, 3 and 5 ppm ClO_2, and 3 ppm ozone (Rodgers et al., 2004). Materon (2003) also reported large population reductions for *E. coli* O157:H7 on inoculated cantaloupes treated with combinations of 200 ppm chlorine + 1.5% lactic acid or 1.5% lactic acid + 1.5% HP. Studies by Shen et al. (2012b) have demonstrated that addition of washing aid T-128 to chlorinated wash solutions and washing with brushing enhanced the inactivation of *S. enterica*, *E. coli* O157:H7, and *Pseudomonas fluorescens* biofilms on cantaloupe rinds.

It is not clear why population reductions were so much greater in some studies than in others. However, studies with *Salmonella* and a nonpathogenic *E. coli* showed that the efficacy of sanitizer treatments for cantaloupe disinfection decreased as the interval between inoculation and treatment increased from 24 to 72 hours (Ukuku and Sapers, 2001; Ukuku et al., 2001). Differences in methodology for melon inoculation, storage, treatment application, and recovery and enumeration of the targeted pathogen can all influence the experimental results (Beuchat and Scouten, 2004; Ukuku and Fett, 2004; Annous et al., 2005). Demonstration of rapid biofilm formation by *Salmonella* on the surface of spot inoculated cantaloupes may explain storage effects on the efficacy of disinfection treatments discussed above (Annous et al., 2005).

The efficacy of sanitizer treatments in inactivating *Salmonella* on cantaloupe surfaces can be enhanced by application at elevated temperatures; for example,

2% lactic acid at 55 to 60°C (Alvarado-Casillas et al., 2007) and 5% HP at 70°C (Ukuku et al., 2004; Ukuku, 2006). Hot-water surface pasteurization of cantaloupes, inoculated with *Salmonella*, achieved population reductions approaching 5 logs with no detrimental effects to melon quality (Ukuku et al., 2004; Annous et al., 2004; Solomon et al., 2006; Ukuku, 2006; Fan, 2008). The rind apparently has sufficient insulating ability to protect the flesh from thermal injury.

Several studies have examined sanitizer treatments to reduce microbial loads on the fresh-cut melon pieces. PAA, HP, and nisin + EDTA were more effective than ClO_2 or chlorine (sodium hypochlorite) treatments in reducing microbial growth on fresh-cut "Galia" melon during storage and delaying softness (Silveira et al., 2008, 2010). Fan et al. (2006) combined hot-water surface pasteurization of whole cantaloupes with low-dose gamma irradiation of the fresh-cut fruit to reduce the population of native microflora while maintaining product quality.

In a study of cantaloupe contamination conducted at four packinghouses located in the Rio Grande Valley of South Texas, Materon (2003) reported that washing resulted in significant reductions (2–3 logs) in populations of aerobic bacteria, total coliforms, and fecal coliforms. In contrast, Gagliardi et al. (2003) reported elevations in bacterial counts of cantaloupes from packinghouses in the Rio Grande Valley, sampled before and after washing and packing. Contamination during processing was traced to a primary wash tank or hydrocooler and may be a reflection of poor water quality, the organic load on melons, and chlorine depletion. It is evident that implementation of Good Agricultural and Manufacturing Practices as well as improvements in disinfection technology are needed to correct such deficiencies.

Apples

Because of the history of *E. coli* O157:H7 outbreaks associated with unpasteurized apple cider, there have been numerous studies of the efficacy of washing and sanitizing agents in disinfecting contaminated apples (Beuchat et al., 1998; Sapers et al., 1999; Wisniewsky et al., 2000; Wright et al., 2000; Achen and Yousef, 2001; Kenney and Beuchat, 2002; Venkitanarayanan et al., 2002; Sapers et al., 2002; Parnell and Harris, 2003; Rodgers et al., 2004). Population reductions on apples inoculated with *E. coli* O157:H7 were generally in the range of 1 to 3 logs for most disinfection treatments. In studies of the efficacy of disinfection treatments for apples, Sapers et al. (2002) identified rapid attachment of bacterial cells to apple surfaces, attachment to inaccessible sites in the stem and calyx areas, and attachment and growth in skin punctures as factors limiting efficacy. The attachment of *E. coli* O157:H7 to surface and internal structures of apples, including discontinuities in the waxy cuticle, lenticels, punctures, the floral tube, seeds, cartilaginous pericarp, and internal trichomes was demonstrated by confocal scanning laser microscopy (Burnett et al., 2000). Rubbing dip-inoculated apples was reported to seal attached cells of *E. coli* O157:H7 within the waxy cutin platelets, thereby protecting them from disinfection (Kenney et al., 2001). Treatment efficacy was poor when apples were treated by spraying in a commercial brush washer (Annous et al., 2001), and efficacy could be improved by immersing apples in the sanitizing solution with good agitation, by heating sanitizer solutions, and by removal of calyx and stem tissue (Sapers et al., 2002). Other studies have demonstrated the localized concentration and survival of microbial contaminants in punctures and the stem and calyx regions of apples (Riordan et al., 2001; Fatemi and Knabel, 2006; Fatemi et al., 2006). Fleischman et al. (2001) and Sapers et al. (2002) demonstrated the efficacy of hot water in disinfecting apple surfaces; however, such treatments resulted in discoloration and softening of surface tissues. Five-log pathogen reductions are now mandated for apple cider (and other fresh juices), but for cider, these reductions must be obtained entirely on the juice, not partly on the whole fruit and the remainder on the juice (FDA, 2001a). Such reductions can be achieved by heat pasteurization or UV treatment of cider.

Although the cider safety issue is no longer dependent on fruit disinfection, potential safety problems remain with fresh-cut apple slices that are vulnerable to contamination with human pathogens in the processing environment; such products may have an extended shelf-life, providing an opportunity for outgrowth of such contaminants. Detection of *L. monocytogenes* in fresh-cut apples resulted in a recall (FDA, 2001b). None of the sanitizing treatments currently available for whole or cut apples can consistently achieve a 5-log reduction in pathogen population without altering product identity. However, conditions allowing contamination of the product in the processing plant environment can be corrected, and survival and growth of human pathogens in the product can be suppressed (Karaibrahimoglu et al., 2004; Pilizota and Sapers, 2004).

Seeds, sprouts, and nuts

Sprouted seeds have a long history of association with outbreaks of foodborne illness from *Salmonella* (FDA, 1999a; CDC, 2009a, 2010, 2011a, 2011c) and Shiga toxin-producing *E. coli* (CDC, 2011d, 2012b). Conditions employed in the production of such products favor the survival and proliferation of human pathogens that are present as contaminants of seeds. Yet interventions for the disinfection of seeds or sprouts are limited in efficacy or detrimental to product yield or quality. In response to an increasing frequency of outbreaks, the FDA issued recommendations in 1999 for disinfecting seeds for sprouting, based on the best available information provided by researchers in the field (FDA, 1999b). Their method calls for soaking the seeds in 20,000 ppm calcium hypochlorite. Yet, even with this treatment, some surviving *E. coli* O157:H7 cells can be detected by enrichment. This may be due in part to differences among seed types in surface roughness, which is negatively correlated with microbial removal (Fransisca and Feng, 2012).

Publications prior to 1999 regarding seed and sprout disinfection are reviewed in the report of the National Advisory Committee on Microbiological Criteria for Foods (FDA, 1999a). In the more recent literature, Kumar et al. (2006) reported that a soak for at least 8 h in a stabilized oxychloro-based sanitizer solution was effective in reducing populations of *E. coli* O157:H7 or *S. enterica* serovars on naturally contaminated mung beans, alfalfa seeds, and other seeds below the limit for detection even with enrichment. However, the treatment failed to inactivate *Salmonella* on damaged mung beans (Hora et al., 2007). A comparison of treatments with gaseous ClO_2, ozone gas, and e-beam irradiation demonstrated that 4- to 5-log reductions in the populations of *S. enterica* and *E. coli* O157:H7 on inoculated tomato, cantaloupe, and lettuce seeds could be obtained (Trinetta et al., 2011). Application of slightly acidic electrolyzed water to inoculated mung bean seeds reduced populations of *E. coli* O15:H7 and *S. enteritidis* by less than 2 logs; reductions on sprouts approached 4 logs (Zhang et al., 2011). It is not clear whether this technology could be used in conjunction with seed disinfection with 20,000 calcium hypochlorite. A related study using the slightly acidic electrolyzed water technology claimed reductions between 2 and 3 logs in populations of *E. coli* and *Salmonella* on a number of ready to eat vegetables and sprouts (Issa-Zacharia et al., 2011). A study by Singla et al. (2011) demonstrated that the combination of malic acid and ozone reduced the population of *Shigella* on radish sprouts by 4.4 logs and on moong bean sprouts by 4.8 logs. An antimicrobial extract of roselle was effective in reducing populations of *S. enterica* on alfalfa sprouts (Jaroni and Ravishankar, 2012).

In common with seeds used for sprout production, tree nuts have a history of human pathogen contamination, resulting in outbreaks of foodborne illness; additionally, contaminated nuts are difficult to disinfect. Salmonellosis outbreaks have been linked to pistachio nuts (CDC, 2009b), almonds (Isaacs et al., 2005; CDC, 2004), and Turkish pine nuts (CDC, 2011e). An *E. coli* O157:H7 outbreak linked to hazelnuts occurred in April, 2011 (CDC, 2011f). Also in April, 2011, an *E. coli* O157:H7 outbreak, associated with shelled walnuts that had been produced in the U.S. and exported to Canada, triggering a Health Hazard Alert and recall from the Canadian Food Inspection Agency (CFIA). A second alert and recall was required in September following detection of this pathogen in raw walnut samples (CFIA, 2011).

Efforts to disinfect raw tree nuts have focused on alternative means of applying thermal energy (see review by Pan et al., 2012). Infrared heating to 104°C, followed by holding for 60 minutes at the elevated temperature, reduced populations of *S.* Enteritidis on inoculated raw almonds by 5.3 logs (Brandl et al., 2008; Yang et al., 2010; Bingol et al., 2011). Du et al. (2010) obtained reductions approaching or exceeding 5 logs for populations of *S.* Enteritidis PT 30 and *S.* Senftenberg 775W by treatment of inoculated almonds with hot oil at 127°C. Harris et al. (2012) submerged almond kernels, inoculated with 5 log cfu/g *S.* Enteritidis PT 30 or *S.* Senftenberg 775W, in hot water at 88°C for 2 min and was unable to recover either *Salmonella* by enrichment. The efficacy of these treatments on other tree nuts has not been reported.

Conclusions

In the absence of a final "kill" step to eliminate spoilage organisms and human pathogens from fresh and fresh-cut fruits and vegetables, the produce industry has depended on avoidance of pre- and postharvest contamination, effective plant sanitation, and use of presumably effective washing and sanitizing agents and methods of sanitizer application to provide consumers with safe products. Yet, in spite of such measures, outbreaks remain a problem. Conventional washing and sanitizing treatments can achieve reductions in the microbial load on fresh produce of 1- to 3-log reductions, good enough to improve quality, control cross-contamination, and reduce the risk of foodborne illness. Yet because the efficacy of conventional disinfection treatments is limited by pathogen survival in inaccessible attachment sites and biofilms, such measures cannot assure safety.

Incremental improvements in washing technology, employing some of the novel agents describe above, can help but will not solve the problem. More potent, penetrating treatments, such as hot water surface pasteurization (where applicable), electron beam irradiation, and exposure to gaseous antimicrobial agents, are required. Since there may be commodities that cannot be irradiated or subjected to other high-energy treatments without loss of quality, other approaches are required as well. This shifts the burden of solving the produce contamination problem from the area of postharvest interventions to preharvest interventions; in other words, by the avoidance of contamination by human pathogens. Unless this fact is fully appreciated, and efforts to reduce produce contamination are comprehensive, not just restricted to washing and sanitizing technology, produce safety cannot be assured.

References

Abadias, M., Usall, J., Oliveira, M., et al., 2008. Efficacy of neutral electrolyzed water (NEW) for reducing microbial contamination on minimally processed vegetables. Int. J. Food Microbiol. 123, 151–158.

Abuladze, T., Li, M., Menetrez, M.Y., et al., 2008. Bacteriophages reduce experimental contamination of hard surfaces, tomato, spinach, broccoli, and ground beef by *Escherichia coli* O157:H7. Appl. Environ. Microbiol. 74, 6230–6238.

Achen, M., Yousef, A.E., 2001. Efficacy of ozone against *Escherichia coli* O157:H7 on apples. J. Food Sci. 66, 1380–1384.

Akbas, M.Y., Olmez, H., 2007. Inactivation of *Escherichia coli* and *Listeria monocytogenes* on iceberg lettuce by dip wash treatments with organic acids. Lett. Appl. Microbiol. 44, 619–624.

Akins, E.D., Harrison, M.A., Hurst, W., 2008. Washing practices on the microflora on Georgia-grown cantaloupes. J. Food Prot. 71, 46–51.

Allende, A., McEvoy, J., Tao, Y., et al., 2009. Antimicrobial effect of acidified sodium chlorite, sodium hypochlorite, and citric acid on *Escherichia coli* O1567:H7 and natural microflora of fresh-cut cilantro. Food Control 20, 230–234.

Allwood, P.B., Malik, Y.S., Hedberg, C.W., et al., 2004. Effect of temperature and sanitizers on the survival of feline calicivirus, *Escherichia coli*, and F-specific coliphage MS2 on leafy salad vegetables. J. Food Prot. 67, 1451–1456.

Alvarado-Casillas, S., Ibarra-Sánchez, S., Rodríguez-García, O., et al., 2007. Comparison of rinsing and sanitizing procedures for reducing bacterial pathogens on fresh cantaloupes and bell peppers. J. Food Prot. 70, 655–660.

Annous, B.A., Burke, A., Sites, J.E., 2004. Surface pasteurization of whole fresh cantaloupes inoculated with *Salmonella* Poona or *Escherichia coli*. J. Food Prot. 67, 1876–1885.

Annous, B.A., Sapers, G.M., Jones, D.M., et al., 2005. Improved recovery procedure for evaluation of sanitizer efficacy in disinfecting contaminated cantaloupes. J. Food Sci. 70, M242–M247.

Annous, B.A., Sapers, G.M., Mattrazzo, A.M., et al., 2001. Efficacy of washing with a commercial flat-bed brush washer, using conventional and experimental washing agents, in reducing populations of *Escherichia coli* on artificially inoculated apples. J. Food Prot. 64, 159–163.

Annous, B.A., Solomon, E.B., Cooke, P.H., et al., 2005. Biofilm formation by *Salmonella* spp. on cantaloupe melons. J. Food Saf. 25, 276–287.

Anon., 2001. Methods to reduce/eliminate pathogens from fresh and fresh-cut produce. Table V-1, Chapter V in Analysis and evaluation of preventive control measures for the control and reduction/elimination of microbial hazards on fresh and fresh-cut produce, U.S. September 30, 2001. Food Drug Adm. Center Food Saf. Appl. Nutr. www.cfsan.fda.gov/~comm/ift3-5.html.

Bae, J.Y., Lee, J.S., Lee, M.H., et al., 2011. Effect of wash treatments on reducing human norovirus on iceberg lettuce and perilla leaf. J. Food Prot. 74, 1908–1911.

Baert, L., Vandekinderen, I., Devlieghere, F., et al., 2009. Efficacy of sodium hypochlorite and peroxyacetic acid to reduce murine norovirus 1, B40-8, *Listeria monocytogenes*, and *Escherichia coli* O157:H7 on shredded iceberg lettuce and in residual wash water. J. Food Prot. 72, 1047–1054.

Barak, J.D., Chue, B., Mills, D.C., 2003. Recovery of surface bacteria from and surface sanitization of cantaloupes. J. Food Prot. 66, 1805–1810.

Bari, M.L., Sabina, Y., Isobe, S., et al., 2003. Effectiveness of electrolyzed acidic water in killing *Escherichia coli* O157:H7, *Salmonella* enteritidis, and *Listeria monocytogenes* on the surface of tomatoes. J. Food Prot. 66, 542–548.

Bartz, J.A., 1982. Infiltration of tomatoes immersed at different temperatures to different depths in suspensions of *Erwinia carotovora* subsp. *carotovora*. Plant Dis. 66, 302–306.

Bartz, J.A., Showalter, R.K., 1981. Infiltration of tomatoes by aqueous bacterial suspensions. Phytopathology 71, 515–518.

Bermúdez-Aguirre, D., Barbosa-Cánovas, G.V., 2013. Disinfection of selected vegetables under nonthermal treatments: chlorine, acid citric, ultraviolet light and ozone. Food Control 29, 82–90.

Beuchat, L.R., 1998. Surface decontamination of fruits and vegetables eaten raw: A review Food Safety Unit. World Health Organization. WHO/FSF/FOS/98.2 www.who.int/foods afety/publications/fs_management/en/surface_decon.pdf.

Beuchat, L.R., Brackett, R.E., 1990. Survival and growth of *Listeria monocytogenes* on lettuce influenced by shredding, chlorine treatment, modified atmosphere packaging and temperature. J. Food Sci. 55, 755–758; 870.

Beuchat, L.R., Nail, B.V., Adler, B.B., et al., 1998. Efficacy of spray application of chlorinated water in killing pathogenic bacteria on raw apples, tomatoes, and lettuce. J. Food Prot. 61, 1305–1311.

Beuchat, L.R., Adler, N.B., Lang, M.M., 2004. Efficacy of chlorine and peroxyacetic acid sanitizer in killing *Listeria monocytogenes* on iceberg and Romaine lettuce using simulated commercial processing conditions. J. Food Prot. 67, 1238–1242.

Beuchat, L.R., Scouten, A.J., 2004. Factors affecting survival, growth, and retrieval of *Salmonella* Poona on intact and wounded cantaloupe rind and stem scar tissue. Food Microbiol. 21, 683–694.

Bialka, K.L., Demirci, A., 2007a. Efficacy of aqueous ozone for the decontamination of *Escherichia coli* O157:H7 and *Salmonella* on raspberries and strawberries. J. Food Prot. 70, 1088–1092.

Bialka, K.L., Demirci, A., 2007b. Decontamination of *Escherichia coli* O157:H7 and *Salmonella enterica* on blueberries using ozone and pulsed UV-light. J. Food Sci. 72, M391–M396.

Bingol, G., Yang, J., Brandl, M., et al., 2011. Infrared pasteurization of raw almonds. J. Food Eng. 104, 387–393.

Blackburn, B.G., Mazurek, J.M., Hlavsa, M., et al., 2006. Cryptosporidiosis associated with ozonated apple cider. Emerg. Infect. Dis. 12 April. Available from www.cdc.gov/ncidod/EID/vol12no04/05-0796.htm.

Block, S.S., 1991. Peroxygen compounds. Chapter 9. In: Block, S.S. (Ed.), Disinfection, sterilization, and preservation, fourth ed. Lea & Febiger, Philadelphia.

Brackett, R.E., 1987. Antimicrobial effect of chlorine on *Listeria monocytogenes*. J. Food Prot. 50, 999–1003.

Brandl, M., Pan, Z., Huynh, S., et al., 2008. Reduction of *Salmonella* Enteritidis population sizes on almond kernels with infrared heat. J. Food Prot. 71, 897–902.

Brown, A.L., Brooks, J.C., Karunasena, E., et al., 2011. Inhibition of *Escherichia coli* O157:H7 and *Clostridium sporogenes* in spinach packaged in modified atmospheres after treatment combined with chlorine and lactic acid bacteria. J. Food Sci. 76, M427–M432.

Burnett, S.L., Chen, J., Beuchat, L.R., 2000. Attachment of *Escherichia coli* O157:H7 to the surfaces and internal structures of apples as detected by confocal scanning laser microscopy. Appl. Environ. Microbiol. 66, 4679–4687.

Caldwell, K.N., Adler, B.B., Anderson, G.L., et al., 2003. Ingestion of *Salmonella enterica* Serotype Poona by a free-living nematode, *Caenorhabditis elegans*, and protection against inactivation by produce sanitizers. Appl. Environ. Microbiol. 69, 4103–4110.

Cannon, J.L., Aydin, A.A., Mann, A.N., et al., 2012. Efficacy of levulinic acid plus sodium dodecyl sulfate-based sanitizer on inactivation of human norovirus surrogates. J. Food Prot. 75, 1532–1535.

Carmichael, I., Harper, I.S., Coventry, M.J., et al., 1999. Bacterial colonization and biofilm development on minimally processed vegetables. J. Appl. Microbiol. Symp. (Suppl. 85), 45S–51S.

Casteel, M.J., Schmidt, C.E., Sobsey, M.D., 2008. Chlorine disinfection of produce to inactivate hepatitis A virus and coliphage MS2. Int. J. Food Microbiol. 125, 267–273.

CDC, 1991. Multistate outbreak of *Salmonella* Poona infections—United States and Canada. MMWR 40, 549–552.

CDC, 2002. Outbreak of *Salmonella* serotype Javiana infections—Orlando, Florida, June 2002. MMWR 51, 683–684.

CDC, 2002a. Multistate outbreaks of *Salmonella* serotype Poona infections associated with eating cantaloupe from Mexico—United States and Canada, 2000–2002. MMWR 51, 1044–1047.

CDC, 2004. Outbreak of *Salmonella* serotype Enteritidis infections associated with raw almonds—United States and Canada, 2003–2004. MMWR 53, 484–487.

CDC, 2005. Outbreaks of *Salmonella* infections associated with eating Roma tomatoes— United States and Canada, 2004. MMWR 54, 325–328.

CDC, 2006. Salmonellosis—Outbreak investigation, October 2006. Division of Bacterial and Mycotic Diseases, Centers for Disease Control and Prevention Press. Release Nov. 23, 2006. www.cdc.gov/ncidod/dbmd/diseaseinfo/salmonellosis2006/outbreak_notice.htm.

CDC, 2008. Investigation of outbreak of infections caused by Salmonella Litchfield. Final update. Posted April 2, 2008.

CDC, 2009a. Investigation of an outbreak of *Salmonella* Saintpaul infections linked to raw alfalfa sprouts. Final update. May 8, 2009.

CDC, 2009b. *Salmonella* in pistachio nuts, 2009. Final update. April 14, 2009.

CDC, 2010. Investigation update: Multistate outbreak of human *Salmonella* Newport infections linked to raw alfalfa sprouts. Final update. June 29, 2010.

CDC, 2011a. Investigation Update: Multistate outbreak of *Salmonella* Panama infections linked to cantaloupe. Final update. June 23, 2011.

CDC, 2011b. Multistate outbreak of Listeriosis linked to whole cantaloupes from Jensen Farms, Colorado. Final update. December 8, 2011.

CDC, 2011c. Investigation update: Multistate outbreak of human *Salmonella* I 4,[5],12:i:- infections linked to alfalfa sprouts. Final update. February 10, 2011.

CDC, 2011d. Investigation update: Outbreak of Shiga toxin-producing *E. coli* O104 (STEC O104:H4) infections associated with travel to Germany. Final update. July 8, 2011.

CDC, 2011e. Multistate outbreak of human *Salmonella* Enteritidis infections linked to Turkish pine nuts. Final update. November 17, 2011.

CDC, 2011f. Investigation update: Multistate outbreak of *E. coli* O157:H7 infections associated with in-shell hazelnuts. Final update. April 7, 2011.

CDC, 2012a. Multistate outbreak of *Salmonella* Typhimurium and *Salmonella* Newport infections linked to cantaloupe. Posted September 13, 2012.

CDC, 2012b. Investigation of Multistate Outbreak of *E. coli* O26 Infections Linked to Consumption of Raw Clover Sprouts. Final update April 5, 2012.

CFIA, 2011. Health hazard alert, Certain bulk and prepackaged raw shelled walnuts may contain *E. coli* O157:H7 bacteria. September 1, 2011.

Chaidez, C., Lopez, J., Vidales, J., et al., 2007. Efficacy of chlorinated and ozonated water in reducing *Salmonella typhimurium* attached to tomato surfaces. Int. J. Environ. Health Res. 17, 311–318.

Chang, T.-L., Streicher, R., Zimmer, H., 1988. The interaction of aqueous solutions of chlorine with malic acid, tartaric acid, and various fruit juices, a source of mutagens. Anal. Lett. 21, 2049–2067.

Chen, W., Jin, T.Z., Gurtler, J.B., et al., 2012. Inactivation of *Salmonella* on whole cantaloupe by application of an antimicrobial coating containing chitosan and allyl isothiocyanate. Int. J. Food Microbiol. 155, 165–170.

Choi, M.-R., Lee, S.-Y., Park, K.H., et al., 2012. Effect of aerosolized malic acid against *Listeria monocytogenes, Salmonella* Typhimurium, and *Escherichia coli* O157:H7 on spinach and lettuce. Food Control 24, 171–176.

Christie, S., 2010a. Turning over a new leaf. Chlorine stabilizer shows promise in wash environment. FreshCUT Mag. 18 (12), 14–15.

Christie, S., 2010b. FreshRinse. 'Groundbreaking' technology is a breakthrough for food safety, Fresh Express says. FreshCUT Mag. 18 (10), 16.

Cummings, K., Barrett, E., Mohle-Boetani, J.C., et al., 2001. A multistate outbreak of *Salmonella enterica* serotype Baildon associated with domestic raw tomatoes 1999. Emerg. Infect. Dis. 7, 1046–1048.

Delaquis, P., Stewart, S., Cazaux, S., et al., 2003. Survival and growth of *Listeria monocytogenes* and *Escherichia coli* O157:H7 in ready-to-eat iceberg lettuce. J. Food Prot. 65, 459–464.

DeWaal, C.S., Barlow, K., 2002. In: Outbreak Alert, seventh ed. Center for Science in the Public Interest, Washington, DC.

Deza, M.A., Araujo, M., Garrido, M.J., 2003. Inactivation of *Escherichia coli* O157:H7, *Salmonella enteritidis* and *Listeria monocytogenes* on the surface of tomatoes by neutral electrolyzed water. Lett. Appl. Microbiol. 37, 482–487.

Ding, T., Dong, Q.-L., Rahman, S.M.E., et al., 2011. Response surface modeling of *Listeria monocytogenes* inactivation on lettuce treated with electrolyzed oxidizing water. J. Food Process. Eng. 34, 1729–1745.

D'Lima, C.B., Linton, R.H., 2002. Inactivation of *Listeria monocytogenes* on lettuce by gaseous and aqueous chlorine dioxide gas and chlorinated water. IFT 2002 Annual Meeting and Food Expo, Anaheim, CA Abstract 15D-4.

Doores, S., 2005. Organic acids. Chapter 4. In: Davidson, P.M., Sofos, J.N., Branen, A.L. (Eds.), Antimicrobials in food. CRC Press, Boca Raton, pp. 91–142.

Drury, J.E., 2011. Inactivation of *Salmonella* enterica on romaine lettuce following spraying with Pro-San™ – a biodegradable foodgrade sanitizer" (2011). Graduate Theses and Dissertations. Paper 10466. http://lib.dr.iastate.edu/etd/10466.

Durak, M.Z., Churey, J.J., Gates, M., et al., 2012. Decontamination of green onions and baby spinach by vaporized ethyl pyruvate. J. Food Prot. 75, 1012–1022.

Durak, M.Z., Churey, J.J., Worobo, R.W., 2012. Efficacy of UV, acidified sodium hypochlorite, and mild heat for decontamination of surfaces and infiltrated *Escherichia coli* O157:H7 on green onions and baby spinach. J. Food Prot. 75, 1198–1206.

Erkman, O., 2010. Antimicrobial effects of hypochlorite on *Escherichia coli* in water and selected vegetables. Foodborne Pathog. Dis. 7, 953–958.

Escudero, M.E., Velázquez, L., Di Genaro, M.S., et al., 1999. Effectiveness of various disinfectants in the elimination of *Yersinia enterocolitica* on fresh lettuce. J. Food Prot. 62, 665–669.

Fallik, E., Aharoni, Y., Grinberg, S., et al., 1994. Postharvest hydrogen peroxide treatment inhibits decay in eggplant and sweet red pepper. Crop Prot. 13, 451–454.

Fan, X., Annous, B.A., Beaulieu, J.C., et al., 2008. Effect of hot water surface pasteurization of whole fruit on shelf life and quality of fresh-cut cantaloupe. J. Food Sci. 73, M91–M98.

Fan, X., Annous, B.A., Keskinen, L.A., et al., 2009. Use of chemical sanitizers to reduce microbial populations and maintain quality of whole and fresh-cut cantaloupe. J. Food Prot. 72, 2453–2460.

Fan, X., Annous, B.A., Sokorai, K.J., et al., 2006. Combination of hot water surface pasteurization of whole fruit and low-dose gamma irradiation of fresh-cut cantaloupe. J. Food Prot. 69, 912–919.

Fan, X., Sokorai, K.J., Engemann, J., et al., 2012. Inactivation of *Listeria innocua, Salmonella* Typhimurium, and *Escherichia coli* O157:H7 on surface and stem scar areas of tomatoes using in-package ozonation. J. Food Prot. 75, 1611–1618.

Fatemi, P., Knabel, S.J., 2006. Evaluation of sanitizer penetration and its effect on destruction of *Escherichia coli* O157:H7 in golden delicious apples. J. Food Prot. 69, 548–555.

Fatemi, P., LaBorde, L.F., Patton, J., et al., 2006. Influence of punctures, cuts, and surface morphologies of Golden Delicious apples on penetration and growth of *Escherichia coli* O157:H7. J. Food Prot. 69, 267–275.

FDA, 1999a. Microbiological safety evaluations and recommendations on sprouted seeds, National Advisory Committee on Microbiological Criteria for Food. www.cfsan.fda.gov/~mow/sprouts2.htm.

FDA, 1999b. Guidance for industry: Reducing microbial food safety hazards for sprouted seeds, U.S. Food and Drug Administration, CFSAN. Oct. 27, 1999.

FDA, 2001a. Hazard Analysis and Critical Control Point (HAACP); Procedures for the Safe and Sanitary Processing and Importing of Juice; Final Rule. Fed. Regist. 66 (13), 6137–6202.

FDA, 2001b. Enforcement report. Recalls and field corrections: Foods—class I. Recall number F-535–1. 20 August 2001. Sliced apples in poly bags. www.fda.gov/bbs/topics/ENFORCE/2001/ENF00708.html.

FDA, 2003. The bad bug book. Foodborne pathogenic microorganisms and natural toxins handbook US Department of Health and Human Services, Food and Drug Administration, Center for Food Safety and Applied Nutrition. Updated: 01–30–2003. www.cfsan.fda.gov/~mow/intro.html.

FDA, 2004. Guidance for industry: Recommendations to processors of apple juice or cider on the use of ozone for pathogen reduction purposes. US Department of Health and Human Services. August 2004. Food Drug Adm./CFSAN. www.cfsan.fda.gov/~dms/juicgu12.html.

FDA, 2007. Guidance for Industry. Guide to minimize microbial food safety hazards of fresh-cut fruits and vegetables. Draft Final Guidance US Department of Health and Human Services, Food and Drug Administration, Center for Food Safety and Applied Nutrition. March 2007. www.cfsan.fda.gov/~tdms/prodgui3.html.

FDA, 2007. FDA News. FDA implementing initiative to reduce tomato-related foodborne illnesses US Department of Health and Human Services, Food and Drug Administration. June 12, 2007. www.fda.gov/bbs/topics/NEWS/2007/NEW01651.htm.

Fett, W.F., 2002. Reduction of native microflora on alfalfa sprouts during propagation by addition of antimicrobial compounds to the irrigation water. Int. J. Food Microbiol. 72, 13–18.

Fisher, K., Phillips, C., McWatt, L., 2009. The use of an antimicrobial citrus vapour to reduce *Enterococcus* sp. on salad products. Int. J. Food Sci. Technol. 44, 1748–1754.

Fleischman, G.J., Batore, C., Merker, R., et al., 2001. Hot water immersion to eliminate *Escherichia coli* O157:H7 on the surface of whole apples: Thermal effects and efficacy. J. Food Prot. 64, 451–455.

Francis, G.A., O'Beirne, D., 1997. Effects of gas atmosphere, antimicrobial dip and temperature on the fate of *Listeria innocua* and *Listeria monocytogenes* on minimally processed lettuce. Int. J. Food Sci. Technol. 32, 141–151.

Fransisca, L., Feng, H., 2012. Effect of surface roughness on inactivation of *Escherichia coli* O157:H7 by new organic acid-surfactant combinations on alfalfa, broccoli and radish seeds. J. Food Prot. 75, 261–269.

Gagliardi, J.V., Millner, P.D., Lester, G., et al., 2003. On-farm and postharvest processing sources of bacterial contamination to melon rinds. J. Food Prot. 66, 82–87.

Ganesh, V., Hettiarachchy, N.S., Griffis, C.L., et al., 2012. Electrostatic spraying of food-grade organic and inorganic acids and plant extracts to decontaminate *Escherichia coli* O157:H7 on spinach and iceberg lettuce. J. Food Sci. 77, M391–M396.

Garcia, A., Mount, J.R., Davidson, P.M., 2003. Ozone and chlorine treatment of minimally processed lettuce. J. Food Sci. 68, 2747–2751.

Garcia, L., Henderson, J., Fabri, M., et al., 2006. Potential sources of microbial contamination in unpasteurized apple cider. J. Food Prot. 69, 137–144.

Gil, M., Selma, M.V., Lopez-Gálvez, F., et al., 2009. Fresh-cut product sanitation and wash water disinfection: problems and solutions. Int. J. Food Microbiol. 134, 37–45.

Gómez-López, V.M., Gobet, J., Selma, M.V., et al., 2013. Operating conditions for the electrolytic disinfection of process wash water from the fresh-cut industry contaminated with *E. coli* O157:H7. Food Control 29, 42–48.

Gonzalez, R.J., Luo, Y., Ruiz-Cruz, S., et al., 2004. Efficacy of sanitizers to inactivate *Escherichia coli* O157:H7 on fresh-cut carrot shreds under simulated process conditions. J. Food Prot. 67, 2375–2380.

Gopal, A., Coventry, J., Wan, J., et al., 2010. Alternative disinfection techniques to extend the shelf life of minimally processed iceberg lettuce. Food Microbiol. 27, 210–219.

Graham, D.M., 1997. Use of ozone for food processing. Food Technol. 51 (6), 72–75.

Guan, W., Huang, L., Fan, X., 2010. Acids in combination with sodium dodecyl sulfate caused quality deterioration of fresh-cut iceberg lettuce during storage in modified atmosphere package. J. Food Sci. 75, S435–S440.

Guentzel, J.L., Liang, L.K., Callan, M.A., et al., 2008. Reduction of bacteria on spinach, lettuce, and surfaces in food service areas using neutral electrolyzed oxidizing water. Food Microbiol. 25, 36–41.

Guo, X., Chen, J., Brackett, R.E., et al., 2002. Survival of *Salmonella* on tomatoes stored at high relative humidity, in soil, and on tomatoes in contact with soil. J. Food Prot. 65, 274–279.

Gurtler, J., Bailey, R., Jin, Z.T., 2012. Inactivation of *E. coli* O157:H7 and *Salmonella enterica* on strawberries by sanitizing solutions. Meeting Abstract. IAFP Meeting, Providence, Rhode Island July 22-25, vol. 1. page 1.

Han, Y., Linton, R.H., Nielson, S.S., et al., 2001. Reduction of *Listeria monocytogenes* on green peppers (*Capsicum annuum* L.) by gaseous and aqueous chlorine dioxide and water washing and its growth at 7 °C. J. Food Prot. 64, 1730–1738.

Hao, J., Lui, H., Liu, R., et al., 2011. Efficacy of slightly acidic electrolyzed water (SAEW) for reducing microbial contamination on fresh-cut cilantro. J. Food Saf. 31, 28–34.

Harris, L.J., Beuchat, L.J., Kajs, T.M., et al., 2001. Efficacy and reproducibility of a produce wash in killing *Salmonella* on the surface of tomatoes assessed with a proposed standard method for produce sanitizers. J. Food Prot. 64, 1477–1482.

Harris, L.J., Uesugi, A.R., Shirin, J.A., et al., 2012. Survival of *Salmonella* Enteritidis PT 30 on inoculated almond kernels in hot water treatments. Food Res. Int. 45, 1093–1098.

Hassenberg, K., Idler, C., Molloy, E., et al., 2007. Use of ozone in a lettuce-washing process: An industrial trial. J. Sci. Food Agric. 87, 914–919.

Hellström, S., Kervinen, R., Lyly, M., et al., 2006. Efficacy of disinfectants to reduce *Listeria monocytogenes* on precut iceberg lettuce. J. Food Prot. 69, 1565–1570.

Hidaka, T., Kirigaya, T., Kamijo, M., et al., 1992. Disappearance of residual chlorine and formation of chloroform in vegetables treated with sodium hypochlorite. J. Food Hyg. Soc. Japan 33, 267–273.

Hilgren, J.D., Salverda, J.A., 2000. Antimicrobial efficacy of a peroxyacetic/octanoic acid mixture in fresh-cut vegetable process waters. J. Food Sci. 65, 1376–1379.

Hirneisen, K.A., Markland, S.M., Kniel, K.E., 2011. Ozone inactivation of norovirus surrogates on fresh produce. J. Food Prot. 74, 836–839.

Grace Ho, K.-L., Luzuriaga, D.A., Rodde, K.M., et al., 2011. Efficacy of a novel sanitizer composed of lactic acid and peroxyacetic acid against single strains of nonpathogenic *Escherichia coli* K-12, *Listeria inocua* and *Lactobacillus plantarum*. J. Food Prot. 74, 1468–1474.

Hora, R., Kumar, M., Kostrzynska, M., et al., 2007. Inactivation of *Escherichia coli* O157:H7 and *Salmonella* on artificially contaminated mung beans (*Vigna radiata* L) using a stabilized oxychloro-based sanitizer. Lett. Appl. Microbiol. 44, 188–193.

Huang, T.-S., Xu, C., Walker, K., et al., 2006. Decontamination efficacy of combined chlorine dioxide with ultrasonification on apples and lettuce. J. Food Sci. 71, M134–M139.

Huang, Y., Chen, H., 2011. Effect of organic acids, hydrogen peroxide and mild heat on inactivation of *Escherichia coli* O157:H7 on baby spinach. Food Control 22, 1178–1183.

Huang, Y., Ye, M., Chen, H., 2012. Efficacy of washing with hydrogen peroxide followed by aerosolized antimicrobials as a novel sanitizing process to inactivate *Escherichia coli* O157:H7 on baby spinach. Int. J. Food Microbiol. 153, 306–313.

Ibarra-Sánchez, L.S., Alvarado-Casillas, S., Rodríguez-García, M.O., et al., 2004. Internalization of bacterial pathogens and their control by selected chemicals. J. Food Prot. 67, 1353–1358.

Inatsu, Y., Bari, M.L., Kawamoto, S., 2007. Application of acidified sodium chlorite prewashing treatment to improve the food hygiene of lightly fermented vegetables. JARQ 41, 17–23.

Isaacs, S., Aramini, J., Ciebin, B., et al., 2005. An international outbreak of salmonellosis associated with raw almonds contaminated with a rare phage type of *Salmonella enteritidis*. J. Food Prot. 68, 191–198.

Issa-Zacharia, A., Kamitani, Y., Miwa, N., et al., 2011. Application of slightly acidic electrolyzed water as a potential non-thermal food sanitizer for decontamination of fresh ready-to-eat vegetables and sprouts. Food Control 22, 601–607.

Izumi, H., 1999. Electrolyzed water as a disinfectant for fresh-cut vegetables. J. Food Sci. 64, 536–539.

Jaroni, S., Ravishankar, D., 2012. Bactericidal effects of roselle (*Hibiscus sabdariffa*) against foodborne pathogens *in vitro* and on romaine lettuce and alfalfa sprouts. Qual. Assur Saf. Crops Foods 4, 33–40.

Jin, T., Gurtler, J.B., 2012. Inactivation of *Salmonella* on tomato stem scars by edible chitosan and organic acid coatings. J. Food Prot. 75, 1368–1372.

Karaca, H., Velioglu, Y.S., 2007. Ozone applications in fruit and vegetable processing. Food Rev. Int. 23, 91–106.

Karaibrahimoglu, Y., Fan, X., Sapers, G.M., et al., 2004. Effect of pH on the survival of *Listeria innocua* in calcium ascorbate solutions and on quality of fresh-cut apples. J. Food Prot. 67, 751–757.

Kenney, S.J., Beuchat, L.R., 2002. Comparison of aqueous commercial cleaners for effectiveness in removing *Escherichia coli* O157:H7 and *Salmonella* Muenchen from the surface of apples. Int. J. Food Microbiol. 74, 47–55.

Kenney, S.J., Burnett, S.L., Beuchat, L.R., 2001. Location of *Escherichia coli* O157:H7 on and in apples as affected by bruising, washing, and rubbing. J. Food Prot. 64, 1328–1333.

Keskinen, L.A., Annous, B.A., 2011. Efficacy of adding detergents to sanitizer solutions for inactivation of *Escherichia coli* O157:H7 on Romaine lettuce. Int. J. Food Microbiol. 147, 157–161.

Keskinen, L.A., Burke, A., Annous, B.A., 2009. Efficacy of chlorine, acidic electrolyzed water and aqueous chlorine dioxide solutions to decontaminate *Escherichia coli* O157:H7 from lettuce leaves. Int. J. Food Microbiol. 132, 134–140.

Kilonzo-Nthenge, A., Chen, F.C., Godwin, S.L., 2006. Efficacy of home washing methods in controlling surface microbial contamination on fresh produce. J. Food Prot. 69, 330–334.

Kim, C., Hung, Y.C., 2012. Inactivation of *E. coli* O157:H7 on blueberries by electrolyzed water, ultraviolet light, and ozone. J. Food Sci. 77, M206–M211.

Kim, J.-G., Yousef, A.E., Chism, G.W., 1999. Use of ozone to inactivate microorganisms on lettuce. J. Food Saf. 19, 17–34.

Kim, Y.-J., Lee, S.-H., Park, J., et al., 2008. Inactivation of *Escherichia coli* O157:H7, *Salmonella typhimurium*, and *Listeria monocytogenes* on stored iceberg lettuce by aqueous chlorine dioxide. J. Food Sci. 73, M418–M422.

Klein, P., Morris, S.C., 2004. Technological breakthrough in postharvest sanitation treatment with the iodine based Isan Process. ISHS Acta Horticulturae 687: International Conference Postharvest Unlimited Downunder 2004. www.actahort.org/members/showpdf?booknra rnr=687_15.

Kondo, N., Murata, M., Isshiki, K., 2006. Efficacy of sodium hypochlorite, fumaric acid, and mild heat in killing native microflora and *Escherichia coli* O157:H7, *Salmonella* Typhimurium DT104 and *Staphylococcus aureus* attached to fresh-cut lettuce. J. Food Prot. 69, 323–329.

Koseki, S., Isobe, S., Itoh, K., 2004. Efficacy of acidic electrolyzed water ice for pathogen control on lettuce. J. Food Prot. 67, 2544–2549.

Koseki, S., Yoshida, K., Kamitami, Y., et al., 2003. Influence of inoculation method, spot inoculation site, and inoculation size on the efficacy of acidic electrolyzed water against pathogens on lettuce. J. Food Prot. 66, 2010–2016.

Kumar, M., Hora, R., Kostrzynska, M., et al., 2006. Inactivation of *Escherichia coli* O157:H7 and *Salmonella* on mung beans, alfalfa, and other seed types destined for sprout production by using an oxychloro-based sanitizer. J. Food Prot. 69, 1571–1578.

Lang, M.M., Harris, L.J., Beuchat, L.R., 2004. Survival and recovery of *Escherichia coli* O157:H7, *Salmonella*, and *Listeria monocytogenes* on lettuce and parsley as affected by method of inoculation, time between inoculation and analysis, and treatment with chlorinated water. J. Food Prot. 67, 1092–1103.

Lee, S.-Y., Dancer, G.I., Chang, S.-S., et al., 2006. Efficacy of chlorine dioxide gas against *Alicyclobacillus acidoterrestris* spores on apple surfaces. Int. J. Food Microbiol. 108, 364–368.

Li, D., Baert, L., De Jonghe, M., et al., 2011. Inactivation of murine norovirus 1, coliphage ΨX174, and *Bacillus fragilis* Phage B40-8. Appl. Environ. Microbiol. 77, 1399–1404.

Liao, C.-H., Shollenberger, L.M., Phillips, J.G., 2003. Lethal and sublethal action of acetic acid on *Salmonella in vitro* and on cut surfaces of apple slices. J. Food Sci. 68, 2793–2798.

Lin, C.-M., Moon, S.S., Doyle, M.P., et al., 2002. Inactivation of *Escherichia coli* O157:H7, *Salmonella enterica* Serotype Enteritidis, and *Listeria monocytogenes* by hydrogen peroxide and lactic acid and by hydrogen peroxide with mild heat. J. Food Prot. 65, 1215–1220.

López-Gálvez, F., Allende, A., Selma, M.V., et al., 2009. Prevention of *Escherichia coli* cross-contamination by different commercial sanitizers during washing of fresh-cut lettuce. Int. J. Food Microbiol. 133, 167–171.

López-Gálvez, F., Allende, A., Truchado, P., et al., 2010. Suitability of aqueous chlorine dioxide versus sodium hypochlorite as an effective sanitizer for preserving quality of fresh-cut lettuce while avoiding by-product formation. Postharvest. Biol. Technol. 55, 53–60.

López-Gálvez, F., Gil, M.I., Truchado, P., et al., 2010. Cross-contamination of fresh-cut lettuce after a short-term exposure during pre-washing cannot be controlled after subsequent washing with chlorine dioxide or sodium hypochlorite. Food Microbiol. 27, 199–204.

Lu, Y., Wu, C., 2010. Reduction of *Salmonella enterica* contamination of grape tomatoes by washing with thyme oil, thymol, and carvacrol as compared with chlorine treatment. J. Food Prot. 73, 2270–2275.

Luo, Y., 2007. Fresh-cut produce wash water re-use affects water quality and packaged product quality and microbial growth in Romaine lettuce. HortScience 42, 1413–1419.

Luo, Y., Nou, X., Yang, Y., et al., 2011. Determination of free chlorine concentrations needed to prevent *Escherichia coli* O157:H7 cross-contamination during fresh-cut produce wash. J. Food Prot. 74, 352–358.

Lukasik, J., Bradley, M.L., Scott, T.M., 2003. Reduction of poliovirus 1, bacteriophages, *Salmonella* Montevideo, and *Escherichia coli* O157:H7 on strawberries by physical and disinfectant washes. J. Food Prot. 66, 188–193.

Mahmoud, B.S.M., Linton, R.H., 2008. Inactivation kinetics of inoculated *Escherichia coli* O157:H7 and *Salmonella enterica* on lettuce by chlorine dioxide gas. Food Microbiol. 25, 244–252.

Mahovic, M., Bartz, J.A., Schneider, K.R., et al., 2009. Chlorine dioxide gas from an aqueous solution: reduction of *Salmonella* in wounds on tomato fruit and movement to sinks in a treatment chamber. J. Food Prot. 72, 952–958.

Materon, L.A., 2003. Survival of *Escherichia coli* O157:H7 applied to cantaloupes and the effectiveness of chlorinated water and lactic acid as disinfectants. World J. Microbiol. Biotechnol 19, 867–873.

Mendonca, A.F., Amoroso, T.L., Knabel, S.J., 1994. Destruction of Gram-negative food-borne pathogens by high pH involves disruption of the cytoplasmic membrane. Appl. Environ. Microbiol. 60, 4009–4014.

Molinos, A.C., Abriouel, H., Omar, N.B., et al., 2005. Effect of immersion solutions containing Enterocin AS-48 on *Listeria monocytogenes* in vegetable foods. Appl. Environ. Microbiol. 71, 7781–7787.

Moore, K.L., Patel, J., Jaroni, D., et al., 2011. Antimicrobial activity of apple, hibiscus, olive, and hydrogen peroxide formulations against *Salmonella enterica* on organic leafy greens. J. Food Prot. 74, 1676–1683.

Nascimento, M.S., Silva, N., Catanozi, M.P.L.M., et al., 2003. Effects of different disinfection treatments on the natural microbiota of lettuce. J. Food Prot. 66, 1697–1700.

Nei, D., Choi, J.-W., Bari, M.L., et al., 2009. Efficacy of chlorine and acidified sodium chlorite on microbial population and quality changes of spinach leaves. Pathog. Dis. 6, 541–546.

Nou, X., Luo, Y., 2010. Whole leaf wash improves chlorine efficiency for microbial reduction and prevents pathogen cross-contamination during fresh-cut lettuce processing. J. Food Sci. 75, M283–M290.

Nou, X., Luo, Y., Hollar, L., et al., 2011. Chlorine stabilizer T-128 enhances efficacy of chlorine against cross-contamination by *E. coli* O157:H7 and *Salmonella* in fresh-cut lettuce processing. J. Food Sci. 76, M218–M224.

Obaidat, M.M., Frank, J.F., 2009. Inactivation of *Salmonella* and *Escherichia coli* O157:H7 on sliced and whole tomatoes by allyl isothiocyanate, carvacrol, and cinnamaldehyde in vapor phase. J. Food Prot. 72, 315–324.

Oh, S.-W., Dancer, G.I., Kang, D.-H., 2005. Efficacy of aerosolized peroxyacetic acid as a sanitizer of lettuce leaves. J. Food Prot. 68, 1743–1747.

Ortega, Y.R., Mann, A., Torres, M.P., et al., 2008. Efficacy of gaseous chlorine dioxide as a sanitizer against *Cryptosporidium parvum*, *Cyclospora cayetanensis*, and *Encephalitozoon intestinalis* on produce. J. Food Prot. 71, 2410–2414.

Ortega, Y.R., Torres, M.P., Tatum, J.M., 2011. Efficacy of levulinic acid-sodium dodecyl sulfate against *Encephalitozoon intestinalis*, *Escherichia coli* O157:H7, and *Cryptosporidium parvum*. J. Food Prot. 74, 140–144.

Osman, M., Janes, M.E., Story, R., et al., 2006. Differential killing activity of cetylpyridinium chloride with or without bacto neutralizing buffer quench against firmly adhered *Salmonella gaminara* and *Shigella sonnei* on cut lettuce stored at 4 degrees C. J. Food Prot. 69, 1286–1291.

Ozer, N.P., Demirci, A., 2006. Electrolyzed oxidizing water treatment for decontamination of raw salmon inoculated with *Escherichia coli* O157:H7 and *Listeria monocytogenes* Scott A and response surface modeling. J. Food Eng. 72, 234–241.

Pan, Z., Bingol, G., McHugh, T., et al., 2012. Review of current technologies for reduction of *Salmonella* on almonds. Food Bioprocess Technol. 5, 2046–2057.

Pangloli, P., Hung, Y.C., Beuchat, L.R., et al., 2009. Reduction of *Escherichia coli* O157:H7 on produce by use of electrolyzed water under simulated food service operation conditions. J. Food Prot. 72, 1854–1861.

Pao, S., Davis, C.L., Kelsey, D.F., 2000. Efficacy of alkaline washing for the decontamination of orange fruit surfaces inoculated with *Escherichia coli*. J. Food Prot. 63, 961–964.

Pao, S., Kelsey, D.F., Khalid, M.F., et al., 2007. Using aqueous chlorine dioxide to prevent contamination of tomatoes with *Salmonella enterica* and *Erwinia carotovora* during fruit washing. J. Food Prot. 70, 629–634.

Paola, C.L., Rocio, C.V., Marcela, M., et al., 2005. Effectiveness of electrolyzed oxidizing water for inactivating *Listeria monocytogenes* in lettuce. Universitas Scientiarum, Revista de la Facultad de Ciencias. Pontificia Univ. Javeriana 10, 97–108.

Parish, M.E., Beuchat, L.R., Suslow, T.V., et al., 2003. Methods to reduce/eliminate pathogens from fresh and fresh-cut produce. Chapter 5 In Compr. Rev. Food Sci. Food Saf. 2 (Suppl.), 161–173.

Park, E.J., Alexander, E., Taylor, G.A., et al., 2008. Effect of electrolyzed water for reduction of foodborne pathogens on lettuce and spinach. J. Food Sci. 73, M268–M272.

Park, E.J., Alexander, E., Taylor, G.A., et al., 2009. The decontaminative effects of acidic electrolyzed water for *Escherichia coli* O157:H7, *Salmonella* Typhimurium, and *Listeria monocytogenes*. Food Microbiol. 26, 386–390.

Park, C.-M., Beuchat, L.R., 1999. Evaluation of sanitizers for killing *Escherichia coli* O157:H7, *Salmonella*; and naturally occurring microorganisms on cantaloupes, honeydew melons and asparagus. Dairy, Food Environ. Sanit. 19, 842–847.

Park, C.-M., Hung, Y.-C., Doyle, M.P., et al., 2001. Pathogen reduction and quality of lettuce treated with electrolyzed oxidizing and acidified chlorinated water. J. Food Sci. 66, 1368–1372.

Park, S.-H., Choi, M.-R., Park, J.-W., et al., 2011. Use of organic acids to inactivate *Escherichia coli* O157:H7, *Salmonella* Typhimurium, and *Listeria monocytogenes* on organic fresh apples and lettuce. J. Food Sci. 76, M293–M298.

Parnell, T.L., Harris, L.J., 2003. Reducing *Salmonella* on apples with wash practices commonly used by consumers. J. Food Prot. 66, 741–747.

Patel, J., Sharma, M., Millner, P., et al., 2010. Inactivation of *Escherichia coli* O157:H7 attached to spinach harvester blade using bacteriophage. Foodborne Pathog. Dis. 8, 541–546.

Pérez-Conesa, D., Cao, J., Chen, L., et al., 2011. Inactivation of *Listeria monocytogenes* and *Escherichia coli* O157:H7 biofilms by micelle-encapsulated eugenol and carvacrol. J. Food Prot. 74, 55–62.

Pilizota, V., Sapers, G.M., 2004. Novel browning inhibitor formulation for fresh-cut apples. J. Food Sci. 69, 140–143.

Pirovani, M.E., Guemes, D.R., Piagnetini, A.M., 2001. Predictive models for available chlorine depletion and total microbial count reduction during washing of fresh-cut spinach. J. Food Sci. 66, 860–864.

Popa, I., Hanson, E.J., Todd, E.C., et al., 2007. Efficacy of chlorine dioxide gas sachets for enhancing the microbiological quality and safety of blueberries. J. Food Prot. 70, 2084–2088.

Predmore, A., Li, J., 2011. Enhanced removal of a human norovirus surrogate from fresh vegetables and fruits by a combination of surfactants and sanitizers. Appl. Environ. Microbiol. 77, 4829–4838.

Rahman, S.M.E., Ding, T., Oh, D.-H., 2010. Inactivation effect of newly developed low concentration electrolyzed water and other sanitizers against microorganisms in spinach. Food Control 21, 1383–1387.

Raiden, R.M., Sumner, S.S., Eifert, J.D., et al., 2003. Efficacy of detergents in removing *Salmonella* and *Shigella* spp. from the surface of fresh produce. J. Food Prot. 66, 2210–2215.

Restaino, L., Frampton, E.W., Hemphill, J.B., et al., 1995. Efficacy of ozonated water against various food-related microorganisms. Appl. Environ. Microbiol. 61, 3471–3475.

Richards, G.M., Beuchat, L.R., 2004. Attachment of *Salmonella* Poona to cantaloupe rind and stem scar tissues as affected by temperature of fruit and inoculum. J. Food Prot. 67, 1359–1364.

Riordan, D.C.R., Sapers, G.M., Annous, B.A., 2000. The survival of *E. coli* O157:H7 in the presence of *Penicillium expansum* and *Glomerella cingulata* in wounds on apple surfaces. J. Food Prot. 63, 1637–1642.

Riordan, D.C., Sapers, G.M., Hankinson, T.H., et al., 2001. A study of U.S. orchards to identify potential sources of *Escherichia coli* O157:H7. J. Food Prot. 64, 1320–1327.

Rodgers, S.L., Cash, J.N., Siddiq, M., et al., 2004. A comparison of different chemical sanitizers for inactivating *Escherichia coli* O157:H7 and *Listeria monocytogenes* in solution and on apples, lettuce, strawberries, and cantaloupe. J. Food Prot. 67, 721–731.

Rodriguez-Garcia, O., González, V.M., Fernández-Escartin, E., 2011. Reduction of *Salmonella enterica*, *Escherichia coli* O157:H7, and *Listeria monocytogenes* with electrolyzed oxidizing water on inoculated Hass avocados (*Persea americana* var. *Hass*). J. Food Prot. 74, 1552–1557.

Rupasinghe, H.P.V., Boulter-Bitzer, J., Odumeru, J.A., 2006. Lactic acid improves the efficacy of anti-microbial washing solutions for apples. J. Food Agr. Environ. 4, 44–48.

Sagong, H.G., Lee, S.Y., Chang, P.S., et al., 2011. Combined effect of ultrasound and organic acids to reduce *Escherichia coli* O157:H7, *Salmonella* Typhimurium, and *Listeria monocytogenes* on organic fresh lettuce. Int. J. Food Microbiol. 145, 287–292.

Samadi, N., Abadian, N., Bakhtiari, D., et al., 2009. Efficacy of detergents and fresh produce disinfectants against microorganisms associated with mixed raw vegetables. J. Food Prot. 72, 1486–1490.

Sampathkumar, B., Khachatourians, G.G., Korber, D.R., 2003. High pH during trisodium phosphate treatment causes membrane damage and destruction of *Salmonella enterica* Serovar Enteriditis. Appl. Environ. Microbiol. 69, 122–129.

Sapers, G.M., 2002. Washing and sanitizing raw materials for minimally processed fruit and vegetable products. Chapter 11. In: Novak, J.S., Sapers, G.M., Juneja, V.K. (Eds.), Microbial Safety of Minimally Processed Foods. CRC Press, Boca Raton, pp. 221–253.

Sapers, G.M., 2003. Hydrogen peroxide as an alternative to chlorine for sanitizing fruits and vegetables, FoodInfo Online Features. IFIS. Publishing: 23 July 2003. http://foodsciencecentral.com/library.html#ifis/12433.

Sapers, G.M., 2005. Washing and sanitizing treatments for fruits and vegetables. Chapter 17. In: Sapers, G.M., Gorny, J.R., Yousef, A.E. (Eds.), Microbiology of Fruits and Vegetables. CRC Press, Boca Raton, pp. 375–400.

Sapers, G.M., Jones, D.M., 2006. Improved sanitizing treatments for fresh tomatoes. J. Food Sci. 71, M252–M256.

Sapers, G.M., Miller, R.L., Annous, B.A., et al., 2002. Improved anti-microbial wash treatments for decontamination of apples. J. Food Sci. 67, 1886–1891.

Sapers, G.M., Miller, R.L., Mattrazzo, A.M., 1999. Effectiveness of sanitizing agents in inactivating *Escherichia coli* in Golden Delicious apples. J. Food Sci. 64, 734–737.

Sapers, G.M., Miller, R.L., Jantschke, M., et al., 2000. Factors limiting the efficacy of hydrogen peroxide washes for decontamination of apples containing *Escherichia coli*. J. Food Sci. 65, 529–532.

Sapers, G.M., Miller, R.L., Pilizota, V., et al., 2001. Shelf-life extension of fresh mushrooms (*Agaricus bisporus*) by application of hydrogen peroxide and browning inhibitors. J. Food Sci. 66, 362–366.

Sapers, G.M., Miller, R.L., Pilizota, V., et al., 2001. Antimicrobial treatments for minimally processed cantaloupe melon. J. Food Sci. 66, 345–349.

Sapers, G.M., Sites, J.E., 2003. Efficacy of 1% hydrogen peroxide wash in decontaminating apples and cantaloupe melons. J. Food Sci. 68, 1793–1797.

Saravacos, G.D., Kostaropoulos, A.E., 2002. Handbook of Food Processing Equipment. Kluwer Boston, Norwell. 256–258.Available on-line at www.Google.com.

Segall, R.H., 1968. Reducing postharvest decay of tomatoes by adding a chlorine source and the surfactant Santormerse F85 to water in field washers. Fla. State Hort. Soc. Proc. 81, 212–214.

Selma, M.V., Allende, A., López-Gálvez, E., et al., 2008c. Disinfection potential of ozone, ultraviolet-C and their combination in wash water for the fresh-cut vegetable industry. Food Microbiol. 25, 809–814.

Selma, M.V., Ibáñez, A.M., A. Allende, M., et al., 2008a. Effect of gaseous ozone and hot water on microbial and sensory quality of cantaloupe and potential transference of *Escherichia coli* O157:H7 during cutting. Food Microbiol. 25, 162–168.

Selma, M.V., Ibáñez, A.M., Cantwell, M., et al., 2008b. Reduction by gaseous ozone of *Salmonella* and microbial flora associated with fresh-cut cantaloupe. Food Microbiol. 25, 558–565.

Sharma, M., Patel, J.R., Conway, W.S., et al., 2009. Effectiveness of bacteriophages in reducing *Escherichia coli* O157:H7 on fresh-cut cantaloupes and lettuce. J. Food Prot. 72, 1481–1485.

Sharma, R., 2005. Ozone decontamination of fresh fruit and vegetables. In: Jongen, W. (Ed.), Improving the safety of fresh fruit and vegetables. Woodhead Publishing, Cambridge, UK.

Sharma, R.R., Demirci, A., Beuchat, L.R., et al., 2002. Inactivation of *Escherichia coli* O157:H7 on inoculated alfalfa seeds with ozonated water and heat treatment. J. Food Prot. 65, 447–451.

Shen, C., Luo, Y., Nou, X., et al., 2013. Dynamic effects of free chlorine concentration, organic load and exposure time on the inactivation of *Salmonella*, *Escherichia coli* O157:H7 and non-O157 Shiga toxin-producing *E. coli* by hypochlorite solutions with high organic loads. J. Food Prot. 76, 386–393.

Shen, C., Luo, Y., Nou, X., et al., 2012a. Fresh produce washing aid, T-128, enhances inactivation of *Salmonella* and *Pseudomonas* biofilms on stainless steel in chlorinated wash solutions. Appl. Environ. Microbiol. 78, 6789–6798.

Shen, C., Luo, Y., Nou, X., et al., 2012b. Enhanced inactivation of *Salmonella*, *Escherichia coli* O157:H7 and *Pseudomonas* biofilms, using fresh produce washing aid, T-128, on cantaloupe rinds with chlorinated wash solutions. Meeting Abstract. IAFP Meeting, Providence, Rhode Island July 23.

Shirron, N., Kisluk, G., Zelikovich, Y., et al., 2009. A comparative study assaying commonly used sanitizers for antimicrobial activity against indicator bacteria and a *Salmonella* Typhimurium strain on fresh produce. J. Food Prot. 72, 2413–2417.

Silveira, A., Aguayo, E., Artes, F., 2010. Emerging sanitizers and clean room packaging for improving the microbial quality of fresh-cut 'Galia' melon. Food Control 21, 863–871.

Silveira, A., Conesa, A., Aguayo, E., et al., 2008. Alternative sanitizers to chlorine for use on fresh-cut 'Galia' (*Cucumis melo* var. *catalupensis*) melon. J. Food Sci. 73, M405–M411.

Singla, R., Ganguli, A., Ghosh, M., 2011. An effective combined treatment using malic acid and ozone inhibits *Shigella* spp. on sprouts. Food Control 22, 1032–1039.

Smilanick, J.L., 2003. Use of ozone in storage and packing facilities. Presented at Washington Tree Fruit Postharvest Conference, December 2–3, 2003, Wenatchee, WA.

Smith, S., Dunbar, M., Tucker, D., et al., 2003. Efficacy of a commercial produce wash on bacterial contamination of lettuce in a food service setting. J. Food Prot. 66, 2359–2361.

Solomon, E.B., Huang, L., Sites, J.E., et al., 2006. Thermal inactivation of *Salmonella* on cantaloupes using hot water. J. Food Sci. 71, M25–M30.

Solomon, E.B., Yaron, S., Matthews, K.R., 2002. Transmission of *Escherichia coli* O157:H7 from contaminated manure and irrigation water to lettuce plant tissue and its subsequent internalization. Appl. Environ. Microbiol. 68, 397–400.

Somers, E.B., Schoeni, J.L., Wong, A.C.L., 1994. Effect of trisodium phosphate on biofilm and planktonic cells of *Campylobacter jejuni*, *Escherichia coli* O157:H7, *Listeria monocytogenes* and *Salmonella* Typhimurium. Int. J. Food Microbiol. 22, 269–276.

Spotts, R.A., 1982. Use of surfactants with chlorine to improve pear decay control. Plant Dis. 66, 725–1982.

Spotts, R.A., Cervantes, L.A., 1992. Effect of ozonated water on postharvest pathogens of pear in laboratory and packinghouse tests. Plant Dis. 76, 256–259.

Steenstrup, L.D., Floros, J.D., 2004. Inactivation of *E. coli* O157:H7 in apple cider by ozone at various temperatures and concentrations. J. Food Proc. Preserv. 28, 103–116.

Stopforth, J.D., Mai, T., Kottapalli, B., et al., 2008. Effect of acidified sodium chlorite, chlorine, and acidic electrolyzed water on *Escherichia coli* O157:H7, *Salmonella* and *Listeria monocytogenes* onto leafy greens. J. Food Prot. 71, 625–628.

Su, X., D'Souza, D.H., 2011. Trisodium phosphate for foodborne virus reduction on produce. Foodborne Pathog. Dis. 8, 713–717.

Suslow, T., 2000. Chlorination in the production and postharvest handling of fresh fruits and vegetables. Chap. 6. In: McLaren, D. (Ed.), Fruit and vegetable processing. Published by the Food Processing Center at the University of Nebraska, Lincoln, NE, pp. 2–15. http://ucce.ucdavis.edu/files/filelibrary/5453/4369.pdf.

Suslow, T., 2000. Postharvest handling for organic crops University of California ANR Communication Services. Publication 7254. http://anrcatalog.ucdavis.edu/pdf/7254.pdf.

Suslow, T.V., 2004. Ozone application for postharvest disinfection of edible horticultural crops University of California Division of Agriculture and Natural Resources. Publication 8133. http://anrcatalog.ucdavis.edu/pdf/8133.pdf.

Sy, K.V., Murray, M.B., Harrison, M.D., et al., 2005. Evaluation of gaseous chlorine dioxide as a sanitizer for killing *Salmonella*, *Escherichia coli* O157:H7, *Listeria monocytogenes*, and yeasts and molds on fresh and fresh-cut produce. J. Food Prot. 68, 1176–1187.

Tian, P., Yang, D., Mandrell, R., 2011. Differences in the binding of human norovirus to and from romaine lettuce and raspberries by water and electrolyzed waters. J. Food Prot. 74, 1364–1369.

Tomas-Callejas, A., Lopez-Velasco, F.G., Sbodio, A., et al., 2012a. Evaluation of current operating standards for chlorine dioxide in disinfection of dump tank and fume for fresh tomatoes. J. Food Prot. 75, 304–313.

Tomas-Callejas, A., Lopez-Galvez, F., Sbodio, A.M., Artes-Hernandez, F., et al., 2012b. Chlorine dioxide and chlorine effectiveness in preventing *Escherichia coli* O157:H7 and *Salmonella* cross-contamination on fresh-cut chard. Food Control 23, 325–332.

Trinetta, V., Vaidya, N., Linton, R., et al., 2011. A comparative study on the effectiveness of chlorine dioxide gas, ozone gas, and e-beam irradiation treatments for inactivation of pathogens inoculated onto tomato, cantaloupe and lettuce seeds. Int. J. Food Microbiol. 146, 203–206.

Ukuku, D.O., 2006. Effect of sanitizing treatments on removal of bacteria from cantaloupe surface, and recontamination with *Salmonella*. Food Microbiol. 23, 289–293.

Ukuku, D.O., Fett, W., 2002. Behavior of *Listeria monocytogenes* inoculated on cantaloupe surfaces and efficacy of washing treatments to reduce transfer from rind to fresh-cut pieces. J. Food Prot. 65, 924–930.

Ukuku, D.O., Fett, W., 2004. Method of applying sanitizers and sample preparation affects recovery of native microflora and *Salmonella* on whole cantaloupe surfaces. J. Food Prot. 67, 999–1004.

Ukuku, D.O., Pilizota, V., Sapers, G.M., 2001. Influence of washing treatment on native microflora and *Escherichia coli* population of inoculated cantaloupes. J. Food Saf. 21, 31–47.

Ukuku, D.O., Pilizota, V., Sapers, G.M., 2004. Effect of hot water and hydrogen peroxide treatments on survival of *Salmonella* and microbial quality of whole and fresh-cut cantaloupe. J. Food Prot. 67, 432–437.

Ukuku, D.O., Sapers, G.M., 2001. Effect of sanitizer treatments on *Salmonella* Stanley attached to the surface of cantaloupe and cell transfer to fresh-cut tissues during cutting practices. J. Food Prot. 64, 1286–1291.

Vadlamudi, S., Taylor, T.M., Blankenburg, C., et al., 2012. Effect of chemical sanitizers on *Salmonella enterica* serovar Poona on the surface of cantaloupe and pathogen contamination of internal tissues as a function of cutting procedure. J. Food Prot. 75, 1766–1773.

Vandekinderen, I., Devlieghere, F., De Meulenaer, B., et al., 2009. Optimization and evaluation of a decontamination step with peroxyacetic acid for fresh-cut produce. Food Microbiol. 26, 882–888.

Venkitanarayanan, K.S., Ezeike, G.O., Hung, Y.-C., et al., 1999a. Efficacy of electrolyzed oxidizing water for inactivating *Escherichia coli* O157:H7, *Salmonella* Enteritidis, and *Listeria monocytogenes*. Appl. Environ. Microbiol. 65, 4276–4279.

Venkitanarayanan, K.S., Ezeike, G.O.I., Hung, Y.-C., et al., 1999b. Inactivation of *Escherichia coli* O157:H7 and *Listeria monocytogenes* on plastic kitchen cutting boards by electrolyzed oxidizing water. J. Food Prot. 62, 857–860.

Venkitanarayanan, K.S., Lin, C.-M., Bailey, H., et al., 2002. Inactivation of *E. coli* O157:H7, *Salmonella* Enteritidis and *Listeria monocytogenes* on apples, oranges, and tomatoes by lactic acid with hydrogen peroxide. J. Food Prot. 65, 100–105.

Venkitanarayanan, K., Zhao, T., Doyle, M.P., 1999. Inactivation of *E. coli* O157:H7 by combination of GRAS chemicals and temperatures. Food Microbiol. 16, 75–82.

Vijayakumar, C., Wolf-Hall, C.E., 2002. Evaluation of household sanitizers for reducing levels of *Escherichia coli* on iceberg lettuce. J. Food Prot. 65, 1646–1650.

Vurma, M., Pandit, R.B., Sastry, S.K., et al., 2009. Inactivation of *Escherichia coli* O157:H7 and natural microbiota on spinach leaves using gaseous ozone during vacuum cooling and simulated transportation. J. Food Prot. 72, 1538–1546.

Wade, W.N., Scouten, A.J., McWatters, K.H., et al., 2003. Efficacy of ozone in killing *Listeria monocytogenes* on alfalfa seeds and sprouts and effects on sensory quality of sprouts. J. Food Prot. 66, 44–51.

Warf, C.C., 2001. The chemistry & modes of action of acidified sodium chlorite. Abstract 91–1. Acidified sodium chlorite—An antimicrobial intervention for the food industry 2001. IFT Annual Meeting, New Orleans, LA.

Wang, H., Feng, H., Luo, Y., 2006. Dual-phasic inactivation of *Escherichia coli* O157:H7 with peroxyacetic acid, acidic electrolyzed water and chlorine on cantaloupes and fresh-cut apples. J. Food Saf. 26, 335–347.

Wang, H., Feng, H., Liang, W., et al., 2009. Effect of surface roughness on retention and removal of *Escherichia coli* O157:H7 on surfaces of selected fruits. J. Food Sci. 74, E8–E15.

Weil, C., Huang, T.S., Kim, J.M., et al., 1995. Growth and survival of *Salmonella* Montevideo on tomatoes and disinfection with chlorinated water. J. Food Prot. 58, 829–836.

Weissinger, W.R., Beuchat, L.R., 2000. Comparison of aqueous chemical treatments to eliminate *Salmonella* on alfalfa seeds. J. Food Prot. 63, 1475–1482.

Wickramanayake, G.B., 1991. Disinfection and sterilization by ozone. Chapter 10. In: Block, S.S. (Ed.), Disinfection, Sterilization, and Preservation, fourth ed. Lea & Febiger, Philadelphia.

Wisniewsky, M.A., Glatz, B.A., Gleason, M.L., et al., 2000. Reduction of *Escherichia coli* O157:H7 counts on whole fresh apples by treatment with sanitizers. J. Food Prot. 63, 703–708.

Wright, J.R., Sumner, S.S., Hackney, C.R., et al., 2000. Reduction of *Escherichia coli* O157:H7 on apples using wash and chemical sanitizer treatments. Dairy, Food Environ. Sanit. 20, 120–126.

Xia, X., Luo, Y., Yang, Y., et al., 2012. Effects of tomato variety, temperature differential, and post-stem removal time on internalization of *Salmonella enterica* Thompson in tomatoes. J. Food Prot. 75, 297–303.

Xu, L., 1999. Use of ozone to improve the safety of fresh fruits and vegetables. Food Technol. 53 (10), 58–63.

Yang, H., Swem, B.L., Li, Y., 2003. The effect of pH on inactivation of pathogenic bacteria on fresh-cut lettuce by dipping treatment with electrolyzed water. J. Food Sci. 68, 1013–1017.

Yang, J., Bingol, G., Pan, Z., et al., 2010. Infrared heating for dry-roasting and pasteurization of almonds. J. Food Eng. 10, 1016.

Yossa, N., Patel, J., Millner, P., et al., 2012. Essential oils reduce *Escherichia coli* O157:H7 and *Salmonella* on spinach leaves. J. Food Prot. 75, 488–496.

Yuk, H.-G., Bartz, J.A., Schneider, K.R., 2005. Effectiveness of individual or combined sanitizer treatments for inactivating *Salmonella* spp. on smooth surface, stem scar, and wounds of tomatoes. J. Food Sci. 70, M409–M414.

Yuk, H.-G., Bartz, J.A., Schneider, K.R., 2006. The effectiveness of sanitizer treatments in inactivation of *Salmonella* spp. from bell pepper, cucumber, and strawberry. J. Food Sci. 71, M95–M99.

Zhang, C., Lu, Z., Li, Y., et al., 2011. Reduction of *Escherichia coli* O157:H7 and *Salmonella enteritidis* on mung bean seeds and sprouts by slightly acidic electrolyzed water. Food Control 22, 792–796.

Zhang, G., Ma, L., Phelan, V.H., et al., 2009. Efficacy of antimicrobial agents in lettuce leaf processing water for control of *Escherichia coli* O157:H7. J. Food Prot. 72, 1392–1397.

Zhang, S., Farber, J.M., 1996. The effects of various disinfectants against *Listeria monocytogenes* on fresh-cut vegetables. Food Microbiol. 13, 311–321.

Zhao, T., Zhao, P., Doyle, M.P., 2009. Inactivation of *Salmonella* and *Escherichia coli* O157:H7 on lettuce and poultry skin by combination of levulinic acid and sodium dodecyl sulfate. J. Food Prot. 72, 928–936.

Zhuang, R.-Y., Beuchat, L.R., 1996. Effectiveness of trisodium phosphate for killing *Salmonella* Montevideo on tomatoes. Lett. Appl. Microbiol. 232, 97–100.

Zhuang, R.-Y., Beuchat, L.R., Angulo, F.J., 1995. Fate of *Salmonella* Montevideo on and in raw tomatoes as affected by temperature and treatment with chlorine. Appl. Environ. Microbiol. 61, 2127–2131.

7CFR205.605 Nonagricultural (nonorganic) substances allowed as ingredients in or on processed products labeled as "organic" or "made with organic (specified ingredients or food group(s)).". *Code of Federal Regulations* 2007 Title 7, Part 205, Section 205.605. Current as of September 27, 2012.

21CFR173.300 Chlorine dioxide. *Code of Federal Regulations* Title 21, Part 173, Section 173.300. Revised as of April 1, 2012.

21CFR173.315 Chemicals used in washing or to assist in the peeling of fruits and vegetables. *Code of Federal Regulations* Title 21, Part 173, Section 173.315. Revised as of April 1, 2012.

21CFR173.325 Acidified sodium chlorite solutions. *Code of Federal Regulations* Title 21, Part 173, Section 173.315. Revised as of April 1, 2012

21CFR173.368 Ozone. *Code of Federal Regulations* Title 21, Part 173, Section 173.368. Revised as of April 1, 2012.

21CFR184.1005 Direct food substances affirmed as generally recognized as safe. Listing of specific substances affirmed as GRAS. Acetic Acid. *Code of Federal Regulations* Title 21, Part 184, Subpart B, Section 184.1005. Revised as of April 1, 2012.

21CFR184.1033 Direct food substances affirmed as generally recognized as safe. Listing of specific substances affirmed as GRAS. Citric Acid. *Code of Federal Regulations* Title 21, Part 184, Subpart B, Section 184.1033. Revised as of April 1, 2012.

21CFR184.1061 Direct food substances affirmed as generally recognized as safe. Listing of specific substances affirmed as GRAS. Lactic Acid. *Code of Federal Regulations* Title 21, Part 184, Subpart B, Section 184.1061. Updated as of Aug. 23, 2012.

21CFR184.1366 Hydrogen peroxide, *Code of Federal Regulations*. Title 21, Part 184, Section 184.1366. Revised as of April 1, 2012.

21CFR184.1769a Direct food substances affirmed as generally recognized as safe. Listing of specific substances affirmed as GRAS. Sodium metasilicate. *Code of Federal Regulations* Title 21, Part 184, Subpart B, Section 184.1769a. Revised as of April 1, 2012.

29CFR 1910.1000 TABLE Z-1 Limits for air contaminants. *Code of Federal Regulations*, Title 29, Part 1910. Section 1910.1000.

40CFR180.1196. Peroxyacetic acid; exemption from the requirement of a tolerance. *Code of Federal Regulations*, Title 40, Part 180, Section 180.1196. As amended, March 4, 2011.

40CFR180.1197. Hydrogen peroxide; exemption from the requirement of a tolerance, *Code of Federal Regulations*, Title 40, Part 180, Section 180.1197. July 1, 2004.

Advanced Technologies for Detection and Elimination of Bacterial Pathogens

18

Brendan A. Niemira[1], Howard Q. Zhang[2]

[1]*Eastern Regional Research Center, U.S. Department of Agriculture, Agricultural Research Service, Wyndmoor, PA*, [2]*Western Regional Research Center, U.S. Department of Agriculture, Agricultural Research Service, Albany, CA*

CHAPTER OUTLINE

Introduction

In recent years, the incidence of foodborne illness (FBI) outbreaks associated with contaminated fruits, vegetables, salads, and juices has increased notably (Sivapalasingam et al., 2004). The Centers for Disease Control and Prevention reported that foodborne outbreaks associated with fresh produce doubled from 1973–1987 to 1988–1992 (Buck et al., 2003; CDC, 2011). More recently, the plant-based commodities accounted for 66% of viral, 32% of bacterial, 25% of chemical, and 30% of parasitic illnesses from 1998 to 2008. Leafy greens were the single food most commonly associated with foodborne illness (Painter et al., 2013). Adherence to established industry standards for Good Agricultural Practices (GAP), Good Manufacturing Practices (GMP), and Good Handling Practices (GHP) can serve to reduce risk. However, by themselves,

The Produce Contamination Problem. http://dx.doi.org/10.1016/B978-0-12-404611-5.00018-X
2014 Published by Elsevier Inc.

these practices have not been able to prevent repeated product recalls and illnesses of exposed consumers. Although various food safety interventions have been proposed as antimicrobial processes that are broadly applicable for produce (a produce "kill step"), barriers to their wide-spread implementation have hampered the food safety efforts of the fresh produce industry (UFPA, 2007; JIFSAN, 2007). Leafy green vegetables such as lettuce and spinach, tomatoes, melons, sprouts, and other fresh produce thus remain vectors for *Escherichia coli* O157:H7, *Salmonella, Listeria,* and other pathogens.

When taken as part of a unified production approach, the GAP, GMP, and GHP protocols in the preharvest, postharvest, and post-processing environments constitute the components of a food safety-oriented Hazard Analysis and Critical Control Point (HACCP) plan. Verification of the efficacy and consistency of the HACCP plan requires detection techniques that are scientifically valid and suitable for commercial production of fresh and fresh-cut produce. This chapter will discuss the latest research in developing rapid, sensitive, and accurate detection technologies, present a summary of the latest research on advanced intervention technologies to inactivate pathogens on produce, and briefly discuss the applicability of new technologies for the production and processing of organic fruits and vegetables.

Detection methods

The benchmarks and standards (e.g., the applicable GAP and GMP and guidance documents) of produce safety are the underpinning of the specific actions taken by growers, processors, shippers, and retailers. In the field, during harvest and processing, after packaging and shipping and in the retail or foodservice environment, the steps taken to ensure the safety of fresh and fresh-cut fruits and vegetables fall into three general categories. These are protocols to *exclude* pathogens from the plants, produce, or packages, and protocols to *contain* or to *eradicate* pathogens (e.g., by application of treatments to suppress growth or to reduce populations sufficiently to assure safety). If completely effective, excluding pathogens will serve to prevent contamination in the first place. Improving these protocols has been the primary focus of much of the industry's effort to date. Although these improvements are achieving positive results, reliance on only one type of control is unlikely to be optimally effective or give complete control. Hence, the focus of this section is on scientific efforts to improve the industry's ability to detect pathogens and subsequently contain contaminated produce.

Standard microbiological testing for quality purposes is common in the produce industry. However, regular testing for contamination by pathogens, including *E. coli* O157:H7, *Salmonella, Shigella,* and *L. monocytogenes*, has historically been less common, although this is becoming more prevalent in the wake of changes in the industry. Relatively slow traditional sampling, enrichment, and enumeration of food-borne pathogens can take 48 to 72 hours, an extremely long time for a commodity class with a shelf-life of only 7 to 14 days. Also, false positives result in needless product recall and expense, while false negatives put the health of the consumer and the viability of the company name and reputation at risk. To be useful for the produce industry, tests need to be rapid, sensitive and accurate, identifying critical limits of viable or potentially viable pathogens in a timely manner.

Immunomagnetic beads and biosensors: separation and concentration

In order to screen high volumes of food material, either from pooled samples or from a flow-through system, a means of concentrating the samples must be employed. Air and water sampling has traditionally used flow-through filtration, with analysis of the filter membrane as the diagnostic step. For bulky, fibrous, or otherwise difficult to filter produce material, this is a less than optimal approach. Recently, advances in the use of immunomagnetic beads have improved the sensitivity and reliability of detection (Tu et al., 2008).

As the name suggests, this technique involves the use of specialized iron beads, coated with plastic. These micron-sized beads are surface treated with antibodies that are specific to the pathogen of interest. The food sample is prepared as a slurry and the beads are added to the suspension. During mixing, the entire population of bacteria in the suspension comes in contact with the beads. Non-target organisms remain in suspension, while the target pathogens bind to the specific antibodies on the bead surface. A powerful magnetic field, either from an electromagnet or from a neodymium boron iron magnet, is used to collect the beads for further analysis. This type of magnetic collection allows for flow-through collection of a large volume of material, and is a significant improvement over older methods that relied on centrifugation to concentrate and collect the immunoattractant material. As the primary basis for the recognition and capture is based on the immunoattraction of the bead surface, this technology can be adapted for a range of purposes, such as concentrating pathogens such as *Salmonella* (Tu et al., 2008), and in detecting non-O157 Shiga toxin-producing *E. coli* in a variety of foods (Wang et al., 2013). This technology can also be used to concentrate and analyze chemical contaminants of interest, such as toxins and adulterants (Gessler et al., 2006).

Immunoattraction has been adapted for use in active biosensors (Nugen and Baeumer, 2008). These are probes for detecting analytes: chemical signals or cell-surface features uniquely characteristic of a single organism. Such probes integrate a biological component, such as a whole bacterium or a biological product (e.g., an enzyme or antibody) with an electronic component to yield a measurable signal. Rather than free-floating beads collected using a physical process, such as centrifugation or magnetic separation, comparably-sized nanofabricated sensor tips are connected to a sensor bank. Upon binding of a bacterium to the immunoattractive sensor tip, the configuration of the sensor is altered, inducing a signal in the attached biosensor. This signal may be a change in electrical conductivity, in optical properties (absorption, reflection, wavelength shift, etc.) or in physical conformation as in a cantilever design. Varying types of biosensors are employed in flow-through or "dipstick" type systems. As with immunomagnetic beads, sample preparation is critical. Sample viscosity, concentration of extraneous or contaminating material, etc., can influence the efficiency and accuracy of the testing. An interesting recent development is the possibility of screen-printed electrodes that would serve as a single-step immunosensor, an approach that would facilitate a broadly acting activation potential when used with a variety of immunomagnetic beads (Volpe et al., 2013)

PCR-based methods

Polymerase chain reaction (PCR) based detection systems are extremely sensitive to the presence of select DNA sequences associated with particular pathogens. Customized probes will bind only to the target sequence, which acts as a template for DNA synthesis by DNA polymerase, during multiple cycles of heating and cooling, leading to rapid amplification. Samples processed using traditional PCR methods undergo many cycles of replication and denaturation. Once the cycling is complete and the target sequence has been fully amplified, the sample is run on an agarose gel. Using a binding marker such as ethidium bromide, the DNA-banding pattern from the samples is read to establish presence/absence, and, to a more limited extent, a quantification of the original copy number and prevalence of a specific organism. A number of tools have been developed in association with this basic procedure to speed up, automate, and increase the sensitivity and accuracy of the process. For example, digital scanners have been applied to read the gels and attempt to quantify the size, position, and intensity of the bands in different lanes.

Traditional PCR methods read the sample after the cycling reactions have been completed. In contrast, real-time PCR methods draw the sample earlier in the reaction process (ABS, 2002; La Paz et al., 2007). This allows for discrimination of the samples during the exponential amplification phase, before the rate of signal increase slows down as all of the polymerase becomes saturated. This improves the sensitivity of the process. Also, real-time PCR systems do not use a gel to read the signal, but analyze the PCR product directly as it is produced. This is done with a photodetector using fluorescent binding dyes that adhere to the DNA. Real-time PCR assays are thus sensitive and specific, and allow detection and accurate quantitation (La Paz et al., 2007). A number of commercial versions of real-time PCR detection technologies are available.

Recently, advances in thermocycler design and improvements in double-labeled probes have increased the utility of this approach. By running multiple fast real-time PCR amplifications simultaneously in conventional 96-well plates, total throughput of samples is improved. Since differential signal chemistries can be used for the different reactions, these reaction products can be analyzed simultaneously.

The conventional and real-time PCR approaches were originally used to isolate and amplify a single DNA or RNA target sequence. Reverse transcriptase-PCR, which amplifies RNA, can be used to differentiate between viable and dead organisms, unlike standard PCR, which amplifies DNA from both. Thus, the analysis and detection focused on a single gene or gene product. Multiplex PCR, however, amplifies multiple genes or gene products in a single reaction, broadening the scope and efficiency of the process (Chang et al., 2008). Combining the multiwell thermocycler design with the techniques of multiplex PCR leads to even greater increases in total throughput and accuracy. When used for molecular epidemiology, multiplex PCR can be used to cross-verify multiple genes against one another, thereby improving the accuracy of detection and identification.

A pre-enrichment step can enhance the ability of detection systems to pick up indications of live pathogens. However, in addition to the time required for pre-enrichment, information regarding the original concentration of bacteria is degraded or lost entirely. The selection and application of real-time PCR protocols, either single target or multiplex, will be driven in part by the intentions of the detection scheme, and the real-world limitations of cost and complexity. Advanced techniques are becoming simpler, easier to use and more widely available (La Paz et al., 2007; Chang et al., 2008). Recently, multiplex PCR-based detection systems have successfully discriminated *Salmonella Typhimurium*, *E. coli* O157:H7, and *L. monocytogenes* in food products (Garrido et al., 2013; Yang et al., 2013). Of course, care must be shown in matching the capabilities of the detection system with the actual needs of the industrial and commercial environment.

Computer/AI optical scanning

Computerized optical scanning is technology that has been under development for some years. Using digital images of produce, it has been possible for some time to use computer software to measure the extent of damage caused by plant pathogens (Niemira et al., 1999). This quantification is based on differences in reflection and light absorption caused by the different physical properties of healthy vs. diseased tissue. More recently, optical scanning in the visible and near-infrared has shown promise in identifying fecal contamination on poultry carcasses (Chao et al., 2008). Using this system, images of the carcass are taken at a number of different wavelengths. Since contamination spots have different spectral characteristics than the underlying skin tissue, subtractive analysis of the various images can enhance the ability of software to identify problem areas.

When applying these techniques to identification of human pathogen contamination on produce, a number of specific hurdles remain. First, it is known that areas of damaged or diseased tissue can increase the harborage of human pathogens (Wells and Butterfield, 1997). Therefore, machine vision, which can be used to identify physical defects (Lee et al., 2008), thus also serves to reduce risk of contamination with human pathogens. It should be noted that the technology has limitations. *Salmonella*, *E. coli* O157:H7 and other human pathogens are often resident on the surfaces of fruits and vegetables without being associated with gross physical defects. Also, while machine vision has proven useful in detecting identifiable fecal contamination, it has been less effective at detecting independent human pathogens not associated with fecal material.

As biofilms represent a potential area of enhanced risk, recent research is developing the ability of machine vision tools to locate and identify biofilm material. Work using *in vitro* biofilms of *E. coli, Pseudomonas pertucinogena, Erwinia chrysanthemi,* and *L. inocula* has shown that excitation with UV-A at 320 to 400 nm enhances the signal from biofilms (Jun et al., 2008). Subtractive analysis of the resulting images collected at various wavelengths provides an actionable level of contrast between biofilm-containing and clear areas. Expansion of the work to *in vivo* biofilms on the surfaces of fruits and vegetables is required. One method of enhancing the optical scanning efficiency is to pre-treat with a pathogen-binding material that alters the reflectivity and/or wavelength scattering of the detector light beam. For example, a study by Sundaram et al. (2013) showed that a silver nanoparticle preparation, coupled with a bioactive polymer, enhanced Raman scattering based on the presence of human pathogens. This type of detection technology remains an area of active technological development.

Antimicrobial intervention technologies

The focus of this section is on the options for eradication of pathogens from fresh and fresh-cut fruits and vegetables. Chemical sanitizers, including a variety of chlorine-based sanitizers, are a standard feature of produce processing. These are primarily intended as a means to prevent cross-contamination, rather than as a true kill step that would eliminate pathogens where they are present in fruits and vegetables. The limited efficacy of conventional liquid-based sanitizing solutions (typically < 2-log reductions) has led to an investigation of a variety of alternatives. Treatments that rely on novel or precision application of chemical sanitizers, such as volatile essential oils, either in liquid phase (Gunduz et al., 2012) or in gas phase (Matan et al., 2005), chlorine dioxide, or ozone (Linton et al., 2006) are the subject of widespread research in order to optimize these treatments for industrial application. One of the most challenging aspects of antimicrobial gas-phase treatments such as chlorine dioxide is maintaining a uniform level of treatment at a high enough concentration to be efficacious. Modified approaches have investigated using lower concentrations renewed over extended treatment times, hours, or even days, such as would be available during shipping. As with shorter duration treatments, uniformity and process control remain obstacles to full implementation.

A number of studies have investigated other promising technologies for sanitizing produce, including use of electrolyzed water (Ayebah et al., 2006), ozonated water (Koseki and Isobe, 2006) ozonated dry ice (Fratamico et al., 2012), and advanced thermal treatments (Annous et al., 2013). These technologies are finding use in a number of different areas in the fresh and fresh-cut produce industry. The remainder of this section will present an overview of several key technologies that are not yet widely implemented in commercial settings. For some of these, there are important technological hurdles to be overcome. For others, the barriers to adoption are regulatory or cost engineering in nature. These technologies will be key areas for research in the future.

Cold plasma

Cold plasma is a promising new sanitizing technology for fresh produce (Niemira, 2012). A number of technological challenges are being addressed in ongoing research. As energy is added to materials, they change state, going from solid to liquid to gas, with large-scale inter-molecular structure breaking down. As additional energy is added, the intra-atomic structures of the components of the gas break down, yielding plasmas – concentrated collections of ions, radical species, and free electrons (Birmingham and Hammerstrom, 2000; Fridman et al., 2005; Gadri et al., 2000; Niemira and Sites, 2008). Therefore, although it is technically a distinct state of matter, cold plasma for all practical purposes may be regarded as energetic form of gas. Cold-plasma technologies used to treat foods have been grouped into three general categories (Niemira and Sites, 2008): *electrode contact* (in which the target is in contact with or between electrodes), *direct treatment* (in which active plasma is deposited directly on the target), and *remote treatment* (in which active plasma is generated at some distance, and plasma is moved to the target).

Electrode contact systems have been shown to achieve reductions as great as 5 logs of *E. coli, Staphylococcus aureus, Bacillus subtillis,* and *Saccharomyces cerevisiae* on foods and inert surfaces (Deng et al., 2007; Kelly-Wintenberg et al., 1999). Cultures of *E. coli* placed within the 1-mm gap spacing of the plasma reactors were reduced by 4.6 and 5.1 log cfu/ml after treatments of 10 s and 60 s, respectively (Sladek and Stoeffels, 2005). As the space between the plasma emitter and the treated culture was increased from 1 mm, antimicrobial efficacy was reduced, until at 10 mm spacing, no reductions were observed at any power level tested. Using the remote treatment reactor, Montie et al. (2000) reduced *E. coli* and *S. aureus* inoculated on polypropylene by 4 or 2 log cfu/ml, respectively, after a 10 s treatment. D-values of 22 s (*Shigella flexneri* and *Vibrio parahaemolyticus*) to 51 s (*E. coli* O157:H7) for pathogens on agar were obtained using the one atmosphere uniform glow discharge plasma system (OAUGDP) (Kayes et al., 2007). Treatment with the OAUGDP for 2 minutes reduced *E. coli* O157:H7 on Red Delicious apples by approximately 3 log cfu, reduced *S.* Enteritidis on cantaloupe by approximately 3 log cfu, and reduced *L. monocytogenes* on iceberg lettuce by approximately 2 log cfu (Critzer et al., 2007).

A gliding arc cold plasma system effectively inactivated *E. coli* O157:H7 and *Salmonella* on agar plates and on the surface of golden delicious apples (Niemira and Sites, 2008). In that study, higher flow rates of plasma (30 or 40 L/min) were more effective than lower flow rates (10 or 20 L/min) in inactivating these pathogens on inoculated apples, and longer exposures were more effective than shorter. At the highest flow rate, treatments of 3 minutes reduced *Salmonella* by 3.4 log cfu, and reduced *E. coli* O157:H7 by 3.5 log cfu. An important area for future research is evaluation of cold plasma treatments to porous surfaces, such as stem scars and fresh-cut surfaces. A study with almonds examined the effect of very rapid treatments using air- or nitrogen-based cold plasma (Niemira, 2012). The system operated at 47 kHz, 524W. Exposure time and distance were varied along with the feed gas. Short treatment with cold plasma significantly reduced *Salmonella* and *E. coli* O157:H7 on inoculated whole almonds. The greatest reduction observed was 1.34 log of *E. coli*

O157:H7 after a 20 s treatment at 6 cm spacing. Nitrogen was generally less effective compared with dry air (Niemira, 2012). This finding agrees with other research that suggests a critical role for oxygen radicals.

The primary modes of action for cold plasma are UV light and reactive chemical products of the ionization process. Various cold plasma systems operate at atmospheric pressures or in low pressure treatment chambers. *Salmonella, Escherichia coli* O157:H7, *Listeria monocytogenes,* and *Staphylococcus aureus* have been reduced by greater than 5 logs, with effective treatment times ranging from 120 s to as little at 3 s, depending on the food treated and the processing conditions. The development of the process will be determined by the optimization of the shape of the applied electrodes, and their method of application (screenprinted, applied, bonded, etc.). Cold plasma is a developing field; research is ongoing to advance the state of the art in cold-plasma emitter design, and to improve the operational application of the technology to fruits and vegetables. Research on sensory impact of the process on the treated fruits and vegetables will be critical for establishing protocols for commercial use.

Irradiation

Irradiation is a nonthermal process in which high-energy electrons or photons are applied to foods, resulting in the inactivation of associated pathogens (Niemira, 2012). Until recently, irradiation of produce was limited to disinfestation and storage life extension. In 2008, the FDA approved the use of irradiation up to 4.0 kGy on fresh lettuce and fresh spinach to kill human pathogens such as *E. coli* O157:H7 and *Salmonella* (FDA, 2008). This intended use, to improve food safety and shelf life, opens new opportunities for implementation of the technology in the arena of lettuce and spinach safety. However, protocols to use the technology effectively in the industrial setting must address matters of cost, consumer acceptance, and retail marketing.

An extensive body of research has demonstrated that this technology is safe and effective. Recently, research has focused on the ability of irradiation to address contamination of pathogens within the interior spaces of a leaf, fruit, or vegetable, which are inaccessible to conventional antimicrobial treatments. Microbiological analysis is made problematic by the inefficient uptake of bacteria via roots and vasculature, complicating the development of a clear risk analysis for this kind of contamination. However, it is clear that penetrating processes such as irradiation may be uniquely suited for dealing with this type of contamination. Nthenge et al. (2007) showed that irradiation eliminated pathogenic bacteria that were internalized within leaf tissues as a result of root uptake. The lettuce plants, grown in hydroponic solutions inoculated with *E. coli* O157:H7, contained the pathogen in the leaf tissue. In that study, irradiation effectively killed the pathogen while a treatment with 200ppm aqueous chlorine was ineffective.

Irradiation was shown to be similarly effective in eliminating internalized *E. coli* O157:H7 from baby spinach and various types of lettuce (Romaine, iceberg, Boston, green leaf, red leaf), while 300 or 600 ppm sodium hypochlorite was generally ineffective (Niemira, 2007; Niemira, 2008). D_{10} values for internalized cells (0.30–0.45 kGy) were 2- to 3-fold higher than for surface associated cells (0.12–0.14 kGy) (Niemira, 2007). The mechanism for this increase in D_{10} value has not been fully described. Pathogen populations within the leaf are generally expected to be very low in a commercial setting, due to the poor efficiency of uptake via the roots. Therefore, near-complete elimination of internalized pathogens may potentially be a practical goal using irradiation doses that do not cause undue sensory damage.

Irradiation is a penetrating process. Along with heat, it is one of the few treatments than can be applied to foods after they are already in the package. Once conventional washes and similar treatments have removed gross contamination, such as adherent residues, foreign matter (insect parts, manure flecks, etc.), more advanced treatments can be used to further reduce risk. These advanced treatments are intended to complement, not to replace, conventional controls and antimicrobial treatments (Fan et al., 2008). Once the produce is as clean as the conventional processes can make it, it is packaged to avoid potential cross-contamination. Irradiation would then be applied to further reduce microbial load. Hence, the packaged produce would then remain untouched by hand or machine during distribution, wholesale, and retail, until it is opened at point-of-consumption by the consumer or by food-service workers. It is expected that irradiation would therefore be used as a terminal process, incorporated into a processing line to be applied post-packaging (Fan et al., 2008). However, the impact of refrigerated storage time on pre- and post-irradiation pathogen recovery and physiology must be considered as part of the entire food handling, food processing chain (Niemira and Boyd, 2013).

Pulsed light

Recent advances in electronics and lighting technology have renewed interest in pulsed light as an antimicrobial process. In addition to improvements in xenon flash lamps and related technologies that produce intense flashes of broad-spectrum light, narrow-spectrum light emitting diode (LED) sources are also of interest. These are being explored for their applicability to fresh produce, and to food contact surfaces in a produce-processing environment.

Applications of broad-spectrum pulsed light for decontamination of surfaces were recently reviewed by Gomez-Lopez et al. (2007). The antimicrobial mode of action of pulsed light is based on the activity of UV-C. Across a range of wavelengths tested, maximum inactivation of *E. coli* was achieved around 270 nm, with antimicrobial efficacy dropping off to zero above 300 nm (Wang et al., 2005). The authors ascribed the majority of antimicrobial efficacy to the 220 to 290 nm range. Total aerobic counts of white cabbage, leek, paprika, carrots, and kale were reduced by 1.6 to 2.6 log cfu/cm^2 following treatment with wide-spectrum pulsed light (Hoornstra et al., 2002). The total luminance was 0.30 J/cm^2, delivered in two pulses. Working with inoculated raspberries exposed to pulsed UV light from a xenon flash lamp, Bialka and Demirci (2008) reduced *E. coli* O157:H7 and *Salmonella* by 3.9 and 3.4 log cfu/g, respectively. A parallel study using inoculated strawberries reported reductions of 2.1 and 2.9 log cfu/g for *E. coli* O157:H7 and *Salmonella*, respectively.

Total luminance for these studies was reported as 25.7 to 72.0 J/cm^2. It should be noted that ensuring uniformity of treatment in a commercial-scale system will be a critical factor in scaling up this technology. The engineering challenges in effectively treating a line throughput of hundreds or thousands of pounds per minute must be thoroughly considered.

Photothermal effects from exposure to pulsed light are a known factor associated with this technology (Gomez-Lopez et al., 2007). In a study of alfalfa seeds treated with pulsed light, the authors noted that treatments were limited by excessive heating of the seeds caused by the intensity of the flash lamp (Sharma and Demirci, 2003). In the case of small fruits treated with pulsed light (Bialka and Demirci, 2008), the temperature of strawberries and raspberries increased to 69°F and 79°F, respectively, during the 60 s treatment time. However, the authors of that study reported that there was no significant effect on the sensory properties of the treated fruits. It may be that future applications of pulsed light could operate in a cold room, or use a post-treatment stream of sterile cold air to remove excessive heat from the treated product.

Recent research has examined a variation on pulsed light technology that uses narrow-spectrum illumination that treats surfaces with light in the visible spectrum. This research derives from the field of medical applications of pulsed light, where an intense UV spectrum cannot be employed without damage to soft tissues of the eye, mouth, or skin. Using selective wavelength filters on a xenon lamp system, Maclean et al. (2008a) demonstrated maximum inactivation of *Staphylococcus aureus*, including methicillin-resistant strains, at 405 nm. A combination treatment of narrow-band light at 405 nm and 880 nm produced by banks of specialized LEDs reduced *S. aureus* and *Pseudomonas aeruginosa* by as much as 1.3 log cfu. Research into the mode of action of blue light has identified oxygen availability as a key factor (Maclean et al., 2008b). The specific chemistry proposed by the authors is related to photoactivation of intracellular porphyrins. Also, broad-spectrum bactericidal effects have been noted for application of 405 nm light, resulting in inactivation of *E. coli, Salmonella, Shigella, Listeria*, and *Mycobacterium* species (Murdoch et al., 2012). These results were obtained in liquid and on food-contact type surfaces. In that study, moist cultures of *L. monocytogenes* were readily inactivated in suspension and on surfaces such as agar, while *S. enterica* was most resistant. After drying *in situ, L. monocytogenes* was less susceptible than *S. enterica* after drying onto PVC and acrylic surfaces. These results suggest that narrow-band pulsed blue light may be a promising area of future research, but that the physiology of the target organism will be a crucial factor in its efficacy.

High-pressure processing

High-pressure processing (HPP) has been successfully applied to a number of liquid and semi-solid vegetable-derived foods such as juices, sauces and guacamole, as well as to meats and seafood products. The process causes a number of physical effects on foods, which can extend shelf-life, as with guacamole, or simplify subsequent processing and preparation at point-of-consumption, as with oysters (Considine et al., 2008).

It also can effectively reduce the microbial load of contaminating microflora and is therefore the subject of consideration from a food safety perspective. Chemical analysis of treated juices, pulps, and similar processed products indicates an acceptable level of sugar retention and enzyme inactivation (Butz et al., 2003). However, the process induces significant changes in protein conformation and cellular structure. For whole fruit or vegetables, these changes can alter the cellular electrolyte balance and enzyme activity, leading to discoloration and off-aromas (Considene et al., 2008). High pressures can shift the critical point of water, leading to glass transitions within the cells of the treated product. This can result in loss of firmness. However, Plaza et al. (2012) showed that 200 MPa increased extracted carotenoid content for astringent persimmons, whereas 400 MPa resulted in no significant differences or even a decrease was observed for non-astringent. The authors concluded that an optimized HPP process improved extraction of potentially health-related compounds and/or modified their bioaccessibility. Therefore, the membrane rupture that can be detrimental in some products can actually enhance value in other cases. Solid frozen fruit products, such as whole berries or sliced fruits, have minimal internal voids and meet different quality criteria than fresh produce products. Therefore, HPP may hold promise as a process for extracable fruit- and/or vegetable-derived products, juices, pulps, and purees, as well as for quick-frozen fruits and vegetables.

Sonication

Ultrasonication is the process of exposing contaminated foods and food contact surfaces to high-frequency sound waves. Recent research has focused on using ultrasound to enhance the antimicrobial efficacy of applied antimicrobial compounds. Huang et al. (2006) found that a 170-kHz ultrasound treatment increased the efficacy of chlorine-dioxide treatments when applied to apples inoculated with *Salmonella* or *E. coli* O157:H7. However, that same study showed that the treatment was generally ineffective at enhancing chemical efficacy for contaminated lettuce. Other studies also have shown that ultrasound does not enhance sanitizer efficacy when applied to lettuce, and can cause sensory damage to leaves with longer treatment times (Ajlouni et al., 2006). Therefore, although ultrasound appears to show potential for improving the treatments applied to some commodities, defining the commercial protocols of its use will require additional information. One possibility is to contribute to the sanitization of wash-water tanks and transport-water flumes to prevent cross-contamination. Further research will define the role that ultrasonication can play in an overall produce-processing system.

Biological controls

Fresh produce supports a varied and complex microflora. Total aerobic populations can range from 10^2 to 10^9 cfu/g and can include bacteria, yeasts, and fungi (Fett, 2006). Complex interactions among the microflora can enhance or detract from the establishment and growth of enteric pathogens (Lund, 1992; Liao et al., 2003).

Utilizing the suppressive and antagonistic effects of native microflora has been a topic of research for a number of years (Beuchat and Bracket, 1990; Nguyen-the and Carlin, 1994; Matos and Garland, 2005).

Liao (2007) demonstrated inhibition of enteric pathogens with native microflora derived from alfalfa seeds and from baby carrot. An isolate of the antagonist *P. fluorescens* (isolate Pf 2-79) suppressed *Salmonella, E. coli,* and *L. monocytogenes* on bell pepper disks. Efficacy of suppression was related to the population ratio of antagonist to pathogen. Where this ratio is 100:1, the pathogen is most effectively suppressed. This antagonist Pf 2-79 was also effective in suppressing *Salmonella* on sprouting seeds (Liao, 2008). Pre-treatment of seeds with Pf 2-79 before sprouting suppressed *Salmonella* growth by 2 to 3 log cfu. These results suggest a role for cultures of compatible antagonists as a dip treatment. Olanya et al. (2013) showed that biocontrol strains of *P. fluorescens* reduced *E.coli* O157:H7 on spinach by 0.5 to 2.1 log cfu/g. In that study, efficacy was significantly affected by storage temperature, and could be improved when combined with other postharvest intervention strategies.

An emerging area of research is the use on produce of bacteriophages as a targeted antimicrobial tool. Anti-*Listeria* bacteriophage treatments for packaged ready-to-eat meat and poultry products were recently approved by the FDA (FDA, 2006). In this application, the phage is applied as a liquid preparation to the surfaces of the food product immediately prior to packaging. It is expected that, much like treatments that use conventional bacterial antagonists, phage-based treatments for fruits and vegetables would be applied as a dip or spray. The specific and most optimal means of usage are still being determined, e.g. in field applications as a preharvest treatment, during postharvest processing, etc. Initial populations of *E. coli* O157:H7 attached to a coupon of stainless steel, a common food contact material for up to 4 days at 4°C, were reduced by 1 to 2 log cfu by bacteriophage KH1; however, populations enmeshed in biofilms were protected (Sharma et al., 2005). When combined with nisin, a phage treatment reduced *L. monocytogenes* on apples and melons by 2.3 and 5.7 log cfu, respectively (Leverentz et al., 2003). In a study of lytic bacteriophages, *E. coli* O157:H7-inoculated lettuce was treated by immersion or spray application (Ferguson et al., 2013). The authors concluded that both methods provided protection from *E. coli* O157:H7 contamination on lettuce, but spray application of lytic bacteriophages to lettuce was more effective in immediately reducing pathogen populations on fresh-cut lettuce.

The challenge of technology development for organic foods

The National Organic Program is a set of regulations governing which technologies can be used during production and processing of organic fruits and vegetables (CFR, 2005). These regulations establish science-based limits on what additives and processes can be used and applied, consistent with the tenets and philosophy of organic production. For example, while chlorine-based sanitizers can be used, they must be limited in use and application so as to result in residues of no more than

4 ppm. High-pressure processing, ohmic heating, and pulsed electric field process-ing of juices is permitted for organic products. In contrast, irradiation is a prohibited process for organic foods, and cannot be used under any circumstances for fresh and fresh-cut fruits and vegetables to be labeled organic.

New technologies may not be specifically addressed by existing governing reg-ulations. The guiding tenets of organic production are applied on a case-by-case basis for approval of new technologies or adaptation of existing technologies. When considering new interventions for which no previous context exists, either organic or conventional, science-based decision making is part of the overall approach to regulation. For example, the FDA has not yet issued a ruling with respect to the applicability of cold plasma for organic foods. In such a circumstance, should the specific constraints on organic produce be taken as constraints on the technology? This is a question with important implications for economics as well as for food tech-nology. While the needs of a specific commodity or market can often be a driver for innovation and technology development, this type of a targeted approach can limit the advancement of an otherwise promising technology. An awareness of the govern-ing regulations for organic or other specialty foods can help to guide development of appropriate technologies, or in the adaptation of existing technologies to organic implementation. The most useful approach is to strike a balance of allowing scientific innovation to generate a host of new tools, and refining their implementation for the specific needs of important commodities or markets.

Acknowledgements

The authors would like to thank Dr. David Kingsley and Dr. Modesto Olanya for their critical reviews of this chapter. Mention of trade names or commercial products in this publication is solely for the purpose of providing specific information and does not imply recommendation or endorsement by the U.S. Department of Agriculture. The USDA is an Equal Opportunity Employer.

References

ABS, AppliedBioSystems, 2002. Real-time PCR vs. Traditional PCR. http://www.appliedbio systems.com/support/tutorials/pdf/rtpcr_vs_tradpcr.pdf (accessed 21.06.13.).

Ajlouni, S., Sibrani, H., Premier, R., Tomkins, B., 2006. Ultrasonication and fresh produce (cos lettuce) preservation. J. Food Sci. 71, M62–M68.

Annous, B.A., Burke, A.M., Sites, J.E., Phillips, J.G., 2013. Commercial thermal process for inactivating Salmonella Poona on surfaces of whole fresh cantaloupes. J. Food Prot. vol 76 (3), 420–428.

Ayebah, B., Hung, Y.C., Kim, C., Frank, J.F., 2006. Efficacy of electrolyzed water in the inactivation of planktonic and biofilm Listeria monocytogenes in the presence of organic matter. J. Food Prot. 69, 2143–2150.

Beuchat, L.R., Brackett, R.E., 1990. Inhibitory effect of raw carrots on Listeria monocytogenes. Appl. Environ. Microbiol. 56, 1734–1742.

Bialka, K.L., Demirci, A., 2008. Efficacy of pulsed UV-light for the decontamination of escherichia coli o157:h7 and salmonella spp. on raspberries and strawberries. J. Food Sci. 73, M201–M207.

Birmingham, J.G., Hammerstrom, D.J., 2000. Bacterial decontamination using ambient pressure nonthermal discharges. IEEE Trans. Plasma Sci. 28, 51–55.

Buck, J.W., Walcott, R.R., Beuchat, L.R., 2003. Recent trends in microbiological safety of fruits and vegetables. Online. Plant Health Progress doi:10.1094/PHP-2003-0121-01-RV.

Butz, P., Fernández García, A., Lindauer, R., Dieterich, S., Bognár, A., Tauscher, B., 2003. Influence of ultra high pressure processing on fruit and vegetable products. J. Food Eng. 56, 233–236.

CDC: Centers for Disease Control and Prevention, 2011. CDC estimates of foodborne illness in the United States. http://www.cdc.gov/foodborneburden/PDFs/FACTSHEET_A_FINDINGS_updated4-13.pdf (accessed 21.06.13.).

CFR, Code of Federal Regulations, 2005. Title 7—Agriculture, Chapter I–Agricultural Marketing Service (Standards, Inspections, Marketing Practices), Department Of Agriculture (Continued) Part 205–National Organic Program. http://www.law.cornell.edu/cfr/text/7/205 (accessed 21.06.13.).

Chang, Y.C., Wang, J.Y., Selvam, A., Kao, S.C., Yang, S.S., Shih, D.Y.C., 2008. Multiplex PCR detection of enterotoxin genes in Aeromonas spp. from suspect food samples in northern Taiwan. J. Food Prot. 71, 2094–2099.

Chao, K., Nou, X., Liu, Y., Kim, M.S., Chan, D.E., Yang, C., Patel, J.R., Sharma, M., 2008. Detection of fecal/ingesta contaminants on poultry processing equipment surfaces by visible and near-infrared reflectance spectroscopy. Appl. Eng. Agric. 24, 49–55.

Critzer, F.A., Kelly-Wintenberg, K., South, S.L., Golden, D.A., 2007. Atmospheric plasma inactivation of foodborne pathogens on fresh produce surfaces. J. Food Prot. 70, 2290–2296.

Considine, K.M., Kelly, A.L., Fitzgerald, G.F., Hill, C., Sleator, R.D., 2008. High-pressure processing – effects on microbial food safety and food quality. FEMS Microbiol. Letters 281, 1–9.

Deng, S.R., Ruan, C.Y., Mok, G., Huang, X.L., Chen, P., 2007. Inactivation of *Escherichia coli* on almonds using nonthermal plasma. J. Food Sci. 72, M62–M66.

Fan, X., Niemira, B.A., Prakash, A., 2008. Ionizing irradiation of fresh and fresh-cut fruits and vegetables. Food Technol. 62, 36–43.

FDA (U.S. Food and Drug Administration), 2006. FDA approval of Listeria-specific bacteriophage preparation on ready-to-eat (RTE) meat and poultry products. http://www.gpo.gov/fdsys/pkg/FR-2006-08-18/html/E6-13621.htm (accessed 21.06.13.).

FDA, 2008. U.S. Food and Drug Administration - Final Rule (73 FR 49593), Irradiation in the Production. In: Processing and Handling of Food 21 CFR Part 179. http://www.gpo.gov/fdsys/pkg/FR-2008-08-22/html/E8-19573.htm (accessed 21.06.13.).

Ferguson, S., Roberts, C., Handy, E., Sharma, M., 2013. Lytic bacteriophages reduce *Escherichia coli* O157:H7 on fresh cut lettuce introduced through cross-contamination. Bacteriophage 3, e24323. 2013. http://dx.doi.org/10.4161/bact.24323.

Fett, W.F., 2006. Inhibition of *Salmonella enterica* by plant-associated pseudomonads in vitro and on sprouting alfalfa seed. J. Food Prot. 69, 719–728.

Fratamico, P.M., Juneja, V., Annous, B.A., Rasanayagam, V., Sundar, M., Braithwaite, D., Fisher, S., 2012. Application of ozonated dry ice (ALIGAL™ Blue Ice) for packaging and transport in the food industry. J. Food Sci. 77, M285–M291.

Fridman, A., Chirokov, A., Gutsol, A., 2005. Non-thermal atmospheric pressure discharges. J. Phys. D. Appl Phys. 38 R1 R24.

Gadri, R.B., Roth, J.R., Montie, T.C., Kelly-Wintenberg, K., Tsai, P., Helfritch, D.J., Feldman, P., Sherman, D.M., Karakaya, F., Chen, Z., 2000. Sterilization and plasma processing of room temperature surfaces with a one atmosphere uniform glow discharge plasma (OAUGDP). Surface Coatings Technol. 131, 528–542.

Garrido, A., Chapela, M.J., Román, B., Fajardo, P., Vieites, J.M., Cabado, A.G., 2013. In-house validation of a multiplex real-time PCR method for simultaneous detection of *Salmonella* spp., *Escherichia coli* O157 and *Listeria monocytogenes*. Int. J. Food Micro. 164, 92–98.

Gessler, F., Hampe, K., Schmidt, M., Bohnel, H., 2006. Immunomagnetic beads assay for the detection of botulinum neurotoxin types C and D. Diag. Microbiol. Inf. Dis. 56, 225–232.

Gomez-Lopez, V.M., Ragaert, P., Debevere, J., Devlieghere, F., 2007. Pulsed light for food decontamination: a review. Food Sci. Technol. 18, 464–473.

Gunduz, T.G., Niemira, B.A., Gonula, S.A., Karapinara, M., 2012. Antimicrobial activity of oregano oil on iceberg lettuce with different attachment conditions. J. Food Sci. 77, M412–M415.

Hoornstra, E., de Jong, G., Notermans, S., 2002. Preservation of vegetables by light. In: Society for Applied Microbiology (Ed.), Frontiers in microbial fermentation and preservation. Wageningen, the Netherlands, pp. 75–77.

Huang, T.-S., Xu, C., Walker, K., West, P., Zhang, S., Weese, J., 2006. Decontamination efficacy of combined chlorine dioxide with ultrasonication on apples and lettuce. J. Food Sci. 71, M134–M139.

JIFSAN (Joint Institute for Food Safety and Applied Nutrition), 2007. Tomato Safety Research Needs Workshop. http://jifsan.umd.edu/events/view/61 (accessed 21.06.13.).

Jun, W., Lee, K., Millner, P.D., Sharma, M., Chao, K., Kim, M.S., 2008. Portable hyperspectral fluorescence imaging system for detection of biofilms on stainless steel surfaces. In: Proceedings of SPIE. Florida, Orlando. March 17–20, 2008.

Kayes, M.M., Critzer, F.J., Kelly-Wintenberg, K., Roth, J.R., Montie, T.C., Golden, D.A., 2007. Inactivation of foodborne pathogens using a one atmosphere uniform glow discharge plasma. Foodborne Path. Dis. 4, 50–59.

Kelly-Wintenberg, K., Hodge, A., Montie, T.C., Eleanu, L.D., Sherman, D., Roth, J.R., Tsai, P., Wadsworth, L., 1999. Use of a one-atmosphere uniform glow discharge plasma to kill a broad spectrum of microorganisms. J. Vac. Sci. Technol. A. 17, 1539–1544.

Koseki, S., Isobe, S., 2006. Effect of ozonated water treatment on microbial control and on browning of iceberg lettuce (*Lactuca sativa* L.). J. Food Prot. 69, 154–160.

La Paz, J.L., Esteve, T., Pla, M., 2007. Comparison of real-time PCR detection chemistries and cycling modes using MON810 event-specific assays as model. J. Agric. Food Chem. 55, 4312–4318.

Lee, K., Kang, S., Delwiche, S.R., Kim, M.S., Noh, S., 2008. Correlation analysis of hyperspectral imagery for multispectral wavelength selection for detection of defects on apples. Sensing IInstr. Food Qual. Saf. 2, 90–96.

Leverentz, B., Conway, W.S., Camp, M.J., Janisiewicz, W.J., Abuladze, T., Yang, M., Saftner, R., Sulakvelidze, A., 2003. Biocontrol of *Listeria monocytogenes* on fresh-cut produce by treatment with lytic bacteriophages and a bacteriocin. Appl. Environ. Microbiol. 69, 4519–4526.

Liao, C.-H., McEvoy, J.L., Smith, J.L., 2003. Control of bacterial soft rot and foodborne human pathogens on fresh fruits and vegetables. In: Huang, H.C., Aharya, S.N. (Eds.), Advances in Plant Disease Management. Research Signpost, Kerala, India, pp. 165–193.

Liao, C.-H., 2007. Inhibition of foodborne pathogens by native microflora recovered from fresh peeled baby carrot and propagated in cultures. J. Food Sci. 72, M134–M139.

Liao, C.-H., 2008. Growth of *Salmonella* on sprouting seeds as affected by the inoculum size, native microbial load and *Pseudomonas fluorescens* 2-79. Lett. Appl. Microbiol. 46, 232–236.

Linton, R.H., Han, Y., Selby, T.L., Nelson, P.E., 2006. Gas-/vapor-phase sanitation (decontamination) treatments. In: Sapers, G.M., Gorny, J.R., Yousef, A.E. (Eds.), Microbiology of fruits and vegetables. Taylor and Francis, New York, NY, pp. 401–436.

Lund, B.M., 1992. Ecosystems in vegetable foods. J. Appl. Bacteriol. Symp. (Suppl.) 73, 115S–126S.

Maclean, M., MacGregor, S.J., Anderson, J.G., Woolsey, G., 2008a. High-intensity narrow-spectrum light inactivation and wavelength sensitivity of Staphylococcus aureus. FEMS Microbiol. Lett. 285, 227–232.

Maclean, M., MacGregor, S.J., Anderson, J.G., Woolsey, G., 2008b. The role of oxygen in the visible-light inactivation of *Staphylococcus aureus*. J. Photochem. Photobiol. B. Biol. 92, 180–184.

Matan, N., Rimkeereea, H., Mawsonb, A.J., Chompreedaa, P., Haruthaithanasana, V., Parker, M., 2005. Antimicrobial activity of cinnamon and clove oils under modified atmosphere conditions. Int. J. Food Microbiol. 107, 180–185.

Matos, A., Garland, J.L., 2005. Effects of community versus single strain of inoculants on the biocontrol of *Salmonella* and microbial community dynamics in alfalfa sprouts. J. Food Prot. 68, 40–48.

Montie, T.C., Kelly-Wintenberg, K., Roth, J.R., 2000. An overview of research using the one atmosphere uniform flow discharge plasma (OAUGDP) for sterilization of surfaces and materials. IEEE Trans. Plasma Sci. 28, 41–50.

Murdoch, L.E., Maclean, M., Endarko, E., MacGregor, S.J., Anderson, J.G., 2012. Bactericidal effects of 405 nm light exposure demonstrated by inactivation of *Escherichia, Salmonella, Shigella, Listeria*, and *Mycobacterium* species in liquid suspensions and on exposed surfaces. Scientific World J. Article. http://dx.doi.org/10.1100/2012/137805 ID 137805.

Nguyen-the, C., Carlin, F., 1994. The microbiology of minimally processed fresh fruits and vegetables. Crit. Rev. Food Sci. Nutr. 34, 371–401.

Niemira, B.A., Kirk, W.W., Stein, J.M., 1999. Screening for late blight susceptibility in potato tubers by digital analysis of cut tuber surfaces. Plant Dis. 83, 469–473.

Niemira, B.A., 2007. Relative efficacy of sodium hypochlorite wash vs. irradiation to inactivate *Escherichia coli* O157:H7 internalized in leaves of romaine lettuce and baby spinach. J. Food Prot. 70, 2526–2532.

Niemira, B.A., 2008. Irradiation vs. chlorination for elimination of *Escherichia coli* O157:H7 internalized in lettuce leaves: influence of lettuce variety. J. Food Sci. 73, M208–M213.

Niemira, B.A., Sites, J., 2008. Cold plasma inactivates *Salmonella* Stanley and *Escherichia coli* O157:H7 inoculated on golden delicious apples. J. Food Prot. 71, 1357–1365.

Niemira, B.A., 2012a. Antimicrobial application of low dose irradiation of fresh and fresh-cut produce. Ch. 14. In: Fan, X., Sommers, C.H. (Eds.), Food Irradiation Research and Technology, second ed. Wiley-Blackwell Publishing, Ames, IA, pp. 253–264.

Niemira, B.A., 2012b. Cold plasma decontamination of foods. Annu. Rev. Food Sci. Technol. 2012 (3), 125–142.

Niemira, B.A., 2012c. Cold plasma reduction of *Salmonella* and *Escherichia coli* O157:H7 on almonds using ambient pressure gases. J. Food Sci. 77, M171–M175.

Niemira, B.A., Boyd, G., 2013. Influence on modified atmosphere and varying time in storage on the irradiation sensitivity of *Salmonella* on sliced roma tomatoes. Rad. Phys. Chem. http://dx.doi.org/10.1016/j.radphyschem.2013.04.021.

Nthenge, A.K., Weese, J.S., Carter, M., Wei, C.I., Huang, T.S., 2007. Efficacy of gamma radiation and aqueous chlorine on *Escherichia coli* O157:H7 in hydroponically grown lettuce plants. J. Food Prot. 70, 748–752.

Nugen, S.R., Baeumner, A.J., 2008. Trends and opportunities in food pathogen detection. Anal. Bioanal. Chem. 391, 451–454.

Olanya, O.M., Ukuku, D.O., Annous, B.A., Niemira, B.A., Sommers, C.H., 2013. Efficacy of *Pseudomonas fluorescens* for biocontrol of *Escherichia coli* O157:H7 on spinach. Int. J. Food Ag. Environ. 11, 86–91.

Painter, J.A., Hoekstra, R.M., Ayers, T., Tauxe, R.V., Braden, C.R., Angulo, F.J., et al., 2013. Attribution of foodborne illnesses, hospitalizations, and deaths to food commodities by using outbreak data, United States, 1998–2008. Emerg. Infect. Dis. http://dx.doi.org/10.3201/eid1903.111866 (accessed 21.06.13.).

Plaza, L., Colina, C., de Ancos, B., Sánchez-Moreno, C., Cano, M.P., 2012. Influence of ripening and astringency on carotenoid content of high-pressure treated persimmon fruit (*Diospyros kaki* L.). Food Chem. 130, 591–597.

Sivapalasingam, S., Friedman, C.R., Cohen, L., Tauxe, R.V., 2004. Fresh produce: a growing cause of outbreaks of foodborne illness in the United States, 1973 through 1997. J. Food Prot. 67, 2342–2353.

Sharma, M., Ryu, J.-H., Beuchat, L.R., 2005. Inactivation of *Escherichia coli* O157:H7 in biofilm on stainless steel by treatment with an alkaline cleaner and a bacteriophage. J. Appl. Microbiol. 99, 449–459.

Sharma, R.R., Demirci, A., 2003. Inactivation of *Escherichia coli* O157:H7 on inoculated alfalfa seeds with pulsed ultraviolet light and response surface modeling. J. Food Sci. 68, 1448–1453.

Sladek, R.E.J., Stoeffels, E., 2005. Deactivation of *Escherichia coli* by the plasma needle. J. Phys. D. Appl. Phys. 38, 1716–1721.

Sundaram, J., Park, B., Kwon, Y., Lawrence, K.C., 2013. Surface enhanced Raman scattering (SERS) with biopolymer encapsulated silver nanosubstrates for rapid detection of foodborne pathogens. Int. J. Food Microbiol. Available online 22 May 2013, ISSN 0168-1605, http://dx.doi.org/10.1016/j.ijfoodmicro.2013.05.013.

Tu, S., Reed, S.A., Gehring, A.G., He, Y., 2008. Detection of *Salmonella enteriditis* from egg components using different immunomagnetic beads and time-resolved fluorescence. Food Anal. Methods 10.1007/s12161-008-9033-4.

UFPA (United Fresh Produce Association), 2007. Leafy Greens Food Safety Research Conference. http://www.unitedfresh.org/newsviews/food_safety_resource_center/research (accessed 21.06.13.).

Volpe, G., Sozzo, U., Piermarini, S., Delibato, E., Palleschi, G., Moscone, D., 2013. Towards the development of a single-step immunosensor based on an electrochemical screen-printed electrode strip coupled with immunomagnetic beads. Anal. Bioanal. Chem. 405, 655–663.

Wang, F., Yang, Q., Kase, J.A., Meng, J., Clotilde, L.M., Lin, A., Ge, B., 2013. Foodborne Path. Dis. Online citation ahead of print. http://dx.doi.org/10.1089/fpd.2012.1448.

Wang, T., MacGregor, S.J., Anderson, J.G., Woolsey, G.A., 2005. Pulsed ultra-violet inactivation spectrum of *Escherichia coli*. Water Res. 39, 2921–2925.

Wells, J.M., Butterfield, J.E., 1997. Salmonella contamination associated with bacterial soft rot of fresh fruits and vegetables in the marketplace. Plant Dis. 81, 867–872.

Yang, Y., Xu, F., Xu, H., Aguilar, Z.P., Niu, R., Yuan, Y., Sun, J., You, X., Lai, W., Xiong, Y., Wan, C., Wei, H., 2013. Magnetic nano-beads based separation combined with propidium monoazide treatment and multiplex PCR assay for simultaneous detection of viable *Salmonella* Typhimurium, *Escherichia coli* O157:H7 and *Listeria monocytogenes* in food products. Food Micro. 34, 418–424.

Conclusions and Recommendations

19

Casey J. Jacob[1], Benjamin J. Chapman[2], Douglas A. Powell[3]

[1]Diagnostic Medicine/Pathobiology, Kansas State University, Manhattan, KS, [2]4-H Youth Development and Family & Consumer Sciences, NC Cooperative Extension Service, North Carolina State University, Raleigh, NC, [3]Vice-president of Communications, IEH Laboratories & Consulting Group, Seattle, WA

CHAPTER OUTLINE

Introduction

From the local market to the local megalomart, the year-round availability of fresh fruits and vegetables has never been greater. Produce is a nutritional superstar, but because many fruits and vegetables are not cooked, anything with which they come into contact is a possible source of contamination. Is the water used for irrigation or rinsing cantaloupe or spinach clean, or is it contaminated with human pathogens? Do the workers who harvest the produce follow strict hygienic practices such as thorough handwashing? What happens to spinach when it is processed and bagged? The possibilities for contamination, as documented in the preceding chapters, are vast.

So what are consumers to think? Consumption of fresh fruits and vegetables is integral to a healthy diet and is a key component of programs designed to address the international epidemic of overweight and obesity. Global public awareness about produce-associated foodborne illness reached a tipping point as a consequence of large-scale outbreaks that occurred in 2005 and 2006. An outbreak of *E. coli*

The Produce Contamination Problem. http://dx.doi.org/10.1016/B978-0-12-404611-5.00019-1

O104:H4 associated with fenugreek sprouts in Germany in 2011 coupled with outbreaks of *Listeria monocytogenes* and *Salmonella* associated with cantaloupe in the United States in the summer of 2011 have focused even greater consumer attention on the safety of produce.

The social and economic impacts of these outbreaks are far-reaching and visible. The challenge lies in how to maximize the benefits of a diet rich in fresh fruits and vegetables while minimizing known risks. The produce contamination problem must be fixed.

Sources of contamination

Contamination begins on the farm in the soil, water, and amendments used to nurture safe, nutritious crops. As noted by Millner (Chapter 4), on-farm conditions and practices are critical determinants of the sanitary condition of fresh produce from both organic and conventional sources. Identification of on-farm pathogen reservoirs and vectors can aid development and use of farm-specific pathogen reduction programs. Research is needed to evaluate the effectiveness of various manure-management practices designed to reduce pathogen loading on-site and minimize pathogen accumulation off-site (via run-off or transport). Some animal viruses, while not zoonotic pathogens, may be suitable indicators of manure treatment efficacy, although research is required to validate their use. Additional research is needed to determine appropriate field management strategies for land areas adjacent to fresh produce crop fields to reduce fugitive enteric pathogen contamination. Many approaches to biological processing of manure exist; however, thermophilic composting remains one of the most cost-effective treatment technologies for manure solids that functions well in a variety of environments. A science-based quality control program (such as a HACCP plan with verification) is required to ensure the safety of compost produced for use on fresh-produce crops.

Millner rather astutely stated that "the non-preferential contamination in fresh-produce–related outbreaks across organic and conventional sources suggests that actual on-site conditions and practices, rather than marketing-based labels, are the critical determinants of the sanitary condition of fresh produce." In other words, microorganisms tend not to preferentially associate with a politically favorable growing regime, exemplified by 2013's outbreak of hepatitis A associated with organically produced frozen berry mix. The sooner the public discussion and buying patterns returns to a basis of biological safety, the better.

Additional microbial risks are found outside of human practices. Although it is thought that wildlife can be reservoirs for human diseases transmitted by contaminated fresh produce, there is, as noted by Rice (Chapter 8), a lack of evidence directly linking contamination by wildlife to outbreaks of human illness. Rice wisely indicates that in most cases there are no economically feasible mechanisms to prevent wildlife from coming into direct contact with produce while being grown. However, awareness – particularly of the surrounding environment – along with some controls can reduce the risk. Clark (Chapter 7) details circumstantial evidence that incriminates wild birds in the contamination of produce at several points throughout production and processing supporting the notion that birds can pose a human health risk by serving as a source of contamination, but can all the birds be killed? This would likely not be possible, nor would it be desirable. There is a need for economically feasible mechanisms to mitigate wildlife-produce interactions.

Commodities at risk

Not all fresh fruits and vegetables are equally susceptible to microbial contamination. Certain commodities – seed sprouts, leafy greens, tomatoes, cantaloupes, green onions, and herbs – are linked to notably more outbreaks of foodborne disease than others. Tree fruits and nuts are rarely associated with such outbreaks.

Matthews (Chapter 9) notes that the economically advantageous practice of processing leafy greens in the field (including harvesting, washing, and packaging) should be examined for potential exposure to microbial contaminants. In addition, hydrocooling processes could have a significant impact on the microbial safety of leafy greens, but research to that end has focused largely on retail environments over processing environments. Matthews also states that guidelines are in place for the microbiological testing of water, soil amendments, and equipment used to grow and harvest leafy greens, as well as the commodities themselves.

The discussion of seed sprouts in Chapter 11 is new to the second edition and raises some important questions concerning microbial safety of this commodity. Recent regulatory changes in the U.S. and EU should move toward improving the microbiological quality of sprouts. Warriner and Smal state that a common theme within the regulations is the responsibility of the seed supplier to provide pathogen-free seeds. But, they also indicate that this may be more difficult for some seed producers (alfalfa) to achieve than others (mung bean). The concept of supplier guarantees coupled with the international movement and lengthy storage time of dried seeds used for sprouts was highlighted by public health authorities during the investigation into 2011's *E. coli* O104 outbreak.

Everyone, including producers, processors, retailers, and consumers, needs to practice good food-safety behavior, based on the best available evidence at the time, and should expend resources to manage their own responsibilities. Therefore, additional measures may include subjecting producers to periodic, unannounced inspections; verifying that GAPs were being used appropriately by growers; and encouraging proper handwashing – a sanitation step – by everyone handling fresh produce or equipment.

In discussing outbreaks associated with tree fruits and nuts, Keller (Chapter 13) says that the occasional association of human pathogens with tree fruits and nuts is facilitated by particular pathogens' tolerance to some extreme conditions (such as desiccation and acidity). The mechanism of contamination of tree fruits involved in foodborne illness outbreaks is often unknown, though poor sanitation and hygiene practices, unsafe harvest or processing methods, internalization of contaminated wash/rinse water facilitated by a temperature gradient, and changes in normal procedures without appropriate attention to food safety (i.e., the lack of a culture of food safety) can lead to growth and survival of human pathogens in or on such fruits, as well as nuts.

Challenges of produce disinfection

When human pathogens are introduced into the production or processing environment, are there ways to minimize the potential for colonization? Yaron (Chapter 2) explains that several bacterial pathogens attach rapidly to produce surfaces and cannot be removed with current washing or agitation regimens. Sapers (Chapter 17) says the efficacy of washing and sanitizing agents for produce is often limited by attachment of microorganisms to inaccessible surfaces, infiltration of microorganisms, or incorporation of microorganisms into resistant biofilms. Cut produce is at even greater risk for bacterial colonization, and sanitizing solutions may be rendered ineffective by organic matter released from cut tissues of produce or by biofilms present on the produce. Some bacterial foodborne pathogens can be internalized by produce when a temperature differential exists between a fruit or vegetable and the fluid it is immersed in, such as in a dump tank. Further, washing equipment can also introduce contamination through the accumulation of soil and microbes in wash water or sanitizing solution. Biofilms and the accumulation of soil have been especially focused on with respect to *Listeria monocytogenes* associated with cantaloupe. This area of study will need to be expanded to help risk managers better understand the concept of transient vs. resident pathogen populations especially in packing facilities.

Despite the proclamations of various consumer advisory groups, washing of produce is of limited use. Sapers notes that the best defense is to minimize contact between human pathogens and fresh produce.

Investigating contamination on the farm

The produce problem needs to be addressed on the farm first. Managing risks on an individual farm basis may yield the most significant reductions in produce-related foodborne illness. As Sapers and Doyle explained in the introduction (Chapter 1), the prevalence of produce contamination by enteric pathogens is generally too low for broadly focused testing, and the understanding of human pathogens in the farm environment is limited. Better detection may help allocate risk reduction resources in the most cost-effective manner. But as described by Niemira and Zhang (Chapter 18), there are lots of technologies but not many near practical applications.

And when problems do happen, the human factor, as pointed out by Farrar and Guzewich (Chapter 3), is often overlooked. Their chapter provided an outstanding overview to improve the investigative process when an outbreak of produce-related foodborne illness occurs, and an urgent call to pay attention to human behavior. For example, Farrar and Guzewich state,

> *Where possible, investigators are strongly encouraged to visually observe*
> *'routine' food preparation/food processing procedures and record objective*
> *measurements of preparation and processing practices, as well as routine*
> *cleaning and sanitation procedures. Often, what is written in procedures*
> *manuals is not what actually occurs in the kitchen or in the food processing*
> *facility. Even with the inevitable bias introduced during physical observation*
> *by investigators, valuable clues are often obtained by simply observing and*
> *documenting a process from start to finish.*

The same advice applies to the farm.

Pre-emptive food safety programs

The safety of the food supply is a global concern requiring the commitment of all countries. A major reason countries import and export food is to satisfy consumer demand. Foodborne illnesses may be linked to the consumption of foods whether grown and manufactured domestically or imported. In January 2011, the FDA's Food Safety Modernization Act (FSMA) was signed into law in the United States. Similar measures aimed at tightening food safety laws have been enacted by other countries. The FSMA in the United States now requires the U.S. FDA to establish science-based minimal standards for the safe production and harvesting of those types of raw fruits and vegetables (e.g., lettuce, tomatoes, cantaloupes) for which standards are necessary to minimize the risk to human health including death. The "Standards for growing, harvesting, packing, and holding of produce for human consumption" will focus on microbiological hazards. Haro et al. (Chapter 15) describes measures implemented in Mexico to improve microbial safety of produce. The "Mexico Calidad Suprema" program is an official mark of identification that guarantees good sanitation, food safety, and a high quality for Mexican products. Johannessen and Cudjoe (Chapter 16) outline changes in EU regulations for sprout production that came about the consequence of the 2011 sprout outbreak. In March 2013, the European Commission published four regulations directed towards the safe production of sprouts. These regulations cover areas from traceability requirements to microbiological criteria for sprouts.

Retailers and consumers are currently driving on-farm food safety program implementation. Clear expectations by these groups can lead producers to reduce the likelihood of illnesses associated with their products through on-farm food-safety programs.

There are several factors that contribute to the successful implementation of sci-entifically validated risk-reduction practices on-farm. To begin, successful on-farm food-safety programs include ongoing support in the form of workshops, documents, and individuals with expertise in food safety to advise on potential risks, implemen-tation of practices, standards to be met, and methods to evaluate any new risks. It is important that any implemented program be constantly revisited and updated with new science, practice developments, and discoveries of risk. An ideal program would also provide rewards to participating producers and be translatable to buyers. Finally, a successful program of on-farm food safety promotes a change in culture on a farm, as opposed to a change of practices in relation to specific risks.

The farm-to-fork approach

The real challenge for food-safety professionals is to garner support for safe food practices in the absence of an outbreak; to create a culture that values microbiologically safe food from farm-to-fork at all times, and not just in response to the glare of the media spotlight. A farm-to-fork approach must be used to target food-safety practices to all food handlers at each stage of food production during typical day-to-day operation. Primary production, processing, distribution/retail, and consumers are the four sectors that comprise the farm-to-fork food continuum. Pathogens with the potential to cause foodborne illness can contaminate food at any point along the continuum. Food safety is, therefore, a shared responsibility among all involved in the food continuum, from producer to consumer, and across all levels of government.

A farm-to-fork approach to fresh produce food safety involves marketing food safety to those involved in the production of safe produce, as well as consumers. The safety of the food supply is a global concern requiring the commitment of all countries. Global food safety standards are required to ensure that food will not be injurious to health regardless of its origin. The purchase of microbiologically safe fresh fruits and vegetables by consumers translates into fewer sick people.

Index

Note: Page numbers followed by "*f*" denote figures; "*t*" tables.

Food Science and Technology International Series

Amerine, M.A., Pangborn, R.M., and Roessler, E.B., 1965. Principles of Sensory Evaluation of Food.

Glicksman, M., 1970. Gum Technology in the Food Industry.

Joslyn, M.A., 1970. Methods in Food Analysis, Second Ed.

Stumbo, C. R., 1973. Thermobacteriology in Food Processing, Second Ed.

Altschul, A.M. (Ed.), New Protein Foods: Volume 1, Technology, Part A—1974. Volume 2, Technology, Part B—1976. Volume 3, Animal Protein Supplies, Part A—1978. Volume 4, Animal Protein Supplies, Part B—1981. Volume 5, Seed Storage Proteins—1985.

Goldblith, S.A., Rey, L., and Rothmayr, W.W., 1975. Freeze Drying and Advanced Food Technology.

Bender, A.E., 1975. Food Processing and Nutrition.

Troller, J.A., and Christian, J.H.B., 1978. Water Activity and Food.

Osborne, D.R., and Voogt, P., 1978. The Analysis of Nutrients in Foods.

Loncin, M., and Merson, R.L., 1979. Food Engineering: Principles and Selected Applications.

Vaughan, J. G. (Ed.), 1979. Food Microscopy.

Pollock, J.R.A. (Ed.), Brewing Science, Volume 1—1979. Volume 2—1980. Volume 3—1987.

Christopher Bauernfeind, J. (Ed.), 1981.Carotenoids as Colorants and Vitamin A Precursors: Technological and Nutritional Applications.

Markakis, P. (Ed.), 1982. Anthocyanins as Food Colors.

Stewart, G.G., and Amerine, M.A. (Eds.), 1982. Introduction to Food Science and Technology, Second Ed.

Iglesias, H.A., and Chirife, J., 1982. Handbook of Food Isotherms: Water Sorption Parameters for Food and Food Components.

Dennis, C. (Ed.), 1983. Post-Harvest Pathology of Fruits and Vegetables.

Barnes, P.J. (Ed.), 1983. Lipids in Cereal Technology.

Pimentel, D., and Hall, C.W. (Eds.), 1984. Food and Energy Resources.

Regenstein, J.M., and Regenstein, C.E., 1984. Food Protein Chemistry: An Introduction for Food Scientists.

Gacula Jr. M.C., and Singh, J., 1984. Statistical Methods in Food and Consumer Research.

Clydesdale, F.M., and Wiemer, K.L. (Eds.), 1985. Iron Fortification of Foods.

Decareau, R.V., 1985. Microwaves in the Food Processing Industry.

Herschdoerfer, S.M. (Ed.), Quality Control in the Food Industry, second edition. Volume 1—1985. Volume 2—1985. Volume 3—1986. Volume 4—1987.

Urbain, W.M., 1986. Food Irradiation.

Bechtel, P.J., 1986. Muscle as Food.

Chan, H.W.-S., 1986. Autoxidation of Unsaturated Lipids.

Cunningham, F.E., and Cox, N.A. (Eds.), 1987. Microbiology of Poultry Meat Products.

McCorkle Jr. C.O., 1987. Economics of Food Processing in the United States.

Japtiani, J., Chan Jr., H.T., and Sakai, W.S., 1987. Tropical Fruit Processing.

Solms, J., Booth, D.A., Dangborn, R.M., and Raunhardt, O., 1987. Food Acceptance and Nutrition.

Macrae, R., 1988. HPLC in Food Analysis, Second Ed.

Pearson, A.M., and Young, R.B., 1989. Muscle and Meat Biochemistry.

Penfield, M.P., and Campbell, A.M., 1990. Experimental Food Science, Third Ed.

Blankenship, L.C., 1991. Colonization Control of Human Bacterial Enteropathogens in Poultry.

Pomeranz, Y., 1991. Functional Properties of Food Components, Second Ed.

Walter, R.H., 1991. The Chemistry and Technology of Pectin.

Stone, H., and Sidel, J.L., 1993. Sensory Evaluation Practices, Second Ed.

Shewfelt, R.L., and Prussia, S.E., 1993. Postharvest Handling: A Systems Approach.

Nagodawithana, T., and Reed, G., 1993. Enzymes in Food Processing, Third Ed.

Hoover, D.G., and Steenson, L.R., 1993. Bacteriocins.

Shibamoto, T., and Bjeldanes, L., 1993. Introduction to Food Toxicology.

Troller, J.A., 1993. Sanitation in Food Processing, Second Ed.

Hafs, D., and Zimbelman, R.G., 1994. Low-fat Meats.

Phillips, L.G., Whitehead, D.M., and Kinsella, J., 1994. Structure-Function Properties of Food Proteins.

Jensen, R.G., 1995. Handbook of Milk Composition.

Roos, Y.H., 1995. Phase Transitions in Foods.

Walter, R.H., 1997. Polysaccharide Dispersions.

Barbosa-Canovas, G.V., Marcela Go´ngora-Nieto, M., Pothakamury, U.R., and Swanson,

B.G., 1999. Preservation of Foods with Pulsed Electric Fields.

Jackson, R.S., 2002. Wine Tasting: A Professional Handbook.

Bourne, M.C., 2002. Food Texture and Viscosity: Concept and Measurement, second ed.

Caballero, B., and Popkin, B.M. (Eds.), 2002. The Nutrition Transition: Diet and Disease in the Developing World.

Cliver, D.O., and Riemann, H.P. (Eds.), 2002. Foodborne Diseases, Second Ed.

Kohlmeier, M., 2003. Nutrient Metabolism.

Stone, H., and Sidel, J.L., 2004. Sensory Evaluation Practices, Third Ed.

Han, J.H., 2005. Innovations in Food Packaging.

Sun, D.-W. (Ed.), 2005. Emerging Technologies for Food Processing.

Riemann, H.P., and Cliver, D.O. (Eds.), 2006. Foodborne Infections and Intoxications, Third Ed.

Arvanitoyannis, I.S., 2008. Waste Management for the Food Industries.

Jackson, R.S., 2008. Wine Science: Principles and Applications, Third Ed.

Sun, D.-W. (Ed.), 2008. Computer Vision Technology for Food Quality Evaluation.

David, K., and Thompson, P., (Eds.), 2008. What Can Nanotechnology Learn From Biotechnology?.

Arendt, E.K., and Bello, F.D. (Eds.), 2008. Gluten-Free Cereal Products and Beverages.

Bagchi, D. (Ed.), 2008. Nutraceutical and Functional Food Regulations in the United States and Around the World.

Singh, R.P., and Heldman, D.R., 2008. Introduction to Food Engineering, Fourth Ed.

Berk, Z., 2009. Food Process Engineering and Technology.

Thompson, A., Boland, M., and Singh, H. (Eds.), 2009. Milk Proteins: From Expression to Food.

Florkowski, W.J., Prussia, S.E., Shewfelt, R.L. and Brueckner, B. (Eds.), 2009. Postharvest Handling, Second Ed.

Gacula Jr., M., Singh, J., Bi, J., and Altan, S., 2009. Statistical Methods in Food and Consumer Research, Second Ed.

Shibamoto, T., and Bjeldanes, L., 2009. Introduction to Food Toxicology, Second Ed.

BeMiller, J. and Whistler, R. (Eds.), 2009. Starch: Chemistry and Technology, Third Ed.

Jackson, R.S., 2009. Wine Tasting: A Professional Handbook, Second Ed.

Sapers, G.M., Solomon, E.B., and Matthews, K.R. (Eds.), 2009. The Produce Contamination Problem: Causes and Solutions.

Heldman, D.R., 2011. Food Preservation Process Design.

Tiwari, B.K., Gowen, A. and McKenna, B. (Eds.), 2011. Pulse Foods: Processing, Quality and Nutraceutical Applications.

Cullen, P.J., Tiwari, B.K., and Valdramidis, V.P. (Eds.), 2012. Novel Thermal and Non-Thermal Technologies for Fluid Foods.

Stone, H., Bleibaum, R., and Thomas, H., 2012. Sensory Evaluation Practices, Fourth Ed.

Kosseva, M.R. and Webb, C. (Eds.), 2013. Food Industry Wastes: Assessment and Recuperation of Commodities.

Morris, J.G. and Potter, M.E. (Eds.), 2013. Foodborne Infections and Intoxications, Fourth Ed.

Berk, Z., 2013. Food Processing Engineering and Technology, Second Ed.

Singh, R.P., and Heldman, D.R., 2014. Introduction to Food Engineering, Fifth Ed.

Han, J.H. (Ed.), 2014. Innovations in Food Packaging, Second Ed.

Madsen, C., Crevel, R., Mills, C., and Taylor, S. (Eds.), 2014. Risk Management for Food Allergy

Jackson, R.S., 2014. Wine Science: Principles and Applications, Fourth Ed.

Printed and bound by CPI Group (UK) Ltd, Croydon, CR0 4YY

08/05/2025

01864858-0002